Implementation of Beam-Type Finite Elements Based on Carrera Unified Formulation

Erasmo Carrera · Gerlando Augello · Riccardo Augello

Implementation of Beam-Type Finite Elements Based on Carrera Unified Formulation

 Springer

Erasmo Carrera
DIMEAS
Politecnico di Torino
Turin, Italy

Gerlando Augello
Turin, Italy

Riccardo Augello
DIMEAS
Politecnico di Torino
Turin, Italy

ISBN 978-3-031-95855-7   ISBN 978-3-031-95856-4 (eBook)
https://doi.org/10.1007/978-3-031-95856-4

This Springer imprint is published by the registered company Springer Nature Switzerland AG
The registered company address is: Gewerbestrasse 11, 6330 Cham, Switzerland

Preface

What prompts the creation of yet another book on the finite element modeling of beam-like structures? The existing literature is rich with contributions from eminent scientists delving into various beam theories. However, a fresh perspective on a classical subject can ignite the curiosity of both seasoned and budding researchers. While this alone might attract interested readers, the authors' motivations for writing a new book on finite elements are distinct. They can be succinctly summarized as follows:

- Traditional one-dimensional (1D) finite elements typically possess a maximum of six degrees of freedom per node. There are various types of 1D finite elements, such as rods, bars, and beams. These elements have different cross-sections, and their primary distinction lies in the loading conditions they can handle. A rod, for instance, is subjected to axial and torsional loads but not bending, exhibiting axial and torsional stiffness without bending stiffness. On the contrary, bars and beams can accommodate various loads, including tension/compression, torsion, bending, and shear. Consequently, the selection of the appropriate type of 1D structure (rod, bar, or beam) depends on the specific external loads and boundary conditions applied to the real component, making this choice highly problem dependent. This classification may pose challenges for beginners and, at times, even experienced users due to potential confusion. The primary objective of this book is to introduce a unified 1D finite element that can emulate the behavior of a rod, bar, or beam without requiring ad-hoc formulations. This approach aims to simplify the modeling process and alleviate the complexities associated with choosing the right structural type, offering a more versatile and user-friendly solution.
- Over the past three decades, the Carrera Unified Formulation (CUF) has emerged in the literature. This formulation was originally proposed by Carrera, and the *unified* aspect of CUF enables the implementation of a 1D mathematical model capable of functioning as a rod, bar, beam, or any other physical representation. This eliminates the need for specific formulations, and users simply need to choose the degree of refinement for the mathematical model. This approach facilitates the

implementation of 1D analyses ranging from low to higher order, encompassing classical Euler-Bernoulli and Timoshenko beams to high-fidelity 1D models.

- Finite elements stand out as versatile and widely employed numerical methods among both researchers and practitioners, serving as a cornerstone for tackling intricate problems in engineering and science. However, the approximations inherent in finite element formulations can sometimes lead to unexpected behaviors under specific conditions. The assumptions made in displacement functions to represent the actual system behavior and the application of numerical integration techniques may result in unpredictable outcomes. One notable challenge in finite element formulations is shear locking, a phenomenon characterized by a significant underestimation of displacements, rendering the structure excessively stiff. The term "locking" signifies that the structure essentially "locks" itself against deformations. The book delves into the intricacies of this problem, exploring in detail potential solutions found in the literature, such as reduced and selectively reduced integration, as well as mixed formulation techniques. These approaches typically revolve around the adjustment of the number of Gauss points used for integrating volumetric and shear strains. The book provides a comprehensive analysis of this issue.

- This book provides the stiffness matrix of a beam structure subjected to various loading conditions, from axial elongation, to bending around transverse directions, and torsion. The determination of the stiffness matrix is approached in each case using the Principle of Virtual Displacement (PVD), a concept rooted in classical mechanics and variational calculus. Within the framework of Lagrangian mechanics, the PVD is instrumental in deriving the equations of motion for a mechanical system. It asserts that the virtual work carried out by internal and external forces acting on a system is zero for any virtual displacement consistent with the system's constraints. What sets this book apart is its introduction of a recursive notation for the evaluation of the stiffness matrix. This innovative approach allows the construction of a matrix whose form is independent of the adopted structural theory.

- An innovative development stemming from CUF is the Node-Dependent Kinematics (NDK) technique, allowing for different kinematics along the length of the 1D mathematical model. This proves valuable when only a specific portion of the 1D structure experiences significant deformation and, consequently, a complex 3D stress state due to simulation requirements. The NDK technique enables refinement exclusively in that portion of the structure, conserving degrees of freedom and, consequently, reducing computational costs. The book will provide a detailed exploration of the NDK concept.

- For each scenario, a dedicated MATLAB script is included in the Appendix. These scripts serve as valuable tools to assist users in systematically deriving the stiffness matrix, offering step-by-step guidance throughout the process.

Turin, Italy

Erasmo Carrera
Gerlando Augello
Riccardo Augello

Competing Interests The authors have no competing interests to declare that are relevant to the content of this manuscript.

Ethics Approval The authors declare that this manuscript does not include primary studies involving animals.

Contents

About the Authors

Erasmo Carrera After earning two degrees (Aeronautics, 1986, and Aerospace Engineering, 1988) at the Politecnico di Torino, Erasmo Carrera received his Ph.D. degree in Aerospace Engineering at the Politecnico di Milano—Politecnico di Torino—Università di Pisa in 1991. He began working as a Researcher at the Department of Aerospace for the Politecnico di Torino in 1992 where he held courses on Missiles and Aerospace Structure Design, Plates and Shells, and the Finite Element Method. He became Associate Professor of Aerospace Structures and Aeroelasticity in 2000, and Full Professor at the Politecnico di Torino in 2011. He has visited the Institute fur Statik und Dynamik, Universitat Stuttgart twice, the first time as a Ph.D. student (six months in 1991) and then as visiting Scientist under a GKKS Grant (18 months in 1995–1996). In the Summer of '96, he was Visiting Professor at the ESM Department of Virginia Tech. His main research topics are: composite materials, finite elements, plates and shells, postbuckling and stability, smart structures, thermal stress, aeroelasticity, multibody dynamics and the design and analysis of non classical lifting systems. He is author of more than 200 articles on these topics, many of which have been published in international journals. He serves as a referee for many journals, such as *Applied Mechanics Reviews*, *AIAA Journal*, *Journal of Sound and Vibration*, *International Journal of Solids and Structures*, *International Journal of Numerical Methods in Engineering* and as contributing editor for Mechanics of Advanced Materials and Structures.

Gerlando Augello is an Italian engineer with over 30 years of experience in mechanical and structural analysis, particularly in the aerospace sector. He earned his Master's Degree in Mechanical Engineering in 1986 and has worked extensively with Thales Alenia Space, holding roles such as Head of Mechanical Computer Aided Engineering (MCAE). His expertise includes finite element analysis, structural optimization, and advanced simulation techniques, with significant contributions to several space projects among which EXOMRAS, Space Rider, EUCLID,

SPACEHAB, COLUMBUS, NODE2 and NODE3. He has authored over 15 technical publications and two books and has served as an adjunct professor at Politecnico di Torino. Recognized as a "Maestro del Lavoro" in 2014, Augello continues to advance methodologies in structural analysis and design.

Riccardo Augello is a Postdoctoral Fellow supported by the prestigious Marie Skłodowska-Curie Actions, a researcher mobility program funded by the European Commission. He earned his Ph.D. in Mechanical Engineering from Politecnico di Torino, Italy, in February 2021. His doctoral research centered on developing advanced mathematical frameworks based on nonlocal mechanics to address the geometrical nonlinear analysis of composite structures. As part of his Ph.D., he collaborated with the City University of Hong Kong, spending six months there conducting research. In July 2022, Augello was appointed Assistant Professor at Politecnico di Torino. He joined California Institute of Technology as visiting postdoc in March 2023 as part of his European research project focussing on advancing mathematical modeling techniques for ultra-thin, deployable, and foldable composite structures, with applications in space technology.

List of Figures

List of Tables

Chapter 1
Introduction

Beam Theories

This book focuses on beam structural theories. Clearly, they prove to be very attractive since they reduce a three-dimensional problem into a set of variables depending only on the longitudinal coordinate. For this reason, beam theories particularly shine when dealing with slender bodies, which engineers frequently encounter in daily practice, such as arches, columns, bridges, rotor blades, and aircraft wings. The classical beam theories are represented by the Euler-Bernoulli Beam Theory (EBBM) [1], Saint-Venant [2], and Timoshenko Beam Theory (TBT) [3, 4]. EBBM, formulated in the 18th century, provides a fundamental understanding of the behavior of slender beams under bending loads. It assumes small deformations and is foundational in the study of beams. On the other hand, Timoshenko's work extended the classical beam theory by incorporating shear deformations. Unlike EBBM, TBT considers the effects of both bending and shear deformation, making it more accurate for certain types of beams. The main limitation of such theories is how they take into account the cross-sectional shear distribution, which is considered null in the former and constant in the latter. Those approximations do not alter the results unless non-conventional effects are present, such as torsion-bending coupling, localized boundary conditions and warping, as highligthed by [5].

To consider those effects, many improvements have been addressed by scientists throughout history, as described in the theory of elasticity book by [6]. The most important contributions are represented by the introduction of shear correction factors (as in the books of [7, 8]). Gruttmann and colleagues [9–11] undertook a comprehensive effort to calculate shear correction factors across various structural scenarios. These included torsional and flexural shearing stresses in prismatic beams, as well as considerations for arbitrary-shaped cross-sections, wide, thin-walled, and bridge-like structures. El Fatmi [12–16] pioneered advancements in displacement models for beam sections by incorporating a warping function, denoted as ϕ, to enhance the representation of normal and shear stress in the beam. The utilization

E. Carrera et al., *Implementation of Beam-Type Finite Elements Based on Carrera Unified Formulation*, https://doi.org/10.1007/978-3-031-95856-4_1

of warping functions as a strategy to refine displacement fields in beams is a well-established approach. The concept of warping functions was initially introduced within the context of the Saint-Venant torsion problem [17, 18]. Early contributions to this approach can be traced back to [19, 20]. The Saint-Venant solution has served as the theoretical foundation for numerous sophisticated beam models. [21] successfully transformed 3D elasticity equations into formulations suitable for beam-like structures. This method involves constructing a beam model by combining a Saint-Venant part with a residual part, allowing its application to a broad range of structures, including thick beams and thin-walled sections.

Asymptotic methods serve as a potent tool for the development of structural models. In the context of beam modeling, early contributions to exploiting the Variational Asymptotic Method (VAM) were made by [22, 23]. These works presented an alternative approach to constructing refined beam theories, utilizing a characteristic parameter (such as the cross-section thickness of a beam) to form an asymptotic series. The retained terms are those exhibiting the same order of magnitude as the parameter when it approaches zero. Significant contributions to asymptotic methods are found in models like the Variational Asymptotic Beam Section (VABS), as demonstrated by [24].

Another methodology to overcome the limitations of classical beam theories is represented by the Generalized Beam Theory (GBT). GBT traces its origins back to Schardt's work [25–27], enhancing classical theories by employing a piece-wise description of thin-walled sections. Silvestre and Camotim, along with their colleagues, extensively applied and extended the GBT in various forms [28]. Numerous higher-order theories, incorporating enhanced displacement fields across the beam cross-section, have been introduced to account for non-classical effects. Washizu [29] delved into considerations on higher-order beam theories. Further refined beam models, covering aspects such as bending, vibration, wave propagation, buckling, and post-buckling, are comprehensively reviewed by [30, 31]. Refined beam models have found extensive application in aeroelastic contexts. Key contributions in this domain include works by [32].

When confronted with complex structural problems, the limitations of classical beam theories must be taken into account. Problem complexity may stem from geometric features (e.g., thin-walled sections), boundary conditions (e.g., localized loads), or material anisotropy (e.g., laminated composites). Each of these factors, individually or in combination, can produce distinctive structural deformations, thereby necessitating a more sophisticated approximation of the structural model's displacement field. As an example, Fig. 1.1 illustrates a double-clamped, thin-walled beam subjected to a uniform transverse pressure.

Here, the challenge arises from the beam's thinness, which causes significant in-plane and out-of-plane deformations that are not captured by classical one-dimensional models like EBBT or TBT. When these effects become critical for design or structural verification, analysts must move beyond elementary approaches. An experienced engineer would instead adopt refined one-dimensional models (e.g., higher-order theories) or higher-dimensional approximations (e.g., plate/shell ele-

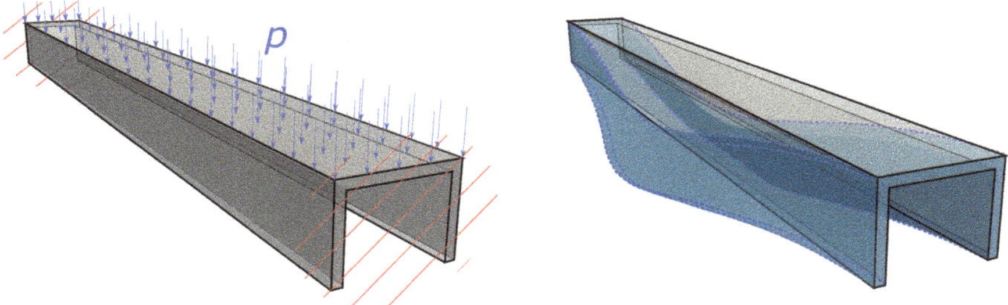

Fig. 1.1 Clamped thin-walled beam subjected to a uniform transverse pressure and related deformation

ments or full three-dimensional elasticity) to ensure accurate predictions of the structural response.

Finite Element Method

Despite the availability of both basic and advanced structural theories, closed-form analytical solutions for most practical engineering problems are seldom available. As a result, modern simulations rely heavily on approximate numerical methods to solve the governing equations of complex structural systems. The most widely used approach, by far, is the Finite Element Method (FEM). Through FEM, the continuum structure is subdivided into smaller finite elements whose stiffness, mass, or damping properties can be calculated individually. By assembling these local contributions, engineers obtain the global response of the structure.

However, one major drawback of FEM (and other numerical methods) is the potentially high computational cost. In FEM, this cost is often proportional to the number of elements in the discretization, which must typically increase to capture complex geometries or sophisticated physical behaviors. Figure 1.2 illustrates a Finite Element (FE) mesh applied to the thin-walled beam example shown in Fig. 1.1, with the nodes colored from green to red to indicate increasing cross-sectional deformation complexity. It is evident that conventional 1D models, such as rods, bars, or beams, ignoring cross-sectional deformation, would produce inaccurate results for this type of slender, thin-walled structure.

To address this shortcoming, users of commercial software may resort to 2D or 3D elements, but thin-walled components require at least 3–4 elements through their thickness to achieve reasonable accuracy, dramatically increasing computational demands. In response to these challenges, the Carrera Unified Formulation (CUF) was introduced. Through CUF, it becomes possible to construct unified beam models that can range from classical, low-order approximations to higher-order, fully

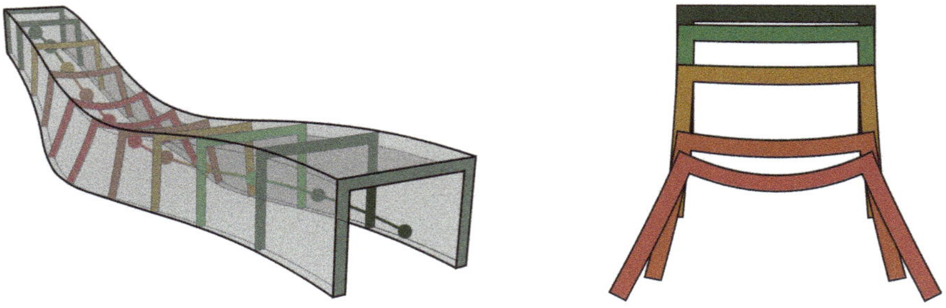

Fig. 1.2 Finite element model application to the study case proposed in Fig. 1.1

refined theories, thereby offering a more efficient and flexible approach to modeling complex beam-like structures.

What This Book Is About

With respect to the earlier book on beam-structure modeling in the CUF framework [33], the present book emphasizes practical guidance by furnishing step-by-step MATLAB scripts for constructing a unified 1D model in straightforward yet realistic scenarios. In addition, it delves into numerical issues commonly encountered in finite element formulations, particularly shear locking, and discusses established remedies. Each proposed solution is paired with a dedicated MATLAB script, thereby streamlining the learning and implementation process.

The concept of 1D Carrera Unified Formulation (CUF) was initially proposed for isotropic compact and thin-walled structures by [34, 35]. These studies employed Taylor-like polynomials to define the cross-sectional displacement field, offering both closed-form and FE solutions. A key advantage of 1D CUF lies in its hierarchical nature, which enables a single formulation to tackle a range of structural challenges. The order of expansion, and hence the refinement level, can be chosen based on a convergence analysis. Subsequent works [36, 37] have compared 1D CUF models with 2D shell formulations for thin-walled, reinforced structures. Their findings highlight the low computational cost of 1D CUF elements and the absence of shear and membrane locking. A novel class of 1D CUF models, employing Lagrange polynomials to characterize the cross-section displacement field, was introduced by [38]. Relying solely on displacement unknowns, these models handle geometric discontinuities and localized boundary conditions more flexibly. The 1D CUF framework has also been extended to free vibration analyses of isotropic beams [39], requiring significantly fewer Degrees of Freedom (DOFs) than shell models. Similar benefits have been observed in buckling [40, 41] and dynamic response [42] analyses.

The investigation of composite structures using 1D CUF has been pursued by [43, 44]. Catapano et al. adopted Taylor beam models with closed-form Navier-type solutions, while Carrera used Lagrange-based beam elements to address more complex systems such as aeronautic longerons. Both approaches demonstrated that 1D CUF can capture 3D stress distributions with 10–100 times fewer DOFs than solid finite elements. Refinement requirements often arise near geometrical and mechanical boundaries, prompting the development of coupling techniques for seamlessly integrating lower- and higher-order models. Biscani [45] employed the Arlequin method to join different beam kinematics along the beam axis, achieving notable computational savings without sacrificing accuracy. Likewise, [46] used Lagrange multipliers with comparable success. Further enhancements in coupling were proposed by [47, 48], who introduced multi-line elements with variable kinematics through the cross-section. This proved especially beneficial for thin-walled and composite beams. Remarkably, 1D CUF allows the node-by-node variation of kinematic refinement, as shown by [49], a topic discussed in detail in this book.

Refined beam models are particularly valuable for aeroelastic applications, where both computational efficiency and high fidelity are crucial. The 1D CUF approach in aeroelasticity was initially investigated by [50], and subsequent work by [51–53] employed the Doublet Lattice Method (DLM) to study unsteady aeroelastic flutter. By combining finite elements with the Doublet Structural Model (DSM), these approaches accurately predicted flutter conditions at a cost far lower than shell-like models. Multifield extensions of 1D CUF include thermomechanical analyses [54, 55], which use closed-form solutions and radial basis functions for temperature fields obtained via the Fourier heat conduction equation. Moreover, piezoelectric structures have been modeled [56–59] by expanding both displacement and electric potential across the cross-section with Lagrange polynomials in a layer-wise fashion. When compared to 3D finite elements, these methods consistently yield high accuracy at a fraction of the computational effort.

Recently, 1D CUF has been further broadened with a Component-Wise (CW) modeling strategy. In CW, each structural component is represented with a dedicated 1D CUF model, and Lagrange polynomials facilitate a straightforward assembly of different parts. This is particularly useful for short, thin-walled beams and typical aerospace components like ribs, stiffeners, and longerons [60, 61]. By leveraging only 1D finite elements, multiscale analysis becomes more direct, since no additional coupling is required to represent layers, fibers, or matrices individually. The CW approach has also demonstrated 3D-like accuracy in failure index computation, achieving up to 100-fold DOF reduction relative to 3D solid elements [62, 63].

Further progress has been made in nonlinear analyses. The works by [64, 65] introduced geometrical nonlinear frameworks for isotropic and composite beams, while [66] extended CUF to material and contact nonlinearities. These advances paved the way for progressive failure studies [67] and multiscale modeling at the microscale [68], underscoring CUF's adaptability and robustness in tackling a wide array of nonlinear problems.

Outline of the Contents

This section offers a concise outline of the book's chapters, highlighting the progression of topics and the level of detail covered.

Chapter 2 Introduces the stiffness matrix of a bar finite element. It is derived by means of the Principle of Virtual Displacement (PVD) and using three approaches: matrix notation, recursive notation against shape functions, and recursive notation against Theory of Structures (TOS). The latter represents the notation used in CUF formalism. The related MATLAB scripts are given in Appendix A.

Chapter 3 Focuses on a beam subjected to flexure, again using the three methods introduced in Chap. 2. Implementation details for the corresponding MATLAB scripts are given in Appendix B. Additionally, this chapter addresses the phenomenon of shear locking, a numerical issue arising from the bending and shear components in the stiffness matrix. To manage integrals related to both bending and shear, the Gauss quadrature method is introduced. Its direct application, referred to as full integration, is discussed at the end of the chapter, with supplementary details in Appendix C.

Chapter 4 Details the possible remedies to overcome the shear locking effect: Uniform Reduced Integration (URI), Selective Reduced Integration (SRI), Mized Interpolation of Tensorial Components (MITC), and the use of higher-order elements Appendix C. The related MATLAB script for the implementation of these remedies are reported in Appendix C. The last part of the chapter provides some numerical examples of the previously described strategies.

Chapters 5, 6 and 7 Present the stiffness matrices for beams under bending, torsion, and combined loading, respectively. Each case uses the three derivation methods introduced earlier. The related MATLAB script are reported in Appendices D, E and F.

Chapter 8 Focuses on the so-called fundamental nucleus, a unified approach based on PVD that facilitates the derivation of the stiffness matrix for a generic beam in a single framework. This represents the basic core of CUF.

Chapter 9 Details the node-dependent kinematic concept, an innovative method allowing displacement unknowns to vary on a node-by-node basis along the beam. The stiffness matrix is constructed using the fundamental nucleus with PVD, independent of cross-sectional geometry, applied loads, or boundary conditions. Moreover, it exploits the unified nature of the fundamental nucleus by changing the shape functions along the beam axis. Focusing on a beam under bending, this chapter derives the corresponding stiffness matrix and demonstrates how varying shape functions can further alleviate shear locking, serving as an additional remedy beyond those introduced in Chap. 4.

Chapter 10 Concludes the book by showcasing the potential of the proposed unified finite-beam approach to handle highly complex structures, including a boat, an industrial building, and a composite launcher.

References

1. Euler, L.: De curvis elasticis. In: Methodus Inveniendi Lineas Curvas Maximi Minimive Proprietate Gaudentes, Sive Solutio Problematis Isoperimetrici Lattissimo Sensu Accepti. Lausanne and Geneva, Bousquet (1744)
2. de Saint-Venant, A.: Mémoire sur la flexion des prismes. Journal de Mathématiques pures et appliquées **1**, 89–189 (1856)
3. Timoshenko, S.P.: On the corrections for shear of the differential equation for transverse vibrations of prismatic bars. Phil. Mag. **41**, 744–746 (1921)
4. Timoshenko, S.P.: On the transverse vibrations of bars of uniform cross section. Phil. Mag. **43**, 125–131 (1922)
5. Mucichescu, D.T.: Bounds for stiffness of prismatic beams. J. Struct. Eng. **110**, 1410–1414 (1984)
6. Novozhilov, V.V.: Theory of Elasticity. Pergamon Press, Oxford (1961)
7. Timoshenko, S.P., Goodier, J.N.: Theory of Elasticity. McGraw-Hill (1970)
8. Sokolnikoff, I.S.: Mathematical Theory of Elasticity. McGrw-Hill (1956)
9. Gruttmann, F., Wagner, W.: Shear correction factors in Timoshenko's beam theory for arbitrary shaped cross-sections. Comput. Mech. **27**, 199–207 (2001)
10. Wagner, W., Gruttmann, F.: A displacement method for the analysis of flexural shear stresses in thin-walled isotropic composite beams. Comput. Struct. **80**, 1843–1851 (2002)
11. Gruttmann, F., Sauer, R., Wagner, W.: Shear stresses in prismatic beams with arbitrary cross-sections. Int. J. Numer. Meth. Eng. **45**, 865–889 (1999)
12. El Fatmi, R.: On the structural behavior and the Saint Venant solution in the exact beam theory. Application to laminated composite beams. Comput. & Struct. **80** 1441–1456 (2002)
13. El Fatmi, R.: Non-uniform warping including the effects of torsion and shear forces. Part I: a general beam theory. Int. J. Solids Struct. **44**(18–19), 5912–5929 (2007)
14. El Fatmi, R.: Non-uniform warping including the effects of torsion and shear forces. Part II: analytical and numerical applications. Int. J. Solids Struct. **44**(18–19), 5930–5952 (2007)
15. El Fatmi, R.: A non-uniform warping theory for beams. C.R. Mec. **335**, 476–474 (2007)
16. El Fatmi, R., Zenzri, H.: A numerical method for the exact elastic beam theory. Applications to homogeneous and composite beams. Int. J. Solids Struct. **41**, 2521–2537 (2004)
17. Ladéveze, P., Simmonds, J.: New concepts for linear beam theory with arbitrary geometry and loading. Eur. J. Mech. A. Solids **17**(3), 377–402 (1998)
18. Lubliner, J.: Plasticity Theory. Macmillan Publishers, London (1990)
19. Volovoi, V.Z.: Thin-Walled Elastic Beams. National Science Foundation, Washington (1961)
20. Benscoter, S.: A theory of Torsion bending for multicell beams. J. Appl. Mech. **21**(1), 25–34 (1954)
21. Ladéveze, P., Simmonds, J.: De nouveaux concepts en théorie des poutres pour des charges et des géométries quelconques. Comptes Rendus Acad. Sci. Paris **332**, 445–462 (1996)
22. Berdichevsky, V.L.: Equations of the theory of anisotropic inhomogeneous rods. Dokl. Akad. Nauk **228**, 558–561 (1976)
23. Berdichevsky, V.L., Armanios, E., Badir. A.: Theory of anisotropic thin-walled closed-cross-section beams. Compos. Eng. **2**(5–7), 411–432 (1992)
24. Volovoi, V.V., Hodges, D.H., Berdichevsky, V.L., Sutyrin, V.G.: Asymptotic theory for static behavior of elastic anisotropic I-beams. Int. J. Solid Struct. **36**, 1017–1043 (1999)
25. Schardt, R.: Eine erweiterung der technischen biegetheorie zur berechnung prismatischer faltwerke. Der Stahlbau **35**, 161–171 (1966)
26. Schardt, R.: Verallgemeinerte technische biegetheorie. Springer (1989)
27. Schardt, R.: Generalized beam theory - an adequate method for coupled stability problems. Thin-Walled Struct. **19**, 161–180 (1994)
28. Silvestre, N., Camotim, D.: First-order generalised beam theory for arbitrary orthotropic materials. Thin-Walled Struct. **40**(9), 791–820 (2002)
29. Washizu, K.: Variational Methods in Elasticity and Plasticity. Pergamon Press, Oxford (1968)

30. Kapania, K., Raciti, S.: Recent advances in analysis of laminated beams and plates, part I: Shear effects and buckling. AIAA J. **27**(7), 923–935 (1989)
31. Kapania, K., Raciti, S.: Recent advances in analysis of laminated beams and plates, part II: Vibrations and wave propagation. AIAA J. **27**(7), 935–946 (1989)
32. Qin, Z., Librescu, L.: On a shear-deformable theory of anisotropic thin-walled beams: further contribution and validations. Compos. Struct. **56**, 345–358 (2002)
33. Carrera, E., Giunta, G., Petrolo, M.: Beam Structures: Classical and Advanced Theories. Wiley (2011)
34. Carrera, E., Giunta, G.: Refined beam theories based on a unified formulation. Int. J. Appl. Mech. **2**(1), 117–143 (2010)
35. Carrera, E., Giunta, G., Nali, P., Petrolo, M.: Refined beam elements with arbitrary cross-section geometries. Comput. Struct. **88**(5–6), 283–293 (2010)
36. Carrera, E., Cinefra, M., Petrolo, M., Nali, P.: Comparisons between 1D (Beam) and 2D (Plate/Shell) Finite Elements to Analyze Thin-Walled Structures. Aerotecnica Missili e Spazio (2014)
37. Carrera, E., Maiaru, M., Petrolo, M., Giunta, G.: A refined 1D element for the structural analysis of single and multiple fiber/matrix cells. Compos. Struct. **96**, 455–468 (2013)
38. Carrera, E., Petrolo, M.: Refined beam elements with only displacement variables and plate/shell capabilities. Meccanica **47**, 537–556 (2012)
39. Carrera, E., Petrolo, M., Nali, P.: Unified formulation applied to free vibrations finite element analysis of beams with arbitrary section. Shock and Vibrations (2010)
40. Ibrahim, S.M., Carrera, E., Petrolo, M., Zappino, E.: Buckling of composite thin walled beams by refined theory. Compos. Struct. **94**(2), 563–570 (2012)
41. Ibrahim, S.M., Carrera, E., Petrolo, M., Zappino, E.: Buckling of thin-walled beams by a refined theory. J. Zhejiang Univ.-Sci. A (Appl. Phy. & Eng.) **13**(10), 747–759 (2012)
42. Carrera, E., Varello, A.: Dynamic response of thin-walled structures by variable kinematic one-dimensional models. J. Sound Vib. **331**(42), 5268–5282 (2012)
43. Catapano, A., Giunta, G., Belouettar, S., Carrera, E.: Static analysis of laminated beams via a unified formulation. Compos. Struct. **94**, 75–83 (2011)
44. Carrera, E., Petrolo, M.: Refined one-dimensional formulations for laminated structure analysis. AIAA J. **50**(1), 176–189 (2012)
45. Biscani, F., Giunta, G., Belouettar, S., Carrera, E., Hu, H.: Variable kinematic beam elements coupled via Arlequin method. Compos. Struct. **93**, 697–708 (2011)
46. Carrera, E., Zappino, E., Filippi, M.: Free vibration analysis of thin-walled cylinders reinforced with longitudinal and transversal stiffeners. J. Vibr. Acous. **135** (2013)
47. Carrera, E., Pagani, A.: Analysis of reinforced and thin-walled structures by multi-line refined 1D/beam models. Int. J. Mech. Sci. **75** (2013)
48. Carrera, E., Pagani, A.: Multi-line enhanced beam model for the analysis of laminated composite structures. Compos.: Part B **57** (2014)
49. Carrera, E., Zappino, E.: One-dimensional finite element formulation with node-dependent kinematics. Comput. & Struct. **192**, 114–125 (2017)
50. Varello, A., Carrera, E., Demasi, L.: Vortex lattice method coupled with advanced one-dimensional structural models. J. Aeroelasticity Struct. Dyn. **2**(2), 53–78 (2011)
51. Pagani, A., Petrolo, M., Carrera, E.: Flutter analysis by refined 1D dynamic stiffness elements and doublet lattice method. Adv. Aircraft Spacecraft Sci. **1**(3), 291–310 (2014)
52. Petrolo, M.: Advanced 1D structural models for flutter analysis of lifting surfaces. Int. J. Aeron. Space Sci. **13**(2), 199–209 (2012)
53. Petrolo, M.: Flutter analysis of composite lifting surfaces by the 1D Carrera Unified Formulation and the doublet lattice method. Compos. Struct. **95**, 539–546 (2013)
54. Giunta, G., Crisafulli, D., Belouettar, S., Carrera, E.: A thermo-mechanical analysis of functionally graded beams via hierarchical modelling. Compos. Struct. **95**, 676–690 (2013)
55. Giunta, G., Metla, N., Belouettar, S., Ferreira, A.J.M., Carrera, E.: A thermo-mechanical analysis of isotropic and composite beams via collocation with radial basis functions. J. Therm. Stress **36** (2013)

56. Giunta, G., Koutsawa, Y., Belouettar, S.: Analysis of Three-Dimensional Piezo-Electric Beams via a Unified Formulation. SMART13: Smart Mat. Struct. (2013)
57. Koutsawa, Y., Giunta, G., Belouettar, S.: Hierarchical FEM modelling of piezo-electric beam structures. Compos. Struct. **95**, 705–718 (2013)
58. Koutsawa, Y., Giunta, G., Belouettar, S.: A free vibration analysis of piezo-electric beams via hierarchical one-dimensional finite elements. J. Intell. Mat. Syst. Struct. (2014)
59. Miglioretti, F., Carrera, E., Petrolo, M.: Variable kinematic beam elements for electro-mechanical analysis. Smart Struct. Syst. **13**(4), 517–546 (2014)
60. Carrere, E., Pagani, A., Petrolo, M.: Use of Lagrange multipliers to combine 1D variable kinematic finite elements. Comput. Struct. **129**, 194–206 (2013)
61. Carrera, E., Pagani, A., Petrolo, M.: Component-wise Method Applied to Vibration of Wing Structures. J. Appl. Mech. **80**(4) (2013)
62. Carrera, E., Maiaru, M., Petrolo, M.: Component-Wise Analysis of Laminated Anisotropic Composites. Int. J. Solids Struct. **49** (2010)
63. Carrera, E., Maiaru, M., Petrolo, M., Giunta, G.: A refined 1D element for the structural analysis of single and multiple fiber/matrix cells. Compos. Struct. **96**, 455–468 (2013)
64. Pagani, A., Carrera, E.: Unified formulation of geometrically nonlinear refined beam theories. Mech. Adv. Mater. Struct. **25**(1), 15–31 (2018)
65. Pagani, A., Carrera, E.: Large-deflection and post-buckling analyses of laminated composite beams by Carrera Unified Formulation. Compos. Struct. **170**, 40–52 (2017)
66. Nagaraj, M.H., Kaleel, I., Carrera, E., Petrolo, M.: Nonlinear analysis of compact and thin-walled metallic structures including localized plasticity under contact conditions. Eng. Struct. **203** (2020)
67. Kaleel, I., Petrolo, M., Waas, A.M., Carrera, E.: Micromechanical progressive failure analysis of fiber-reinforced composite using refined beam models. J. Appl. Mech. **85**(2) (2018)
68. Kaleel, I., Petrolo, M., Carrera, E.: Elastoplastic and progressive failure analysis of fiber-reinforced composites via an efficient nonlinear microscale model. Aerotecnica Missili & Spazio **97**, 103–110 (2018)

Chapter 2
Stiffness Matrix of a Bar Element

Graphical Abstract

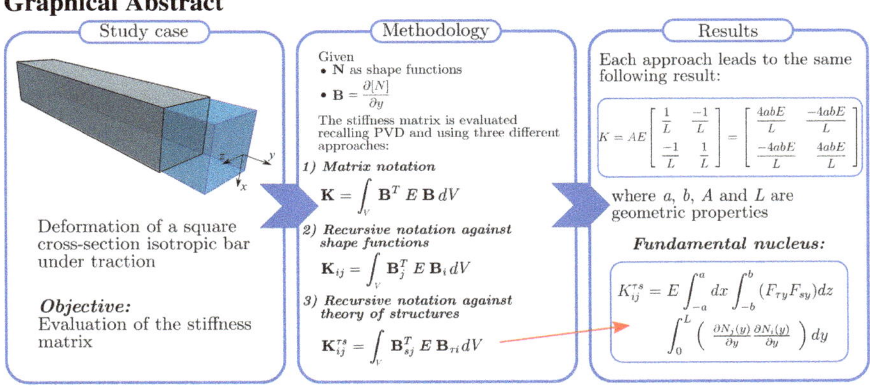

Overview

In this chapter, the stiffness matrix of a two-node bar element is derived. The bar is a structural element with axial and torsional stiffness, but only the axial stiffness is considered in this derivation. The methodology behind this derivation is presented in Sect. 2.1. Section 2.2 details the assumptions underlying the finite element bar definition and specifies the associated geometric data. The derivation of the stiffness matrix employs the Principle of Virtual Displacements (PVD), a concept grounded in classical mechanics and variational calculus. PVD serves as a core tool for formulating equations of motion in mechanical systems. A virtual displacement is an infinitesimal, hypothetical displacement consistent with the system's constraints. According to PVD, the total virtual work of both external and internal forces equals zero for any such permissible virtual displacement when a system is in equilibrium. Three distinct approaches are used throughout this chapter to evaluate the stiffness matrix:

E. Carrera et al., *Implementation of Beam-Type Finite Elements Based on Carrera Unified Formulation*, https://doi.org/10.1007/978-3-031-95856-4_2

- Matrix notation (Sect. 2.4);
- Recursive notation based on shape functions (Sect. 2.5);
- Recursive notation derived from the Theory of Structures (TOS) (Sect. 2.6).

Section 2.6 introduces the generic expansion functions F_τ and F_s, which express the relevant variables and unknowns in expansion series along the bar's main axis and across its cross-section. This concept underpins the Carrera Unified Formulation (CUF). Appendix A provides the dedicated MATLAB script for the study case presented in this chapter.

2.1 Structural Model of a Bar

Figure 2.1 shows the deformed shape of a bar subjected to a tensile load. This structural element has axial and torsional stiffness. For the purposes of this case, only the axial stiffness is considered, allowing the bar to undergo axial loads, whether in tension or compression.

Figure 2.2 depicts the displacement of a generic point P, which, under the influence of an external tensile load, shifts to position P' along the longitudinal axis of the bar. The displacement is denoted as $v^0(y)$, and its magnitude varies based on the location along the longitudinal axis of the bar, i.e. it is function of y.

Regarding the displacement field, a tensile load induces elongation in the bar, represented by $v^0(y)$. The displacement remains constant for all points within the

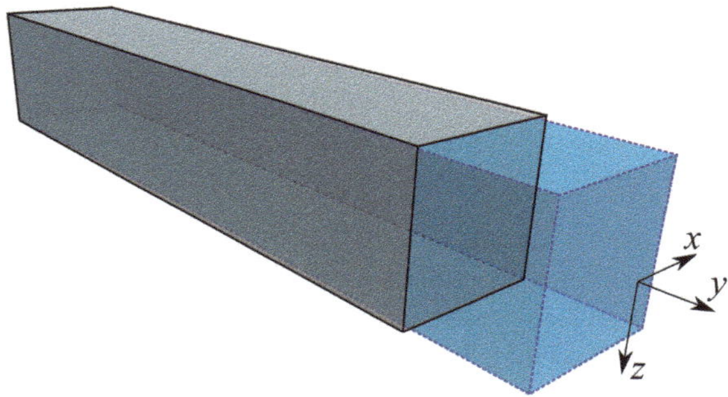

Fig. 2.1 Deformation of a bar under traction

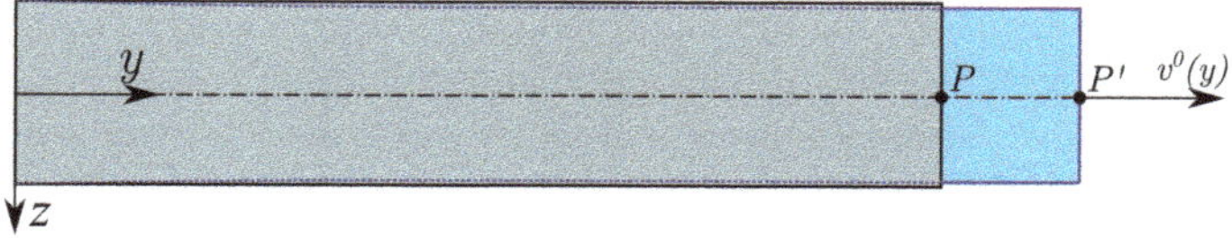

Fig. 2.2 1D displacement of a bar under traction

same cross-section of the bar but varies along the length of the bar. The displacement field, in a Cartesian coordinate system (x, y, z), is expressed as follows:

$$
\begin{aligned}
u(x, y, z) &= 0 \\
v(x, y, z) &= v^0(y) \\
w(x, y, z) &= 0
\end{aligned}
\tag{2.1}
$$

where the exponent 0 refers to the degree of the displacement function over the cross-section of the bar, and therefore it means that the displacement in this domain is constant. The geometric relations between the strain ϵ_{yy} and the displacement $v^0(y)$ is the following

$$
\epsilon_{yy} = \frac{\partial v^0(y)}{\partial y}
\tag{2.2}
$$

From the previous equation, the differential operator b can be introduced

$$
b = \frac{\partial}{\partial y}
\tag{2.3}
$$

Then, Eq. (2.2) can be written as

$$
\epsilon_{yy} = b\, v^0(y)
\tag{2.4}
$$

The constitutive relation between the stress and the strain occurring in the bar during the application of a tensile (or compressive) load is obtained employing three-dimensional (3D) generalized Hooke's law, considering a linear elastic isotropic materials. The only stress component in this case is σ_{yy}. The stress-strain relation can be written as

$$
\sigma_{yy} = C_{11}\, \epsilon_{yy}
\tag{2.5}
$$

The material parameter C_{11} can be expressed as a function of the material properties so that

$$
C_{11} = 2G + \lambda
\tag{2.6}
$$

where

$$
G = \frac{E}{2(1+v)}, \qquad \lambda = \frac{vE}{(1+v)(1-2v)}
\tag{2.7}
$$

In the previous expression, v is the Poisson's ratio, E is the Young's modulus and G is the shear modulus. The terms λ and G are also named Lamé constants.

Assuming the displacements to be in the plane of the bar cross-section, Poisson effects can be omitted, then v in Eq. (2.7) is zero. Then, one has

$$
G = \frac{E}{2}, \quad \lambda = 0, \quad C_{11} = E
\tag{2.8}
$$

Finally, in the case of a bar under tensile load, the stress-strain relation becomes

$$\sigma_{yy} = E\,\epsilon_{yy}. \tag{2.9}$$

2.2   Finite Element Approximation

The finite element approximation of the bar under traction is considered here. The example refers to a two-node bar element as reported in the following red box. The bar's coordinate system is defined with the y axis aligned along its length, while the x and z axes define the cross-section. v_1^0 and v_2^0 represent the displacements along the y direction for nodes 1 and 2 of the bar, respectively. The subscripts 1 and 2 denote the bar nodes, and the exponent 0 corresponds to the notation in Eq. (2.1).

Geometry of the bar

The following dimensions are considered to evaluate the stiffness matrix of the bar under axial displacement.

- $L = 500$ mm;
- y of node 1 = 0 mm
  y of node 2 = 500 mm.

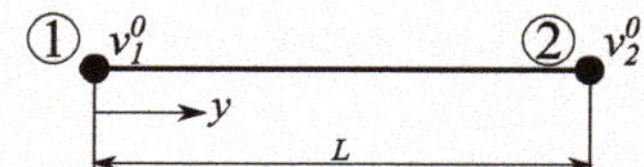

The reference frame is located at the center of a square cross-section. Consequently, the coordinates of the nodes on the axis and the points of the cross-section are

- $a = b = 10$ mm;
- $P_1 = (x = -10, z = 10)$ mm
  $P_2 = (x = -10, z = -10)$ mm
  $P_3 = (x = 10, z = -10)$ mm
  $P_4 = (x = 10, z = 10)$ mm.

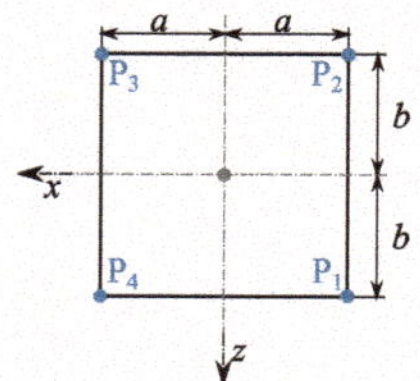

The following quantities can be calculated

- $dA = dx\,dz, \quad A = \int_A dA = \int_{-a}^{a} dx \int_{-b}^{b} dz = 4ab;$
- $dV = dA\,dy = dx\,dz\,dy, \quad V = \int_{-a}^{a} dx \int_{-b}^{b} dz \int_{0}^{L} dy = 4abL.$

2.3 Principle of Virtual Displacement

PVD is here recalled to derive the governing equation of a structural problem. PVD can be written in its static case as

$$\delta L_{\text{int}} = \delta L_{\text{ext}} \tag{2.10}$$

where δL_{int} and δL_{ext} are the virtual variation of the internal strain energy and the external work, respectively, with δ denoting the variation. The internal work can be expressed in matrix notation as as

$$\delta L_{\text{int}} = \int_V \delta \boldsymbol{\epsilon}^T \boldsymbol{\sigma} \, dV \tag{2.11}$$

The same equation can be written in explicit form:

$$\delta L_{\text{int}} = \int_V (\sigma_{xx} \delta \epsilon_{xx} + \sigma_{yy} \delta \epsilon_{yy} + \sigma_{zz} \delta \epsilon_{zz} + \sigma_{xz} \delta \epsilon_{xz} + \sigma_{yz} \delta \epsilon_{yz} + \sigma_{xy} \delta \epsilon_{xy}) \, dV \tag{2.12}$$

In this chapter, the only non-zero stress and strain components are those along the y direction. Thus, Eq. (2.12) becomes

$$\delta L_{\text{int}} = \int_V (\sigma_{yy} \delta \epsilon_{yy}) \, dV \tag{2.13}$$

Introducing Eqs. (2.4) and (2.5), the previous relation can be written as

$$\delta L_{\text{int}} = \int_V E \, b \, v^0 \, \delta(b \, v^0) dV. \tag{2.14}$$

2.4 Matrix Notation

In this section the stiffness matrix of the bar is calculated starting from Eq. (2.14). The variation of the displacement $v^0(y)$ along the bar length can be expressed as a function of the displacements of the bar nodes 1 and 2, i.e. v_1^0 and v_2^0 respectively, and the shape functions. The shape functions used for this purpose are based on Lagrange polynomials and are expressed as

$$N_1(y) = 1 - \frac{y}{L}$$
$$N_2(y) = \frac{y}{L} \tag{2.15}$$

Then, the expression which describes the variation of the displacement along the bar, is the following

$$v^0(y) = N_1(y)v_1^0 + N_2(y)v_2^0 \tag{2.16}$$

Equation 2.16 can be written as

$$v^0(y) = [N_1(y) \quad N_2(y)] \begin{bmatrix} v_1^0 \\ v_2^0 \end{bmatrix} = [N_1(y) \quad N_2(y)]\, \mathbf{S}^0 \tag{2.17}$$

Equation 2.17 introduces $\mathbf{S}^0$, which is the unknowns vector of the problem.

Introducing the *displacement differentiation matrix* $\mathbf{B}$, it is obtained by deriving the shape functions used to describe the bar displacements (Eq. (2.15)). It is expressed in the following

$$\mathbf{B} = b\,[N_1(y) \quad N_2(y)] = \left[\frac{\partial N_1(y)}{\partial y} \quad \frac{\partial N_2(y)}{\partial y} \right] \tag{2.18}$$

Introducing the defined $\mathbf{B}$ into Eq. (2.14), and rearranging the parameters, it becomes

$$\delta L_{\text{int}} = \int_V \delta \mathbf{S}^0 \, \mathbf{B}^T E \, \mathbf{B} \, \mathbf{S}^0 \, dV \tag{2.19}$$

where the stiffness matrix $\mathbf{K}$ of the bar can be introduced as

$$\mathbf{K} = \int_V \mathbf{B}^T E \mathbf{B} \, dV \tag{2.20}$$

Equation (2.35) can be written as:

$$\mathbf{K} = \int_V \begin{pmatrix} \frac{\partial N_1(y)}{\partial y} \\ \frac{\partial N_2(y)}{\partial y} \end{pmatrix} E \left(\frac{\partial N_1(y)}{\partial y} \quad \frac{\partial N_2(y)}{\partial y} \right) dV \tag{2.21}$$

The volume integral of Eq. (2.21) must be solved. If the following equivalences are considered,

$$dV = dA\,dy, \quad dA = dx\,dz \tag{2.22}$$

Equation (2.21) becomes

$$\mathbf{K} = \int_A dA \int_y \begin{pmatrix} \frac{\partial N_1(y)}{\partial y} \\ \frac{\partial N_2(y)}{\partial y} \end{pmatrix} E \left(\frac{\partial N_1(y)}{\partial y} \quad \frac{\partial N_2(y)}{\partial y} \right) dy \tag{2.23}$$

In the Eq. (2.23) the surface integral $\int_A dA$ has to be evaluated over the bar cross-section, whereas the line integral $\int_L dy$ along the bar length. Since E is a constant and by computing the product inside the line integral, Eq. (2.23) can written as:

$$\mathbf{K} = E \int_{-a}^{a} dx \int_{-b}^{b} dz \int_{0}^{L} \begin{pmatrix} \frac{\partial N_1(y)}{\partial y}\frac{\partial N_1(y)}{\partial y} & \frac{\partial N_1(y)}{\partial y}\frac{\partial N_2(y)}{\partial y} \\ \frac{\partial N_2(y)}{\partial y}\frac{\partial N_1(y)}{\partial y} & \frac{\partial N_2(y)}{\partial y}\frac{\partial N_2(y)}{\partial y} \end{pmatrix} dy \tag{2.24}$$

The derivatives of the shape functions which are in Eq. (2.24) can be calculated by their expression in Eq. (2.15). These are

$$\frac{\partial N_1(y)}{\partial y} = \frac{\partial}{\partial y}\left(1 - \frac{y}{L}\right) = \frac{-1}{L}$$

$$\frac{\partial N_2(y)}{\partial y} = \frac{\partial}{\partial y}\left(\frac{y}{L}\right) = \frac{1}{L} \tag{2.25}$$

By introducing the results of Eq. (2.25) into Eqs. (2.18) and (2.24), the following equations are obtained

$$\mathbf{B} = \begin{bmatrix} \dfrac{-1}{L} & \dfrac{1}{L} \end{bmatrix} \tag{2.26}$$

$$\mathbf{K} = E \int_{-a}^{a} dx \int_{-b}^{b} dz \int_{0}^{L} \begin{pmatrix} \dfrac{1}{L^2} & \dfrac{-1}{L^2} \\ \dfrac{-1}{L^2} & \dfrac{1}{L^2} \end{pmatrix} dy \tag{2.27}$$

Equation (2.27) can be written as follows

$$\mathbf{K} = E \int_{-a}^{a} dx \int_{-b}^{b} dz \begin{pmatrix} \int_{0}^{L}\dfrac{1}{L^2}dy & \int_{0}^{L}\dfrac{-1}{L^2}dy \\ \int_{0}^{L}\dfrac{-1}{L^2}dy & \int_{0}^{L}\dfrac{1}{L^2}dy \end{pmatrix} \tag{2.28}$$

and then

$$\mathbf{K} = 2a2bE \begin{pmatrix} \dfrac{1}{L} & \dfrac{-1}{L} \\ \dfrac{-1}{L} & \dfrac{1}{L} \end{pmatrix} = 4abE \begin{pmatrix} \dfrac{1}{L} & \dfrac{-1}{L} \\ \dfrac{-1}{L} & \dfrac{1}{L} \end{pmatrix} = \begin{bmatrix} \dfrac{4abE}{L} & \dfrac{-4abE}{L} \\ \dfrac{-4abE}{L} & \dfrac{4abE}{L} \end{bmatrix} \tag{2.29}$$

Considering that $4ab = A$, the stiffness matrix of the bar is the following

$$\mathbf{K} = \begin{bmatrix} \dfrac{4abE}{L} & \dfrac{-4abE}{L} \\[2ex] \dfrac{-4abE}{L} & \dfrac{4abE}{L} \end{bmatrix} = \begin{bmatrix} \dfrac{AE}{L} & \dfrac{-AE}{L} \\[2ex] \dfrac{-AE}{L} & \dfrac{AE}{L} \end{bmatrix} \tag{2.30}$$

In order to obtain the numerical values of the matrix shown in Eq. (2.30), the numerical data are considered. Considering a steel alloy with $E = 210000 \frac{\text{N}}{\text{mm}^2}$, the numerical matrix of the bar is:

$$\mathbf{K} = \begin{bmatrix} \dfrac{4\cdot10\cdot10\cdot210000}{500} & \dfrac{-4\cdot10\cdot10\cdot210000}{500} \\[2ex] \dfrac{-4\cdot10\cdot10\cdot210000}{500} & \dfrac{4\cdot10\cdot10\cdot210000}{500} \end{bmatrix} = \begin{bmatrix} 168000 & -168000 \\[2ex] -168000 & 168000 \end{bmatrix} \tag{2.31}$$

The comprehensive procedure to derive the matrix presented in Eq. (2.31) is detailed in Appendix A.1, and the step-by-step implementation is provided as a MAT-LAB script.

2.5   Recursive Notation Against Shape Functions

In this section the stiffness matrix of the bar is calculated using a recursive notation against shape functions. The displacements field of Eq. (2.1) can be written as follows:

$$v^0(y) = N_i(y)v_i^0 \tag{2.32}$$

with i ranging from 1 to 2, as the number of nodes of the beam. The $\mathbf{B}$ matrix of Eq. (2.18) can be written using the recursive notation as:

$$\mathbf{B}_i = \frac{\partial}{\partial y}\left(N_i(y)\right) \tag{2.33}$$

In order to distinguish the variables from their virtual variation, the second index j is introduced.

$$\mathbf{B}_j = \frac{\partial}{\partial y}\left(N_j(y)\right) \tag{2.34}$$

Equation 2.35 can be then rewritten using the recursive notation of Eq. 2.34.

$$\mathbf{K} = \int_V \mathbf{B}_j^T E \mathbf{B}_i \, dV \tag{2.35}$$

Introducing the definition of $\mathbf{B}_i$ and $\mathbf{B}_j$ into Eq. (2.35), and looping the indexes i and j from 1 to 2, i.e. the nodes of the element, it has the following form

$$\mathbf{K} = \int_V \begin{pmatrix} \frac{\partial N_1(y)}{\partial y} \\ \frac{\partial N_2(y)}{\partial y} \end{pmatrix} E \begin{pmatrix} \frac{\partial N_1(y)}{\partial y} & \frac{\partial N_2(y)}{\partial y} \end{pmatrix} dV \tag{2.36}$$

As a result, Eq. (2.36) is the same as Eq. 2.21. Considering the notation of the stiffness matrix as:

$$\mathbf{K} = \begin{bmatrix} K_{11} & K_{12} \\ & \\ K_{21} & K_{22} \end{bmatrix} \tag{2.37}$$

it becomes as follows

$$\mathbf{K} = AE \begin{bmatrix} \int_0^L \left(\frac{\partial}{\partial y}(N_1(y)) \frac{\partial}{\partial y}(N_1(y)) \right) dy & \int_0^L \left(\frac{\partial}{\partial y}(N_1(y)) \frac{\partial}{\partial y}(N_2(y)) \right) dy \\ \int_0^L \left(\frac{\partial}{\partial y}(N_2(y)) \frac{\partial}{\partial y}(N_1(y)) \right) dy & \int_0^L \left(\frac{\partial}{\partial y}(N_2(y)) \frac{\partial}{\partial y}(N_2(y)) \right) dy \end{bmatrix} \tag{2.38}$$

Introducing the derivatives of the shape functions, it is obtained

$$\mathbf{K} = AE \begin{bmatrix} \int_0^L \left(\frac{-1}{L} \frac{-1}{L} \right) dy & \int_0^L \left(\frac{-1}{L} \frac{1}{L} \right) dy \\ \int_0^L \left(\frac{1}{L} \frac{-1}{L} \right) dy & \int_0^L \left(\frac{1}{L} \frac{1}{L} \right) dy \end{bmatrix} \tag{2.39}$$

Note that Eq. (2.39) is equal to the equation (2.28). The stiffness matrix obtained using the recursive notation against shape function, introducing the numerical data, can be written as

$$\mathbf{K} = AE \begin{bmatrix} \frac{1}{L} & \frac{-1}{L} \\ \frac{-1}{L} & \frac{1}{L} \end{bmatrix} = \begin{bmatrix} \frac{4 \cdot 10 \cdot 10 \cdot 210000}{L} & \frac{-4 \cdot 10 \cdot 10 \cdot 210000}{L} \\ \frac{-4 \cdot 10 \cdot 10 \cdot 210000}{L} & \frac{4 \cdot 10 \cdot 10 \cdot 210000}{L} \end{bmatrix} \tag{2.40}$$

and

$$
\mathbf{K} = \begin{bmatrix} \dfrac{4\cdot10\cdot10\cdot210000}{500} & \dfrac{-4\cdot10\cdot10\cdot210000}{500} \\[2ex] \dfrac{-4\cdot10\cdot10\cdot210000}{500} & \dfrac{4\cdot10\cdot10\cdot210000}{500} \end{bmatrix} = \begin{bmatrix} 168000 & -168000 \\[2ex] -168000 & 168000 \end{bmatrix}
\tag{2.41}
$$

The comprehensive procedure to derive the matrix presented in Eq. (2.41) is detailed in Appendix A.2, and the step-by-step implementation is provided as a MAT-LAB script.

2.6 Recursive Notation Against the Theory of Structures

In this section the stiffness matrix of the bar is calculated using a recursive notation against TOS. Introducing $s_n = s_x, s_y, s_z$, Eq. (2.1) becomes

$$
\begin{aligned}
s_x &= u(x, y, z) = 0 \\
s_y &= v(x, y, z) = v^0(y) \\
s_z &= w(x, y, z) = 0
\end{aligned}
\tag{2.42}
$$

Let us now introduce the *so-called* expansion functions F. They are introduced so that

$$
\begin{aligned}
s_x &= F_{1_x} \times 0 \\
s_y &= F_{1_y} \times v^0(y) \\
s_z &= F_{1_z} \times 0
\end{aligned}
\tag{2.43}
$$

In this case

$$
F_{1_x} = F_{1_z} = 0, \qquad F_{1_y} = 1
\tag{2.44}
$$

Equation (6.38) can be expressed in a generic form introducing the index τ which ranges from 1 to the number of the terms in the expansion functions F. This case can then expressed as

$$
s_n = F_{\tau_n} s_{\tau_n}
\tag{2.45}
$$

with the generic index n introduced for the directions x, y and z and

$$
s_{1_x} = s_{1_z} = 0, \qquad s_{1_y} = v^0(y)
\tag{2.46}
$$

The same procedure can be used for the virtual variation, by introducing the index s

$$
\delta s_n = F_{s_n} s_{s_n}
\tag{2.47}
$$

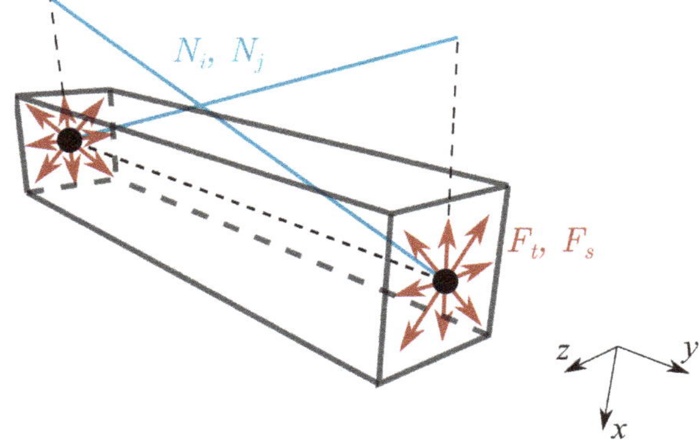

Fig. 2.3 Representation of the axial approximation (using N_i, N_j), and the cross-section expansion (using F_τ, F_s) of the bar using a two-node finite element

By applying the recursive notation against TOS, $\mathbf{B}_{\tau i}$ can be written as follows

$$\mathbf{B}_{\tau i} = F_{\tau y}\frac{\partial N_i(y)}{\partial y} \tag{2.48}$$

and its virtual variation

$$\mathbf{B}_{sj} = F_{sy}\frac{\partial N_j(y)}{\partial y} \tag{2.49}$$

Figure 2.3 shows the two approximations. The shape functions N_i and N_j expand the solution from the nodes to the axis. Expansion F_τ and F_s expand the solution from the nodes to the cross-section of the bar.

Finally, the stiffness matrix can be expressed as follows

$$\mathbf{K}_{ij}^{\tau s} = \int_V \mathbf{B}_{sj}^T E \mathbf{B}_{\tau i}\, dV = \int_V \left(F_{sy}\frac{\partial N_j(y)}{\partial y} \right) E \left(F_{\tau y}\frac{\partial N_i(y)}{\partial y} \right) dV \tag{2.50}$$

and

$$\mathbf{K}_{ij}^{\tau s} = E \int_A F_{\tau y} F_{sy}\, dA \int_y \frac{\partial N_j(y)}{\partial y}\frac{\partial N_i(y)}{\partial y}\, dy \tag{2.51}$$

Equation (2.51) can be written as follow

$$\mathbf{K}_{ij}^{\tau s} = E \int_{-a}^{a} dx \int_{-b}^{b} (F_{\tau y} F_{sy})\, dz \int_0^L \frac{\partial N_j(y)}{\partial y}\frac{\partial N_i(y)}{\partial y}\, dy \tag{2.52}$$

Considering that $\tau = 1$ and $s = 1$ and that both the indexes i and j can vary from 1 to 2, following the same notation already presented in Eq. (2.37)

$$\mathbf{K}^{11} = \begin{bmatrix} K_{11}^{11} & K_{12}^{11} \\ \\ K_{21}^{11} & K_{22}^{11} \end{bmatrix} \tag{2.53}$$

Iterating the indices i and j from 1 to 2, the resulting terms are as follows:

$$K_{11}^{11} = E \int_{-a}^{a} dx \int_{-b}^{b} (F_{1y}F_{1y})dz \int_{0}^{L} \left[\frac{\partial}{\partial y}(N_1(y))\frac{\partial}{\partial y}(N_1(y)) \right] dy$$

$$K_{12}^{11} = E \int_{-a}^{a} dx \int_{-b}^{b} (F_{1y}F_{1y})dz \int_{0}^{L} \left[\frac{\partial}{\partial y}(N_1(y))\frac{\partial}{\partial y}(N_2(y)) \right] dy$$

$$K_{21}^{11} = E \int_{-a}^{a} dx \int_{-b}^{b} (F_{1y}F_{1y})dz \int_{0}^{L} \left[\frac{\partial}{\partial y}(N_2(y))\frac{\partial}{\partial y}(N_1(y)) \right] dy \tag{2.54}$$

$$K_{22}^{11} = E \int_{-a}^{a} dx \int_{-b}^{b} (F_{1y}F_{1y})dz \int_{0}^{L} \left[\frac{\partial}{\partial y}(N_2(y))\frac{\partial}{\partial y}(N_2(y)) \right] dy$$

Since

$$\int_{A} F_{\tau y}F_{sy}\,dA = \int_{A} F_{1y}F_{1y}\,dA = \int_{-a}^{-a} dx \int_{-b}^{-b} F_{1y}F_{1y}dz \tag{2.55}$$

Equation (2.54) then become as follows

$$K_{11}^{11} = AE \int_{0}^{L} \left(\frac{\partial}{\partial y}(N_1(y))\frac{\partial}{\partial y}(N_1(y)) \right) dy$$

$$K_{11}^{12} = AE \int_{0}^{L} \left(\frac{\partial}{\partial y}(N_1(y))\frac{\partial}{\partial y}(N_2(y)) \right) dy$$

$$K_{11}^{21} = AE \int_{0}^{L} \left(\frac{\partial}{\partial y}(N_2(y))\frac{\partial}{\partial y}(N_1(y)) \right) dy \tag{2.56}$$

$$K_{11}^{22} = AE \int_{0}^{L} \left(\frac{\partial}{\partial y}(N_2(y))\frac{\partial}{\partial y}(N_2(y)) \right) dy$$

The final expression of the stiffness matrix is equal to that of Eq. (2.40) and reported below.

$$\mathbf{K} = AE \begin{bmatrix} \dfrac{1}{L} & \dfrac{-1}{L} \\[2mm] \dfrac{-1}{L} & \dfrac{1}{L} \end{bmatrix} = \begin{bmatrix} \dfrac{4abE}{L} & \dfrac{-4abE}{L} \\[2mm] \dfrac{-4abE}{L} & \dfrac{4abE}{L} \end{bmatrix} \tag{2.57}$$

In order to obtain the numerical values for the matrix of Eq. (2.57) the same geometric and material data are considered. The numerical matrix of a bar, calculated by using the recursive notation against TOS results the following

$$\mathbf{K} = \begin{bmatrix} \dfrac{4 \cdot 10 \cdot 10 \cdot 210000}{500} & \dfrac{-4 \cdot 10 \cdot 10 \cdot 210000}{500} \\[3mm] \dfrac{-4 \cdot 10 \cdot 10 \cdot 210000}{500} & \dfrac{4 \cdot 10 \cdot 10 \cdot 210000}{500} \end{bmatrix} = \begin{bmatrix} 168000 & -168000 \\[3mm] -168000 & 168000 \end{bmatrix} \tag{2.58}$$

The comprehensive procedure to derive the matrix presented in Eq. (2.58) is detailed in Appendix A.3, and the step-by-step implementation is provided as a MATLAB script.

Chapter 3
Stiffness Matrix of a Beam Bending Around z Axis

Graphical Abstract

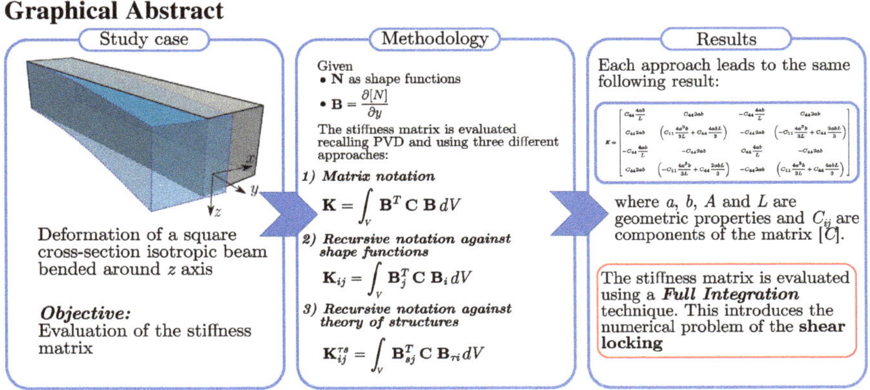

Overview

In this chapter, the stiffness matrix of a two-node beam element is derived. The beam is a structural element with axial, torsional and bending stiffness, but only the bending stiffness is considered in this derivation. The methodology behind this derivation is presented in Sect. 3.1. Section 3.2 details the assumptions underlying the finite element beam definition and specifies the associated geometric data. The stiffness matrix derivation utilizes the PVD approach, as discussed in the preceding chapter. The same techniques are employed, including:

- Matrix notation (Sect. 3.4);
- Recursive notation against the shape functions (Sect. 3.5);
- Recursive notation against the Theory of Structures (TOS) (Sect. 3.6).

Section 3.7 discusses the shear locking phenomenon, a well-documented numerical issue that can significantly compromise the accuracy of beam displacement predictions. In the standard finite element formulation of beams, shear stiffness tends to be

E. Carrera et al., *Implementation of Beam-Type Finite Elements Based on Carrera Unified Formulation*, https://doi.org/10.1007/978-3-031-95856-4_3

overestimated while bending stiffness is concurrently underestimated. This imbalance causes external loads to be primarily resisted by the erroneously high shear stiffness rather than the intended bending mechanism, leading to overly stiff beam responses and inaccurate displacement fields.

3.1 Structural Model of a Beam

Figure 3.1 shows the deformed shape of a beam bending around the z axis. Three displacement components are involved in the deformation of the beam: the displacement u along the direction x, the displacement v along the direction y, the displacement w along the direction z.

The bending of a beam around z axis produces displacements in the xy plane. Then, the displacement w along the z axis can be assumed as zero. Figure 3.2 shows a two-dimensional representation of deformed beam. In this figure, $u^0(y)$ refers to the displacement along direction x as a function of the beam axis y and $\phi_z(y)$ is the rotation angle around the z axis.

Given a specific point P, identified in the figure at a distance x from the midpoint of the beam cross-section, it has a displacement v_P along the longitudinal y axis. Assuming small rotations (i.e., $tan(\phi_z(y)) \simeq \phi_z(y)$), this displacement is given by $x\phi_z(y)$.

The displacement field is the following:

Fig. 3.1 Bending around z axis

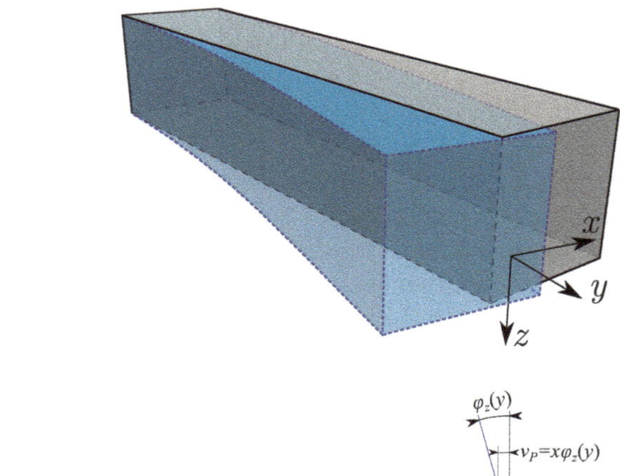

Fig. 3.2 Two-dimensional assumptions of bending around z axis

$$
\begin{aligned}
u(x, y, z) &= -u^0(y) \\
v(x, y, z) &= x\phi_z(y) \\
w(x, y, z) &= 0
\end{aligned}
\tag{3.1}
$$

For the case under study, the strain components of interest are ϵ_{yy} and ϵ_{xy} and. The geometric relations between these strain components and the displacements are:

$$
\epsilon_{yy} = \frac{\partial v}{\partial y}, \quad \epsilon_{xy} = \frac{\partial u}{\partial y} + \frac{\partial v}{\partial x}
\tag{3.2}
$$

Given the displacement components of Eq. (3.1), the following expressions of the strains are obtained

$$
\epsilon_{yy} = x\frac{\partial \phi_z(y)}{\partial y} \quad \epsilon_{xy} = -\frac{\partial u^0(y)}{\partial y} + \phi_z
\tag{3.3}
$$

In a matrix form, Eq. (3.3) reads as:

$$
\epsilon = \left\{ \begin{matrix} \epsilon_{yy} \\ \epsilon_{xy} \end{matrix} \right\} = \begin{bmatrix} 0 & x\dfrac{\partial}{\partial y} \\ -\dfrac{\partial}{\partial y} & 1 \end{bmatrix} \left\{ \begin{matrix} u^0(y) \\ \phi_z(y) \end{matrix} \right\}
\tag{3.4}
$$

Introducing the differential operator matrix $\mathbf{b}$, Eq. (3.4) can be expressed as:

$$
\mathbf{b} = \begin{bmatrix} 0 & x\dfrac{\partial}{\partial y} \\ -\dfrac{\partial}{\partial y} & 1 \end{bmatrix}, \quad \epsilon = \mathbf{b} \left\{ \begin{matrix} u^0(y) \\ \phi_z(y) \end{matrix} \right\} = \mathbf{b}\, \mathbf{s}^0
\tag{3.5}
$$

where $\mathbf{s}^0$ equals to

$$
\mathbf{s}^0 = \left\{ \begin{matrix} u^0(y) \\ \phi_z(y) \end{matrix} \right\}
\tag{3.6}
$$

The constitutive relations are based on the material coefficients and the Hooke's law. The constitutive relations of interest are the following

$$
\begin{aligned}
\sigma_{yy} &= C_{11}\epsilon_{yy} \\
\sigma_{xy} &= C_{44}\epsilon_{xy}
\end{aligned}
\tag{3.7}
$$

The expression of C_{11} is given in Eqs. (2.6) and (2.7), while the expression of C_{44} is equal to

$$
C_{44} = G
\tag{3.8}
$$

Eq. (3.7) can be written in matrix form, reading as

$$\boldsymbol{\sigma} = \mathbf{C}\boldsymbol{\epsilon} \tag{3.9}$$

where

$$\boldsymbol{\sigma} = \left\{ \begin{array}{c} \sigma_{yy} \\ \sigma_{xy} \end{array} \right\} \tag{3.10}$$

$$\boldsymbol{\epsilon} = \left\{ \begin{array}{c} \epsilon_{yy} \\ \epsilon_{xy} \end{array} \right\} \tag{3.11}$$

$$\mathbf{C} = \left[\begin{array}{cc} C_{11} & 0 \\ 0 & C_{44} \end{array} \right]. \tag{3.12}$$

3.2 Finite Element Approximation

The finite element approximation of the beam under bending around z axis is considered here. This example refers to a two-node beam element, as shown in Fig. 3.3, where subscripts 1 and 2 refer to node 1 and node 2 of the beam.

The coordinate system of the beam is also indicated in Fig. 3.3: the y axis lays along the element length, and as a consequence the x and z axes define the cross-section of the beam.

u_1^0 and u_2^0 are the displacements along the coordoinate x, while ϕ_{z1} and ϕ_{z2} refer to the rotations of the cross-section of the beam around the z axis at nodes 1 and 2. The geometric characteristics and the dimensions of the beam are the same as those of the bar reported in the Sect. 2.2.

3.3 Principle of Virtual Displacement

PVD is used to derive the stiffness matrix of the beam. In this chapter, σ_{yy} and σ_{xy} are the non-zero component of the stress tensor. Consequently, Eq. (2.12) becomes

$$\delta L_{\text{int}} = \int_V (\sigma_{yy}\delta\epsilon_{yy}) + (\sigma_{xy}\delta\epsilon_{xy}) \, dV \tag{3.13}$$

Introducing Eq. (3.9) into Eq. (3.13), it becomes:

Fig. 3.3 Schematic representation of a beam with 2 nodes

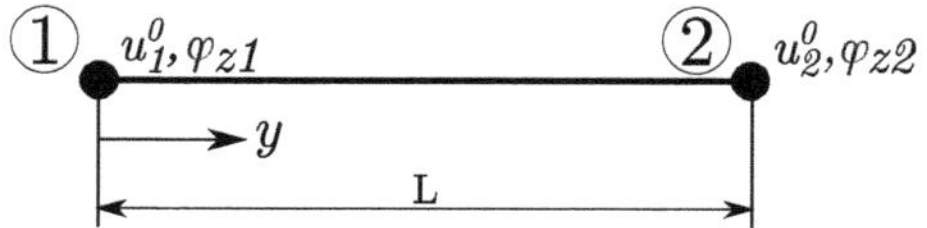

$$\delta L_{\text{int}} = \int_V \delta \epsilon^T \mathbf{C} \epsilon \, dV \tag{3.14}$$

Considering Eqs. (3.5), (3.14) is written as follows:

$$\delta L_{\text{int}} = \int_V \delta \left(\mathbf{bs}^0\right)^T \mathbf{C} \, \mathbf{bs}^0 \, dV. \tag{3.15}$$

3.4 Matrix Notation

The variation, along the beam length, of the displacement $u^0(y)$ can be expressed as a function of the displacements of the beam nodes 1 and 2, i.e. u_1^0 and u_2^0 respectively, and the shape functions. In the same manner, the variation of the rotation angle ϕ_z, which defines the $v^0(y)$ displacement, can be described using the rotations ϕ_{z1} and ϕ_{z2} of the beam nodes 1 and 2 and the shape functions. The shape functions used for this purpose are based on Lagrange polynomials and are the following

$$\begin{aligned} N_1(y) &= 1 - \frac{y}{L} \\ N_2(y) &= \frac{y}{L} \end{aligned} \tag{3.16}$$

The expressions that describe the variation, along the beam, of the displacement $u^0(y)$ and the rotation ϕ_z, are the following

$$u^0(y) = N_1(y)u_1^0 + N_2(y)u_2^0 \tag{3.17}$$

$$\phi_z(y) = N_1(y)\phi_{z1} + N_2(y)\phi_{z2}$$

The matrix form of Eq. (3.17) is:

$$\mathbf{s}^0 = \begin{bmatrix} N_1 & 0 & N_2 & 0 \\ 0 & N_1 & 0 & N_2 \end{bmatrix} \begin{bmatrix} u_1^0 \\ \phi_{z1} \\ u_2^0 \\ \phi_{z2} \end{bmatrix} = \begin{bmatrix} N_1 & 0 & N_2 & 0 \\ 0 & N_1 & 0 & N_2 \end{bmatrix} \mathbf{S}^0 \tag{3.18}$$

In Eq. (3.18), $\mathbf{S}^0$ is introduced to express the unknowns of the problem. Introducing the shape functions matrix $\mathbf{N}$, it is

$$\mathbf{N} = \begin{bmatrix} N_1 & 0 & N_2 & 0 \\ 0 & N_1 & 0 & N_2 \end{bmatrix} \tag{3.19}$$

Introducing the *displacement differentiation matrix* **B**, it is obtained by deriving the shape functions used to describe the beam displacements. It is expressed in the following

$$\mathbf{B} = \mathbf{b}^*\mathbf{N} = \begin{bmatrix} 0 & x\frac{\partial}{\partial y} \\ -\frac{\partial}{\partial y} & 1 \end{bmatrix} \begin{bmatrix} N_1 & 0 & N_2 & 0 \\ 0 & N_1 & 0 & N_2 \end{bmatrix} =$$

$$= \begin{bmatrix} 0 & xN_{1,y} & 0 & xN_{2,y} \\ -N_{1,y} & N_1 & -N_{2,y} & N_2 \end{bmatrix}$$

(3.20)

where

$$N_{1,y} = \frac{\partial N_1(y)}{\partial y}$$

$$N_{2,y} = \frac{\partial N_2(y)}{\partial y}$$

(3.21)

Introducing Eqs. (3.18) and (3.20), Eq. (3.15) becomes

$$\delta L_{\text{int}} = \int_V \delta \mathbf{S}^0 \mathbf{B}^T \mathbf{C} \, \mathbf{B} \, \mathbf{S}^0 \, dV$$

(3.22)

where the stiffness matrix **K** of the beam is

$$\mathbf{K} = \int_V \mathbf{B}^T \mathbf{C} \mathbf{B}$$

(3.23)

Using the previous definitions expressed in Eqs. (3.12) and (3.20) the explicit form of the stiffness matrix **K** becomes

$$\mathbf{K} = \int_V \begin{bmatrix} 0 & xN_{1,y} & 0 & xN_{2,y} \\ -N_{1,y} & N_1 & -N_{2,y} & N_2 \end{bmatrix}^T \begin{bmatrix} C_{11} & 0 \\ 0 & C_{44} \end{bmatrix} \begin{bmatrix} 0 & xN_{1,y} & 0 & xN_{2,y} \\ -N_{1,y} & N_1 & -N_{2,y} & N_2 \end{bmatrix} dV$$

(3.24)

$$\mathbf{K} = \int_V \begin{bmatrix} 0 & -N_{1,y} \\ xN_{1,y} & N_1 \\ 0 & -N_{2,y} \\ xN_{2,y} & N_2 \end{bmatrix} \begin{bmatrix} 0 & C_{11}xN_{1,y} & 0 & C_{11}xN_{2,y} \\ -C_{44}N_{1,y} & C_{44}N_1 & -C_{44}N_{2,y} & C_{44}N_2 \end{bmatrix} dV$$

(3.25)

By performing the double matrix product of Eq. (3.25), the resultant stiffness matrix becomes

$$\mathbf{K} = \int_V \begin{bmatrix} C_{44}N_{1,y}N_{1,y} & -C_{44}N_1N_{1,y} & C_{44}N_{1,y}N_{2,y} & -C_{44}N_2N_{1,y} \\ -C_{44}N_1N_{1,y} & C_{11}x^2N_{1,y}N_{1,y} + C_{44}N_1N_1 & -C_{44}N_1N_{2,y} & C_{11}x^2N_{1,y}N_{2,y} + C_{44}N_1N_2 \\ C_{44}N_{2,y}N_{1,y} & -C_{44}N_1N_{2,y} & C_{44}N_{2,y}N_{2,y} & -C_{44}N_{2,y}N_2 \\ -C_{44}N_2N_{1,y} & C_{11}x^2N_{1,y}N_{2,y} + C_{44}N_2N_1 & -C_{44}N_2N_{2,y} & C_{11}x^2N_{2,y}N_{2,y} + C_{44}N_2N_2 \end{bmatrix} dV$$

$$= \begin{bmatrix} k_{11} & k_{12} & k_{13} & k_{14} \\ k_{21} & k_{22} & k_{23} & k_{24} \\ k_{31} & k_{32} & k_{33} & k_{34} \\ k_{41} & k_{42} & k_{43} & k_{44} \end{bmatrix}$$

$$(3.26)$$

Subsequently, the volume integral of each term of matrix $\mathbf{K}$ of Eq. (3.26) can be evaluated separately considering the following main geometrical relations and shape functions.

Geometrical relations and shape functions

$$I_z = \int_A x^2\, dA \qquad\qquad I_z = \int_{-a}^{a} x^2\, dx \int_{-b}^{b} dz = \frac{4a^3b}{3}$$

$$N_1(y) = 1 - \frac{y}{L} \qquad\qquad N_2(y) = \frac{y}{L}$$

$$N_{1,y} = \frac{\partial}{\partial y}\left(1 - \frac{y}{L}\right) = -\frac{1}{L} \qquad N_{2,y} = \frac{\partial}{\partial y}\left(\frac{y}{L}\right) = \frac{1}{L}$$

- $L = 500$ mm;
- $a = b = 10$ mm.

$$
\begin{aligned}
k_{11} &= \int_V (C_{44} N_{1,y} N_{1,y}) dV = C_{44} \int_{-a}^{a} dx \int_{-b}^{b} dz \int_0^L N_{1,y} N_{1,y} dy \\
&= C_{44} \int_{-a}^{a} dx \int_{-b}^{b} dz \int_0^L \frac{1}{L^2} dy \\
&= C_{44} \frac{A}{L} = C_{44} \frac{4ab}{L}
\end{aligned}
$$

$$
\begin{aligned}
k_{12} &= \int_V (-C_{44} N_1 N_{1,y}) dV = -C_{44} \int_{-a}^{a} dx \int_{-b}^{b} dz \int_0^L N_1 N_{1,y} dy \\
&= -C_{44} \int_{-a}^{a} dx \int_{-b}^{b} dz \int_0^L \left[\left(1 - \frac{y}{L}\right)\left(-\frac{1}{L}\right) \right] dy = \\
&= -C_{44} \int_{-a}^{a} dx \int_{-b}^{b} dz \int_0^L \left(-\frac{1}{L} + \frac{y}{L^2}\right) dy \\
&= C_{44} \frac{A}{2} = C_{44} \frac{4ab}{2} = C_{44} 2ab
\end{aligned}
$$

$$
\begin{aligned}
k_{13} &= \int_V (C_{44} N_{1,y} N_{2,y}) dV = C_{44} \int_{-a}^{a} dx \int_{-b}^{b} dz \int_0^L N_{1,y} N_{2,y} dy \\
&= C_{44} \int_{-a}^{a} dx \int_{-b}^{b} dz \int_0^L \left[\left(-\frac{1}{L}\right)\left(\frac{1}{L}\right) \right] dy = C_{44} \int_{-a}^{a} dx \int_{-b}^{b} dz \int_0^L -\frac{1}{L^2} dy \\
&= -C_{44} \frac{A}{L} = -C_{44} \frac{4ab}{L}
\end{aligned}
$$

$$
\begin{aligned}
k_{14} &= \int_V (-C_{44} N_2 N_{1,y}) dV = -C_{44} \int_{-a}^{a} dx \int_{-b}^{b} dz \int_0^L N_2 N_{1,y} dy \\
&= -C_{44} \int_{-a}^{a} dx \int_{-b}^{b} dz \int_0^L \left[\left(\frac{y}{L}\right)\left(-\frac{1}{L}\right) \right] dy = \\
&= -C_{44} \int_{-a}^{a} dx \int_{-b}^{b} dz \int_0^L \left(-\frac{y}{L^2}\right) dy \\
&= C_{44} \frac{A}{2} = C_{44} \frac{4ab}{2} = C_{44} 2ab
\end{aligned}
$$

$$
\begin{aligned}
k_{22} &= \int_V \left(C_{11} x^2 N_{1,y} N_{1,y} + C_{44} N_1 N_1 \right) dV \\
&= C_{11} \int_{-a}^{a} x^2 dx \int_{-b}^{b} dz \int_0^L N_{1,y} N_{1,y} dy + C_{44} \int_{-a}^{a} dx \int_{-b}^{b} dz \int_0^L N_1 N_1 dy \\
&= C_{11} \int_{-a}^{a} x^2 dx \int_{-b}^{b} dz \int_0^L \left[\left(-\frac{1}{L}\right)\left(-\frac{1}{L}\right) \right] dy + \\
&\quad + C_{44} \int_{-a}^{a} dx \int_{-b}^{b} dz \int_0^L \left(1 - \frac{y}{L}\right)\left(1 - \frac{y}{L}\right) dy \\
&= C_{11} \int_{-a}^{a} x^2 dx \int_{-b}^{b} dz \int_0^L \left(\frac{1}{L^2}\right) dy + \\
&\quad + C_{44} \int_{-a}^{a} dx \int_{-b}^{b} dz \int_0^L \left(1 + \frac{y^2}{L^2} - 2\frac{y}{L}\right) dy \\
&= C_{11} \frac{I_z}{L} + C_{44} \frac{AL}{3} = C_{11} \frac{4a^3 b}{3L} + C_{44} \frac{4abL}{3}
\end{aligned}
$$

$$k_{23} = \int_V (-C_{44}N_1 N_{2,y})dV = C_{44}\int_{-a}^{a}dx\int_{-b}^{b}dz\int_{0}^{L}N_1 N_{2,y}dy$$

$$= -C_{44}\int_{-a}^{a}dx\int_{-b}^{b}dz\int_{0}^{L}\left[\left(1-\frac{y}{L}\right)\left(\frac{1}{L}\right)\right]dy =$$

$$= -C_{44}\int_{-a}^{a}dx\int_{-b}^{b}dz\int_{0}^{L}\left(\frac{1}{L}-\frac{y}{L^2}\right)dy$$

$$= -C_{44}\frac{A}{2} = -C_{44}\frac{4ab}{2} = -C_{44}2ab$$

$$k_{24} = \int_V \left(C_{11}x^2 N_{1,y}N_{2,y} + C_{44}N_1 N_2\right)dV$$

$$= C_{11}\int_{-a}^{a}x^2 dx\int_{-b}^{b}dz\int_{0}^{L}N_{1,y}N_{2,y}dy + C_{44}\int_{-a}^{a}dx\int_{-b}^{b}dz\int_{0}^{L}N_1 N_2 dy$$

$$= C_{11}\int_{-a}^{a}x^2 dx\int_{-b}^{b}dz\int_{0}^{L}\left[\left(-\frac{1}{L}\right)\left(\frac{1}{L}\right)\right]dy+$$

$$+C_{44}\int_{-a}^{a}dx\int_{-b}^{b}dz\int_{0}^{L}\left(1-\frac{y}{L}\right)\left(\frac{y}{L}\right)dy$$

$$= C_{11}\int_{-a}^{a}x^2 dx\int_{-b}^{b}dz\int_{0}^{L}\left(-\frac{1}{L^2}\right)dy + C_{44}\int_{-a}^{a}dx\int_{-b}^{b}dz\int_{0}^{L}\left(\frac{y}{L}-\frac{y^2}{L^2}\right)dy$$

$$= -C_{11}\frac{I_z}{L} + C_{44}\frac{AL}{6} = -C_{11}\frac{4a^3 b}{3L} + C_{44}\frac{4abL}{6} = -C_{11}\frac{4a^3 b}{3L} + C_{44}\frac{2abL}{3}$$

$$k_{33} = \int_V (C_{44}N_{2,y}N_{2,y})dV = C_{44}\int_{-a}^{a}dx\int_{-b}^{b}dz\int_{0}^{L}N_{2,y}N_{2,y}dy$$

$$= C_{44}\int_{-a}^{a}dx\int_{-b}^{b}dz\int_{0}^{L}\left[\left(\frac{1}{L}\right)\left(\frac{1}{L}\right)\right]dy = C_{44}\int_{-a}^{a}dx\int_{-b}^{b}dz\int_{0}^{L}\frac{1}{L^2}dy$$

$$= C_{44}\frac{A}{L} = C_{44}\frac{4ab}{L}$$

$$k_{34} = \int_V (-C_{44}N_2 N_{2,y})dV = -C_{44}\int_{-a}^{a}dx\int_{-b}^{b}dz\int_{0}^{L}N_2 N_{1,y}dy$$

$$= -C_{44}\int_{-a}^{a}dx\int_{-b}^{b}dz\int_{0}^{L}\left[\left(\frac{y}{L}\right)\left(-\frac{1}{L}\right)\right]dy =$$

$$= -C_{44}\int_{-a}^{a}dx\int_{-b}^{b}dz\int_{0}^{L}\left(-\frac{y}{L}+\frac{y}{L^2}\right)dy$$

$$= -C_{44}\frac{A}{2} = -C_{44}\frac{4ab}{2} = -C_{44}2ab$$

$$k_{44} = \int_V \left(C_{11}x^2 N_{2,y}N_{2,y} + C_{44}N_2 N_2\right)dV$$

$$= C_{11}\int_{-a}^{a}x^2 dx\int_{-b}^{b}dz\int_{0}^{L}N_{2,y}N_{2,y}dy + C_{44}\int_{-a}^{a}dx\int_{-b}^{b}dz\int_{0}^{L}N_2 N_2 dy$$

$$= C_{11}\int_{-a}^{a}x^2 dx\int_{-b}^{b}dz\int_{0}^{L}\left[\left(\frac{1}{L}\right)\left(\frac{1}{L}\right)\right]dy+$$

$$+C_{44}\int_{-a}^{a}dx\int_{-b}^{b}dz\int_{0}^{L}\left(\frac{y}{L}\right)\left(\frac{y}{L}\right)dy$$

$$= C_{11}\int_{-a}^{a}x^2 dx\int_{-b}^{b}dz\int_{0}^{L}\left(\frac{1}{L^2}\right)dy + C_{44}\int_{-a}^{a}dx\int_{-b}^{b}dz\int_{0}^{L}\left(\frac{y^2}{L^2}\right)dy$$

$$= C_{11}\frac{I_z}{L} + C_{44}\frac{AL}{3} = C_{11}\frac{4a^3 b}{3L} + C_{44}\frac{4abL}{3}$$

Considering that the matrix **K** is symmetric, i.e. $K_{21} = K_{21}$, $K_{31} = K_{13}$, $K_{32} = K_{23}$, $K_{41} = K_{14}$, $K_{42} = K_{24}$ and $K_{43} = K_{34}$, its complete form results in the following

$$\mathbf{K} = \begin{bmatrix} C_{44}\dfrac{4ab}{L} & C_{44}2ab & -C_{44}\dfrac{4ab}{L} & C_{44}2ab \\[2ex] C_{44}2ab & \left(C_{11}\dfrac{4a^3b}{3L}+C_{44}\dfrac{4abL}{3}\right) & -C_{44}2ab & \left(-C_{11}\dfrac{4a^3b}{3L}+C_{44}\dfrac{2abL}{3}\right) \\[2ex] -C_{44}\dfrac{4ab}{L} & -C_{44}2ab & C_{44}\dfrac{4ab}{L} & -C_{44}2ab \\[2ex] C_{44}2ab & \left(-C_{11}\dfrac{4a^3b}{3L}+C_{44}\dfrac{2abL}{3}\right) & -C_{44}2ab & \left(C_{11}\dfrac{4a^3b}{3L}+C_{44}\dfrac{4abL}{3}\right) \end{bmatrix} \tag{3.27}$$

Introducing the numerical data, the matrix of the beam is:

$$\mathbf{K} = \begin{bmatrix} 84000 & 21000000 & -84000 & 21000000 \\ 21000000 & 7005600000 & -21000000 & 3494400000 \\ -84000 & -21000000 & 84000 & -21000000 \\ 21000000 & 3494400000 & -21000000 & 7005600000 \end{bmatrix} \tag{3.28}$$

The comprehensive procedure for deriving the matrix presented in Eq. (3.28) is detailed in Appendix B.1, and the step-by-step implementation is provided as a MATLAB script.

3.5 Recursive Notation Against Shape Functions

In this section the stiffness matrix of the beam is calculated using a recursive notation against the shape functions. The displacements field of Eq. (3.17) can be written as follows.

$$u^0(y) = N_i(y)u_i^0$$
$$\phi_z(y) = N_i(y)\phi_{zi} \tag{3.29}$$

where i is either 1 or 2, according to which node of the beam is considered. $\mathbf{B}$ matrix of Eq. (3.20) can be expressed as

$$\mathbf{B} = \begin{bmatrix} B_1 & B_2 \end{bmatrix} \tag{3.30}$$

where

$$\mathbf{B}_1 = \begin{bmatrix} 0 & xN_{1,y} \\ -N_{1,y} & N_1 \end{bmatrix} \tag{3.31}$$

and

$$\mathbf{B}_2 = \begin{bmatrix} 0 & x N_{2,y} \\ -N_{2,y} & N_2 \end{bmatrix} \tag{3.32}$$

and using the recursive notation the $\mathbf{B}$ it becomes

$$\mathbf{B}_i = \begin{bmatrix} 0 & x N_{i,y} \\ -N_{i,y} & N_i \end{bmatrix} \tag{3.33}$$

In order to distinguish the variables from their virtual variation, the second index j is introduced.

$$\mathbf{B}_j = \begin{bmatrix} 0 & x N_{j,y} \\ -N_{j,y} & N_j \end{bmatrix} \tag{3.34}$$

The subscripts i and j are either 1 or 2. The stiffness matrix can be written as follows

$$\mathbf{K}_{ij} = \int_V \mathbf{B}_j^T \mathbf{C} \mathbf{B}_i dV = \int_V \begin{bmatrix} 0 & -N_{j,y} \\ x N_{j,y} & N_j \end{bmatrix} \begin{bmatrix} C_{11} & 0 \\ 0 & C_{44} \end{bmatrix} \begin{bmatrix} 0 & x N_{i,y} \\ -N_{i,y} & N_i \end{bmatrix} \tag{3.35}$$

By computing the triple product, the previous matrix becomes

$$\mathbf{K}_{ij} = \int_V \begin{bmatrix} C_{44} N_{i,y} N_{j,y} & -C_{44} N_i N_{j,y} \\ -C_{44} N_j N_{i,y} & \left(C_{11} x^2 N_{i,y} N_{j,y} + C_{44} N_i N_j \right) \end{bmatrix} dV \tag{3.36}$$

Solving the volume integral of Eq. (3.36), the following matrices are obtained

$$\mathbf{K}_{ij} = \begin{bmatrix} C_{44} \int_A dA \int_L N_{i,y} N_{j,y} dL & -C_{44} \int_A dA \int_L N_i N_{j,y} dL \\ -C_{44} \int_A dA \int_L N_j N_{i,y} dL & \begin{aligned} & C_{11} \int_A x^2 dA \int_L N_{i,y} N_{j,y} dL + \\ & + C_{44} \int_A dA \int_L N_i N_j dL \end{aligned} \end{bmatrix} \tag{3.37}$$

$$\mathbf{K}_{ij} = \begin{bmatrix} C_{44} \int_{-a}^{a} dx \int_{-b}^{b} dz \int_{0}^{L} N_{i,y}N_{j,y}dy & -C_{44} \int_{-a}^{a} dx \int_{-b}^{b} dz \int_{0}^{L} N_i N_{j,y}dy \\[4ex] -C_{44} \int_{-a}^{a} dx \int_{-b}^{b} dz \int_{0}^{L} N_j N_{i,y}dy & \begin{aligned} & C_{11} \int_{-a}^{a} x^2 dx \int_{-b}^{b} dz \int_{0}^{L} N_{i,y}N_{j,y}dy+ \\ & +C_{44} \int_{-a}^{a} dx \int_{-b}^{b} dz \int_{0}^{L} N_i N_j dy \end{aligned} \end{bmatrix} \quad (3.38)$$

By looping the indexes i and j from 1 to 2 the following submatrices of the global element of to the beam under consideration can be obtained

$$\mathbf{K}_{11} = \begin{bmatrix} C_{44} \int_{-a}^{a} dx \int_{-b}^{b} dz \int_{0}^{L} N_{1,y}N_{1,y}dy & -C_{44} \int_{-a}^{a} dx \int_{-b}^{b} dz \int_{0}^{L} N_1 N_{1,y}dy \\[4ex] -C_{44} \int_{-a}^{a} dx \int_{-b}^{b} dz \int_{0}^{L} N_1 N_{1,y}dy & \begin{aligned} & C_{11} \int_{-a}^{a} x^2 dx \int_{-b}^{b} dz \int_{0}^{L} N_{1,y}N_{1,y}dy+ \\ & +C_{44} \int_{-a}^{a} dx \int_{-b}^{b} dz \int_{0}^{L} N_1 N_1 dy \end{aligned} \end{bmatrix}$$
$$(3.39)$$

$$\mathbf{K}_{12} = \begin{bmatrix} C_{44} \int_{-a}^{a} dx \int_{-b}^{b} dz \int_{0}^{L} N_{1,y}N_{2,y}dy & -C_{44} \int_{-a}^{a} dx \int_{-b}^{b} dz \int_{0}^{L} N_1 N_{2,y}dy \\[4ex] -C_{44} \int_{-a}^{a} dx \int_{-b}^{b} dz \int_{0}^{L} N_2 N_{1,y}dy & \begin{aligned} & C_{11} \int_{-a}^{a} x^2 dx \int_{-b}^{b} dz \int_{0}^{L} N_{1,y}N_{2,y}dy+ \\ & +C_{44} \int_{-a}^{a} dx \int_{-b}^{b} dz \int_{0}^{L} N_1 N_2 dy \end{aligned} \end{bmatrix}$$
$$(3.40)$$

$$\mathbf{K}_{21} = \begin{bmatrix} C_{44} \int_{-a}^{a} dx \int_{-b}^{b} dz \int_{0}^{L} N_{2,y}N_{1,y}dy & -C_{44} \int_{-a}^{a} dx \int_{-b}^{b} dz \int_{0}^{L} N_2 N_{1,y}dy \\[4ex] -C_{44} \int_{-a}^{a} dx \int_{-b}^{b} dz \int_{0}^{L} N_1 N_{2,y}dy & \begin{aligned} & C_{11} \int_{-a}^{a} x^2 dx \int_{-b}^{b} dz \int_{0}^{L} N_{2,y}N_{1,y}dy+ \\ & +C_{44} \int_{-a}^{a} dx \int_{-b}^{b} dz \int_{0}^{L} N_2 N_1 dy \end{aligned} \end{bmatrix}$$
$$(3.41)$$

$$\mathbf{K}_{22} = \begin{bmatrix} C_{44} \int_{-a}^{a} dx \int_{-b}^{b} dz \int_{0}^{L} N_{2,y}N_{2,y}dy & -C_{44} \int_{-a}^{a} dx \int_{-b}^{b} dz \int_{0}^{L} N_2 N_{2,y}dy \\[4ex] -C_{44} \int_{-a}^{a} dx \int_{-b}^{b} dz \int_{0}^{L} N_2 N_{2,y}dy & \begin{aligned} & C_{11} \int_{-a}^{a} x^2 dx \int_{-b}^{b} dz \int_{0}^{L} N_{2,y}N_{2,y}dy+ \\ & +C_{44} \int_{-a}^{a} dx \int_{-b}^{b} dz \int_{0}^{L} N_2 N_2 dy \end{aligned} \end{bmatrix}$$
$$(3.42)$$

The matrix $\mathbf{K}$ can be written as follows

$$\mathbf{K} = \begin{bmatrix} \mathbf{K}_{11} & \mathbf{K}_{12} \\ \mathbf{K}_{21} & \mathbf{K}_{22} \end{bmatrix} \tag{3.43}$$

Each component expressed from Eq. (3.39) to Eq. (3.42) can be written as follows

$$\mathbf{K}_{11} = \begin{bmatrix} k_{11} & k_{12} \\ k_{21} & k_{22} \end{bmatrix} \tag{3.44}$$

$$\mathbf{K}_{12} = \begin{bmatrix} k_{13} & k_{14} \\ k_{23} & k_{24} \end{bmatrix} \tag{3.45}$$

$$\mathbf{K}_{21} = \begin{bmatrix} k_{31} & k_{32} \\ k_{41} & k_{42} \end{bmatrix} \tag{3.46}$$

$$\mathbf{K}_{22} = \begin{bmatrix} k_{33} & k_{34} \\ k_{43} & k_{44} \end{bmatrix} \tag{3.47}$$

Equation (3.43), considering Eqs. from (3.44) to (3.47), results equal to Eq. (3.27). The resultant stiffness matrix is then

$$\mathbf{K} = \begin{bmatrix} C_{44}\dfrac{4ab}{L} & C_{44}2ab & -C_{44}\dfrac{4ab}{L} & C_{44}2ab \\[2mm] C_{44}2ab & \left(C_{11}\dfrac{4a^3 b}{3L} + C_{44}\dfrac{4abL}{3}\right) & -C_{44}2ab & \left(-C_{11}\dfrac{4a^3 b}{3L} + C_{44}\dfrac{2abL}{3}\right) \\[2mm] -C_{44}\dfrac{4ab}{L} & -C_{44}2ab & C_{44}\dfrac{4ab}{L} & -C_{44}2ab \\[2mm] C_{44}2ab & \left(-C_{11}\dfrac{4a^3 b}{3L} + C_{44}\dfrac{2abL}{3}\right) & -C_{44}2ab & \left(C_{11}\dfrac{4a^3 b}{3L} + C_{44}\dfrac{4abL}{3}\right) \end{bmatrix} \tag{3.48}$$

Introducing the numerical data, the matrix of the beam, reads as

$$\mathbf{K} = \begin{bmatrix} 84000 & 21000000 & -84000 & 21000000 \\ 21000000 & 7005600000 & -21000000 & 3494400000 \\ -84000 & -21000000 & 84000 & -21000000 \\ 21000000 & 3494400000 & -21000000 & 7005600000 \end{bmatrix} \tag{3.49}$$

The comprehensive procedure for deriving the matrix presented in Eq. (3.5) is detailed in Appendix B.2, and the step-by-step implementation is provided as a MATLAB script.

3.6 Recursive Notation Against the Theory of Structure

In this section the stiffness matrix of the beam is calculated using a recursive notation against TOS. Introducing $s_n = s_x, s_y, s_z$, Eq. (3.1) becomes

$$\begin{aligned} s_x &= u(x, y, z) = -u^0(y) \\ s_y &= v(x, y, z) = x\phi_z(y) \\ s_z &= w(x, y, z) = 0 \end{aligned} \tag{3.50}$$

Then, the expansion functions F are introduced, so that

$$\begin{aligned} s_x &= F_{1_x} \times u^0(y) \\ s_y &= F_{1_y} \times \phi_z(y) \\ s_z &= F_{1_z} \times 0 \end{aligned} \tag{3.51}$$

In this case

$$F_{1_x} = -1 \qquad F_{1_y} = x, \qquad F_{1_z} = 0 \tag{3.52}$$

Equation (2.44) can be expressed in a generic form introducing the index τ which ranges from 1 to the number of the terms in the expansion functions F. This case can then expressed as

$$s_n = F_{\tau_n} s_{\tau_n} \tag{3.53}$$

with the generic index n introduced for the directions x, y and z and

$$s_{1_x} = u^0(y) \qquad s_{1_y} = \phi_z(y), \qquad s_{1_z} = 0 \tag{3.54}$$

The same procedure can be used for the virtual variation, by introducing the index s for the virtual variation.

$$\delta s_n = F_{s_n} s_{s_n} \tag{3.55}$$

By applying the recursive notation against TOS, $\mathbf{B}_{\tau i}$ and $\mathbf{B}_{sj}$ can be written as follows

$$\mathbf{B}_{\tau i} = \begin{bmatrix} 0 & F_{\tau y} N_{i,y} \\ F_{\tau x} N_{i,y} & F_{\tau y,x} N_i \end{bmatrix} \tag{3.56}$$

$$\mathbf{B}_{sj} = \begin{bmatrix} 0 & F_{sy} N_{j,y} \\ F_{sx} N_{j,y} & F_{sy,x} N_j \end{bmatrix} \tag{3.57}$$

The stiffness matrix can be expressed as follows

$$\mathbf{K}_{ij}^{\tau s} = \int_V \mathbf{B}_{sj}^T \mathbf{C} \mathbf{B}_{\tau i} dV =$$
$$= \int_V \begin{bmatrix} 0 & F_{sx} N_{j,y} \\ F_{sy} N_{j,y} & F_{sy,x} N_j \end{bmatrix} \begin{bmatrix} C_{11} & 0 \\ 0 & C_{44} \end{bmatrix} \begin{bmatrix} 0 & F_{\tau y} N_{i,y} \\ F_{\tau x} N_{i,y} & F_{\tau y,x} N_i \end{bmatrix} \tag{3.58}$$

Performing the triple product, the previous matrix becomes

$$\mathbf{K}_{ij}^{\tau s} = \int_V \begin{bmatrix} C_{44} F_{sx} N_{j,y} F_{\tau x} N_{i,y} & C_{44} F_{sx} N_{j,y} F_{\tau y,x} N_i \\ C_{44} F_{sy,x} N_j F_{\tau x} N_{i,y} & C_{11} F_{sy} N_{j,y} F_{\tau y} N_{i,y} + \\ & + C_{44} F_{sy,x} N_j F_{\tau y,x} N_i \end{bmatrix} dV \tag{3.59}$$

Solving the volume integral of Eq. (3.59), the following matrices are obtained

$$\mathbf{K}_{ij}^{\tau s} = \begin{bmatrix} C_{44} \int_A F_{sx} F_{\tau x} dA \int_L N_{i,y} N_{j,y} dL & C_{44} \int_A F_{sx} F_{\tau y,x} dA \int_L N_i N_{j,y} dL \\ C_{44} \int_A F_{sy,x} F_{\tau x} dA \int_L N_j N_{i,y} dL & C_{11} \int_A F_{sy} F_{\tau y} dA \int_L N_{i,y} N_{j,y} dL + \\ & + C_{44} \int_A F_{sy,x} F_{\tau y,x} dA \int_L N_i N_j dL \end{bmatrix} \tag{3.60}$$

Considering the values of the products between the expansion functions and their derivatives

Expansion functions operations

$$F_{sx} F_{\tau x} = (-1)(-1) = 1$$

$$F_{sx} F_{\tau y,x} = (-1)(1) = -1$$

$$F_{sy,x} F_{\tau x} = (1)(-1) = -1 \qquad i = 1, 2 \qquad \tau = 1, 2$$

$$F_{sy} F_{\tau y} = (x)(x) = x^2$$

$$F_{sy,x} F_{\tau y,x} = (1)(1) = 1$$

the stiffness matrix of Eq. (3.60) becomes the following

$$\mathbf{K}_{ij}^{\tau s} = \begin{bmatrix} C_{44} \int_{-a}^{a} 1 dx \int_{-b}^{b} dz \int_{0}^{L} [N_{i,y} N_{j,y}] dy & C_{44} \int_{-a}^{a} -1 dx \int_{-b}^{b} dz \int_{0}^{L} [N_{i} N_{j,y}] dy \\ C_{44} \int_{-a}^{a} -1 dx \int_{-b}^{b} dz \int_{0}^{L} [N_{j} N_{i,y}] dy & \begin{aligned} C_{11} \int_{-a}^{a} x^2 dx \int_{-b}^{b} dz \int_{0}^{L} [N_{i,y} N_{j,y}] dy + \\ + C_{44} \int_{-a}^{a} 1 dx \int_{-b}^{b} dz \int_{0}^{L} [N_{i} N_{j}] dy \end{aligned} \end{bmatrix}$$

$$(3.61)$$

The stiffness matrix for the beam then results the same as the matrix of Eq. (3.48) and it is reported below

$$\mathbf{K} = \begin{bmatrix} C_{44} \dfrac{4ab}{L} & C_{44} 2ab & -C_{44} \dfrac{4ab}{L} & C_{44} 2ab \\ C_{44} 2ab & \left(C_{11} \dfrac{4a^3 b}{3L} + C_{44} \dfrac{4abL}{3} \right) & -C_{44} 2ab & \left(-C_{11} \dfrac{4a^3 b}{3L} + C_{44} \dfrac{2abL}{3} \right) \\ -C_{44} \dfrac{4ab}{L} & -C_{44} 2ab & C_{44} \dfrac{4ab}{L} & -C_{44} 2ab \\ C_{44} 2ab & \left(-C_{11} \dfrac{4a^3 b}{3L} + C_{44} \dfrac{2abL}{3} \right) & -C_{44} 2ab & \left(C_{11} \dfrac{4a^3 b}{3L} + C_{44} \dfrac{4abL}{3} \right) \end{bmatrix}$$

$$(3.62)$$

The numerical matrix of the beam, calculated by using the recursive notation against TOS, is:

$$
\mathbf{K} = \begin{bmatrix} 84000 & 21000000 & -84000 & 21000000 \\ 21000000 & 7005600000 & -21000000 & 3494400000 \\ -84000 & -21000000 & 84000 & -21000000 \\ 21000000 & 3494400000 & -21000000 & 7005600000 \end{bmatrix} \tag{3.63}
$$

The comprehensive procedure for deriving the matrix presented in Eq. (3.6) is detailed in Appendix B.3, and the step-by-step implementation is provided as a MATLAB script.

3.7 Locking Phenomena

The adoption of a priori kinematic assumptions in Finite Element (FE) models can give rise to locking phenomena, which have a profound impact on the accuracy and convergence of numerical solutions. When a linear element is subjected to bending, artificially introduced shear deformations can cause an excessive increase in the stiffness of the element. This increased stiffness prevents the solution from converging to the correct value and leads to distortions in the predicted structural response. Two principal types of locking deserve particular attention: Poisson locking and shear locking. Poisson locking emerges from deformation coupling associated with the Poisson ratio and may arise in classical structural theories such as those by Euler–Bernoulli, Timoshenko, when transverse strains are not modeled accurately. By contrast, shear locking is an inherent numerical artifact frequently attributable to low-order interpolation schemes. In beam elements, shear locking manifests as an overestimation of shear stiffness coupled with an underestimation of bending stiffness. The resulting numerical solution often depicts deflections that are significantly smaller than the actual physical displacement, thereby hindering convergence to the correct solution. Given its prevalence and potential to corrupt results, shear locking is a focal point of discussion in this chapter, along with strategies to mitigate its effects.

3.7.1 The Shear Locking

The shear locking phenomenon in a beam element with linear shape functions arises from an incorrect representation of the shear deformation component in the beam's total strain energy. In particular, the linear element employed to describe bending about the z can exhibit shear locking, leading to inaccurate displacement predictions when using the stiffness matrix presented in (3.62). This matrix, formulated with local coordinates x and z in the cross-sectional plane, and y along the beam's length,

Fig. 3.4 Local coordinates along a beam with 2 nodes

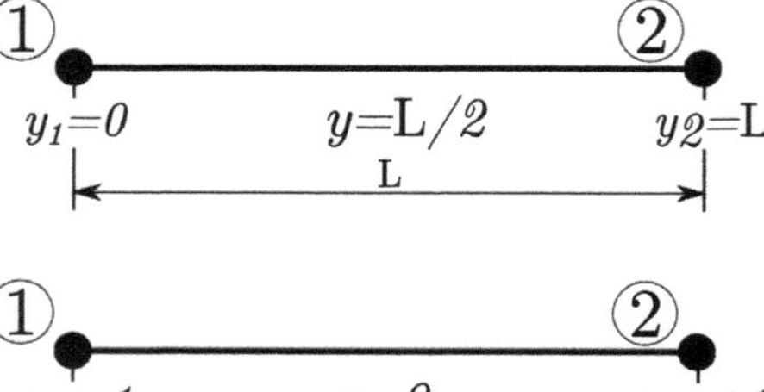

Fig. 3.5 Natural coordinates along a beam with 2 nodes

includes terms evaluated through a triple integration process. An alternative approach involves deriving the stiffness matrix in natural coordinates (ξ, η, r), where ξ and η span the cross-section of the beam, and r runs along its length. In this system, r varies between -1 at $y = 0$ and $+1$ at $y = L$. Figures 3.4 and 3.5 illustrate the relationship between the local coordinate system (x, y, z) and the natural coordinate system (ξ, η, r).

The relation between local and natural coordinates is expressed in the following

$$y = \frac{L}{2}(1 + r)$$
$$r = \frac{2y}{L} - 1 \tag{3.64}$$

Substituting Eq. (3.64) into the equations of the shape functions expressed in terms of local coordinate (Eq. (3.16))

$$N_1(y) = 1 - \frac{y}{L}$$
$$N_2(y) = \frac{y}{L} \tag{3.65}$$

the shape functions expressed in terms of the natural coordinate r are obtained.

$$N_1(r) = \frac{1}{2}(1 - r)$$
$$N_2(r) = \frac{1}{2}(1 + r) \tag{3.66}$$

The Jacobian matrix is introduced now. This matrix, denoted as **J**, plays a crucial role in this element formulation, which is denoted as *isoparametric*. It serves as a *scaling* factor between the derivatives of the natural and local coordinate systems. **J** can be determined through the following computation, using the expression of Eq. (3.64):

$$\mathbf{J} = \left[\frac{\partial y}{\partial r} \right] = \left[\frac{\partial}{\partial r} \left(\left(\frac{1}{2}(1+r) \right) L \right) \right] = \left[\frac{L}{2} \right] \tag{3.67}$$

The determinant of the previously computed Jacobian Matrix $\mathbf{J}$ is

$$det\, \mathbf{J} = det \left[\frac{L}{2} \right] = \frac{L}{2} \tag{3.68}$$

Considering Eq. (3.64), the following relation applies

$$dy = \frac{L}{2} dr = det\, \mathbf{J}\, dr \tag{3.69}$$

The inverse of the Jacobian matrix $\mathbf{J}$ is

$$\mathbf{J}^{-1} = \left[\frac{L}{2} \right]^{-1} = \left[\frac{2}{L} \right] \tag{3.70}$$

As a result

$$\frac{d}{dy} = \mathbf{J}^{-1} \frac{d}{dr}. \tag{3.71}$$

3.7.2 Bending and Shear Stiffness Matrix Parts

The stiffness matrix of a beam can be split into two parts, the bending and shear components. With reference to Fig. 3.3 and considering the shape functions related to the natural coordinates of Eq. (3.66), the beam nodes displacements $u^0(y)$ and $\phi_z(y)$ in terms of the natural coordinates can be written as

$$u^0(r) = N_1(r)u_1^0 + N_2(r)u_2^0$$

$$\phi_z(r) = N_1(r)\phi_{z1} + N_2(r)\phi_{z2} \tag{3.72}$$

and

$$u^0(r) = \frac{1}{2}(1-r)\, u_1^0 + \frac{1}{2}(1+r)\, u_2^0$$

$$\phi_z(r) = \frac{1}{2}(1-r)\, \phi_{z1} + \frac{1}{2}(1+r)\, \phi_{z2} \tag{3.73}$$

Equation (3.73) can be written in matrix form, becoming

$$\left\{ \begin{array}{c} u^0(r) \\ \phi_x(r) \end{array} \right\} = \begin{bmatrix} \dfrac{1}{2}(1-r) & 0 & \dfrac{1}{2}(1+r) & 0 \\ 0 & \dfrac{1}{2}(1-r) & 0 & \dfrac{1}{2}(1+r) \end{bmatrix} \left\{ \begin{array}{c} u_1^0 \\ \phi_{z1} \\ u_2^0 \\ \phi_{z2} \end{array} \right\} \tag{3.74}$$

Introducing Eq. (3.74) into Eq. (3.4) the following expression is obtained

$$\boldsymbol{\epsilon} = \left\{ \begin{array}{c} \epsilon_{yy} \\ \epsilon_{xy} \end{array} \right\} =$$

$$= \begin{bmatrix} 0 & F_{\tau y}\mathbf{J}^{-1}\dfrac{\partial}{\partial r} \\ F_{\tau x}\mathbf{J}^{-1}\dfrac{\partial}{\partial r} & F_{\tau y,x} \end{bmatrix} \begin{bmatrix} \dfrac{1}{2}(1-r) & 0 & \dfrac{1}{2}(1+r) & 0 \\ 0 & \dfrac{1}{2}(1-r) & 0 & \dfrac{1}{2}(1+r) \end{bmatrix} \left\{ \begin{array}{c} u_1^0 \\ \phi_{z1} \\ u_2^0 \\ \phi_{z2} \end{array} \right\} \tag{3.75}$$

Generalizing Eq. (3.75), it can be written

$$\boldsymbol{\epsilon} = \left\{ \begin{array}{c} \epsilon_{yy} \\ \epsilon_{xy} \end{array} \right\} =$$

$$= \begin{bmatrix} 0 & F_{\tau y}\mathbf{J}^{-1}\dfrac{\partial}{\partial r} \\ F_{\tau x}\mathbf{J}^{-1}\dfrac{\partial}{\partial r} & F_{\tau y,x} \end{bmatrix} \begin{bmatrix} N_1(r) & 0 & N_2(r) & 0 \\ 0 & N_1(r) & 0 & N_2(r) \end{bmatrix} \left\{ \begin{array}{c} u_1^0 \\ \phi_{z1} \\ u_2^0 \\ \phi_{z2} \end{array} \right\} \tag{3.76}$$

and evaluating the derivatives, it has

$$\boldsymbol{\epsilon} = \left\{ \begin{array}{c} \epsilon_{yy} \\ \epsilon_{xy} \end{array} \right\} =$$

$$= \begin{bmatrix} 0 & F_{\tau y}\mathbf{J}^{-1}N_{1,r}(r) & 0 & F_{\tau y}\mathbf{J}^{-1}N_{2,r}(r) \\ F_{\tau x}\mathbf{J}^{-1}N_{1,r}(r) & F_{\tau y,x}N_1(r) & F_{\tau x}\mathbf{J}^{-1}N_{2,r}(r) & F_{\tau y,x}N_2(r) \end{bmatrix} \left\{ \begin{array}{c} u_1^0 \\ \phi_{z1} \\ u_2^0 \\ \phi_{z2} \end{array} \right\} \tag{3.77}$$

The previous expression can be split into two parts, one related to the longitudinal strains ϵ_{yy}, which is the bending part, and one related to the shear strains ϵ_{xy}, which is the shear part. By operating this partition, it is obtained

$$\epsilon_{yy} = \begin{bmatrix} 0 & F_{\tau y}\mathbf{J}^{-1}N_{1,r}(r) & 0 & F_{\tau y}\mathbf{J}^{-1}N_{2,r}(r) \end{bmatrix} \left\{ \begin{array}{c} u_1^0 \\ \phi_{z1} \\ u_2^0 \\ \phi_{z2} \end{array} \right\} \tag{3.78}$$

and

$$\epsilon_{xy} = \begin{bmatrix} F_{\tau x}\mathbf{J}^{-1}N_{1,r}(r) & F_{\tau y,x}N_1(r) & F_{\tau x}\mathbf{J}^{-1}N_{2,r}(r) & F_{\tau y,x}N_2(r) \end{bmatrix} \left\{ \begin{array}{c} u_1^0 \\ \phi_{z1} \\ u_2^0 \\ \phi_{z2} \end{array} \right\} \tag{3.79}$$

From the previous two equations, the following matrices $\mathbf{B}_{b_{\tau i}}$ (related to the bending part) and the matrix $\mathbf{B}_{s_{\tau i}}$ (related to the shear part) can be introduced

$$\mathbf{B}_{b_{\tau i}} = \left[0 \; F_{\tau y} \mathbf{J}^{-1} N_{1,r}(r) \; 0 \; F_{\tau y} \mathbf{J}^{-1} N_{2,r}(r) \right] \tag{3.80}$$

$$\mathbf{B}_{s_{\tau i}} = \left[F_{\tau x} \mathbf{J}^{-1} N_{1,r}(r) \; F_{\tau y,x} N_1(r) \; F_{\tau x} \mathbf{J}^{-1} N_{2,r}(r) \; F_{\tau y,x} N_2(r) \right] \tag{3.81}$$

Consequently, their variation can be introduced.

$$\mathbf{B}_{b_{sj}} = \left[0 \; F_{sy} \mathbf{J}^{-1} N_{1,r}(r) \; 0 \; F_{sy} \mathbf{J}^{-1} N_{2,r}(r) \right] \tag{3.82}$$

$$\mathbf{B}_{s_{sj}} = \left[F_{sx} \mathbf{J}^{-1} N_{1,r}(r) \; F_{sy,x} N_1(r) \; F_{sx} \mathbf{J}^{-1} N_{2,r}(r) \; F_{sy,x} N_2(r) \right] \tag{3.83}$$

Then, the bending part of the stiffness matrix $\mathbf{K}_b$ can be calculated as follows

$$\mathbf{K}_b = \int_V \mathbf{B}_{b_{sj}}^T \mathbf{C}_s \mathbf{B}_{b_{\tau i}} dV \tag{3.84}$$

whereas for the shear part $\mathbf{K}_s$ one has

$$\mathbf{K}_s = \int_V \mathbf{B}_{s_{sj}}^T \mathbf{C}_s \mathbf{B}_{s_{\tau i}} dV \tag{3.85}$$

In the previous equations the matrices $\mathbf{C}_b$ and $\mathbf{C}_s$ are the bending and shear terms of the material coefficients matrix $\mathbf{C}$, respectively (see (3.12)). These values are the following

$$C_b = C_{11} \qquad C_s = C_{44} \tag{3.86}$$

Using Eqs. (3.80) and (3.82) the bending part $\mathbf{K}_b$ of the stiffness matrix (see (3.84)) becomes

$$\mathbf{K}_b = \int_V \begin{bmatrix} 0 \\ F_{sy} \mathbf{J}^{-1} N_{1,r}(r) \\ 0 \\ F_{sy} \mathbf{J}^{-1} N_{2,r}(r) \end{bmatrix} C_{11} \left[0 \; F_{\tau y} \mathbf{J}^{-1} N_{1,r}(r) \; 0 \; F_{\tau y} \mathbf{J}^{-1} N_{2,r}(r) \right] dV \tag{3.87}$$

Using Eqs. (3.81) and (3.83) the shear part $\mathbf{K}_s$ of the stiffness matrix (see (3.85)) becomes

$$\mathbf{K}_s = \int_V \begin{bmatrix} F_{sx} \mathbf{J}^{-1} N_{1,r}(r) \\ F_{sy,x} N_1(r) \\ F_{sx} \mathbf{J}^{-1} N_{2,r}(r) \\ F_{sy,x} N_2(r) \end{bmatrix} C_{44} \left[F_{\tau x} \mathbf{J}^{-1} N_{1,r}(r) \; F_{\tau y,x} N_1(r) \; F_{\tau x} \mathbf{J}^{-1} N_{2,r}(r) \; F_{\tau y,x} N_2(r) \right] dV \tag{3.88}$$

Performing the triple product on the volume integral of Eq. (3.88), the following relation is obtained

$$\mathbf{K}_s = \int_V \begin{bmatrix} C_{44}F_{sx}\mathbf{J}^{-1}N_{1,r}(r) & C_{44}F_{sx}\mathbf{J}^{-1}N_{1,r}(r) & C44F_{sx}\mathbf{J}^{-1}N_{1,r}(r) & C_{44}F_{sx}\mathbf{J}^{-1} \\ F_{\tau x}\mathbf{J}^{-1}N_{1,r}(r) & F_{\tau y,x}N_1(r) & F_{\tau x}\mathbf{J}^{-1}N_{2,r}(r) & N_{1,r}(r)F_{\tau y,x}N_2(r) \\ C_{44}F_{sy,x}N_1(r) & C_{44}F_{sy,x}N_1(r) & C_{44}F_{sy,x}N_1(r) & C_{44}F_{sy,x}N_1(r) \\ F_{\tau x}\mathbf{J}^{-1}N_{1,r}(r) & F_{\tau y,x}N_1(r) & F_{\tau x}\mathbf{J}^{-1}N_{2,r}(r) & F_{\tau y,x}N_2(r) \\ C_{44}F_{sx}\mathbf{J}^{-1}N_{2,r}(r) & C_{44}F_{sx}\mathbf{J}^{-1} & C_{44}F_{sx}\mathbf{J}^{-1} & C_{44}F_{sx}\mathbf{J}^{-1} \\ F_{\tau x}\mathbf{J}^{-1}N_{1,r}(r) & N_{2,r}(r)F_{\tau y,x}N_1(r) & N_{2,r}(r)F_{\tau x}\mathbf{J}^{-1}N_{2,r}(r) & N_{2,r}(r)F_{\tau y,x}N_2(r) \\ C_{44}F_{sy,x}N_2(r) & C_{44}F_{sy,x}N_2(r) & C_{44}F_{sy,x}N_2(r) & C_{44}F_{sy,x}N_2(r) \\ F_{\tau x}\mathbf{J}^{-1}N_{1,r}(r) & F_{\tau y,x}N_1(r) & F_{\tau x}\mathbf{J}^{-1}N_{2,r}(r) & F_{\tau y,x}N_2(r) \end{bmatrix} dV$$

$$\mathbf{K}_s = \begin{bmatrix} \int_V (C_{44}F_{sx}F_{\tau x}\mathbf{J}^{-1}N_{1,r}(r)\mathbf{J}^{-1}N_{1,r}(r))dV & \int_V (C_{44}F_{sx}F_{\tau y,x}\mathbf{J}^{-1}N_{1,r}(r)N_1(r))dV & \int_V (C_{44}F_{sx}F_{\tau x}\mathbf{J}^{-1}N_{1,r}(r)\mathbf{J}^{-1}N_{2,r}(r))dV & \int_V (C_{44}F_{sx}F_{\tau y,x}\mathbf{J}^{-1}N_{1,r}(r)N_2(r))dV \\ \int_V (C44C_{44}F_{sy,x}F_{\tau x}N_1(r)\mathbf{J}^{-1}N_{1,r}(r))dV & \int_V (CC_{44}F_{sy,x}F_{\tau y,x}N_1(r)N_1(r))dV & \int_V (C_{44}F_{sy,x}F_{\tau x}N_1(r)\mathbf{J}^{-1}N_{2,r}(r))dV & \int_V (CC_{44}F_{sy,x}F_{\tau y,x}N_1(r)N_2(r))dV \\ \int_V (C_{44}F_{sx}F_{\tau x}\mathbf{J}^{-1}N_{2,r}(r)\mathbf{J}^{-1}N_{1,r}(r))dV & \int_V (C_{44}F_{sx}F_{\tau y,x}\mathbf{J}^{-1}N_{2,r}(r)N_1(r))dV & \int_V (C_{44}F_{sx}F_{\tau x}\mathbf{J}^{-1}N_{2,r}(r)\mathbf{J}^{-1}N_{2,r}(r))dV & \int_V (C_{44}F_{sx}F_{\tau y,x}\mathbf{J}^{-1}N_{2,r}(r)N_2(r))dV \\ \int_V (C_{44}F_{sy,x}F_{\tau x}N_2(r)\mathbf{J}^{-1}N_{1,r}(r))dV & \int_V (C_{44}F_{sy,x}F_{\tau y,x}N_2(r)N_1(r))dV & \int_V (C_{44}F_{sy,x}F_{\tau x}N_2(r)\mathbf{J}^{-1}N_{2,r}(r))dV & \int_V (C_{44}F_{sy,x}F_{\tau y,x}N_2(r)N_2(r))dV \end{bmatrix} \tag{3.89}$$

Computing the triple product on the volume integral of Eq. (3.87), it is obtained

$$\mathbf{K}_b = \int_V \begin{bmatrix} 0 & 0 & 0 & 0 \\ 0 & C_{11}F_{sy}F_{\tau y}\mathbf{J}^{-1}N_{1,r}(r)\mathbf{J}^{-1}N_{1,r}(r) & 0 & C_{11}F_{sy}F_{\tau y}\mathbf{J}^{-1}N_{1,r}(r)\mathbf{J}^{-1}N_{2,r}(r) \\ 0 & 0 & 0 & 0 \\ 0 & C_{11}F_{sy}F_{\tau y}\mathbf{J}^{-1}N_{2,r}(r)\mathbf{J}^{-1}N_{1,r}(r) & 0 & C_{11}F_{sy}F_{\tau y}\mathbf{J}^{-1}N_{2,r}(r)\mathbf{J}^{-1}N_{2,r}(r) \end{bmatrix} dV ==$$
$$\begin{bmatrix} 0 & 0 & 0 & 0 \\ 0 & \int_V (C_{11}F_{sy}F_{\tau y}\mathbf{J}^{-1}N_{1,r}(r)\mathbf{J}^{-1}N_{1,r}(r))dV & 0 & \int_V (C_{11}F_{sy}F_{\tau y}\mathbf{J}^{-1}N_{1,r}(r)\mathbf{J}^{-1}N_{2,r}(r))dV \\ 0 & 0 & 0 & 0 \\ 0 & \int_V (C_{11}F_{sy}F_{\tau y}\mathbf{J}^{-1}N_{2,r}(r)\mathbf{J}^{-1}N_{1,r}(r))dV & 0 & \int_V (C_{11}F_{sy}F_{\tau y}\mathbf{J}^{-1}N_{2,r}(r)\mathbf{J}^{-1}N_{2,r}(r))dV \end{bmatrix} dV. \tag{3.90}$$

3.7.3 Gauss Quadrature Technique

Gauss quadrature is a numerical method used to evaluate line, surface, and volume integrals by converting them into summations. It involves three main steps: (1) evaluating the integrand at selected Sampling Points (SPs), (2) multiplying each evaluated value by a corresponding weight, and (3) summing all the weighted values. In this study case, Gauss quadrature is applied only along the y axis, because the focus is on shear locking, which depends on how strains vary in this direction. To address other locking phenomena such as volumetric locking, Gauss quadrature would also need to be applied along the x and z directions. Building upon these steps, the assessment of a line integral can be described as follows:

$$\int_{-1}^{1} F(r)dr = \sum_{i=1}^{n_{SP}} (w_i)\,(F(SP_i)) \tag{3.91}$$

where F is the integrating function, w_i is the i_{th} weight of the i_{th} SP and n_{SP} is the total number of the SPs. For example, if 2 SPs (i.e. $n_{SP} = 2$), are considered, Eq. (3.91) becomes

$$\int_{-1}^{1} F(r)dr = (w_1)\,(F(SP_1)) + (w_2)\,(F(SP_2)) \tag{3.92}$$

If F is constant ($F(r) = F_{const}$), then it is applied for every $F(SP_i)$ in Eq. (3.92). In this scenario, Eq. (3.92) simplifies to the following:

$$\int_{-1}^{1} F_{const}dr = (w_1)\,(F_{const}) + (w_2)\,(F_{const}) \tag{3.93}$$

In the case of a surface integral, it has the following form

$$\int_{-1}^{1}\int_{-1}^{1} F(\xi, \eta)d\xi d\eta = \sum_{i,j=1}^{n_{SP}} (w_i)\,(w_j)\,\big(F(SP_i, SP_j)\big) \tag{3.94}$$

In case of $n_{SP} = 2$ it is

$$\int_{-1}^{1}\int_{-1}^{1} F(\xi, \eta)d\xi d\eta = (w_1)\,(w_1)\,(F(SP_1, SP_1)) + (w_1)\,(w_2)\,(F(SP_1, SP_2)) +$$
$$+ (w_2)\,(w_1)\,(F(SP_2, SP_1)) + (w_2)\,(w_2)\,(F(SP_2, SP_2)) \tag{3.95}$$

In Eq. (3.95), the integrand is $F(\xi, \eta)$. For example, to evaluate $F(SP_1, SP_1)$ the coordinate value associated with SP_1 is substituted for both ξ and η. Similarly, $F(SP_1, SP_2)$ is obtained by substituting the coordinate value of SP_1 into ξ and that of SP_2 into η. If F depends only on ξ or only on η, then the function simplifies to $F(\xi)$ or $F(\eta)$, respectively, and only a single coordinate value is needed. In such cases, Eq. (3.95) is modified accordingly. For instance, if F depends solely on η, it reduces to the following form

$$\int_{-1}^{1}\int_{-1}^{1} F(\eta)d\xi d\eta = (w_1)\,(w_1)\,(F(SP_1)) + (w_1)\,(w_2)\,(F(SP_1)) +$$
$$+ (w_2)\,(w_1)\,(F(SP_2)) + (w_2)\,(w_2)\,(F(SP_2)) \tag{3.96}$$

If the function F in the (3.94) has a constant value ($F(\xi, \eta) = F_{const}$), then this constant value is used for every $F(SP_i)$ in the Eq. (3.95). In this last case the Eq. (3.95) becomes the following

$$\int_{-1}^{1}\int_{-1}^{1} F(\eta)d\xi d\eta = (w_1)(w_1)(F_{const}) + (w_1)(w_2)(F_{const}) + (w_2)(w_1)(F_{const}) + \\ + (w_2)(w_2)(F_{const}) \tag{3.97}$$

In the case of a volume integral, it is

$$\int_{-1}^{1}\int_{-1}^{1}\int_{-1}^{1} F(\xi,\eta,r)d\xi d\eta dr = \sum_{i,j,k=1}^{SP} (w_i)(w_j)(w_k)\big(F(SP_i,SP_j,SP_k)\big) \tag{3.98}$$

In the case of $n_{SP} = 2$, it is

$$\int_{-1}^{1}\int_{-1}^{1}\int_{-1}^{1} F(\xi,\eta,r)d\xi d\eta dr = (w_1)(w_1)(w_1)(F(SP_1))(F(SP_1))(F(SP_1)) \\ + (w_1)(w_1)(w_2)(F(SP_1,SP_1,SP_2)) + \\ + (w_1)(w_2)(w_2)(F(SP_1,SP_2,SP_2)) + \\ + (w_1)(w_2)(w_1)(F(SP_1,SP_2,SP_1)) + \\ + (w_2)(w_1)(w_1)(F(SP_2,SP_1,SP_1)) + \\ + (w_2)(w_1)(w_2)(F(SP_2,SP_1,SP_2)) + \\ + (w_2)(w_2)(w_1)(F(SP_2,SP_2,SP_1)) + \\ + (w_2)(w_2)(w_2)(F(SP_2,SP_2,SP_2)) \tag{3.99}$$

The same clarifications as those shown in Eqs. (3.96) and (3.97) can be also applied in the case of volume integral.

3.7.4 Full Integration

The outcome achieved through Gauss quadrature is generally an approximation of the exact result. There exists a rule which asserts that if the integrating function is a polynomial, then the degree of the polynomial n can be exactly evaluated if the following relation holds:

$$n \leqslant 2n_{SP} - 1 \tag{3.100}$$

If this relation applies, then the polynomial is exactly evaluated and in this case it is referred to as Full Integration (FI) technique. If the number of SPs used is such as the results of $2n_{SP} - 1$ is smaller than the degree of the integrating polynomial, then the polynomial is evaluated approximately and in this case it is referred to as *reduced integration*.

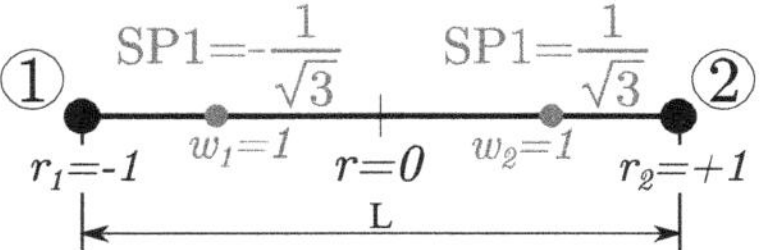

Fig. 3.6 Position of 2 SPs along the two-node beam

Regarding FE analysis, it is crucial to use a suitable integration order to achieve an accurate evaluation of integrals through Gauss quadrature. When the appropriate integration order is employed, Gauss quadrature yields Full Integration (FI), ensuring that matrices, such as the stiffness matrix, are precisely evaluated (i.e., obtaining the same value as analytical integration).

Concerning the two-node linear beam element, the stiffness matrix calculation is exact when using a 2 Sampling Points (SPs) Gauss quadrature (SP = 2). The positions of these 2 SPs along the beam, i.e., the coordinate r of these SPs, along with their corresponding weights, are illustrated in the Fig. 3.6.

Note that the coordinates for the SPs are the following

$$SP_1 : r = -\frac{1}{\sqrt{3}}, \quad SP_2 : r = \frac{1}{\sqrt{3}}$$

The related weights in this case are $w_1 = 1$ and $w_2 = 1$. Using the rules and the definitions established previously for the Gauss quadrature technique, the volume integrals can be computed.

3.7.5 Full Integration Applied to the Bending Part of the Stiffness Matrix

The bending part of the stiffness matrix is written as follows

$$\mathbf{K}_{b_{FULL}} = \begin{bmatrix} 0 & 0 & 0 & 0 \\ 0 & K_{b22} & 0 & K_{b24} \\ 0 & 0 & 0 & 0 \\ 0 & K_{b42} & 0 & K_{b44} \end{bmatrix} \tag{3.101}$$

The term K_{b22} of the matrix of Eq. (3.101) can be written as follows

$$
\begin{aligned}
K_{b22} &= \int_V \left(C_{11} F_{sy} F_{\tau y} \mathbf{J}^{-1} N_{1,r}(r) \mathbf{J}^{-1} N_{1,r}(r) \right) dV \\
&= \int_A \left(C_{11} F_{sy} F_{\tau y} \right) dA \int_L \left(\mathbf{J}^{-1} N_{1,r}(r) \mathbf{J}^{-1} N_{1,r}(r) \right) dL \\
&= \int_{-a}^{a} \int_{-b}^{b} \left(C_{11} F_{sy} F_{\tau y} \right) dx dz \int_0^L \left(\mathbf{J}^{-1} N_{1,r}(r) \mathbf{J}^{-1} N_{1,r}(r) \right) dy
\end{aligned}
\tag{3.102}
$$

Below are some relations derived from the previously provided definitions (see (3.64) and (3.69))

$$
\begin{aligned}
\text{if } \ y = 0 \quad & r = -1 \\
\text{if } \ y = L \quad & r = 1 \\
dy = det \mathbf{J} dr
\end{aligned}
\tag{3.103}
$$

By applying the relations from Eq. (3.103), the term K_{b22} can be expressed as follows:

$$
K_{b22} = \int_{-a}^{a} \int_{-b}^{b} \left(C_{11} F_{sy} F_{\tau y} \right) dx dz \int_{-1}^{1} \left(\mathbf{J}^{-1} N_{1,r}(r) \mathbf{J}^{-1} N_{1,r}(r) \right) det \ \mathbf{J} dr
\tag{3.104}
$$

The red part of the term K_{b22} in Eq. (3.104) is a surface integral and it is solved analytically. The solution of the surface integral can be done as follows

$$
\begin{aligned}
\int_{-a}^{a} \int_{-b}^{b} \left(C_{11} F_{sy} F_{\tau y} \right) dx dz &= \int_{-a}^{a} \int_{-b}^{b} \left(C_{11} x \cdot x \right) dx dz = \\
&= C_{11} \int_{-a}^{a} x^2 dx \int_{-b}^{b} dz = \frac{4}{3} C_{11} a^3 b
\end{aligned}
\tag{3.105}
$$

The blue part of the term K_{b22} of in Eq. (3.104), constitutes a line integral expressed in terms of natural coordinates r. The solution to this integral will be obtained using Gauss quadrature and the principles outlined in Eqs. (3.91), (3.92), and (3.93) are employed. To compute the integral, a Gauss quadrature is employed, utilizing two SPs ($n_{SP} = 2$). This choice leads to a FI Gauss quadrature. The subsequent operations between shape functions are utilized for the evolution of the stiffness matrix:

Operations between shape functions

$$\mathbf{J}^{-1}N_{1,r}(r)\mathbf{J}^{-1}N_{1,r}(r) = \frac{2}{L}\frac{d}{dr}\left(\frac{1}{2}(1-r)\right)\frac{2}{L}\frac{d}{dr}\left(\frac{1}{2}(1-r)\right) = \frac{1}{L^2}$$

$$\mathbf{J}^{-1}N_{1,r}(r)\mathbf{J}^{-1}N_{2,r}(r) = \frac{2}{L}\frac{d}{dr}\left(\frac{1}{2}(1-r)\right)\frac{2}{L}\frac{d}{dr}\left(\frac{1}{2}(1+r)\right) = -\frac{1}{L^2}$$

$$\mathbf{J}^{-1}N_{2,r}(r)\,[\mathbf{J}]^{-1}N_{2,r}(r) = \frac{2}{L}\frac{d}{dr}\left(\frac{1}{2}(1+r)\right)\frac{2}{L}\frac{d}{dr}\left(\frac{1}{2}(1+r)\right) = \frac{1}{L^2}$$

$$\mathbf{J}^{-1}N_{1,r}(r)N_1(r) = \frac{2}{L}\frac{d}{dr}\left(\frac{1}{2}(1-r)\right)\left(\frac{1}{2}(1-r)\right) = -\frac{1}{2L}(1-r)$$

$$\mathbf{J}^{-1}N_{2,r}(r)N_1(r) = \frac{2}{L}\frac{d}{dr}\left(\frac{1}{2}(1+r)\right)\left(\frac{1}{2}(1-r)\right) = \frac{1}{2L}(1-r)$$

$$N_1(r)N_1(r) = \left(\frac{1}{2}(1-r)\right)\left(\frac{1}{2}(1-r)\right) = \left(\frac{1}{2}(1-r)\right)^2 = \left(\frac{1}{4}+\frac{r^2}{4}-\frac{r}{2}\right)$$

$$N_1(r)\mathbf{J}^{-1}N_{2,r}(r) = \left(\frac{1}{2}(1-r)\right)\frac{2}{L}\frac{d}{dr}\left(\frac{1}{2}(1+r)\right) = -\frac{1}{2L}(1-r)$$

$$N_1(r)N_2(r) = \left(\frac{1}{2}(1-r)\right)\left(\frac{1}{2}(1+r)\right) = \left(\frac{1}{4}(1-r^2)\right)$$

$$\mathbf{J}^{-1}N_{2,r}(r)N_2(r) = \frac{2}{L}\frac{d}{dr}\left(\frac{1}{2}(1+r)\right)\left(\frac{1}{2}(1+r)\right) = \frac{1}{2L}(1-r)$$

$$N_2(r)N_2(r) = \left(\frac{1}{2}(1+r)\right)\left(\frac{1}{2}(1+r)\right) = \left(\frac{1}{2}(1+r)\right)^2 = \left(\frac{1}{4}+\frac{r^2}{4}+\frac{r}{2}\right)$$

The FI Gauss quadrature of the line integral is shown below

$$\int_{-1}^{1}\left(\mathbf{J}^{-1}N_{1,r}(r)\mathbf{J}^{-1}N_{1,r}(r)\right)det\mathbf{J}dr = [(w_1)(F(SP_1)) + (w_2)(F(SP_2))]\,det\mathbf{J}$$

$$= [(w_1)(F_{const}) + (w_2)(F_{const})]\,det\mathbf{J}$$

$$= \left[(1)\left(\frac{1}{L^2}\right) + (1)\left(\frac{1}{L^2}\right)\right]\frac{L}{2} \tag{3.106}$$

$$= \frac{1}{L}$$

Using the results from Eqs. (3.105) and (3.106), the term K_{22} of the bending part of the stiffness matrix can be evaluated as shown below

$$K_{b22} = \int_{-a}^{a} \int_{-b}^{b} \left(C_{11} F_{sy} F_{\tau y} \right) dx dz \int_{-1}^{1} \left(\mathbf{J}^{-1} N_{1,r}(r) \mathbf{J}^{-1} N_{1,r}(r) \right) det \mathbf{J} dr =$$

$$= \left(C_{11} \frac{4a^3 b}{3} \right) \left(\frac{1}{L} \right) = C_{11} \frac{4a^3 b}{3L}$$

$$(3.107)$$

The term K_{b24} can be written as follows

$$K_{b24} = \int_{V} \left(C_{11} F_{sy} F_{\tau y} \mathbf{J}^{-1} N_{1,r}(r) \mathbf{J}^{-1} N_{2,r}(r) \right) dV$$

$$= \int_{A} \left(C_{11} F_{sy} F_{\tau y} \right) dA \int_{L} \left(\mathbf{J}^{-1} N_{1,r}(r) \mathbf{J}^{-1} N_{2,r}(r) \right) dL \qquad (3.108)$$

$$= \int_{-a}^{a} \int_{-b}^{b} \left(C_{11} F_{sy} F_{\tau y} \right) dx dy \int_{0}^{L} \left(\mathbf{J}^{-1} N_{1,r}(r) \mathbf{J}^{-1} N_{2,r}(r) \right) dx$$

If the relations of Eq. (3.103) are applied, the term K_{b24} can be expressed as reported below

$$K_{b24} = \int_{-a}^{a} \int_{-b}^{b} \left(C_{11} F_{sy} F_{\tau y} \right) dx dy \int_{-1}^{1} \left(\mathbf{J}^{-1} N_{1,r}(r) \mathbf{J}^{-1} N_{2,r}(r) \right) det \mathbf{J} dr$$

$$(3.109)$$

The red part of the matrix term K_{b24} is equal to the surface integral of Eq. (3.105).

The blue part of the matrix term K_{b24} is a line integral and then the rules of Eqs. (3.91), (3.92) and (3.93) apply for it.

The FI Gauss quadrature of the line integral is shown below

$$\int_{-1}^{1} \left(\mathbf{J}^{-1} N_{1,r}(r) \mathbf{J}^{-1} N_{1,r}(r) \right) det \mathbf{J} dr = \left[(w_1) \left(F(SP_1) \right) + (w_2) \left(F(SP_2) \right) \right] det \mathbf{J}$$

$$= \left[(w_1) \left(F_{const} \right) + (w_2) \left(F_{const} \right) \right] det \mathbf{J}$$

$$= \left[(1) \left(-\frac{1}{L^2} \right) + (1) \left(-\frac{1}{L^2} \right) \right] \frac{L}{2} \qquad (3.110)$$

$$= -\frac{1}{L}$$

The term K_{24} of the bending part of the stiffness matrix is shown below

$$K_{b24} = \int_{-a}^{a} \int_{-b}^{b} \left(C_{11} F_{sy} F_{\tau y} \right) dx dy \int_{-1}^{1} \left(\mathbf{J}^{-1} N_{1,r}(r) \mathbf{J}^{-1} N_{2,r}(r) \right) det \mathbf{J} dr =$$

$$= \left(C_{11} \frac{4a^3 b}{3} \right) \left(-\frac{1}{L} \right) = -C_{11} \frac{4a^3 b}{3L}$$

$$(3.111)$$

The term K_{b42} can be written as follows

$$K_{b42} = \int_V \left(C_{11} F_{sy} F_{\tau y} \mathbf{J}^{-1} N_{2,r}(r) \mathbf{J}^{-1} N_{1,r}(r) \right) dV$$
$$= \int_A \left(C_{11} F_{sy} F_{\tau y} \right) dA \int_L \left(\mathbf{J}^{-1} N_{2,r}(r) \mathbf{J}^{-1} N_{1,r}(r) \right) dL \qquad (3.112)$$
$$= \int_{-a}^{a} \int_{-b}^{b} \left(C_{11} F_{sy} F_{\tau y} \right) dx\,dy \int_{0}^{L} \left(\mathbf{J}^{-1} N_{2,r}(r) \mathbf{J}^{-1} N_{1,r}(r) \right) dx$$

If the relations of Eq. (3.103) are employed, the term K_{b42} can be expressed as reported below

$$K_{b42} = \int_{-a}^{a} \int_{-b}^{b} \left(C_{11} F_{sy} F_{\tau y} \right) dx\,dy \int_{-1}^{1} \left(\mathbf{J}^{-1} N_{2,r}(r) \mathbf{J}^{-1} N_{1,r}(r) \right) det\mathbf{J}\,dr$$
$$(3.113)$$

The red part of the matrix term K_{b42} is equal to the surface integral of Eq. (3.105).

The blue part of the matrix term K_{b42} written in Eq. (3.113) is a line integral and it is equal to the term of Eq. (3.110) .

As a result the term K_{42} of the bending part of the stiffness matrix is calculated as follows

$$K_{b42} = \int_{-a}^{a} \int_{-b}^{b} \left(C_{11} F_{sy} F_{\tau y} \right) dx\,dy \int_{-1}^{1} \left(\mathbf{J}^{-1} N_{2,r}(r) \mathbf{J}^{-1} N_{1,r}(r) \right) det\mathbf{J}\,dr =$$
$$= \left(C_{11} \frac{4a^3 b}{3} \right) \left(-\frac{1}{L} \right) = -C_{11} \frac{4a^3 b}{3L}$$
$$(3.114)$$

The term K_{b44} can be written as follows

$$K_{b44} = \int_V \left(C_{11} F_{sy} F_{\tau y} \mathbf{J}^{-1} N_{2,r}(r) \mathbf{J}^{-1} N_{2,r}(r) \right) dV$$
$$= \int_A \left(C_{11} F_{sy} F_{\tau y} \right) dA \int_L \left(\mathbf{J}^{-1} N_{2,r}(r) \mathbf{J}^{-1} N_{2,r}(r) \right) dL \qquad (3.115)$$
$$= \int_{-a}^{a} \int_{-b}^{b} \left(C_{11} F_{sy} F_{\tau y} \right) dx\,dy \int_{0}^{L} \left(\mathbf{J}^{-1} N_{2,r}(r) \mathbf{J}^{-1} N_{2,r}(r) \right) dx$$

If relations of Eq. (3.103) are applied, the term K_{b44} can be expressed as reported below

$$K_{b44} = \int_{-a}^{a} \int_{-b}^{b} \left(C_{11} F_{sy} F_{\tau y} \right) dx\,dy \int_{-1}^{1} \left(\mathbf{J}^{-1} N_{2,r}(r) \mathbf{J}^{-1} N_{2,r}(r) \right) det\mathbf{J}\,dr$$
$$(3.116)$$

The red part of the matrix term K_{b44} is equal to the surface integral of Eq. (3.105).

The blue part of the matrix term K_{b44} written in Eq. (3.116) is a line integral and then the rules of Eqs. (3.91), (3.92) and (3.93) apply for it.

In order to calculate this integral, a Gauss quadrature is used and then two SPs are used ($n_{SP} = 2$). This results in a FI Gauss quadrature.

The integrating function is

$$\mathbf{J}^{-1} N_{2,r}(r) \mathbf{J}^{-1} N_{2,r}(r) = \frac{2}{L}\frac{d}{dr}\left(\frac{1}{2}(1+r)\right)\frac{2}{L}\frac{d}{dr}\left(\frac{1}{2}(1+r)\right) = \frac{1}{L^2} \tag{3.117}$$

The result of the line integral is

$$\int_{-1}^{1}\left(\mathbf{J}^{-1} N_{2,r}(r)\mathbf{J}^{-1} N_{2,r}(r)\right) det\mathbf{J}\, dr = [(w_1)(F(SP_1)) + (w_2)(F(SP_2))]\, det\mathbf{J}$$

$$= [(w_1)(F_{const}) + (w_2)(F_{const})]\, det\mathbf{J}$$

$$= \left[(1)\left(\frac{1}{L^2}\right) + (1)\left(\frac{1}{L^2}\right)\right]\frac{L}{2} \tag{3.118}$$

$$= \frac{1}{L}$$

Using the previous computation of Eqs. (3.105) and (3.118), the term K_{44} of the bending part of the stiffness matrix can be calculated as shown below

$$K_{b44} = \int_{-1}^{1}\int_{-1}^{1}\left(C_{11} F_{sy} F_{\tau y}\right) det\mathbf{J}_c d\xi\, d\eta \int_{-1}^{1}\left(\mathbf{J}^{-1} N_{2,r}(r)\mathbf{J}^{-1} N_{2,r}(r)\right) det\mathbf{J}\, dr =$$

$$= \left(C_{11}\frac{4a^3b}{3}\right)\left(\frac{1}{L}\right) = C_{11}\frac{4a^3b}{3L} \tag{3.119}$$

Considering the previous results obtained for the terms K_{b22}, K_{b24}, K_{b42}, and K_{b44}, the benign part of the stiffness matrix, computed using the FI Gauss quadrature, is equal to

$$\mathbf{K}_{b_{FULL}} = \begin{bmatrix} 0 & 0 & 0 & 0 \\ 0 & \dfrac{4C_{11}a^3b}{3L} & 0 & -\dfrac{4C_{11}a^3b}{3L} \\ 0 & 0 & 0 & 0 \\ 0 & -\dfrac{4C_{11}a^3b}{3L} & 0 & \dfrac{4C_{11}a^3b}{3L} \end{bmatrix}. \tag{3.120}$$

3.7.6 Full Integration Applied to the Shear Part of the Stiffness Matrix

The shear part of the stiffness matrix is written in the following generic form

$$
\mathbf{K}_{S_{FULL}} =
\begin{bmatrix}
K_{s11} & K_{s12} & K_{s13} & K_{s14} \\
K_{s21} & K_{s22} & K_{s23} & K_{s24} \\
K_{s31} & K_{s32} & K_{s33} & K_{s34} \\
K_{s41} & K_{s42} & K_{s43} & K_{s44}
\end{bmatrix}
\tag{3.121}
$$

The terms of the matrix of Eq. (3.121) correspond to the following expressions

$$
\begin{aligned}
K_{s11} &= \int_V \left(C_{44} F_{sx} F_{\tau x} \mathbf{J}^{-1} N_{1,r}(r) \mathbf{J}^{-1} N_{1,r}(r) \right) dV \\
&= \int_A (C_{44} F_{sx} F_{\tau x})\, dA \int_L \left(\mathbf{J}^{-1} N_{1,r}(r) \mathbf{J}^{-1} N_{1,r}(r) \right) dL \\
&= \int_{-a}^{a} \int_{-b}^{b} (C_{44} F_{sx} F_{\tau x})\, dx dz \int_{0}^{L} \left(\mathbf{J}^{-1} N_{1,r}(r) \mathbf{J}^{-1} N_{1,r}(r) \right) dx \\
&= \int_{-a}^{a} \int_{-b}^{b} (C_{44} F_{sx} F_{\tau x})\, dx dz \int_{-1}^{1} \left(\mathbf{J}^{-1} N_{1,r}(r) \mathbf{J}^{-1} N_{1,r}(r) \right) det \mathbf{J} dr
\end{aligned}
$$

$$
\begin{aligned}
K_{s12} &= \int_V \left(C_{44} F_{sx} F_{\tau y,x} \mathbf{J}^{-1} N_{1,r}(r) N_1(r) \right) dV \\
&= \int_A \left(C_{44} F_{sx} F_{\tau y,x} \right) dA \int_L \left(\mathbf{J}^{-1} N_{1,r}(r) N_1(r) \right) dL \\
&= \int_{-a}^{a} \int_{-b}^{b} \left(C_{44} F_{sx} F_{\tau y,x} \right) dx dz \int_{0}^{L} \left(\mathbf{J}^{-1} N_{1,r}(r) N_1(r) \right) dx \\
&= \int_{-a}^{a} \int_{-b}^{b} \left(C_{44} F_{sx} F_{\tau y,x} \right) dx dz \int_{-1}^{1} \left(\mathbf{J}^{-1} N_{1,r}(r) N_1(r) \right) det \mathbf{J} dr
\end{aligned}
$$

$$
\begin{aligned}
K_{s13} &= \int_V \left(C_{44} F_{sx} F_{\tau x} \mathbf{J}^{-1} N_{1,r}(r) \mathbf{J}^{-1} N_{2,r}(r) \right) dV \\
&= \int_A (C_{44} F_{sx} F_{\tau x})\, dA \int_L \left(\mathbf{J}^{-1} N_{1,r}(r) \mathbf{J}^{-1} N_{2,r}(r) \right) dL \\
&= \int_{-a}^{a} \int_{-b}^{b} (C_{44} F_{sx} F_{\tau x})\, dx dz \int_{0}^{L} \left(\mathbf{J}^{-1} N_{1,r}(r) \mathbf{J}^{-1} N_{2,r}(r) \right) dx \\
&= \int_{-a}^{a} \int_{-b}^{b} (C_{44} F_{sx} F_{\tau x})\, dx dz \int_{-1}^{1} \left(\mathbf{J}^{-1} N_{1,r}(r) \mathbf{J}^{-1} N_{2,r}(r) \right) det \mathbf{J} dr
\end{aligned}
$$

$$
\begin{aligned}
K_{s14} &= \int_V \left(C_{44} F_{sx} F_{\tau y,x} \mathbf{J}^{-1} N_{1,r}(r) N_2(r) \right) dV \\
&= \int_A \left(C_{44} F_{sx} F_{\tau y,x} \right) dA \int_L \left(\mathbf{J}^{-1} N_{1,r}(r) N_2(r) \right) dL
\end{aligned}
$$

$$= \int_{-a}^{a} \int_{-b}^{b} \left(C_{44} F_{sx} F_{\tau y,x} \right) dx dz \int_{0}^{L} \left(\mathbf{J}^{-1} N_{1,r}(r) N_2(r) \right) dx$$

$$= \int_{-a}^{a} \int_{-b}^{b} \left(C_{44} F_{sx} F_{\tau y,x} \right) dx dz \int_{-1}^{1} \left(\mathbf{J}^{-1} N_{1,r}(r) N_2(r) \right) det \mathbf{J} dr$$

$$K_{s21} = \int_{V} \left(C_{44} F_{sy,x} F_{\tau x} N_1(r) \mathbf{J}^{-1} N_{1,r}(r) \right) dV$$

$$= \int_{A} \left(C_{44} F_{sy,x} F_{\tau x} \right) dA \int_{L} \left(N_1(r) \mathbf{J}^{-1} N_{1,r}(r) \right) dL$$

$$= \int_{-a}^{a} \int_{-b}^{b} \left(C_{44} F_{sy,x} F_{\tau x} \right) dx dz \int_{0}^{L} \left(N_1(r) \mathbf{J}^{-1} N_{1,r}(r) \right) dx$$

$$= \int_{-a}^{a} \int_{-b}^{b} \left(C_{44} F_{sy,x} F_{\tau x} \right) dx dz \int_{-1}^{1} \left(N_1(r) \mathbf{J}^{-1} N_{1,r}(r) \right) det \mathbf{J} dr$$

$$K_{s22} = \int_{V} \left(C_{44} F_{sy,x} F_{\tau y,x} N_1(r) N_1(r) \right) dV$$

$$= \int_{A} \left(C_{44} F_{sy,x} F_{\tau y,x} \right) dA \int_{L} \left(N_1(r) N_1(r) \right) dL$$

$$= \int_{-a}^{a} \int_{-b}^{b} \left(C_{44} F_{sy,x} F_{\tau y,x} \right) dx dz \int_{0}^{L} \left(N_1(r) N_1(r) \right) dx$$

$$= \int_{-a}^{a} \int_{-b}^{b} \left(C_{44} F_{sy,x} F_{\tau y,x} \right) dx dz \int_{-1}^{1} \left(N_1(r) N_1(r) \right) det \mathbf{J} dr$$

$$K_{s23} = \int_{V} \left(C_{44} F_{sy,x} F_{\tau x} N_1(r) \mathbf{J}^{-1} N_{2,r}(r) \right) dV$$

$$= \int_{A} \left(C_{44} F_{sy,x} F_{\tau x} \right) dA \int_{L} \left(N_1(r) \mathbf{J}^{-1} N_{2,r}(r) \right) dL$$

$$= \int_{-a}^{a} \int_{-b}^{b} \left(C_{44} F_{sy,x} F_{\tau x} \right) dx dz \int_{0}^{L} \left(N_1(r) \mathbf{J}^{-1} N_{2,r}(r) \right) dx$$

$$= \int_{-a}^{a} \int_{-b}^{b} \left(C_{44} F_{sy,x} F_{\tau x} \right) dx dz \int_{-1}^{1} \left(N_1(r) \mathbf{J}^{-1} N_{2,r} \right) det \mathbf{J} dr$$

$$K_{s24} = \int_{V} \left(C_{44} F_{sy,x} F_{\tau y,x} N_1(r) N_2(r) \right) dV$$

$$= \int_{A} \left(C_{44} F_{sy,x} F_{\tau y,x} \right) dA \int_{L} \left(N_1(r) N_2(r) \right) dL$$

$$= \int_{-a}^{a} \int_{-b}^{b} \left(C_{44} F_{sy,x} F_{\tau y,x} \right) dx dz \int_{0}^{L} \left(N_1(r) N_2(r) \right) dx$$

$$= \int_{-a}^{a} \int_{-b}^{b} \left(C_{44} F_{sy,x} F_{\tau y,x} \right) dx dz \int_{-1}^{1} \left(N_1(r) N_2(r) \right) det \mathbf{J} dr$$

$$
\begin{aligned}
K_{s31} &= \int_V \left(C_{44} F_{sx} F_{\tau x} \mathbf{J}^{-1} N_{2,r}(r) \mathbf{J}^{-1} N_{1,r}(r) \right) dV \\
&= \int_A \left(C_{44} F_{sx} F_{\tau x} \right) dA \int_L \left(\mathbf{J}^{-1} N_{2,r}(r) \mathbf{J}^{-1} N_{1,r}(r) \right) dL \\
&= \int_{-a}^{a} \int_{-b}^{b} \left(C_{44} F_{sx} F_{\tau x} \right) dx dz \int_0^L \left(\mathbf{J}^{-1} N_{2,r}(r) \mathbf{J}^{-1} N_{1,r}(r) \right) dx \\
&= \int_{-a}^{a} \int_{-b}^{b} \left(C_{44} F_{sx} F_{\tau x} \right) dx dz \int_{-1}^{1} \left(\mathbf{J}^{-1} N_{2,r}(r) \mathbf{J}^{-1} N_{1,r}(r) \right) det\mathbf{J} dr
\end{aligned}
$$

$$
\begin{aligned}
K_{s32} &= \int_V \left(C_{44} F_{sx} F_{\tau y,x} \mathbf{J}^{-1} N_{2,r}(r) N_1(r) \right) dV \\
&= \int_A \left(C_{44} F_{sx} F_{\tau y,x} \right) dA \int_L \left(\mathbf{J}^{-1} N_{2,r}(r) N_1(r) \right) dL \\
&= \int_{-a}^{a} \int_{-b}^{b} \left(C_{44} F_{sx} F_{\tau y,x} \right) dx dz \int_0^L \left(\mathbf{J}^{-1} N_{2,r}(r) N_1(r) \right) dx \\
&= \int_{-a}^{a} \int_{-b}^{b} \left(C_{44} F_{sx} F_{\tau y,x} \right) dx dz \int_{-1}^{1} \left(\mathbf{J}^{-1} N_{2,r}(r) N_1(r) \right) det\mathbf{J} dr
\end{aligned}
$$

$$
\begin{aligned}
K_{s33} &= \int_V \left(C_{44} F_{sx} F_{\tau x} \mathbf{J}^{-1} N_{2,r}(r) \mathbf{J}^{-1} N_{2,r}(r) \right) dV \\
&= \int_A \left(C_{44} F_{sx} F_{\tau x} \right) dA \int_L \left(\mathbf{J}^{-1} N_{2,r}(r) \mathbf{J}^{-1} N_{2,r}(r) \right) dL \\
&= \int_{-a}^{a} \int_{-b}^{b} \left(C_{44} F_{sx} F_{\tau x} \right) dx dz \int_0^L \left(\mathbf{J}^{-1} N_{2,r}(r) \mathbf{J}^{-1} N_{2,r}(r) \right) dx \\
&= \int_{-a}^{a} \int_{-b}^{b} \left(C_{44} F_{sx} F_{\tau x} \right) dx dz \int_{-1}^{1} \left(\mathbf{J}^{-1} N_{2,r}(r) \mathbf{J}^{-1} N_{2,r}(r) \right) det\mathbf{J} dr
\end{aligned}
$$

$$
\begin{aligned}
K_{s34} &= \int_V \left(C_{44} F_{sx} F_{\tau y,x} \mathbf{J}^{-1} N_{2,r}(r) N_2(r) \right) dV \\
&= \int_A \left(C_{44} F_{sx} F_{\tau y,x} \right) dA \int_L \left(\mathbf{J}^{-1} N_{2,r}(r) N_2(r) \right) dL \\
&= \int_{-a}^{a} \int_{-b}^{b} \left(C_{44} F_{sx} F_{\tau y,x} \right) dx dz \int_0^L \left(\mathbf{J}^{-1} N_{2,r}(r) N_2(r) \right) dx \\
&= \int_{-a}^{a} \int_{-b}^{b} \left(C_{44} F_{sx} F_{\tau y,x} \right) dx dz \int_{-1}^{1} \left(\mathbf{J}^{-1} N_{2,r}(r) N_2(r) \right) det\mathbf{J} dr
\end{aligned}
$$

$$
\begin{aligned}
K_{s41} &= \int_V \left(C_{44} F_{sy,x} F_{\tau x} N_2(r) \mathbf{J}^{-1} N_{1,r}(r) \right) dV \\
&= \int_A \left(C_{44} F_{sy,x} F_{\tau x} \right) dA \int_L \left(N_2(r) \mathbf{J}^{-1} N_{1,r}(r) \right) dL \\
&= \int_{-a}^{a} \int_{-b}^{b} \left(C_{44} F_{sy,x} F_{\tau x} \right) dx dz \int_0^L \left(N_2(r) \mathbf{J}^{-1} N_{1,r}(r) \right) dx \\
&= \int_{-a}^{a} \int_{-b}^{b} \left(C_{44} F_{sy,x} F_{\tau x} \right) dx dz \int_{-1}^{1} \left(N_2(r) \mathbf{J}^{-1} N_{1,r}(r) \right) det\mathbf{J} dr
\end{aligned}
$$

$$
\begin{aligned}
K_{s42} &= \int_V \left(C_{44} F_{sy,x} F_{\tau y,x} N_2(r) N_1(r) \right) dV \\
&= \int_A \left(C_{44} F_{sy,x} F_{\tau y,x} \right) dA \int_L \left(N_2(r) N_1(r) \right) dL \\
&= \int_{-a}^{a} \int_{-b}^{b} \left(C_{44} F_{sy,x} F_{\tau y,x} \right) dx dz \int_0^L \left(N_2(r) N_1(r) \right) dx \\
&= \int_{-a}^{a} \int_{-b}^{b} \left(C_{44} F_{sy,x} F_{\tau y,x} \right) dx dz \int_{-1}^{1} \left(N_2(r) N_1(r) \right) \det \mathbf{J} dr
\end{aligned}
$$

$$
\begin{aligned}
K_{s43} &= \int_V \left(C_{44} F_{sy,x} F_{\tau x} N_2(r) \mathbf{J}^{-1} N_{2,r}(r) \right) dV \\
&= \int_A \left(C_{44} F_{sy,x} F_{\tau x} \right) dA \int_L \left(N_2(r) \mathbf{J}^{-1} N_{2,r}(r) \right) dL \\
&= \int_{-a}^{a} \int_{-b}^{b} \left(C_{44} F_{sy,x} F_{\tau x} \right) dx dz \int_0^L \left(N_2(r) \mathbf{J}^{-1} N_{2,r}(r) \right) dx \\
&= \int_{-a}^{a} \int_{-b}^{b} \left(C_{44} F_{sy,x} F_{\tau x} \right) dx dz \int_{-1}^{1} \left(N_2(r) \mathbf{J}^{-1} N_{2,r}(r) \right) \det \mathbf{J} dr
\end{aligned}
$$

$$
\begin{aligned}
K_{s44} &= \int_V \left(C_{44} F_{sy,x} F_{\tau y,x} N_2(r) N_2(r) \right) dV \\
&= \int_A \left(C_{44} F_{sy,x} F_{\tau y,x} \right) dA \int_L \left(N_2(r) N_2(r) \right) dL \\
&= \int_{-a}^{a} \int_{-b}^{b} \left(C_{44} F_{sy,x} F_{\tau y,x} \right) dx dz \int_0^L \left(N_2(r) N_2(r) \right) dx \\
&= \int_{-a}^{a} \int_{-b}^{b} \left(C_{44} F_{sy,x} F_{\tau y,x} \right) dx dz \int_{-1}^{1} \left(N_2(r) N_2(r) \right) \det \mathbf{J} dr
\end{aligned}
$$

Obtaining the shear terms of the stiffness matrix requires evaluating the relevant integrals. It is important to note the following: the red-colored integrands for each matrix term (corresponding to the surface integrals) are computed analytically, whereas the blue-colored integrands (corresponding to the line integrals) are evaluated via Gauss quadrature. These latter functions either remain constant or reduce to polynomials of degree at most 2. A second-degree polynomial is fully integrated by a 2-point Gauss quadrature. In fact, the highest polynomial degree that can be exactly integrated using two sampling points (SPs) is given by $2 \times SP - 1 = 2 \times 2 - 1 = 3$. When evaluating the blue-colored integrands, it is necessary to take into account the shape function definitions given in Eq. (3.7.5). The evaluation of each term in $\mathbf{K}_{S_{FULL}}$ is provided below.

$$K_{s11} = \int_{-a}^{a} \int_{-b}^{b} (C_{44} F_{sx} F_{\tau x}) \, dx \, dz \int_{-1}^{1} \left(\mathbf{J}^{-1} N_{1,r}(r) \mathbf{J}^{-1} N_{1,r}(r) \right) det \mathbf{J} \, dr$$

$$F_{sx} F_{\tau x} = (-1)(-1) = 1$$

$$\int_{-a}^{a} \int_{-b}^{b} (C_{44} F_{sx} F_{\tau x}) \, dx \, dz \quad = \quad \int_{-a}^{a} \int_{-b}^{b} ((C_{44})(1)) \, dx \, dz = \quad C_{44} \int_{-a}^{a} dx \int_{-b}^{b} dz =$$

$$C_{44} 4ab$$

$$\mathbf{J}^{-1} N_{1,r}(r) \mathbf{J}^{-1} N_{1,r}(r) = \frac{2}{L} \frac{d}{dr} \left(\frac{1}{2}(1-r) \right) \frac{2}{L} \frac{d}{dr} \left(\frac{1}{2}(1-r) \right) = \frac{1}{L^2}$$

$$\int_{-1}^{1} \left(\mathbf{J}^{-1} N_{1,r}(r) \mathbf{J}^{-1} N_{1,r}(r) \right) det \mathbf{J} \, dr \qquad = \int_{-1}^{1} \left(\frac{1}{L^2} \right) \left(\frac{L}{2} \right) dr \qquad =$$

$$\left[(1) \left(\frac{1}{L^2} \right) + (1) \left(\frac{1}{L^2} \right) \right] \frac{L}{2} \quad \frac{1}{L}$$

$$K_{s11} = C_{44} 4ab \, \frac{1}{L} = \frac{4 C_{44} ab}{L}$$

$$K_{s12} = \int_{-a}^{a} \int_{-b}^{b} (C_{44} F_{sx} F_{\tau y,x}) \, dx \, dz \int_{-1}^{1} \left(\mathbf{J}^{-1} N_{1,r}(r) N_1(r) \right) det \mathbf{J} \, dr$$

$$F_{sx} F_{\tau y,x} = (-1)(1) = -1$$

$$\int_{-a}^{a} \int_{-b}^{b} (C_{44} F_{sx} F_{\tau y,x}) \, dx \, dz \quad = \quad \int_{-a}^{a} \int_{-b}^{b} ((C_{44})(-1)) \, dx \, dz = \quad -C_{44} \int_{-a}^{a} dx \int_{-b}^{b} dz =$$

$$-C_{44} 4ab$$

$$\mathbf{J}^{-1} N_{1,r}(r) N_1(r) = \frac{2}{L} \frac{d}{dr} \left(\frac{1}{2}(1-r) \right) \left(\frac{1}{2}(1-r) \right) = -\frac{1}{2L}(1-r)$$

$$\int_{-1}^{1} \left(\mathbf{J}^{-1} N_{1,r}(r) N_1(r) \right) det \mathbf{J} \, dr =$$

$$\int_{-1}^{1} \left(-\frac{1}{2L}(1-r) \right) \left(\frac{L}{2} \right) dr \quad = \left(-\frac{1}{2L} \right) \left[(1) \left(1 - \left(-\frac{1}{\sqrt{3}} \right) \right) + (1) \left(1 - \left(+\frac{1}{\sqrt{3}} \right) \right) \right] \frac{L}{2} =$$

$$-\frac{1}{2}$$

$$K_{s12} = -C_{44} 4ab \, -\frac{1}{2} = 2 C_{44} ab$$

$$K_{s13} = \int_{-a}^{a} \int_{-b}^{b} (C_{44} F_{sx} F_{\tau x})\, dx\, dz \int_{-1}^{1} \left(\mathbf{J}^{-1} N_{1,r}(r) \mathbf{J}^{-1} N_{2,r}(r) \right) det\mathbf{J}\, dr$$

$$F_{sx} F_{\tau x} = (-1)(-1) = 1$$

$$\int_{-a}^{a} \int_{-b}^{b} (C_{44} F_{sx} F_{\tau x})\, dx\, dz \;=\; \int_{-a}^{a} \int_{-b}^{b} ((C_{44})(1))\, dx\, dz = \; C_{44} \int_{-a}^{a} dx \int_{-b}^{b} dz = $$

$$C_{44} 4ab$$

$$\mathbf{J}^{-1} N_{1,r}(r) \mathbf{J}^{-1} N_{2,r}(r) = \frac{2}{L} \frac{d}{dr}\left(\frac{1}{2}(1-r) \right) \frac{2}{L} \frac{d}{dr}\left(\frac{1}{2}(1+r) \right) = -\frac{1}{L^2}$$

$$\int_{-1}^{1} \left(\mathbf{J}^{-1} N_{1,r}(r) \mathbf{J}^{-1} N_{2,r}(r) \right) det\mathbf{J}\, dr = \int_{-1}^{1} \left(-\frac{1}{L^2} \right)\left(\frac{L}{2} \right) dr \;=\;$$

$$\left[(1)\left(-\frac{1}{L^2} \right) + (1)\left(-\frac{1}{L^2} \right) \right] \frac{L}{2} = \; -\frac{1}{L}$$

$$K_{s13} = \; = C_{44} 4ab \; -\frac{1}{L} = \; -\frac{4C_{44}ab}{L}$$

$$K_{s14} = \int_{-a}^{a} \int_{-b}^{b} (C_{44} F_{sx} F_{\tau y,x})\, dx\, dz \int_{-1}^{1} \left(\mathbf{J}^{-1} N_{1,r}(r) N_2(r) \right) det\mathbf{J}\, dr$$

$$F_{sx} F_{\tau y,x} = (-1)(1) = -1$$

$$\int_{-a}^{a} \int_{-b}^{b} (C_{44} F_{sx} F_{\tau y,x})\, dx\, dz \;=\; \int_{-a}^{a} \int_{-b}^{b} ((C_{44})(-1))\, dx\, dz = \; -C_{44} \int_{-a}^{a} dx \int_{-b}^{b} dz = $$

$$-C_{44} 4ab$$

$$\mathbf{J}^{-1} N_{1,r}(r) N_2(r) = \frac{2}{L} \frac{d}{dr}\left(\frac{1}{2}(1-r) \right)\left(\frac{1}{2}(1+r) \right) = -\frac{1}{2L}(1+r)$$

$$\int_{-1}^{1} \left(\mathbf{J}^{-1} N_{1,r}(r) N_2(r) \right) det\mathbf{J}\, dr = $$

$$\int_{-1}^{1} \left(-\frac{1}{2L}(1+r) \right)\left(\frac{L}{2} \right) dr \;=\; \left(-\frac{1}{2L} \right)\left[(1)\left(1 + \left(-\frac{1}{\sqrt{3}} \right) \right) + (1)\left(1 + \left(+\frac{1}{\sqrt{3}} \right) \right) \right] \frac{L}{2} = $$

$$-\frac{1}{2}$$

$$K_{s14} = \; = -C_{44} 4ab \; -\frac{1}{2} = \; 2C_{44}ab$$

$$K_{s21} = \int_{-a}^{a} \int_{-b}^{b} \left(C_{44} F_{sy,x} F_{\tau x} \right) dx\,dz \int_{-1}^{1} \left(\mathbf{J}^{-1} N_{1,r}(r) N_1(r) \right) det\mathbf{J}\,dr$$

$$F_{sy,x} F_{\tau x} = (1)(-1) = -1$$

$$\int_{-a}^{a} \int_{-b}^{b} \left(C_{44} F_{sy,x} F_{\tau x} \right) dx\,dz \;=\; \int_{-a}^{a} \int_{-b}^{b} \left((C_{44})(-1) \right) dx\,dz = \;-C_{44} \int_{-a}^{a} dx \int_{-b}^{b} dz =$$

$$-C_{44} 4ab$$

$$\mathbf{J}^{-1} N_{1,r}(r) N_1(r) = \frac{2}{L} \left[\frac{d}{dr} \left(\frac{1}{2}(1-r) \right) \right] \left(\frac{1}{2}(1-r) \right) = -\frac{1}{2L}(1-r)$$

$$\int_{-1}^{1} \left(\mathbf{J}^{-1} N_{1,r}(r) N_1(r) \right) det\mathbf{J}\,dr =$$

$$\int_{-1}^{1} \left(-\frac{1}{2L}(1-r) \right) \left(\frac{L}{2} \right) dr \;=\; \left(-\frac{1}{2L} \right) \left[(1)\left(1 - \left(-\frac{1}{\sqrt{3}} \right) \right) + (1)\left(1 - \left(+\frac{1}{\sqrt{3}} \right) \right) \right] \frac{L}{2} =$$

$$-\frac{1}{2}$$

$$K_{s21} = \; = -C_{44} 4ab -\frac{1}{2} = \mathbf{2C_{44} ab}$$

$$K_{s22} = \int_{-a}^{a} \int_{-b}^{b} \left(C_{44} F_{sy,x} F_{\tau y,x} \right) dx\,dz \int_{-1}^{1} \left(N_1(r) N_1(r) \right) det\mathbf{J}\,dr$$

$$F_{sy,x} F_{\tau y,x} = (1)(1) = 1$$

$$\int_{-a}^{a} \int_{-b}^{b} \left(C_{44} F_{sy,x} F_{\tau y,x} \right) dx\,dz = \int_{-a}^{a} \int_{-b}^{b} \left((C_{44})(1) \right) dx\,dz = C_{44} \int_{-a}^{a} dx \int_{-b}^{b} dz = C_{44} 4ab$$

$$N_1(r) N_1(r) = \left(\frac{1}{2}(1-r) \right) \left(\frac{1}{2}(1-r) \right) = \left(\frac{1}{2}(1-r) \right)^2 = \left(\frac{1}{4} + \frac{r^2}{4} - \frac{r}{2} \right)$$

$$\int_{-1}^{1} \left(N_1(r) N_1(r) \right) det\mathbf{J}\,dr =$$

$$\int_{-1}^{1} \left(\frac{1}{4} + \frac{r^2}{4} - \frac{r}{2} \right) \left(\frac{L}{2} \right) dr = \left[(1)\left(\frac{1}{4} + \frac{(-\frac{1}{\sqrt{3}})^2}{4} - \frac{(-\frac{1}{\sqrt{3}})}{2} \right) + (1)\left(\frac{1}{4} + \frac{(+\frac{1}{\sqrt{3}})^2}{4} - \frac{(+\frac{1}{\sqrt{3}})}{2} \right) \right] \frac{L}{2} =$$

$$\frac{L}{3}$$

$$K_{s22} = \; = C_{44} 4ab \, \frac{L}{3} = \mathbf{\frac{4C_{44} Lab}{3}}$$

$$K_{s23} = \int_{-a}^{a} \int_{-b}^{b} \left(C_{44} F_{sy,x} F_{\tau x} \right) dx dz \int_{-1}^{1} \left(\mathbf{J}^{-1} N_{2,r}(r) N_1(r) \right) det\mathbf{J} dr$$

$$F_{sy,x} F_{\tau x} = (1)(-1) = -1$$

$$\int_{-a}^{a} \int_{-b}^{b} \left(C_{44} F_{sy,x} F_{\tau x} \right) dx dz \quad = \quad \int_{-a}^{a} \int_{-b}^{b} \left((C_{44})(-1) \right) dx dz = \quad -C_{44} \int_{-a}^{a} dx \int_{-b}^{b} dz =$$

$$-C_{44} 4ab$$

$$\mathbf{J}^{-1} N_{2,r}(r) N_1(r) = \frac{2}{L} \left[\frac{d}{dr} \left(\frac{1}{2}(1+r) \right) \right] \left(\frac{1}{2}(1-r) \right) = \frac{1}{2L}(1-r)$$

$$\int_{-1}^{1} \left(\mathbf{J}^{-1} N_{2,r}(r) N_1(r) \right) det\mathbf{J} dr =$$

$$\int_{-1}^{1} \left(\frac{1}{2L}(1-r) \right) \left(\frac{L}{2} \right) dr = \left[(1) \left(\frac{1}{2L}(1-(-\frac{1}{\sqrt{3}})) \right) + (1) \left(\frac{1}{2L}(1-(\frac{1}{\sqrt{3}})) \right) \right] \frac{L}{2} = \frac{1}{2}$$

$$K_{s23} = = -C_{44} 4ab \, \frac{1}{2} = -2C_{44} ab$$

$$K_{s24} = \int_{-a}^{a} \int_{-b}^{b} \left(C_{44} F_{sy,x} F_{\tau y,x} \right) dx dz \int_{-1}^{1} \left(N_1(r) N_2(r) \right) det\mathbf{J} dr$$

$$F_{sy,x} F_{\tau y,x} = (1)(1) = 1$$

$$\int_{-a}^{a} \int_{-b}^{b} \left(C_{44} F_{sy,x} F_{\tau y,x} \right) dx dz = \int_{-a}^{a} \int_{-b}^{b} \left((C_{44})(1) \right) dx dz = C_{44} \int_{-a}^{a} dx \int_{-b}^{b} dz = C_{44} 4ab$$

$$N_1(r) N_2(r) = \left(\frac{1}{2}(1-r) \right) \left(\frac{1}{2}(1+r) \right) = \left(\frac{1}{4}(1-r^2) \right)$$

$$\int_{-1}^{1} \left(N_1(r) N_2(r) \right) det\mathbf{J} dr$$

$$= \int_{-1}^{1} \left(\frac{1}{4}(1-r^2) \right) \left(\frac{L}{2} \right) dr = \left[(1) \left(\frac{1}{4} \left(1 - \left(-\frac{1}{\sqrt{3}} \right)^2 \right) \right) + (1) \left(\frac{1}{4} \left(1 - \left(+\frac{1}{\sqrt{3}} \right)^2 \right) \right) \right] \frac{L}{2} =$$

$$\frac{L}{6}$$

$$K_{s24} = = C_{44} 4ab \, \frac{L}{6} = \frac{2C_{44} Lab}{3}$$

$$K_{31} = \int_{-a}^{a} \int_{-b}^{b} (C_{44} F_{sx} F_{\tau x}) \, dx dz \int_{-1}^{1} \left(\mathbf{J}^{-1} N_{1,r}(r) \mathbf{J}^{-1} N_{2,r}(r) \right) det\mathbf{J} dr$$

$$F_{sx} F_{\tau x} = (-1)(-1) = 1$$

$$\int_{-a}^{a} \int_{-b}^{b} (C_{44} F_{sx} F_{\tau x}) \, dx dz = \int_{-a}^{a} \int_{-b}^{b} ((C_{44})(1)) \, dx dz = C_{44} \int_{-a}^{a} dx \int_{-b}^{b} dz = C_{44} 4ab$$

$$\mathbf{J}^{-1} N_{1,r}(r) \mathbf{J}^{-1} N_{2,r}(r) = \frac{2}{L} \frac{d}{dr} \left(\frac{1}{2}(1-r) \right) \frac{2}{L} \frac{d}{dr} \left(\frac{1}{2}(1+r) \right) = -\frac{1}{L^2}$$

$$\int_{-1}^{1} \left(\mathbf{J}^{-1} N_{1,r}(r) \mathbf{J}^{-1} N_{2,r}(r) \right) det\mathbf{J} dr =$$

$$= \int_{-1}^{1} \left(-\frac{1}{L^2} \right) \left(\frac{L}{2} \right) dr = \left[(1) \left(-\frac{1}{L^2} \right) + (1) \left(-\frac{1}{L^2} \right) \right] \frac{L}{2} = -\frac{1}{L}$$

$$K_{s31} = = C_{44} 4ab - \frac{1}{L} = -\frac{4C_{44} ab}{L}$$

$$K_{32} = \int_{-a}^{a} \int_{-b}^{b} (C_{44} F_{sy,x} F_{\tau x}) \, dx dz \int_{-1}^{1} \left(\mathbf{J}^{-1} N_{2,r}(r) N_1(r) \right) det\mathbf{J} dr$$

$$F_{sy,x} F_{\tau x} = (1)(-1) = -1$$

$$\int_{-a}^{a} \int_{-b}^{b} (C_{44} F_{sy,x} F_{\tau x}) \, dx dz = \int_{-a}^{a} \int_{-b}^{b} ((C_{44})(-1)) \, dx dz = -C_{44} \int_{-a}^{a} dx \int_{-b}^{b} dz =$$
$$-C_{44} 4ab$$

$$\mathbf{J}^{-1} N_{2,r}(r) N_1(r) = \frac{2}{L} \left[\frac{d}{dr} \left(\frac{1}{2}(1+r) \right) \right] \left(\frac{1}{2}(1-r) \right) = \frac{1}{2L}(1-r)$$

$$\int_{-1}^{1} \left(\mathbf{J}^{-1} N_{2,r}(r) N_1(r) \right) det\mathbf{J} dr =$$

$$\int_{-1}^{1} \left(\frac{1}{2L}(1-r) \right) \left(\frac{L}{2} \right) dr = \left[(1) \left(\frac{1}{2L}(1-(-\frac{1}{\sqrt{3}})) \right) + (1) \left(\frac{1}{2L}(1-(\frac{1}{\sqrt{3}})) \right) \right] \frac{L}{2} = \frac{1}{2}$$

$$K_{s32} = = -C_{44} 4ab \, \frac{1}{2} = -2C_{44} ab$$

$$K_{33} = \int_{-a}^{a} \int_{-b}^{b} \left(C_{44} F_{sx} F_{\tau x} \right) dx dz \int_{-1}^{1} \left(\mathbf{J}^{-1} N_{2,r}(r) \mathbf{J}^{-1} N_{2,r}(r) \right) det \mathbf{J} dr$$

$$F_{sx} F_{\tau x} = (-1)(-1) = 1$$

$$\int_{-a}^{a} \int_{-b}^{b} \left(C_{44} F_{sx} F_{\tau x} \right) dx dz = \int_{-a}^{a} \int_{-b}^{b} \left((C_{44})(1) \right) dx dz = C_{44} \int_{-a}^{a} dx \int_{-b}^{b} dz = C_{44} 4ab$$

$$\mathbf{J}^{-1} N_{2,r}(r) \mathbf{J}^{-1} N_{2,r}(r) = \frac{2}{L} \frac{d}{dr} \left(\frac{1}{2}(1+r) \right) \frac{2}{L} \frac{d}{dr} \left(\frac{1}{2}(1+r) \right) = \frac{1}{L^2}$$

$$\int_{-1}^{1} \left(\mathbf{J}^{-1} N_{2,r}(r) \mathbf{J}^{-1} N_{2,r}(r) \right) det \mathbf{J} dr = \int_{-1}^{1} \left(\frac{1}{L^2} \right) \left(\frac{L}{2} \right) dr = \left[(1) \left(\frac{1}{L^2} \right) + (1) \left(\frac{1}{L^2} \right) \right] \frac{L}{2} = \frac{1}{L}$$

$$K_{s33} = = C_{44} 4ab \, \frac{1}{L} = \frac{4 C_{44} ab}{L}$$

$$K_{34} = \int_{-a}^{a} \int_{-b}^{b} \left(C_{44} F_{sx} F_{\tau y,x} \right) dx dz \int_{-1}^{1} \left(\mathbf{J}^{-1} N_{2,r}(r) N_2(r) \right) det \mathbf{J} dr$$

$$F_{sx} F_{\tau y,x} = (-1)(1) = -1$$

$$\int_{-a}^{a} \int_{-b}^{b} \left(C_{44} F_{sx} F_{\tau y,x} \right) dx dz = \int_{-a}^{a} \int_{-b}^{b} \left((C_{44})(-1) \right) dx dz = -C_{44} \int_{-a}^{a} dx \int_{-b}^{b} dz = -C_{44} 4ab$$

$$\mathbf{J}^{-1} N_{2,r}(r) N_2(r) = \frac{2}{L} \left[\frac{d}{dr} \left(\frac{1}{2}(1+r) \right) \right] \left(\frac{1}{2}(1+r) \right) = \frac{1}{2L}(1+r)$$

$$\int_{-1}^{1} \left(\mathbf{J}^{-1} N_{2,r}(r) N_2(r) \right) det \mathbf{J} dr =$$

$$\int_{-1}^{1} \left(\frac{1}{2L}(1+r) \right) \left(\frac{L}{2} \right) dr = \left[(1) \left(\frac{1}{2L}(1 + (-\frac{1}{\sqrt{3}})) \right) + (1) \left(\frac{1}{2L}(1 + (\frac{1}{\sqrt{3}})) \right) \right] \frac{L}{2} = \frac{1}{2}$$

$$K_{s34} = = -C_{44} 4ab \, \frac{1}{2} = -2 C_{44} ab$$

$$K_{41} = \int_{-a}^{a} \int_{-b}^{b} \left(C_{44} F_{sy,x} F_{\tau x} \right) dx dz \int_{-1}^{1} \left(\mathbf{J}^{-1} N_{1,r}(r) N_2(r) \right) det \mathbf{J} dr$$

$$F_{sy,x} F_{\tau x} = (1)(-1) = -1$$

$$\int_{-a}^{a} \int_{-b}^{b} \left(C_{44} F_{sy,x} F_{\tau x} \right) dx dz \quad = \quad \int_{-a}^{a} \int_{-b}^{b} \left((C_{44})(-1) \right) dx dz = \quad -C_{44} \int_{-a}^{a} dx \int_{-b}^{b} dz =$$

$$-C_{44} 4ab$$

$$\mathbf{J}^{-1} N_{1,r}(r) N_2(r) = \frac{2}{L} \frac{d}{dr} \left(\frac{1}{2}(1-r) \right) \left(\frac{1}{2}(1+r) \right) = -\frac{1}{2L}(1+r)$$

$$\int_{-1}^{1} \left(\mathbf{J}^{-1} N_{1,r}(r) N_2(r) \right) det \mathbf{J} dr =$$

$$\int_{-1}^{1} \left(-\frac{1}{2L}(1+r) \right) \left(\frac{L}{2} \right) dr \quad = \left(-\frac{1}{2L} \right) \left[(1) \left(1 + \left(-\frac{1}{\sqrt{3}} \right) \right) + (1) \left(1 + \left(+\frac{1}{\sqrt{3}} \right) \right) \right] \frac{L}{2} =$$

$$-\frac{1}{2}$$

$$K_{s41} = = -C_{44} 4ab -\frac{1}{2} = 2C_{44} ab$$

$$K_{42} = \int_{-a}^{a} \int_{-b}^{b} \left(C_{44} F_{sy,x} F_{\tau y,x} \right) dx dz \int_{-1}^{1} \left(N_1(r) N_2(r) \right) det \mathbf{J} dr$$

$$F_{sy,x} F_{\tau y,x} = (1)(1) = 1$$

$$\int_{-a}^{a} \int_{-b}^{b} \left(C_{44} F_{sy,x} F_{\tau y,x} \right) dx dz = \int_{-a}^{a} \int_{-b}^{b} \left((C_{44})(1) \right) dx dz = C_{44} \int_{-a}^{a} dx \int_{-b}^{b} dz = C_{44} 4ab$$

$$N_1(r) N_2(r) = \left(\frac{1}{2}(1-r) \right) \left(\frac{1}{2}(1+r) \right) = \left(\frac{1}{4}(1-r^2) \right)$$

$$\int_{-1}^{1} \left(N_1(r) N_2(r) \right) det \mathbf{J} dr =$$

$$\int_{-1}^{1} \left(\frac{1}{4}(1-r^2) \right) \left(\frac{L}{2} \right) dr = \left[(1) \left(\frac{1}{4} \left(1 - \left(-\frac{1}{\sqrt{3}} \right)^2 \right) \right) + (1) \left(\frac{1}{4} \left(1 - \left(+\frac{1}{\sqrt{3}} \right)^2 \right) \right) \right] \frac{L}{2} =$$

$$\frac{L}{6}$$

$$K_{s42} = = C_{44} 4ab \; \frac{L}{6} = \frac{2C_{44} L ab}{3}$$

$$K_{43} = \int_{-a}^{a} \int_{-b}^{b} \left(C_{44} F_{sy,x} F_{\tau x} \right) dx dz \int_{-1}^{1} \left(\mathbf{J}^{-1} N_{2,r}(r) N_2(r) \right) \det \mathbf{J} dr$$

$$F_{sy,x} F_{\tau x} = (1)(-1) = -1$$

$$\int_{-a}^{a} \int_{-b}^{b} \left(C_{44} F_{sy,x} F_{\tau x} \right) dx dz = \int_{-a}^{a} \int_{-b}^{b} \left((C_{44})(-1) \right) dx dz = -C_{44} \int_{-a}^{a} dx \int_{-b}^{b} dz = \; -C_{44} 4ab$$

$$\mathbf{J}^{-1} N_{2,r}(r) N_2(r) = \frac{2}{L} \left[\frac{d}{dr} \left(\frac{1}{2}(1+r) \right) \right] \left(\frac{1}{2}(1+r) \right) = \frac{1}{2L}(1+r)$$

$$\int_{-1}^{1} \left(\mathbf{J}^{-1} N_{2,r}(r) N_2(r) \right) \det \mathbf{J} dr =$$

$$\int_{-1}^{1} \left(\frac{1}{2L}(1+r) \right) \left(\frac{L}{2} \right) dr = \left[(1) \left(\frac{1}{2L}(1+(-\frac{1}{\sqrt{3}})) \right) + (1) \left(\frac{1}{2L}(1+(\frac{1}{\sqrt{3}})) \right) \right] \frac{L}{2} = \; \frac{1}{2}$$

$$K_{s43} = = -C_{44} 4ab \, \frac{1}{2} = -2C_{44} ab$$

$$K_{44} = \int_{-a}^{a} \int_{-b}^{b} \left(C_{44} F_{sy,x} F_{\tau y,x} \right) dx dz \int_{-1}^{1} \left(N_2(r) N_2(r) \right) \det \mathbf{J} dr$$

$$F_{sy,x} F_{\tau y,x} = (1)(1) = 1$$

$$\int_{-a}^{a} \int_{-b}^{b} \left(C_{44} F_{sy,x} F_{\tau y,x} \right) dx dz = \int_{-a}^{a} \int_{-b}^{b} \left((C_{44})(1) \right) dx dz = C_{44} \int_{-a}^{a} dx \int_{-b}^{b} dz = \; C_{44} 4ab$$

$$N_2(r) N_2(r) = \left(\frac{1}{2}(1+r) \right) \left(\frac{1}{2}(1+r) \right) = \left(\frac{1}{2}(1+r) \right)^2 = \left(\frac{1}{4} + \frac{r^2}{4} + \frac{r}{2} \right)$$

$$\int_{-1}^{1} \left(N_2(r) N_2(r) \right) \det \mathbf{J} dr =$$

$$\int_{-1}^{1} \left(\frac{1}{4} + \frac{r^2}{4} + \frac{r}{2} \right) \left(\frac{L}{2} \right) dr \; = \left[(1) \left(\frac{1}{4} + \frac{(-\frac{1}{\sqrt{3}})^2}{4} + \frac{(-\frac{1}{\sqrt{3}})}{2} \right) + (1) \left(\frac{1}{4} + \frac{(+\frac{1}{\sqrt{3}})^2}{4} + \frac{(+\frac{1}{\sqrt{3}})}{2} \right) \right] \frac{L}{2} =$$

$$\frac{L}{3}$$

$$K_{s44} = = C_{44} 4ab \, \frac{L}{3} = \frac{4 C_{44} L ab}{3}.$$

3.7.7 Full Integration: The Complete Stiffness Matrix

The shear part of the stiffness matrix, computed using the FI Gauss quadrature, is the following

$$
\mathbf{K}_{SFULL} =
\begin{bmatrix}
\dfrac{4C_{44}ab}{L} & 2C_{44}ab & -\dfrac{4C_{44}ab}{L} & 2C_{44}ab \\[2mm]
2C_{44}ab & \dfrac{4C_{44}abL}{3} & -2C_{44}ab & \dfrac{2C_{44}abL}{3} \\[2mm]
-\dfrac{4C_{44}ab}{L} & -2C_{44}ab & \dfrac{4C_{44}ab}{L} & -2C_{44}ab \\[2mm]
2C_{44}ab & \dfrac{2C_{44}abL}{3} & -2C_{44}ab & \dfrac{4C_{44}abL}{3}
\end{bmatrix}
\tag{3.122}
$$

The bending part of the stiffness matrix, computed previously using the FI Gauss quadrature, is reported below

$$
\mathbf{K}_{bFULL} =
\begin{bmatrix}
0 & 0 & 0 & 0 \\[2mm]
0 & \dfrac{4C_{11}a^3b}{3L} & 0 & -\dfrac{4C_{11}a^3b}{3L} \\[2mm]
0 & 0 & 0 & 0 \\[2mm]
0 & -\dfrac{4C_{11}a^3b}{3L} & 0 & \dfrac{4C_{11}a^3b}{3L}
\end{bmatrix}
\tag{3.123}
$$

The complete stiffness matrix for a two-node element computed using the FI Gauss quadrature, can be obtained summing the two previous matrices

$$
\mathbf{K}_{FULL} =
\begin{bmatrix}
\dfrac{4C_{44}ab}{L} & 2C_{44}ab & -\dfrac{4C_{44}ab}{L} & 2C_{44}ab \\[3mm]
2C_{44}ab & \dfrac{4C_{44}abL}{3}+\dfrac{4C_{11}a^3b}{3L} & -2C_{44}ab & \dfrac{2C_{44}abL}{3}-\dfrac{4C_{11}a^3b}{3L} \\[3mm]
-\dfrac{4C_{44}ab}{L} & -2C_{44}ab & \dfrac{4C_{44}ab}{L} & -2C_{44}ab \\[3mm]
2C_{44}ab & \dfrac{2C_{44}abL}{3}-\dfrac{4C_{11}a^3b}{3L} & -2C_{44}ab & \dfrac{4C_{44}abL}{3}+\dfrac{4C_{11}a^3b}{3L}
\end{bmatrix}
\tag{3.124}
$$

3.7.8 Full Integration: Considerations

In Eq. (3.124), the stiffness matrix is derived using a 2-point Gauss quadrature, which corresponds to FI for integrand functions with a maximum polynomial degree of 2. Consequently, this matrix precisely matches the one presented in Eq. (3.62), obtained analytically. However, when this matrix is used to compute displacements in a structural analysis (for instance, a cantilever beam), it fails to predict accurate results. Specifically, it overestimates the shear stiffness while underestimating the bending stiffness, leading to displacement values that are significantly smaller than expected. Furthermore, the two component matrices that must be summed, namely, the bending component from Eq. (3.123) and the shear component from Eq. (3.122), exhibit different ranks (1 and 2, respectively). This discrepancy in rank indicates that the corresponding finite element is unlikely to produce reliable results. The MATLAB script for the derivation of the FI stiffness matrix is reported in Appendix C.1.

Chapter 4
Numerical Remedies Against Shear Locking

Graphical Abstract

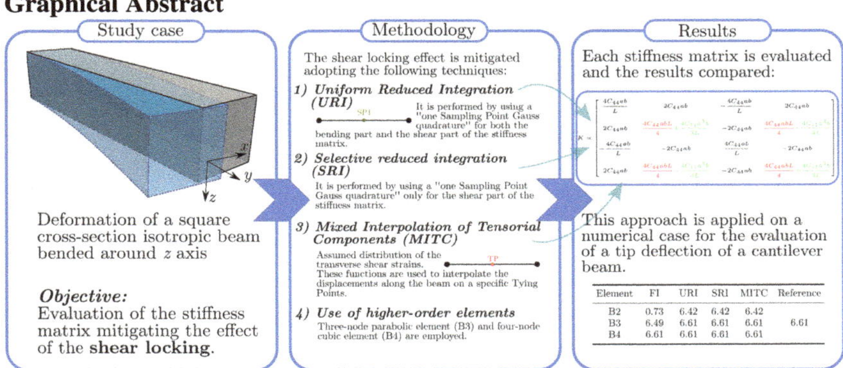

Element	FI	URI	SRI	MITC	Reference
B2	0.73	6.42	6.42	6.42	
B3	6.49	6.61	6.61	6.61	6.61
B4	6.61	6.61	6.61	6.61	

Introduction

This chapter provides a comprehensive review of four key strategies designed to mitigate shear locking in two-node beam elements under bending: Uniform Reduced Integration (URI), Selective Reduced Integration (SRI), Mixed Interpolation of Tensorial Components (MITC), and the use of higher-order Finite Elements (FEs).

- Section 4.1 introduces URI, which applies reduced Gauss integration (one sampling point for Gauss quadrature) uniformly to both the bending and shear components of the stiffness matrix.
- Section 4.2 discusses SRI, where reduced integration (one sampling point) is used exclusively for the shear component, while Full Integration (FI) (two sampling points) is retained for the bending component.
- Section 4.3 presents MITC, an approach that assumes a distribution of transverse shear strains to develop shear locking-free displacement-based 1D beam models. In particular, the stiffness matrix for a B2 beam element is derived by applying the principle of virtual displacements to this assumed shear strain distribution.

E. Carrera et al., *Implementation of Beam-Type Finite Elements Based on Carrera Unified Formulation*, https://doi.org/10.1007/978-3-031-95856-4_4

- Section 4.4 details the fourth strategy: employing higher-order 1D elements (e.g., three-node B3 elements and four-node B4 elements) to alleviate shear locking.

Subsequent sections, including Sect. 4.5 and the concluding parts of the chapter, illustrate the efficacy of these methods through numerical examples. These examples highlight key differences between FI and URI approaches, offering insights into the relative performance of each technique in addressing shear locking.

4.1   Uniform Reduced Integration

In the context of URI, the Gauss quadrature using one SP is employed instead of the Gauss quadrature with two SPs, for both the bending and shear components of the stiffness matrix. Figure 4.1 illustrates the positions of the sampling point along the beam (i.e., the coordinate r of the sampling point) and the corresponding weight. The coordinate of the sampling point is $r = 0$ and the related weight are $w_1 = 2$.

4.1.1   Uniform Reduced Integration Applied to the Bending Part of the Stiffness Matrix

Regarding the bending part of the stiffness matrix, URI has to be applied to the terms K_{b22}, K_{b24}, K_{b42}, K_{b44}, outlined in Eq. (3.101). The K_{b22} term is investigated hereafter, and URI must be applied to the line integral of the expression that represents the K_{b22} term.

$$K_{b22} = \int_{-a}^{a} \int_{-b}^{b} \left(C_{11} F_{sy} F_{\tau y} \right) dx dz \int_{-1}^{1} \left(\mathbf{J}^{-1} N_{1,r}(r) \mathbf{J}^{-1} N_{1,r}(r) \right) det\,\mathbf{J}\,dr \tag{4.1}$$

The portion of the term K_{b22} highlighted in red corresponds to a surface integral, and it is solved analytically. The solution of this surface integral is obtained through the following steps:

Fig. 4.1 Position of 1 sampling points along the two-node beam

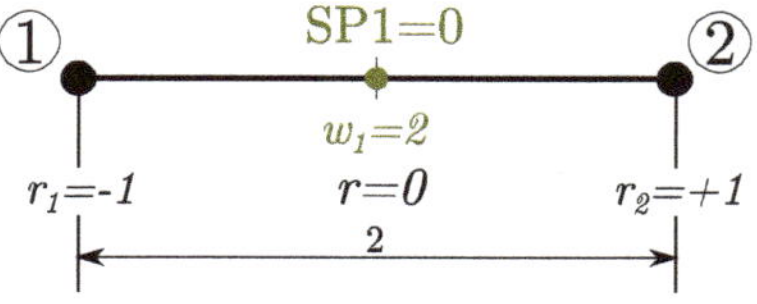

$$\int_{-a}^{a}\int_{-b}^{b}\left(C_{11}F_{sy}F_{\tau y}\right)dxdz \;=\; \int_{-a}^{a}\int_{-b}^{b}\left(C_{11}x\cdot x\right)dxdz \;=$$
$$= C_{11}\int_{-a}^{a}x^{2}dx\int_{-b}^{b}dz \;=\; \frac{4}{3}C_{11}a^{3}b \tag{4.2}$$

The blue part of the bending matrix term K_{b22} of Eq. (4.1) is solved using one sampling point according to URI.

$$\mathbf{J}^{-1}N_{1,r}(r)\mathbf{J}^{-1}N_{1,r}(r) = \frac{2}{L}\frac{d}{dr}\left(\frac{1}{2}(1-r)\right)\frac{2}{L}\frac{d}{dr}\left(\frac{1}{2}(1-r)\right) = \frac{1}{L^{2}} \tag{4.3}$$

Note that the integrating function is constant. By applying URI, the following result is derived

$$\int_{-1}^{1}\left(\mathbf{J}^{-1}N_{1,r}(r)\mathbf{J}^{-1}N_{1,r}(r)\right)det\mathbf{J}dr = \left[(w_{1})\left(F(SP_{1})\right)\right]det\mathbf{J} =$$
$$= \left[(w_{1})\left(F_{const}\right)\right]det\mathbf{J} = \tag{4.4}$$
$$= \left[(2)\left(\frac{1}{L^{2}}\right)\right]\frac{L}{2} = \frac{1}{L}$$

Using the previous computation of Eqs. (4.2) and (4.4), the value of the term K_{b22} of the bending part of the stiffness matrix is evaluated as presented below:

$$K_{b22} = \int_{-a}^{a}\int_{-b}^{b}\left(C_{11}F_{sy}F_{\tau y}\right)dxdz\int_{-1}^{1}\left(\mathbf{J}^{-1}N_{1,r}(r)\mathbf{J}^{-1}N_{1,r}(r)\right)det\mathbf{J}dr =$$
$$= \left(C_{11}\frac{4a^{3}b}{3}\right)\left(\frac{1}{L}\right) = C_{11}\frac{4a^{3}b}{3L} \tag{4.5}$$

By following the same approach employed for K_{b22}, the remaining terms of the bending matrix can be obtained. The expression for the term K_{b24} of the bending matrix is as follows:

$$K_{b24} = \int_{-a}^{a}\int_{-b}^{b}\left(C_{11}F_{sy}F_{\tau y}\right)dxdy\int_{-1}^{1}\left(\mathbf{J}^{-1}N_{1,r}(r)\mathbf{J}^{-1}N_{2,r}(r)\right)det\mathbf{J}dr \tag{4.6}$$

The red part is computed analytically and the following result is obtained

$$\int_{-a}^{a}\int_{-b}^{b} \left(C_{11}\, F_{sy}\, F_{\tau y}\right) dx\, dz = \int_{-a}^{a}\int_{-b}^{b} \left(C_{11}x \cdot x\right) dx\, dz =$$
$$= C_{11} \int_{-a}^{a} x^2\, dx \int_{-b}^{b} dz = \frac{4}{3} C_{11} a^3 b \tag{4.7}$$

Analyzing the blue part, it is evident that the integrating function can be expressed as follows:

$$\mathbf{J}^{-1} N_{1,r}(r)\, \mathbf{J}^{-1} N_{2,r}(r) = \frac{2}{L}\frac{d}{dr}\left(\frac{1}{2}(1-r)\right)\frac{2}{L}\frac{d}{dr}\left(\frac{1}{2}(1+r)\right) = -\frac{1}{L^2} \tag{4.8}$$

Using the previous results, the following result is obtained

$$\int_{-1}^{1} \left(\mathbf{J}^{-1} N_{1,r}(r)\, \mathbf{J}^{-1} N_{2,r}(r)\right) det\mathbf{J}\, dr = [(w_1)\,(F(SP_1))]\, det\mathbf{J} =$$
$$= [(w_1)\,(F_{const})]\, det\mathbf{J} = \tag{4.9}$$
$$= \left[(1)\left(-\frac{1}{L^2}\right)\right]\frac{L}{2} = -\frac{1}{L}$$

Using the previous computation of Eqs. (4.7) and (4.9), the value of the term K_{24} in the bending part of the stiffness matrix can be determined, as illustrated below:

$$K_{b24} = \int_{-a}^{a}\int_{-b}^{b} \left(C_{11}\, F_{sy}\, F_{\tau y}\right) dx\, dy \int_{-1}^{1} \left(\mathbf{J}^{-1} N_{1,r}(r)\, \mathbf{J}^{-1} N_{2,r}(r)\right) det\mathbf{J}\, dr =$$
$$= \left(C_{11} \frac{4a^3 b}{3}\right)\left(-\frac{1}{L}\right) = -C_{11}\frac{4a^3 b}{3L} \tag{4.10}$$

The term K_{b42} of the bending matrix is the following

$$K_{b42} = \int_{-a}^{a}\int_{-b}^{b} \left(C_{11}\, F_{sy}\, F_{\tau y}\right) dx\, dy \int_{-1}^{1} \left(\mathbf{J}^{-1} N_{2,r}(r)\, \mathbf{J}^{-1} N_{1,r}(r)\right) det\mathbf{J}\, dr \tag{4.11}$$

The red part is computed analytically and the following result is obtained

$$\int_{-a}^{a}\int_{-b}^{b} \left(C_{11}\, F_{sy}\, F_{\tau y}\right) dx\, dz = \int_{-a}^{a}\int_{-b}^{b} \left(C_{11}x \cdot x\right) dx\, dz =$$
$$C_{11} \int_{-a}^{a} x^2\, dx \int_{-b}^{b} dz = \frac{4}{3} C_{11} a^3 b \tag{4.12}$$

Examining the blue part of the Eq. (4.11), it is evident that the integrating function can be expressed as follows:

$$\mathbf{J}^{-1} N_{2,r}(r) \mathbf{J}^{-1} N_{1,r}(r) = \frac{2}{L} \frac{d}{dr}\left(\frac{1}{2}(1-r)\right) \frac{2}{L} \frac{d}{dr}\left(\frac{1}{2}(1+r)\right) = -\frac{1}{L^2} \tag{4.13}$$

Building on the previously derived result and applying the URI technique, the following outcome is obtained:

$$\int_{-1}^{1} \left(\mathbf{J}^{-1} N_{2,r}(r)\mathbf{J}^{-1} N_{1,r}(r)\right) det\,\mathbf{J}\,dr = \left[(w_1)\left(F(SP_1)\right)\right] det\,\mathbf{J} =$$

$$= \left[(w_1)\left(F_{const}\right)\right] det\,\mathbf{J} = \tag{4.14}$$

$$= \left[(1)\left(-\frac{1}{L^2}\right)\right]\frac{L}{2} = -\frac{1}{L}$$

Using the previous computation of Eqs. (4.12) and (4.14), the value of the term K_{42} in the bending part of the stiffness matrix can be calculated, as described below:

$$K_{b42} = \int_{-a}^{a} \int_{-b}^{b} \left(C_{11} F_{sy} F_{\tau y}\right) dxdy \int_{-1}^{1} \left(\mathbf{J}^{-1} N_{2,r}(r)\mathbf{J}^{-1} N_{1,r}(r)\right) det\,\mathbf{J}\,dr =$$

$$= \left(C_{11}\frac{4a^3 b}{3}\right)\left(-\frac{1}{L}\right) = -C_{11}\frac{4a^3 b}{3L}$$

$$\tag{4.15}$$

The term K_{b44} can be expressed as reported below

$$K_{b44} = \int_{-a}^{a} \int_{-b}^{b} \left(C_{11} F_{sy} F_{\tau y}\right) dxdy \int_{-1}^{1} \left(\mathbf{J}^{-1} N_{2,r}(r)\mathbf{J}^{-1} N_{2,r}(r)\right) det\,\mathbf{J}\,dr \tag{4.16}$$

The red part of the previous equation is computed analytically and the following result is obtained

$$\int_{-a}^{a} \int_{-b}^{b} \left(C_{11} F_{sy} F_{\tau y}\right) dxdz = \int_{-a}^{a} \int_{-b}^{b} \left(C_{11} x \cdot x\right) dxdz =$$

$$= C_{11} \int_{-a}^{a} x^2 dx \int_{-b}^{b} dz = \frac{4}{3}C_{11}a^3 b \tag{4.17}$$

Considering the blue of the previous relation, the integrating function can be expressed as follows:

$$\mathbf{J}^{-1}N_{2,r}(r)\mathbf{J}^{-1}N_{2,r}(r) = \frac{2}{L}\frac{d}{dr}\left(\frac{1}{2}(1+r)\right)\frac{2}{L}\frac{d}{dr}\left(\frac{1}{2}(1+r)\right) = \frac{1}{L^2} \tag{4.18}$$

Building upon the previously derived result and employing the URI method, the outcome is as follows:

$$\int_{-1}^{1}\left(\mathbf{J}^{-1}N_{2,r}(r)\mathbf{J}^{-1}N_{2,r}(r)\right)det\,\mathbf{J}dr = [(w_1)\,(F(SP_1))]\,det\,\mathbf{J} =$$

$$= [(w_1)\,(F_{const})]\,det\,\mathbf{J} = \tag{4.19}$$

$$= \left[(1)\left(\frac{1}{L^2}\right)\right]\frac{L}{2} = \frac{1}{L}$$

By employing the calculations from the previous Eq. (4.17) and (4.19), the value of the term K_{44} in the bending part of the stiffness matrix can be determined, as illustrated below:

$$K_{b44} = \int_{-a}^{a}\int_{-b}^{b}\left(C_{11}F_{sy}F_{\tau y}\right)dxdy\int_{-1}^{1}\left(\mathbf{J}^{-1}N_{2,r}(r)\mathbf{J}^{-1}N_{2,r}(r)\right)det\,\mathbf{J}dr =$$

$$= \left(C_{11}\frac{4a^3b}{3}\right)\left(\frac{1}{L}\right) = C_{11}\frac{4a^3b}{3L}$$

$$\tag{4.20}$$

Considering the previous results obtained for the terms K_{b22}, K_{b24}, K_{b42}, and K_{b44}, the bending part of the stiffness matrix calculated using URI is the following

$$\mathbf{K}_{bURI} = \begin{bmatrix} 0 & 0 & 0 & 0 \\ 0 & \dfrac{4C_{11}a^3b}{3L} & 0 & -\dfrac{4C_{11}a^3b}{3L} \\ 0 & 0 & 0 & 0 \\ 0 & -\dfrac{4C_{11}a^3b}{3L} & 0 & \dfrac{4C_{11}a^3b}{3L} \end{bmatrix} \tag{4.21}$$

4.1.2 Uniform Reduced Integration Applied to the Shear Part of the Stiffness Matrix

The shear part of the stiffness matrix is shown here

$$\mathbf{K}_{SURI} = \begin{bmatrix} K_{s11} & K_{s12} & K_{s13} & K_{s14} \\ K_{s21} & K_{s22} & K_{s23} & K_{s24} \\ K_{s31} & K_{s32} & K_{s33} & K_{s34} \\ K_{s41} & K_{s42} & K_{s43} & K_{s44} \end{bmatrix} \tag{4.22}$$

The evaluation of each term of $\mathbf{K}_{SURI}$ are reported in the following

$$K_{s11} = \int_{-a}^{a} \int_{-b}^{b} (C_{44} F_{sx} F_{\tau x}) \, dx \, dz \int_{-1}^{1} \left(\mathbf{J}^{-1} N_{1,r}(r) \mathbf{J}^{-1} N_{1,r}(r) \right) \det \mathbf{J} \, dr$$

$$\int_{-a}^{a} \int_{-b}^{b} (C_{44} F_{sx} F_{\tau x}) \, dx \, dz = \int_{-a}^{a} \int_{-b}^{b} ((C_{44})(1)) \, dx \, dz = C_{44} \int_{-a}^{a} dx \int_{-b}^{b} dz = C_{44} 4ab$$

$$\mathbf{J}^{-1} N_{1,r}(r) \mathbf{J}^{-1} N_{1,r}(r) = \frac{2}{L} \frac{d}{dr} \left(\frac{1}{2}(1-r) \right) \frac{2}{L} \frac{d}{dr} \left(\frac{1}{2}(1-r) \right) = \frac{1}{L^2}$$

$$\int_{-1}^{1} \left(\mathbf{J}^{-1} N_{1,r}(r) \mathbf{J}^{-1} N_{1,r}(r) \right) \det \mathbf{J} \, dr = \int_{-1}^{1} \left(\frac{1}{L^2} \right) \left(\frac{L}{2} \right) dr = \left[(2) \left(\frac{1}{L^2} \right) \right] \frac{L}{2} = \frac{1}{L}$$

$$K_{s11} = = C_{44} 4ab \frac{1}{L} = \frac{4C_{44}ab}{L}$$

$$K_{s12} = \int_{-a}^{a} \int_{-b}^{b} \left(C_{44} F_{sx} F_{\tau y,x} \right) dx dz \int_{-1}^{1} \left(\mathbf{J}^{-1} N_{1,r}(r) N_1(r) \right) det\, \mathbf{J} dr$$

$$\int_{-a}^{a} \int_{-b}^{b} \left(C_{44} F_{sx} F_{\tau y,x} \right) dx dz = \int_{-a}^{a} \int_{-b}^{b} \left((C_{44})(-1) \right) dx dz = -C_{44} \int_{-a}^{a} dx \int_{-b}^{b} dz =$$
$$-C_{44} 4ab$$

$$\mathbf{J}^{-1} N_{1,r}(r) N_1(r) = \frac{2}{L} \frac{d}{dr} \left(\frac{1}{2}(1-r) \right) \left(\frac{1}{2}(1-r) \right) = -\frac{1}{2L}(1-r)$$

$$\int_{-1}^{1} \left(\mathbf{J}^{-1} N_{1,r}(r) N_1(r) \right) det\, \mathbf{J} dr = \qquad \int_{-1}^{1} \left(-\frac{1}{2L}(1-r) \right) \left(\frac{L}{2} \right) dr \qquad =$$
$$\left(-\frac{1}{2L} \right) [(2)(1-0)] \frac{L}{2} = -\frac{1}{2}$$

$$K_{s12} = = -C_{44} 4ab \left(-\frac{1}{2} \right) = 2C_{44}ab$$

$$K_{s13} = \int_{-a}^{a} \int_{-b}^{b} \left(C_{44} F_{sx} F_{\tau x} \right) dx dz \int_{-1}^{1} \left(\mathbf{J}^{-1} N_{1,r}(r) \mathbf{J}^{-1} N_{2,r}(r) \right) det\, \mathbf{J} dr$$

$$\int_{-a}^{a} \int_{-b}^{b} \left(C_{44} F_{sx} F_{\tau x} \right) dx dz = \int_{-a}^{a} \int_{-b}^{b} \left((C_{44})(1) \right) dx dz = C_{44} \int_{-a}^{a} dx \int_{-b}^{b} dz = C_{44} 4ab$$

$$\mathbf{J}^{-1} N_{1,r}(r) \mathbf{J}^{-1} N_{2,r}(r) = \frac{2}{L} \frac{d}{dr} \left(\frac{1}{2}(1-r) \right) \frac{2}{L} \frac{d}{dr} \left(\frac{1}{2}(1+r) \right) = -\frac{1}{L^2}$$

$$\int_{-1}^{1} \left(\mathbf{J}^{-1} N_{1,r}(r) \mathbf{J}^{-1} N_{2,r}(r) \right) det\, \mathbf{J} dr = \int_{-1}^{1} \left(-\frac{1}{L^2} \right) \left(\frac{L}{2} \right) dr = \left[(2) \left(-\frac{1}{L^2} \right) \right] \frac{L}{2}$$
$$= -\frac{1}{L}$$

$$K_{s13} = = C_{44} 4ab \left(-\frac{1}{L} \right) = -\frac{4C_{44}ab}{L}$$

$$K_{s14} = \int_{-a}^{a} \int_{-b}^{b} \left(C_{44} F_{sx} F_{\tau y,x} \right) dxdz \int_{-1}^{1} \left(\mathbf{J}^{-1} N_{1,r}(r) N_2(r) \right) det\, \mathbf{J} dr$$

$$\int_{-a}^{a} \int_{-b}^{b} \left(C_{44} F_{sx} F_{\tau y,x} \right) dxdz = \int_{-a}^{a} \int_{-b}^{b} \left((C_{44})(-1) \right) dxdz = -C_{44} \int_{-a}^{a} dx \int_{-b}^{b} dz = -C_{44} 4ab$$

$$\mathbf{J}^{-1} N_{1,r}(r) N_2(r) = \frac{2}{L} \frac{d}{dr} \left(\frac{1}{2}(1-r) \right) \left(\frac{1}{2}(1+r) \right) = -\frac{1}{2L}(1+r)$$

$$\int_{-1}^{1} \left(\mathbf{J}^{-1} N_{1,r}(r) N_2(r) \right) det\, \mathbf{J} dr = \int_{-1}^{1} \left(-\frac{1}{2L}(1+r) \right) \left(\frac{L}{2} \right) dr =$$

$$\left(-\frac{1}{2L} \right) [(2)(1+0)] \frac{L}{2} = -\frac{1}{2}$$

$$K_{s14} = = -C_{44} 4ab \left(-\frac{1}{2} \right) = 2C_{44} ab$$

$$K_{s21} = \int_{-a}^{a} \int_{-b}^{b} \left(C_{44} F_{sy,x} F_{\tau x} \right) dxdz \int_{-1}^{1} \left(\mathbf{J}^{-1} N_{1,r}(r) N_1(r) \right) det\, \mathbf{J}] dr$$

$$\int_{-a}^{a} \int_{-b}^{b} \left(C_{44} F_{sy,x} F_{\tau x} \right) dxdz = \int_{-a}^{a} \int_{-b}^{b} \left((C_{44})(-1) \right) dxdz = -C_{44} \int_{-a}^{a} dx \int_{-b}^{b} dz = -C_{44} 4ab$$

$$\mathbf{J}^{-1} N_{1,r}(r) N_1(r) = \frac{2}{L} \left[\frac{d}{dr} \left(\frac{1}{2}(1-r) \right) \right] \left(\frac{1}{2}(1-r) \right) = -\frac{1}{2L}(1-r)$$

$$\int_{-1}^{1} \left(\mathbf{J}^{-1} N_{1,r}(r) N_1(r) \right) det\, \mathbf{J} dr = \int_{-1}^{1} \left(-\frac{1}{2L}(1-r) \right) \left(\frac{L}{2} \right) dr =$$

$$\left(-\frac{1}{2L} \right) [(2)(1-0) +] \frac{L}{2} = -\frac{1}{2}$$

$$K_{s21} = = -C_{44} 4ab \left(-\frac{1}{2} \right) = 2C_{44} ab$$

$$K_{s22} = \int_{-a}^{a} \int_{-b}^{b} \left(C_{44} F_{sy,x} F_{\tau y,x} \right) dx\,dz \int_{-1}^{1} \left(N_1(r) N_1(r) \right) \det \mathbf{J}\,dr$$

$$\int_{-a}^{a} \int_{-b}^{b} \left(C_{44} F_{sy,x} F_{\tau x} \right) dx\,dz = \int_{-a}^{a} \int_{-b}^{b} \left((C_{44})(-1) \right) dx\,dz = -C_{44} \int_{-a}^{a} dx \int_{-b}^{b} dz =$$
$$-C_{44}4ab$$

$$N_1(r) N_1(r) = \left(\frac{1}{2}(1-r) \right)\left(\frac{1}{2}(1-r) \right) = \left(\frac{1}{2}(1-r) \right)^2 = \left(\frac{1}{4} + \frac{r^2}{4} - \frac{r}{2} \right)$$

$$\int_{-1}^{1} \left(N_1(r) N_1(r) \right) \det \mathbf{J}\,dr = \qquad \int_{-1}^{1} \left(\frac{1}{4} + \frac{r^2}{4} - \frac{r}{2} \right) \left(\frac{L}{2} \right) dr \qquad =$$
$$\left[(2)\left(\frac{1}{4} + 0 - 0 \right) \right] \frac{L}{2} = \frac{L}{4}$$

$$K_{s22} = = C_{44}4ab\left(\frac{L}{4} \right) = \frac{4C_{44}Lab}{4}$$

$$K_{s23} = \int_{-a}^{a} \int_{-b}^{b} \left(C_{44} F_{sy,x} F_{\tau x} \right) dx\,dz \int_{-1}^{1} \left(\mathbf{J}^{-1} N_{2,r}(r) N_1(r) \right) \det \mathbf{J}\,dr$$

$$\int_{-a}^{a} \int_{-b}^{b} \left(C_{44} F_{sy,x} F_{\tau x} \right) dx\,dz \qquad = \qquad \int_{-a}^{a} \int_{-b}^{b} \left((C_{44})(-1) \right) dx\,dz =$$
$$-C_{44} \int_{-a}^{a} dx \int_{-b}^{b} dz = -C_{44}4ab$$

$$\mathbf{J}^{-1} N_{2,r}(r) N_1(r) = \frac{2}{L} \left[\frac{d}{dr} \left(\frac{1}{2}(1+r) \right) \right] \left(\frac{1}{2}(1-r) \right) = \frac{1}{2L}(1-r)$$

$$\int_{-1}^{1} \left(\mathbf{J}^{-1} N_{2,r}(r) N_1(r) \right) \det \mathbf{J}\,dr = \qquad \int_{-1}^{1} \left(\frac{1}{2L}(1-r) \right) \left(\frac{L}{2} \right) dr \qquad =$$
$$\left(\frac{1}{2L} \right) [(2)(1-0)] \frac{L}{2} = \frac{1}{2}$$

$$K_{s23} = = -C_{44}4ab\left(\frac{1}{2} \right) = -2C_{44}ab$$

$$K_{s24} = \int_{-a}^{a} \int_{-b}^{b} \left(C_{44} F_{sy,x} F_{\tau y,x} \right) dx dz \int_{-1}^{1} \left(N_1(r) N_2(r) \right) det\,\mathbf{J} dr$$

$$\int_{-a}^{a} \int_{-b}^{b} \left(C_{44} F_{sy,x} F_{\tau y,x} \right) dx dz \qquad = \qquad \int_{-a}^{a} \int_{-b}^{b} \left((C_{44})(1) \right) dx dz =$$

$$C_{44} \int_{-a}^{a} dx \int_{-b}^{b} dz = C_{44} 4ab$$

$$N_1(r) N_2(r) = \left(\frac{1}{2}(1-r) \right) \left(\frac{1}{2}(1+r) \right) = \left(\frac{1}{4}(1-r^2) \right)$$

$$\int_{-1}^{1} \left(N_1(r) N_2(r) \right) det\,\mathbf{J} dr = \int_{-1}^{1} \left(\frac{1}{4}(1-r^2) \right) \left(\frac{L}{2} \right) dr = \left[(2) \left(\frac{1}{4}(1-0) \right) \right] \frac{L}{2} =$$

$$\frac{L}{4}$$

$$K_{s24} = = C_{44} 4ab \left(\frac{L}{4} \right) = \frac{4 C_{44} Lab}{4}$$

$$K_{s31} = \int_{-a}^{a} \int_{-b}^{b} \left(C_{44} F_{sx} F_{\tau x} \right) dx dz \int_{-1}^{1} \left(\mathbf{J}^{-1} N_{1,r}(r) \mathbf{J}^{-1} N_{2,r}(r) \right) det\,\mathbf{J} | dr$$

$$\int_{-a}^{a} \int_{-b}^{b} \left(C_{44} F_{sx} F_{\tau x} \right) dx dz = \int_{-a}^{a} \int_{-b}^{b} \left((C_{44})(1) \right) dx dz = C_{44} \int_{-a}^{a} dx \int_{-b}^{b} dz = C_{44} 4ab$$

$$\mathbf{J}^{-1} N_{1,r}(r) \mathbf{J}^{-1} N_{2,r}(r) = \frac{2}{L} \frac{d}{dr} \left(\frac{1}{2}(1-r) \right) \frac{2}{L} \frac{d}{dr} \left(\frac{1}{2}(1+r) \right) = -\frac{1}{L^2}$$

$$\int_{-1}^{1} \left(\mathbf{J}^{-1} N_{1,r}(r) \mathbf{J}^{-1} N_{2,r}(r) \right) det\,\mathbf{J} dr =$$

$$= \int_{-1}^{1} \left(-\frac{1}{L^2} \right) \left(\frac{L}{2} \right) dr = \left[(2) \left(-\frac{1}{L^2} \right) \right] \frac{L}{2} = -\frac{1}{L}$$

$$K_{s31} = = C_{44} 4ab \left(-\frac{1}{L} \right) = -\frac{4 C_{44} ab}{L}$$

$$K_{s32} = \int_{-a}^{a} \int_{-b}^{b} \left(C_{44} F_{sy,x} F_{\tau x} \right) dx\,dz \int_{-1}^{1} \left(\mathbf{J}^{-1} N_{2,r}(r) N_1(r) \right) det\,\mathbf{J}dr$$

$$\int_{-a}^{a} \int_{-b}^{b} \left(C_{44} F_{sy,x} F_{\tau x} \right) dx\,dz = \int_{-a}^{a} \int_{-b}^{b} \left((C_{44})(-1) \right) dx\,dz = -C_{44} \int_{-a}^{a} dx \int_{-b}^{b} dz =$$
$$-C_{44} 4ab$$

$$\mathbf{J}^{-1} N_{2,r}(r) N_1(r) = \frac{2}{L} \left[\frac{d}{dr} \left(\frac{1}{2}(1+r) \right) \right] \left(\frac{1}{2}(1-r) \right) = \frac{1}{2L}(1-r)$$

$$\int_{-1}^{1} \left(\mathbf{J}^{-1} N_{2,r}(r) N_1(r) \right) det\,\mathbf{J}dr = \qquad \int_{-1}^{1} \left(\frac{1}{2L}(1-r) \right) \left(\frac{L}{2} \right) dr \qquad =$$
$$\left(\frac{1}{2L} \right) \left[(2)(1-0) \right] \frac{L}{2} = \frac{1}{2}$$

$$K_{s32} = = -C_{44} 4ab \left(\frac{1}{2} \right) = -2C_{44}ab$$

$$K_{s33} = \int_{-a}^{a} \int_{-b}^{b} \left(C_{44} F_{sx} F_{\tau x} \right) dx\,dz \int_{-1}^{1} \left(\mathbf{J}^{-1} N_{2,r}(r) \mathbf{J}^{-1} N_{2,r}(r) \right) det\,\mathbf{J}dr$$

$$\int_{-a}^{a} \int_{-b}^{b} \left(C_{44} F_{sx} F_{\tau x} \right) dx\,dz \quad = \quad \int_{-a}^{a} \int_{-b}^{b} \left((C_{44})(1) \right) dx\,dz = \quad C_{44} \int_{-a}^{a} dx \int_{-b}^{b} dz =$$
$$C_{44} 4ab$$

$$\mathbf{J}^{-1} N_{2,r}(r) \mathbf{J}^{-1} N_{2,r}(r) = \frac{2}{L} \frac{d}{dr} \left(\frac{1}{2}(1+r) \right) \frac{2}{L} \frac{d}{dr} \left(\frac{1}{2}(1+r) \right) = \frac{1}{L^2}$$

$$\int_{-1}^{1} \left(\mathbf{J}^{-1} N_{2,r}(r) \mathbf{J}^{-1} N_{2,r}(r) \right) det\,\mathbf{J}dr = \int_{-1}^{1} \left(\frac{1}{L^2} \right) \left(\frac{L}{2} \right) dr = \left[(2) \left(\frac{1}{L^2} \right) \right] \frac{L}{2} =$$
$$\frac{1}{L}$$

$$K_{s33} = = C_{44} 4ab \frac{1}{L} = \frac{4C_{44}ab}{L}$$

$$K_{s34} = \int_{-a}^{a} \int_{-b}^{b} \left(C_{44} F_{sx} F_{\tau y,x} \right) dx\, dz \int_{-1}^{1} \left(\mathbf{J}^{-1} N_{2,r}(r) N_2(r) \right) \det \mathbf{J}\, dr$$

$$\int_{-a}^{a} \int_{-b}^{b} \left(C_{44} F_{sx} F_{\tau y,x} \right) dx\, dz = \int_{-a}^{a} \int_{-b}^{b} \left((C_{44})(-1) \right) dx\, dz = -C_{44} \int_{-a}^{a} dx \int_{-b}^{b} dz = -C_{44} 4ab$$

$$\mathbf{J}^{-1} N_{2,r}(r) N_2(r) = \frac{2}{L} \left[\frac{d}{dr} \left(\frac{1}{2}(1+r) \right) \right] \left(\frac{1}{2}(1+r) \right) = \frac{1}{2L}(1+r)$$

$$\int_{-1}^{1} \left(\mathbf{J}^{-1} N_{2,r}(r) N_2(r) \right) \det \mathbf{J}\, dr = \int_{-1}^{1} \left(\frac{1}{2L}(1+r) \right) \left(\frac{L}{2} \right) dr =$$

$$\left(\frac{1}{2L} \right) [(2)(1+0)] \frac{L}{2} = \frac{1}{2}$$

$$K_{s34} = = -C_{44} 4ab \left(\frac{1}{2} \right) = -2C_{44} ab$$

$$K_{s41} = \int_{-a}^{a} \int_{-b}^{b} \left(C_{44} F_{sy,x} F_{\tau x} \right) dx\, dz \int_{-1}^{1} \left(\mathbf{J}^{-1} N_{1,r}(r) N_2(r) \right) \det \mathbf{J}\, dr$$

$$\int_{-a}^{a} \int_{-b}^{b} \left(C_{44} F_{sy,x} F_{\tau x} \right) dx\, dz = \int_{-a}^{a} \int_{-b}^{b} \left((C_{44})(-1) \right) dx\, dz = -C_{44} \int_{-a}^{a} dx \int_{-b}^{b} dz = -C_{44} 4ab$$

$$\mathbf{J}^{-1} N_{1,r}(r) N_2(r) = \frac{2}{L} \frac{d}{dr} \left(\frac{1}{2}(1-r) \right) \left(\frac{1}{2}(1+r) \right) = -\frac{1}{2L}(1+r)$$

$$\int_{-1}^{1} \left(\mathbf{J}^{-1} N_{1,r}(r) N_2(r) \right) \det \mathbf{J}\, dr = \int_{-1}^{1} \left(-\frac{1}{2L}(1+r) \right) \left(\frac{L}{2} \right) dr =$$

$$\left(-\frac{1}{2L} \right) [(2)(1+0)] \frac{L}{2} = -\frac{1}{2}$$

$$K_{s41} = = -C_{44} 4ab \left(-\frac{1}{2} \right) = 2C_{44} ab$$

$$K_{s42} = \int_{-a}^{a} \int_{-b}^{b} \left(C_{44} F_{sy,x} F_{\tau y,x} \right) dx dz \int_{-1}^{1} \left(N_1(r) N_2(r) \right) \det \mathbf{J} dr$$

$$\int_{-a}^{a} \int_{-b}^{b} \left(C_{44} F_{sy,x} F_{\tau y,x} \right) dx dz \;=\; \int_{-a}^{a} \int_{-b}^{b} \left((C_{44})(1) \right) dx dz = \; C_{44} \int_{-a}^{a} dx \int_{-b}^{b} dz = \; C_{44} 4ab$$

$$N_1(r) N_2(r) = \left(\frac{1}{2}(1-r) \right) \left(\frac{1}{2}(1+r) \right) = \left(\frac{1}{4}(1-r^2) \right)$$

$$\int_{-1}^{1} \left(N_1(r) N_2(r) \right) \det \mathbf{J} dr = \; \int_{-1}^{1} \left(\frac{1}{4}(1-r^2) \right) \left(\frac{L}{2} \right) dr \; = \left[(2) \left(\frac{1}{4}(1-0) \right) \right] \frac{L}{2} = \; \frac{L}{4}$$

$$K_{s42} = \; = C_{44} 4ab \left(\frac{L}{4} \right) = \frac{4 C_{44} L ab}{4}$$

$$K_{s43} = \int_{-a}^{a} \int_{-b}^{b} \left(C_{44} F_{sy,x} F_{\tau x} \right) dx dz \int_{-1}^{1} \left(\mathbf{J}^{-1} N_{2,r}(r) N_2(r) \right) \det \mathbf{J} dr$$

$$\int_{-a}^{a} \int_{-b}^{b} \left(C_{44} F_{sy,x} F_{\tau x} \right) dx dz \;=\; \int_{-a}^{a} \int_{-b}^{b} \left((C_{44})(-1) \right) dx dz = \; -C_{44} \int_{-a}^{a} dx \int_{-b}^{b} dz = \; -C_{44} 4ab$$

$$\mathbf{J}^{-1} N_{2,r}(r) N_2(r) = \frac{2}{L} \left[\frac{d}{dr} \left(\frac{1}{2}(1+r) \right) \right] \left(\frac{1}{2}(1+r) \right) = \frac{1}{2L}(1+r)$$

$$\int_{-1}^{1} \left(\mathbf{J}^{-1} N_{2,r}(r) N_2(r) \right) \det \mathbf{J} dr = \; \int_{-1}^{1} \left(\frac{1}{2L}(1+r) \right) \left(\frac{L}{2} \right) dr \; = $$

$$\left(\frac{1}{2L} \right) [(2)(1+0)] \frac{L}{2} = \; \frac{1}{2}$$

$$K_{s43} = \; = -C_{44} 4ab \left(\frac{1}{2} \right) = -2 C_{44} ab$$

$$K_{s44} = \int_{-a}^{a}\int_{-b}^{b} (C_{44}\, F_{sy,x}\, F_{\tau y,x})\, dx\, dz \int_{-1}^{1} (N_2(r)N_2(r))\, det\, \mathbf{J}\, dr$$

$$\int_{-a}^{a}\int_{-b}^{b} (C_{44}\, F_{sy,x}\, F_{\tau y,x})\, dx\, dz \;=\; \int_{-a}^{a}\int_{-b}^{b} ((C_{44})(1))\, dx\, dz \;=\; C_{44}\int_{-a}^{a} dx \int_{-b}^{b} dz \;=\; C_{44}4ab$$

$$N_2(r)N_2(r) = \left(\frac{1}{2}(1+r)\right)\left(\frac{1}{2}(1+r)\right) = \left(\frac{1}{2}(1+r)\right)^2 = \left(\frac{1}{4}+\frac{r^2}{4}+\frac{r}{2}\right)$$

$$\int_{-1}^{1} (N_2(r)N_2(r))\, det\, \mathbf{J}\, dr = \int_{-1}^{1}\left(\frac{1}{4}+\frac{r^2}{4}+\frac{r}{2}\right)\left(\frac{L}{2}\right) dr \;=\;$$

$$\left[(2)\left(\frac{1}{4}+0+0\right)\right]\frac{L}{2} = \frac{L}{4}$$

$$K_{s44} = = C_{44}4ab\left(\frac{L}{4}\right) = \frac{4C_{44}Lab}{4}$$

$$K_{S_{URI}} = \begin{bmatrix} \dfrac{4C_{44}ab}{L} & 2C_{44}ab & -\dfrac{4C_{44}ab}{L} & 2C_{44}ab \\[2ex] 2C_{44}ab & \dfrac{4C_{44}abL}{4} & -2C_{44}ab & \dfrac{4C_{44}abL}{4} \\[2ex] -\dfrac{4C_{44}ab}{L} & -2C_{44}ab & \dfrac{4C_{44}ab}{L} & -2C_{44}ab \\[2ex] 2C_{44}ab & \dfrac{4C_{44}abL}{4} & -2C_{44}ab & \dfrac{4C_{44}abL}{4} \end{bmatrix} \qquad (4.23)$$

4.1.3 Uniform Reduced Integration: Considerations on Bending Part and Shear Part of the Stiffness Matrix

When evaluating the bending stiffness matrices using FI and URI, they are similar. However, a discrepancy arises in the shear components, specifically in the red terms K_{s22}, K_{s24}, K_{s42}, K_{s44}. These differences can be expressed as follows:

$$K_{s22FI} = \frac{1}{3} \cdot C_{44}AL = 0.333 \cdot C_{44}AL \text{ vs } K_{s22URI} = \frac{1}{4} \cdot C_{44}AL = 0.25 \cdot C_{44}AL$$

$$K_{s24FI} = \frac{1}{6} \cdot C_{44}AL = 0.167 \cdot C_{44}AL \text{ vs } K_{s24URI} = \frac{1}{4} \cdot C_{44}AL = 0.25 \cdot C_{44}AL$$

$$K_{s42FI} = \frac{1}{6} \cdot C_{44}AL = 0.167 \cdot C_{44}AL \text{ vs } K_{s42URI} = \frac{1}{4} \cdot C_{44}AL = 0.25 \cdot C_{44}AL$$

$$K_{s44FI} = \frac{1}{3} \cdot C_{44}AL = 0.333 \cdot C_{44}AL \text{ vs } K_{s44URI} = \frac{1}{4} \cdot C_{44}AL = 0.25 \cdot C_{44}AL$$

$$(4.24)$$

The differences in the coefficients of FI and URI shear parts range from $+0.083$ to -0.083. Although these variations may seem minor, they have a substantial impact on the predicted bending stiffness and, consequently, on the displacement. This effect will be demonstrated in the numerical examples presented in Sect. 4.5.2.

4.1.4 Uniform Reduced Integration: The Complete Stiffness Matrix

The complete stiffness matrix for a two-node element using URI, by summing the two matrices of Eqs. (4.21) and (4.24) is

$$
K_{URI} =
\begin{bmatrix}
\dfrac{4C_{44}ab}{L} & 2C_{44}ab & -\dfrac{4C_{44}ab}{L} & 2C_{44}ab \\[2mm]
2C_{44}ab & \dfrac{4C_{44}abL}{4}+\dfrac{4C_{11}a^3b}{3L} & -2C_{44}ab & \dfrac{4C_{44}abL}{4}-\dfrac{4C_{11}a^3b}{3L} \\[2mm]
-\dfrac{4C_{44}ab}{L} & -2C_{44}ab & \dfrac{4C_{44}ab}{L} & -2C_{44}ab \\[2mm]
2C_{44}ab & \dfrac{4C_{44}abL}{4}-\dfrac{4C_{11}a^3b}{3L} & -2C_{44}ab & \dfrac{4C_{44}abL}{4}+\dfrac{4C_{11}a^3b}{3L}
\end{bmatrix}
$$

$$(4.25)$$

The MATLAB script for the derivation of the FI stiffness matrix is reported in Appendix C.2.

4.2 Selective Reduced Integration

In SRI, a one sampling point Gauss quadrature is applied only to the shear component of the stiffness matrix, while two sampling points are used for the bending component, thereby achieving FI. Figure 4.1 illustrates the position of the sampling point along the beam (i.e., its coordinate r) and the corresponding weight.

4.2.1 Selective Reduced Integration Applied to the Bending Part of the Stiffness Matrix

Implementing a two sampling points Gauss quadrature for the bending portion of the stiffness matrix follows the procedure outlined in Sect. 3.7.5. As a result, the bending matrix of SRI matches the one in Eq. (3.120), which is reported below:

$$K_{b_{SRI}} = \begin{bmatrix} 0 & 0 & 0 & 0 \\ 0 & \dfrac{4C_{11}a^3b}{3L} & 0 & -\dfrac{4C_{11}a^3b}{3L} \\ 0 & 0 & 0 & 0 \\ 0 & -\dfrac{4C_{11}a^3b}{3L} & 0 & \dfrac{4C_{11}a^3b}{3L} \end{bmatrix} \tag{4.26}$$

4.2.2 Selective Reduced Integration Applied to the Shear Part of the Stiffness Matrix

The application of a one sampling point Gauss quadrature to the shear part of the stiffness matrix aligns with the description provided in Sect. 4.1.2. Consequently, the shear part matrix for SRI is equivalent to the matrix presented in equation (4.26). This matrix is reported below:

$$
K_{S_{SRI}} =
\begin{bmatrix}
\dfrac{4C_{44}ab}{L} & 2C_{44}ab & -\dfrac{4C_{44}ab}{L} & 2C_{44}ab \\[2ex]
2C_{44}ab & \dfrac{4C_{44}abL}{4} & -2C_{44}ab & \dfrac{4C_{44}abL}{4} \\[2ex]
-\dfrac{4C_{44}ab}{L} & -2C_{44}ab & \dfrac{4C_{44}ab}{L} & -2C_{44}ab \\[2ex]
2C_{44}ab & \dfrac{4C_{44}abL}{4} & -2C_{44}ab & \dfrac{4C_{44}abL}{4}
\end{bmatrix}
\tag{4.27}
$$

4.2.3   *Selective Reduced Integration: The Complete Stiffness Matrix*

The complete stiffness matrix of a two-node element evaluated using SRI, can be obtained by summing the two previous contributions of Eqs. (4.21) and (4.27)

$$
K_{SRI} =
\begin{bmatrix}
\dfrac{4C_{44}ab}{L} & 2C_{44}ab & -\dfrac{4C_{44}ab}{L} & 2C_{44}ab \\[2ex]
2C_{44}ab & \dfrac{4C_{44}abL}{4}+\dfrac{4C_{11}a^3b}{3L} & -2C_{44}ab & \dfrac{4C_{44}abL}{4}-\dfrac{4C_{11}a^3b}{3L} \\[2ex]
-\dfrac{4C_{44}ab}{L} & -2C_{44}ab & \dfrac{4C_{44}ab}{L} & -2C_{44}ab \\[2ex]
2C_{44}ab & \dfrac{4C_{44}abL}{4}-\dfrac{4C_{11}a^3b}{3L} & -2C_{44}ab & \dfrac{4C_{44}abL}{4}+\dfrac{4C_{11}a^3b}{3L}
\end{bmatrix}
\tag{4.28}
$$

Note that the stiffness matrices of Eqs. (4.25) and (4.28) are the same. The MATLAB script for the derivation of the FI stiffness matrix is reported in Appendix C.3.

4.3 Mixed Interpolations of Tensorial Components

Mixed Interpolation of Tensorial Components (MITC) is a family of finite element formulations created to mitigate shear and membrane locking in plate and shell analyses. Early research on mixed or assumed-strain approaches, pioneered by [1], laid the groundwork for such methods. However, the specific acronym "MITC" and the systematic framework for these elements were introduced by [2, 3]. Their work formalized the MITC approach, using carefully chosen interpolations for stress or strain fields to reduce spurious constraints, and led to popular elements such as MITC4, MITC9, and more. These elements strike a balance between computational efficiency and accuracy and remain a cornerstone of robust plate and shell finite element modeling. In this book, the same approach is employed for a two-node beam finite element. From the assumed distribution, the stiffness matrix is derived using the principle of virtual work. The shape functions, defined in the local coordinate y, are presented below:

$$N_1(y) = \left(1 - \frac{y}{L}\right)$$

$$N_2(y) = \left(\frac{y}{L}\right)$$

$$(4.29)$$

These functions are used to interpolate the displacements along the beam. In order to proceed with the interpolation of the shear strains, an *assumed interpolation function* is introduced

$$\bar{N}_m \qquad\qquad m = 1, ..., n_{node} - 1 \qquad\qquad (4.30)$$

where m denotes a number of points named *tying points*. These points are used to tie the interpolations of the displacements with the assumed strains. The number of *tying points* ranges from 1 to the number of element FE nodes minus 1. As far as the two-node element is concerned, m is equal to 1 and then there exists only 1 *tying point*. The *a-priori assumed interpolation function* is the following

$$\bar{N}_1 = 1 \qquad\qquad (4.31)$$

This assumed interpolation function has a constant value equal to 1. The considered *tying point* has the following coordinate value

$$y_T = \frac{L}{2} \qquad\qquad (4.32)$$

Figure 4.2 shows the position of the *tying point* along the beam.

In order to compute the overall stiffness matrix for the two-node element using MITC, the bending stiffness and the shear stiffness must be computed.

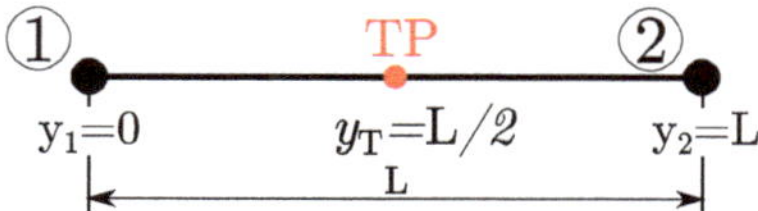

Fig. 4.2 Tying point position in local coordinates along a beam with 2 nodes

4.3.1 Bending Stiffness Matrix Using the Mixed Interpolations of Tensorial Components

The bending stiffness is computed by following the approach described in Sect. 3.7.2. In MITC formulation, local coordinates are utilized, which means $N_{1,y}$ replaces $\mathbf{J}^{-1}N_{1,r}(r)$, $N_{2,y}$ replaces $\mathbf{J}^{-1}N_{2,r}(r)$, $N_1(y)$ and $N_2(y)$ take the place of $N_1(r)$ and $N_2(r)$, resepctively. Consequently, it has:

$$\mathbf{K}_{b_{MITC}} = \begin{bmatrix} 0 & 0 & 0 & 0 \\ 0 & \int_V (C_{11}F_{sy}F_{\tau y} \, N_{1,y}(y)N_{1,y}(y))dV & 0 & \int_V (C_{11}F_{sy}F_{\tau y} \, N_{1,y}(y)N_{2,y}(y))dV \\ 0 & 0 & 0 & 0 \\ 0 & \int_V (C1C_{11}1F_{sy}F_{\tau y} \, N_{2,y}(y)N_{1,y}(y))dV & 0 & \int_V (C_{11}F_{sy}F_{\tau y} \, N_{2,y}(y)N_{2,y}(y))dV \end{bmatrix} dV \tag{4.33}$$

The previous matrix can be written, in a more generic way, as follows

$$\mathbf{K}_{b_{MITC}} = \begin{bmatrix} 0 & 0 & 0 & 0 \\ 0 & k_{22_{b_{MITC}}} & 0 & k_{24_{b_{MITC}}} \\ 0 & 0 & 0 & 0 \\ 0 & k_{42_{b_{MITC}}} & 0 & k_{44_{b_{MITC}}} \end{bmatrix} dV \tag{4.34}$$

In order to compute the terms of the matrix of Eq. (4.34) the following relations apply (see also relations of Eq. (4.29))

> **Expansion and shape functions**
>
> $$F_{\tau y} = x, \qquad F_{sy} = x$$
>
> $$F_{sy} F_{\tau y} = (x)(x) = x^2$$
>
> $$N_{1,y}(y) = \frac{\partial N_1(y)}{\partial y} = \frac{\partial}{\partial y}\left(1 - \frac{y}{L}\right) = -\frac{1}{L}$$
>
> $$N_{2,y}(y) = \frac{\partial N_2(y)}{\partial y} = \frac{\partial}{\partial y}\left(\frac{y}{L}\right) = \frac{1}{L}$$
>
> $$N_{1,y}(y)N_{1,y}(y) = \left(-\frac{1}{L}\right)\left(-\frac{1}{L}\right) = \frac{1}{L^2}$$
>
> $$N_{1,y}(y)N_{2,y}(y) = \left(-\frac{1}{L}\right)\left(\frac{1}{L}\right) = -\frac{1}{L^2}$$
>
> $$N_{2,y}(y)N_{1,y}(y) = \left(\frac{1}{L}\right)\left(-\frac{1}{L}\right) = -\frac{1}{L^2}$$
>
> $$N_{2,y}(y)N_{2,y}(y) = \left(\frac{1}{L}\right)\left(\frac{1}{L}\right) = \frac{1}{L^2}$$

Based on the previous expressions, the terms of the matrix of Eq. (4.34) can be then computed as follows

$$
\begin{aligned}
k_{22_{b_{MITC}}} &= \int_V \left(C_{11} F_{sy} F_{\tau y} N_{1,y}(y) N_{1,y}(y)\right) dV = \int_A \left(C_{11} F_{sy} F_{\tau y}\right) dA \int_L \left(N_{1,y}(y) N_{1,y}(y)\right) dL \\
&= C_{11} \int_{-a}^{a} \left(x^2\right) dx \int_{-b}^{b} dz \int_{0}^{L} \frac{1}{L^2} dy = C_{11} \left(\frac{2}{3}a^3\right)(2b)\left(\frac{1}{L}\right) = C_{11}\frac{4a^3 b}{3L}
\end{aligned}
$$

$$
\begin{aligned}
k_{24_{b_{MITC}}} &= \int_V \left(C_{11} F_{sy} F_{\tau y} N_{1,y}(y) N_{2,y}(y)\right) dV = \int_A \left(C_{11} F_{sy} F_{\tau y}\right) dA \int_L \left(N_{1,y}(y) N_{2,y}(y)\right) dL \\
&= C_{11} \int_{-a}^{a} \left(x^2\right) dx \int_{-b}^{b} dz \int_{0}^{L} -\frac{1}{L^2} dy = C_{11} \left(\frac{2}{3}a^3\right)(2b)\left(-\frac{1}{L}\right) = -C_{11}\frac{4a^3 b}{3L}
\end{aligned}
$$

$$
\begin{aligned}
k_{42_{b_{MITC}}} &= \int_V \left(C_{11} F_{sy} F_{\tau y} N_{2,y}(y) N_{1,y}(y)\right) dV = \int_A \left(C_{11} F_{sy} F_{\tau y}\right) dA \int_L \left(N_{1,y}(y) N_{1,y}(y)\right) dL \\
&= C_{11} \int_{-a}^{a} \left(x^2\right) dx \int_{-b}^{b} dz \int_{0}^{L} -\frac{1}{L^2} dy = C_{11} \left(\frac{2}{3}a^3\right)(2b)\left(-\frac{1}{L}\right) = -C_{11}\frac{4a^3 b}{3L}
\end{aligned}
$$

$$k_{44_{b_{MITC}}} = \int_V \left(C_{11} F_{sy} F_{\tau y} N_{2,y}(y) N_{2,y}(y) \right) dV = \int_A \left(C_{11} F_{sy} F_{\tau y} \right) dA \int_L \left(N_{2,y}(y) N_{2,y}(y) \right) dL$$

$$= C_{11} \int_{-a}^{a} \left(x^2 \right) dx \int_{-b}^{b} dz \int_0^L \frac{1}{L^2} dy = C_{11} \left(\frac{2}{3} a^3 \right) (2b) \left(\frac{1}{L} \right) = C_{11} \frac{4a^3 b}{3L}$$

The bending stiffness matrix using MITC becomes

$$\mathbf{K}_{b_{MITC}} = \begin{bmatrix} 0 & 0 & 0 & 0 \\ 0 & C_{11}\dfrac{4a^3 b}{3L} & 0 & -C_{11}\dfrac{4a^3 b}{3L} \\ 0 & 0 & 0 & 0 \\ 0 & -C_{11}\dfrac{4a^3 b}{3L} & 0 & C_{11}\dfrac{4a^3 b}{3L} \end{bmatrix} \tag{4.35}$$

Clearly, the bending stiffness matrix using MITC is equal to those obtained with FI, URI and SRI.

4.3.2   The Shear Stiffness Matrix Using the Mixed Interpolations of Tensorial Components

To compute the shear stiffness matrix part, the shear strains are interpolated using the assumed interpolation function $\bar{N}_1$. These shear strains, originally computed in the natural coordinate system in Eq. (3.79), are presented below.

$$\epsilon_{xy} = \begin{bmatrix} F_{\tau x} N_{1,y}(y) & F_{\tau y,x} N_1(y) & F_{\tau x} N_{2,y}(y) & F_{\tau y,x} N_2(y) \end{bmatrix} \begin{Bmatrix} u_1^0 \\ \phi_{z1} \\ u_2^0 \\ \phi_{z2} \end{Bmatrix} \tag{4.36}$$

The interpolation of the shear strains is performed following

$$\bar{\epsilon}_{xy} = \bar{N}_m \epsilon_{xy_m} \tag{4.37}$$

Since m is equal to 1, one can write

$$\bar{\epsilon}_{xy} = \bar{N}_1 \epsilon_{xy_1} \tag{4.38}$$

where $\bar{\epsilon}_{xy}$ are the *assumed strains* or the *interpolated strains*, the $\bar{N}_1$ is the *assumed interpolation function* equal to 1 and ϵ_{xy_1} are the shear strains computed in the *tying point* 1. Using the previous Eq. (4.36) for the shear strains, Eq. (4.38) becomes

$$\bar{\epsilon}_{xy} = \bar{N}_1 \epsilon_{xy_1} =$$

$$\left[\bar{N}_1 F_{\tau x}(N_{1,y}(y))_1 \ \bar{N}_1 F_{\tau y,x}(N_1(y))_1 \ \bar{N}_1 F_{\tau x}(N_{2,y}(y))_1 \ \bar{N}_1 F_{\tau y,x}(N_2(y))_1 \right] \left\{ \begin{array}{c} u_1^0 \\ \phi_{z1} \\ u_2^0 \\ \phi_{z2} \end{array} \right\}$$

$$(4.39)$$

In the previous expression the terms $(N_1(y))_1$ and $(N_2(y))_1$ indicate the *displacements interpolation functions* computed at the *tying point* 1, whereas the terms $\left(N_{1,y}(y)\right)_1$ and $\left(N_{2,y}(y)\right)_1$ indicate the derivatives of the *displacements interpolation functions* computed at the *tying point* 1.

Considering the computations of Sect. 3.7.2 it can be written

$$\bar{\mathbf{B}}_{\tau i} = \left[\bar{N}_1 F_{\tau x} \left(N_{1,y}(y)\right)_1 \ \bar{N}_1 F_{\tau y,x} \left(N_1(y)\right)_1 \ \bar{N}_1 F_{\tau x} \left(N_{2,y}(y)\right)_1 \ \bar{N}_1 F_{\tau y,x} \left(N_2(y)\right)_1 \right]$$

$$(4.40)$$

$$\bar{\mathbf{B}}_{sj} = \left[\bar{N}_1 F_{sx} \left(N_{1,y}(y)\right)_1 \ \bar{N}_1 F_{sy,x} \left(N_1(y)\right)_1 \ \bar{N}_1 F_{sx} \left(N_{2,y}(y)\right)_1 \ \bar{N}_1 F_{sy,x} \left(N_2(y)\right)_1 \right]$$

$$(4.41)$$

where $\bar{\mathbf{B}}_{\tau i}$ and $\bar{\mathbf{B}}_{sj}$ are matrices related to MITC approach. The shear stiffness matrix part using MITC approach can be calculated using the following equation

$$\mathbf{K}_s = \int_V \bar{\mathbf{B}}_{sj}^T \mathbf{C}_s \bar{\mathbf{B}}_{\tau i} dV \qquad (4.42)$$

Introducing Eqs. 4.40 and 4.41 into Eq. 4.42, the following relation can be written

$$
\mathbf{K}_{S_{MITC}} = \int_V
\begin{bmatrix}
\bar{N}_1 F_{sx}\,(N_{1,y}(y))_1 \\[4pt]
\bar{N}_1 F_{sy,x}\,(N_1(y))_1 \\[4pt]
\bar{N}_1 F_{sx}\,(N_{2,y}(y))_1 \\[4pt]
\bar{N}_1 F_{sy,x}\,(N_2(y))_1
\end{bmatrix}
C_{44}
\begin{bmatrix}
\bar{N}_1 F_{\tau x} & \bar{N}_1 F_{\tau y,x} & \bar{N}_1 F_{\tau x} & \bar{N}_1 F_{\tau y,x} \\
(N_{1,y}(y))_1 & (N_1(y))_1 & (N_{2,y}(y))_1 & (N_2(y))_1
\end{bmatrix}
dV \qquad (4.43)
$$

Performing the triple product on the volume integral of Eq. (4.43), it can be written as

$$
\mathbf{K}_{S_{MITC}} = C_{44} \int_V
\begin{bmatrix}
\begin{array}{cccc}
\bar{N}_1 F_{sx}\,(N_{1,y}(y))_1 & \bar{N}_1 F_{sx}\,(N_{1,y}(y))_1 & \bar{N}_1 F_{sx}\,(N_{1,y}(y))_1 & \bar{N}_1 F_{sx}\,(N_{1,y}(y))_1 \\
\bar{N}_1 F_{\tau x}\,(N_{1,y}(y))_1 & \bar{N}_1 F_{\tau y,x}\,(N_1(y))_1 & \bar{N}_1 F_{\tau x}\,(N_{2,y}(y))_1 & \bar{N}_1 F_{\tau y,x}\,(N_2(y))_1 \\[18pt]
\bar{N}_1 F_{sy,x}\,(N_1(y))_1 & \bar{N}_1 F_{sy,x}\,(N_1(y))_1 & \bar{N}_1 F_{sy,x}\,(N_1(y))_1 & \bar{N}_1 F_{sy,x}\,(N_1(y))_1 \\
\bar{N}_1 F_{\tau x}\,(N_{1,y}(y))_1 & \bar{N}_1 F_{\tau y,x}\,(N_1(y))_1 & \bar{N}_1 F_{\tau x}\,(N_{2,y}(y))_1 & \bar{N}_1 F_{\tau y,x}\,(N_2(y))_1 \\[18pt]
\bar{N}_1 F_{sx}\,(N_{2,y}(y))_1 & \bar{N}_1 F_{sx}\,(N_{2,y}(y))_1 & \bar{N}_1 F_{sx}\,(N_{2,y}(y))_1 & \bar{N}_1 F_{sx}\,(N_{2,y}(y))_1 \\
\bar{N}_1 F_{\tau x}\,(N_{1,y}(y))_1 & \bar{N}_1 F_{\tau y,x}\,(N_1(y))_1 & \bar{N}_1 F_{\tau x}\,(N_{2,y}(y))_1 & \bar{N}_1 F_{\tau y,x}\,(N_2(y))_1 \\[18pt]
\bar{N}_1 F_{sy,x}\,(N_2(y))_1 & \bar{N}_1 F_{sy,x}\,(N_2(y))_1 & \bar{N}_1 F_{sy,x}\,(N_2(y))_1 & \bar{N}_1 F_{sy,x}\,(N_2(y))_1 \\
\bar{N}_1 F_{\tau x}\,(N_{1,y}(y))_1 & \bar{N}_1 F_{\tau y,x}\,(N_1(y))_1 & \bar{N}_1 F_{\tau x}\,(N_{2,y}(y))_1 & \bar{N}_1 F_{\tau y,x}\,(N_2(y))_1
\end{array}
\end{bmatrix}
dV
$$

$$(4.44)$$

$$
\mathbf{K}_{S_{MITC}} = C_{44}
\begin{bmatrix}
\begin{array}{cccc}
\int_y (\bar{N}_1 F_{sx}\,(N_{1,y}(y))_1 & \int_y (\bar{N}_1 F_{sx}\,(N_{1,y}(y))_1 & \int_y (\bar{N}_1 F_{sx}\,(N_{1,y}(y))_1 & \int_y (\bar{N}_1 F_{sx}\,(N_{1,y}(y))_1 \\
N_1 F_{\tau x}\,(N_{1,y}(y))_1)\,dV & N_1 F_{\tau y,x}\,(N_1(y))_1)\,dV & N_1 F_{\tau x}\,(N_{2,y}(y))_1)\,dV & N_1 F_{\tau y,x}\,(N_2(y))_1)\,dV \\[18pt]
\int_y (\bar{N}_1 F_{sy,x}\,(N_1(y))_1 & \int_y (\bar{N}_1 F_{sy,x}\,(N_1(y))_1 & \int_y (\bar{N}_1 F_{sy,x}\,(N_1(y))_1 & \int_y (\bar{N}_1 F_{sy,x}\,(N_1(y))_1 \\
N_1 F_{\tau x}\,(N_{1,y}(y))_1)\,dV & N_1 F_{\tau y,x}\,(N_1(y))_1)\,dV & N_1 F_{\tau x}\,(N_{2,y}(y))_1)\,dV & N_1 F_{\tau y,x}\,(N_2(y))_1)\,dV \\[18pt]
\int_y (\bar{N}_1 F_{sx}\,(N_{2,y}(y))_1 & \int_y (\bar{N}_1 F_{sx}\,(N_{2,y}(y))_1 & \int_y (\bar{N}_1 F_{sx}\,(N_{2,y}(y))_1 & \int_y (\bar{N}_1 F_{sx}\,(N_{2,y}(y))_1 \\
N_1 F_{\tau x}\,(N_{1,y}(y))_1)\,dV & N_1 F_{\tau y,x}\,(N_1(y))_1)\,dV & N_1 F_{\tau x}\,(N_{2,y}(y))_1)\,dV & N_1 F_{\tau y,x}\,(N_2(y))_1)\,dV \\[18pt]
\int_y (\bar{N}_1 F_{sy,x}\,(N_2(y))_1 & \int_y (\bar{N}_1 F_{sy,x}\,(N_2(y))_1 & \int_y (\bar{N}_1 F_{sy,x}\,(N_2(y))_1 & \int_y (\bar{N}_1 F_{sy,x}\,(N_2(y))_1 \\
N_1 F_{\tau x}\,(N_{1,y}(y))_1)\,dV & N_1 F_{\tau y,x}\,(N_1(y))_1)\,dV & N_1 F_{\tau x}\,(N_{2,y}(y))_1)\,dV & N_1 F_{\tau y,x}\,(N_2(y))_1)\,dV
\end{array}
\end{bmatrix}
$$

$$(4.45)$$

In order to calculate the terms of the matrix of Eq. (4.45) the following relations must be taken into account

Expansion and shape functions

$$F_{\tau x} = -1; \quad F_{\tau y,x} = 1; \quad F_{sx} = -1; \quad F_{sy,x} = 1$$

$$F_{\tau x} F_{sx} = 1; \quad F_{\tau y,x} F_{sx} = -1; \quad F_{\tau x} F_{sy,x} = -1; \quad F_{\tau y,x} F_{sy,x} = 1$$

$$\bar{N}_1 = 1$$

$$(N_1(y))_1 = \left(1 - \frac{y}{L}\right)_1 = \left(1 - \frac{y_T}{L}\right) = \left(1 - \frac{\frac{L}{2}}{L}\right) = \frac{1}{2}$$

$$(N_{1,y}(y))_1 = \left(\frac{\partial}{\partial y}(1 - \frac{y}{L})\right)_1 = \left(-\frac{1}{L}\right)_1 = -\frac{1}{L}$$

$$(N_2(y))_1 = \left(\frac{y}{L}\right)_1 = \left(\frac{y_T}{L}\right) = \left(\frac{\frac{L}{2}}{L}\right) = \frac{1}{2}$$

$$(N_{2,y}(y))_1 = \left(\frac{\partial}{\partial y}(\frac{y}{L})\right)_1 = \left(\frac{1}{L}\right)_1 = \frac{1}{L}$$

Following are the calculations of the terms of the shear stiffness matrix part using the MITC approach.

$$K_{11_{s_{MITC}}} = C_{44} \int_V \left(\bar{N}_1 F_{sx} \left(N_{1,y}(y)\right)_1 \bar{N}_1 F_{\tau x} \left(N_{1,y}(y)\right)_1 \right) dV$$

$$= C_{44} \int_A F_{\tau x} F_{sx} dA \int_L \bar{N}_1 \left(N_{1,y}(y)\right)_1 \bar{N}_1 \left(N_{1,y}(y)\right)_1 dL$$

$$= C_{44} \int_{-a}^{a} 1 dx \int_{-b}^{b} dz \int_0^L 1 \left(-\frac{1}{L}\right) 1 \left(-\frac{1}{L}\right) dy = C_{44} \frac{4ab}{L}$$

$$K_{12_{s_{MITC}}} = C_{44} \int_V \left(\bar{N}_1 F_{sx} \left(N_{1,y}(y)\right)_1 \bar{N}_1 F_{\tau y,x} \left(N_1(y)\right)_1 \right) dV$$

$$= C_{44} \int_A F_{\tau y,x} F_{sx} dA \int_L \bar{N}_1 \left(N_{1,y}(y)\right)_1 \bar{N}_1 \left(N_1(y)\right)_1 dL$$

$$= C_{44} \int_{-a}^{a} -1 dx \int_{-b}^{b} dz \int_0^L 1 \left(-\frac{1}{L}\right) 1 \left(\frac{1}{2}\right) dy = C_{44} 2ab$$

$$K_{13_{sMITC}} = C_{44} \int_V \left(\bar{N}_1 F_{sx} \left(N_{1,y}(y) \right)_1 \bar{N}_1 F_{\tau x} \left(N_{2,y}(y) \right)_1 \right) dV$$

$$= C_{44} \int_A F_{\tau x} F_{sx} dA \int_L \bar{N}_1 \left(N_{1,y}(y) \right)_1 \bar{N}_1 \left(N_{2,y}(y) \right)_1 dL$$

$$= C_{44} \int_{-a}^{a} 1 dx \int_{-b}^{b} dz \int_0^L 1 \left(-\frac{1}{L} \right) 1 \left(\frac{1}{L} \right) dy = -C_{44} \frac{4ab}{L}$$

$$K_{14_{sMITC}} = C_{44} \int_V \left(\bar{N}_1 F_{sx} \left(N_{1,y}(y) \right)_1 \bar{N}_1 F_{\tau y,x} \left(N_2(y) \right)_1 \right) dV$$

$$= C_{44} \int_A F_{\tau y,x} F_{sx} dA \int_L \bar{N}_1 \left(N_{1,y}(y) \right)_1 \bar{N}_1 \left(N_2(y) \right)_1 dL$$

$$= C_{44} \int_{-a}^{a} -1 dx \int_{-b}^{b} dz \int_0^L 1 \left(-\frac{1}{L} \right) 1 \left(\frac{1}{2} \right) dy = C_{44} 2ab$$

$$K_{21_{sMITC}} = C_{44} \int_V \left(\bar{N}_1 F_{sy,x} \left(N_{1,y}(y) \right)_1 \bar{N}_1 F_{\tau x} \left(N_{1,y}(y) \right)_1 \right) dV$$

$$= C_{44} \int_A F_{sy,x} F_{\tau x} dA \int_L \bar{N}_1 \left(N_{1,y}(y) \right)_1 \bar{N}_1 \left(N_{1,y}(y) \right)_1 dL$$

$$= C_{44} \int_{-a}^{a} -1 dx \int_{-b}^{b} dz \int_0^L 1 \left(\frac{1}{2} \right) 1 \left(-\frac{1}{L} \right) dy = C_{44} 2ab$$

$$K_{22_{sMITC}} = C_{44} \int_V \left(\bar{N}_1 F_{sy,x} \left(N_1(y) \right)_1 \bar{N}_1 F_{\tau y,x} \left(N_1(y) \right)_1 \right) dV$$

$$= C_{44} \int_A F_{\tau y,x} F_{sy,x} dA \int_L \bar{N}_1 \left(N_1(y) \right)_1 \bar{N}_1 \left(N_1(y) \right)_1 dL$$

$$= C_{44} \int_{-a}^{a} 1 dx \int_{-b}^{b} dz \int_0^L 1 \left(\frac{1}{2} \right) 1 \left(\frac{1}{2} \right) dy = C_{44} abL$$

$$K_{23_{sMITC}} = C_{44} \int_V \left(\bar{N}_1 F_{sy,x} \left(N_1(y) \right)_1 \bar{N}_1 F_{\tau x} \left(N_{2,y}(y) \right)_1 \right) dV$$

$$= C_{44} \int_A F_{\tau x} F_{sy,x} dA \int_L \bar{N}_1 \left(N_1(y) \right)_1 \bar{N}_1 \left(N_{2,y}(y) \right)_1 dL$$

$$= C_{44} \int_{-a}^{a} -1 dx \int_{-b}^{b} dz \int_0^L 1 \left(\frac{1}{2} \right) 1 \left(\frac{1}{L} \right) dy = -C_{44} 2ab$$

$$K_{24_{sMITC}} = C_{44} \int_V \left(\bar{N}_1 F_{sy,x} \left(N_1(y) \right)_1 \bar{N}_1 F_{\tau y,x} \left(N_2(y) \right)_1 \right) dV$$

$$= C_{44} \int_A F_{\tau y,x} F_{sy,x} dA \int_L \bar{N}_1 \left(N_{1,y}(y) \right)_1 \bar{N}_1 \left(N_2(y) \right)_1 dL$$

$$= C_{44} \int_{-a}^{a} 1 dx \int_{-b}^{b} dz \int_0^L 1 \left(\frac{1}{2} \right) 1 \left(\frac{1}{2} \right) dy = C_{44} abL$$

$$K_{31_{sMITC}} = C_{44} \int_V \left(\bar{N}_1 F_{sx} \left(N_{2,y}(y) \right)_1 \bar{N}_1 F_{\tau x} \left(N_{1,y}(y) \right)_1 \right) dV$$

$$= C_{44} \int_A F_{\tau x} F_{sx} dA \int_L \bar{N}_1 \left(N_{2,y}(y) \right)_1 \bar{N}_1 \left(N_{1,y}(y) \right)_1 dL$$

$$= C_{44} \int_{-a}^{a} 1 dx \int_{-b}^{b} dz \int_0^L 1 \left(\frac{1}{L} \right) 1 \left(-\frac{1}{L} \right) dy = -C_{44} \frac{4ab}{L}$$

$$K_{32_{sMITC}} = C_{44} \int_V \left(\bar{N}_1 F_{sx} \left(N_{2,y}(y) \right)_1 \bar{N}_1 F_{\tau y,x} \left(N_1(y) \right)_1 \right) dV$$

$$= C_{44} \int_A F_{\tau y,x} F_{sx} dA \int_L \bar{N}_1 \left(N_{2,y}(y) \right)_1 \bar{N}_1 \left(N_1(y) \right)_1 dL$$

$$= C_{44} \int_{-a}^{a} -1 dx \int_{-b}^{b} dz \int_0^L 1 \left(\frac{1}{L} \right) 1 \left(\frac{1}{2} \right) dy = -C_{44} 2ab$$

$$K_{33_{sMITC}} = C_{44} \int_V \left(\bar{N}_1 F_{sx} \left(N_{2,y}(y) \right)_1 \bar{N}_1 F_{\tau x} \left(N_{2,y}(y) \right)_1 \right) dV$$

$$= C_{44} \int_A F_{\tau x} F_{sx} dA \int_L \bar{N}_1 \left(N_{2,y}(y) \right)_1 \bar{N}_1 \left(N_{2,y}(y) \right)_1 dL$$

$$= C_{44} \int_{-a}^{a} 1 dx \int_{-b}^{b} dz \int_0^L 1 \left(\frac{1}{L} \right) 1 \left(\frac{1}{L} \right) dy = C_{44} \frac{4ab}{L}$$

$$K_{34_{sMITC}} = C_{44} \int_V \left(\bar{N}_1 F_{sx} \left(N_{2,y}(y) \right)_1 \bar{N}_1 F_{\tau y,x} \left(N_2(y) \right)_1 \right) dV$$

$$= C_{44} \int_A F_{\tau y,x} F_{sx} dA \int_L \bar{N}_1 \left(N_{2,y}(y) \right)_1 \bar{N}_1 \left(N_2(y) \right)_1 dL$$

$$= C_{44} \int_{-a}^{a} -1 dx \int_{-b}^{b} dz \int_0^L 1 \left(\frac{1}{L} \right) 1 \left(\frac{1}{2} \right) dy = -C_{44} 2ab$$

$$K_{41_{s_{MITC}}} = C_{44} \int_V \left(\bar{N}_1 F_{sy,x} \left(N_2(y) \right)_1 \bar{N}_1 F_{\tau x} \left(N_{1,y}(y) \right)_1 \right) dV$$

$$= C_{44} \int_A F_{\tau x} F_{sy,x} \, dA \int_L \bar{N}_1 \left(N_2(y) \right)_1 \bar{N}_1 \left(N_{1,y}(y) \right)_1 dL$$

$$= C_{44} \int_{-a}^{a} -1 \, dx \int_{-b}^{b} dz \int_0^L 1 \left(\frac{1}{2} \right) 1 \left(-\frac{1}{L} \right) dy = C_{44} 2ab$$

$$K_{42_{s_{MITC}}} = C_{44} \int_V \left(\bar{N}_1 F_{sy,x} \left(N_2(y) \right)_1 \bar{N}_1 F_{\tau y,x} \left(N_1(y) \right)_1 \right) dV$$

$$= C_{44} \int_A F_{\tau y,x} F_{sy,x} \, dA \int_L \bar{N}_1 \left(N_{2,y}(y) \right)_1 \bar{N}_1 \left(N_1(y) \right)_1 dL$$

$$= C_{44} \int_{-a}^{a} 1 \, dx \int_{-b}^{b} dz \int_0^L 1 \left(\frac{1}{2} \right) 1 \left(\frac{1}{2} \right) dy = C_{44} abL$$

$$K_{43_{s_{MITC}}} = C_{44} \int_V \left(\bar{N}_1 F_{sy,x} \left(N_2(y) \right)_1 \bar{N}_1 F_{\tau x} \left(N_{2,y}(y) \right)_1 \right) dV$$

$$= C_{44} \int_A F_{\tau x} F_{sy,x} \, dA \int_L \bar{N}_1 \left(N_2(y) \right)_1 \bar{N}_1 \left(N_{2,y}(y) \right)_1 dL$$

$$= C_{44} \int_{-a}^{a} -1 \, dx \int_{-b}^{b} dz \int_0^L 1 \left(\frac{1}{2} \right) 1 \left(\frac{1}{L} \right) dy = -C_{44} 2ab$$

$$K_{44_{s_{MITC}}} = C_{44} \int_V \left(\bar{N}_1 F_{sy,x} \left(N_2(y) \right)_1 \bar{N}_1 F_{\tau y,x} \left(N_2(y) \right)_1 \right) dV$$

$$= C_{44} \int_A F_{\tau y,x} F_{sy,x} \, dA \int_L \bar{N}_1 \left(N_2(y) \right)_1 \bar{N}_1 \left(N_2(y) \right)_1 dL$$

$$= C_{44} \int_{-a}^{a} 1 \, dx \int_{-b}^{b} dz \int_0^L 1 \left(\frac{1}{2} \right) 1 \left(\frac{1}{2} \right) dy = C_{44} abL$$

$$\mathbf{K}_{s_{MITC}} = \begin{bmatrix} \dfrac{4C_{44}ab}{L} & 2C_{44}ab & -\dfrac{4C_{44}ab}{L} & 2C_{44}ab \\[2ex] 2C_{44}ab & C_{44}abL & -C_{44}2ab & C_{44}abL \\[2ex] \dfrac{-4C_{44}ab}{L} & 2 - C_{44}ab & \dfrac{4C_{44}ab}{L} & 2 - C_{44}ab \\[2ex] C_{44}2ab & C_{44}abL & -C_{44}2ab & C_{44}abL \end{bmatrix} \tag{4.46}$$

The shear stiffness matrix part using MITC approach is equal to those obtained with URI (see Eq. (4.23)) and SRI (see Eq. (4.28)) approaches.

4.3.2.1 Mixed Interpolations of Tensorial Components: The Complete Stiffness Matrix

The complete stiffness matrix of a two-node element computed using MITC approach, can be obtained summing the two previous matrices of Eqs. (4.35) and (4.46)

$$
\mathbf{K}_{MITC} =
\begin{bmatrix}
\dfrac{4C_{44}ab}{L} & 2C_{44}ab & -\dfrac{4C_{44}ab}{L} & 2C_{44}ab \\[2mm]
2C_{44}ab & C_{44}abL+C_{11}\dfrac{4a^3b}{3L} & -C_{44}2ab & C_{44}abL-C_{11}\dfrac{4a^3b}{3L} \\[2mm]
\dfrac{-4C_{44}ab}{L} & 2C_{44}ab & \dfrac{4C_{44}ab}{L} & 2C_{44}ab \\[2mm]
C_{44}2ab & C_{44}abL-C_{11}\dfrac{4a^3b}{3L} & -C_{44}2ab & C_{44}abL+C_{11}\dfrac{4a^3b}{3L}
\end{bmatrix}
\tag{4.47}
$$

The MATLAB script for the derivation of the FI stiffness matrix is reported in Appendix C.4.

4.4 Use of Higher-Order Elements

The use of higher-order elements can result in a relieving of the shear locking effects. In the following sections, the three-node and the four-node elements are considered for this purpose. They are referred to as B3 and B4, and they make use of quadratic and cubic interpolation, respectively.

4.4.1 Quadratic B3 Element

Figure 4.3 shows the representation of the quadratic B3 element. The shape functions related to a quadratic B3 element are reported below using the natural coordinate.

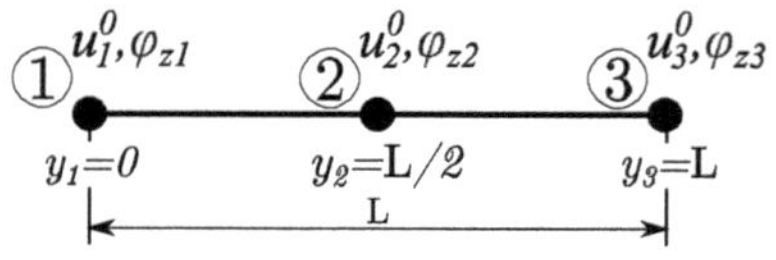

Fig. 4.3 Schematic representation of a three-node beam (B3 element)

$$N_1(r) = \frac{1}{2}r\,(r-1)$$
$$N_2(r) = (1+r)\,(1-r) \tag{4.48}$$
$$N_3(r) = \frac{1}{2}r\,(r+1)$$

The Jacobian $\mathbf{J}$ related to the B3 can be computed as follow (using the expression of Eq. (3.64))

$$\mathbf{J} = \left[\frac{\partial y}{\partial r}\right] = \left[\frac{\partial}{\partial r}\left(\left(\frac{1}{2}(1+r)\right)L\right)\right] = \left[\frac{L}{2}\right] \tag{4.49}$$

The determinant of the previously computed Jacobian $\mathbf{J}$ is

$$det\,\mathbf{J} = det\left[\frac{L}{2}\right] = \frac{L}{2} \tag{4.50}$$

Considering Eq. (3.64), the following relation applies

$$dy = \frac{L}{2}dr = det\,\mathbf{J}dr \tag{4.51}$$

The inverse of the Jacobian $\mathbf{J}$ is

$$\mathbf{J}^{-1} = \left[\frac{L}{2}\right]^{-1} = \left[\frac{2}{L}\right] \tag{4.52}$$

With reference to the Fig. 4.3 and considering the shape functions related to the natural coordinates of Eq. (4.48), the beam nodes displacements $u^0(y)$ and $\phi_z(y)$ can be expressed as follows

$$u^0(y) = N_1(r)u_1^0 + N_2(r)u_2^0 + N_3(r)u_3^0$$

$$\tag{4.53}$$

$$\phi_z(y) = N_1(r)\phi_{z1} + N_2(r)\phi_{z2} + N_3(r)\phi_{z3}$$

Introducing the values of the shape functions

$$u^0(y) = \frac{1}{2}r\,(r-1)\,u_1^0 + (1+r)\,(1-r)\,u_2^0 + \frac{1}{2}r\,(r+1)\,u_3^0$$

$$\phi_z(y) = \frac{1}{2}r\,(r-1)\,\phi_{z1} + (1+r)\,(1-r)\,\phi_{z2} + \frac{1}{2}r\,(r+1)\,\phi_{z3} \tag{4.54}$$

Equation (4.54), can be written in matrix form as

$$\left\{\begin{array}{c} u^0(y) \\ \phi_x(y) \end{array}\right\} = \begin{bmatrix} \frac{1}{2}r\,(r-1) & 0 & (1+r)\,(1-r) & 0 & \frac{1}{2}r\,(r+1) & 0 \\ 0 & \frac{1}{2}r\,(r-1) & 0 & (1+r)\,(1-r) & 0 & \frac{1}{2}r\,(r+1) \end{bmatrix} \left\{\begin{array}{c} u_1^0 \\ \phi_{z1} \\ u_2^0 \\ \phi_{z2} \\ u_3^0 \\ \phi_{z3} \end{array}\right\} \tag{4.55}$$

The geometrical relation is the following

$$\epsilon = \left\{\begin{array}{c} \epsilon_{yy} \\ \epsilon_{xy} \end{array}\right\} = \begin{bmatrix} 0 & F_{\tau y}\dfrac{\partial}{\partial y} \\ F_{\tau x}\dfrac{\partial}{\partial y} & F_{\tau y,x} \end{bmatrix} \left\{\begin{array}{c} u^0(y) \\ \phi_x(y) \end{array}\right\} \tag{4.56}$$

Introducing Eq. (4.56) into Eq. (4.55), the following expression is obtained

$$\epsilon = \left\{\begin{array}{c} \epsilon_{yy} \\ \epsilon_{xy} \end{array}\right\} = \begin{bmatrix} 0 & F_{\tau y}\dfrac{\partial}{\partial y} \\ F_{\tau x}\dfrac{\partial}{\partial y} & F_{\tau y,x} \end{bmatrix} \begin{bmatrix} \frac{1}{2}r\,(r-1) & 0 & (1+r)\,(1-r) & 0 & \frac{1}{2}r\,(r+1) & 0 \\ 0 & \frac{1}{2}r\,(r-1) & 0 & (1+r)\,(1-r) & 0 & \frac{1}{2}r\,(r+1) \end{bmatrix} \left\{\begin{array}{c} U_1^0 \\ \phi_{z1} \\ U_2^0 \\ \phi_{z2} \\ U_3^0 \\ \phi_{z3} \end{array}\right\} \tag{4.57}$$

Recalling the relation $\dfrac{\partial}{\partial y} = \mathbf{J}^{-1}\dfrac{\partial}{\partial r}$, Eq. (4.57) can be written as

$$\epsilon = \left\{\begin{array}{c} \epsilon_{yy} \\ \epsilon_{xy} \end{array}\right\} = \begin{bmatrix} 0 & F_{\tau y}\mathbf{J}^{-1}\dfrac{\partial}{\partial r} \\ F_{\tau x}\mathbf{J}^{-1}\dfrac{\partial}{\partial r} & F_{\tau y,x} \end{bmatrix} \begin{bmatrix} \frac{1}{2}r\,(r-1) & 0 & (1+r)\,(1-r) & 0 & \frac{1}{2}r\,(r+1) & 0 \\ 0 & \frac{1}{2}r\,(r-1) & 0 & (1+r)\,(1-r) & 0 & \frac{1}{2}r\,(r+1) \end{bmatrix} \left\{\begin{array}{c} u_1^0 \\ \phi_{z1} \\ u_2^0 \\ \phi_{z2} \\ u_3^0 \\ \phi_{z3} \end{array}\right\} \tag{4.58}$$

Generalizing Eq. (4.58), it is

$$\boldsymbol{\epsilon} = \left\{ \begin{matrix} \epsilon_{yy} \\ \epsilon_{xy} \end{matrix} \right\} = \begin{bmatrix} 0 & F_{\tau y}\mathbf{J}^{-1}\dfrac{\partial}{\partial r} \\[2ex] F_{\tau x}\mathbf{J}^{-1}\dfrac{\partial}{\partial r} & F_{\tau y,x} \end{bmatrix} \begin{bmatrix} N_1(r) & 0 & N_2(r) & 0 & N_3(r) & 0 \\[1ex] 0 & N_1(r) & 0 & N_2(r) & 0 & N_3(r) \end{bmatrix} \left\{ \begin{matrix} u_1^0 \\ \phi_{z1} \\ u_2^0 \\ \phi_{z2} \\ u_3^0 \\ \phi_{z3} \end{matrix} \right\}$$

$$(4.59)$$

and performing the derivatives

$$\boldsymbol{\epsilon} = \left\{ \begin{matrix} \epsilon_{yy} \\ \epsilon_{xy} \end{matrix} \right\} = \begin{bmatrix} 0 & \begin{matrix} F_{\tau y}\mathbf{J}^{-1} \\ N_{1,r}(r) \end{matrix} & 0 & \begin{matrix} F_{\tau y}\mathbf{J}^{-1} \\ N_{2,r}(r) \end{matrix} & 0 & \begin{matrix} F_{\tau y}\mathbf{J}^{-1} \\ N_{3,r}(r) \end{matrix} \\[3ex] \begin{matrix} F_{\tau x}\mathbf{J}^{-1} \\ N_{1,r}(r) \end{matrix} & \begin{matrix} F_{\tau y,x} \\ N_1(r) \end{matrix} & \begin{matrix} F_{\tau x}\mathbf{J}^{-1} \\ N_{2,r}(r) \end{matrix} & \begin{matrix} F_{\tau y,x} \\ N_2(r) \end{matrix} & \begin{matrix} F_{\tau x}\mathbf{J}^{-1} \\ N_{3,r}(r) \end{matrix} & \begin{matrix} F_{\tau y,x} \\ N_3(r) \end{matrix} \end{bmatrix} \left\{ \begin{matrix} u_1^0 \\ \phi_{z1} \\ u_2^0 \\ \phi_{z2} \\ u_3^0 \\ \phi_{z3} \end{matrix} \right\}$$

$$(4.60)$$

The previous expression can be split into two parts: one related to the longitudinal strains ϵ_{yy}, which is the bending contribution, and one related to the shear strains ϵ_{xy}, which is the shear contribution.

$$\epsilon_{yy} = \begin{bmatrix} 0 & F_{\tau y}\mathbf{J}^{-1}N_{1,r}(r) & 0 & F_{\tau y}\mathbf{J}^{-1}N_{2,r}(r) & 0 & F_{\tau y}\mathbf{J}^{-1}N_{3,r}(r) \end{bmatrix} \left\{ \begin{matrix} u_1^0 \\ \phi_{z1} \\ u_2^0 \\ \phi_{z2} \\ u_3^0 \\ \phi_{z3} \end{matrix} \right\} \qquad (4.61)$$

and

$$\epsilon_{xy} = \begin{bmatrix} F_{\tau x}\mathbf{J}^{-1} & F_{\tau y,x} & F_{\tau x}\mathbf{J}^{-1} & F_{\tau y,x} & F_{\tau x}\mathbf{J}^{-1} & F_{\tau y,x} \\ N_{1,r}(r) & N_1(r) & N_{2,r}(r) & N_2(r) & N_{3,r}(r) & N_3(r) \end{bmatrix} \left\{ \begin{matrix} u_1^0 \\ \phi_{z1} \\ u_2^0 \\ \phi_{z2} \\ u_3^0 \\ \phi_{z3} \end{matrix} \right\} \qquad (4.62)$$

The following matrices $\mathbf{B}_{bi}$ (related to the bending part) and the matrix $\mathbf{B}_{si}$ (related to the shear part) can be introduced

$$\mathbf{B}_{b\tau i} = \begin{bmatrix} 0 & F_{\tau y}\mathbf{J}^{-1}N_{1,r}(r) & 0 & F_{\tau y}\mathbf{J}^{-1}N_{2,r}(r) & 0 & F_{\tau y}\mathbf{J}^{-1}N_{3,r}(r) \end{bmatrix} \qquad (4.63)$$

$$\mathbf{B}_{S_{\tau i}} = \begin{bmatrix} F_{\tau x}\mathbf{J}^{-1} & F_{\tau y,x} & F_{\tau x}\mathbf{J}^{-1} & F_{\tau y,x} & F_{\tau x}\mathbf{J}^{-1} & F_{\tau y,x} \\ N_{1,r}(r) & N_1(r) & N_{2,r}(r) & N_2(r) & N_{3,r}(r) & N_3(r) \end{bmatrix} \quad (4.64)$$

Introducing the expansion functions F_{sx} and F_{sy} for the virtual variations

$$\mathbf{B}_{b_{sj}} = \begin{bmatrix} 0 & F_{sy}\mathbf{J}^{-1}N_{1,r}(r) & 0 & F_{sy}\mathbf{J}^{-1}N_{2,r}(r) & 0 & F_{sy}\mathbf{J}^{-1}N_{3,r}(r) \end{bmatrix} \quad (4.65)$$

$$\mathbf{B}_{S_{sj}} = \begin{bmatrix} F_{sx}\mathbf{J}^{-1} & F_{sy,x} & F_{sx}\mathbf{J}^{-1} & F_{sy,x} & F_{sx}\mathbf{J}^{-1} & F_{sy,x} \\ N_{1,r}(r) & N_1(r) & N_{2,r}(r) & N_2(r) & N_{3,r}(r) & N_3(r) \end{bmatrix} \quad (4.66)$$

The bending part of the stiffness matrix $\mathbf{K}_{b_{B3}}$ can be calculated as follows, using the previous expressions for $\mathbf{B}_{b_{sj}}$ and $\mathbf{B}_{b_{\tau i}}$

$$\mathbf{K}_{b_{B3}} = \int_V \mathbf{B}_{bj}^T \mathbf{C}_b \mathbf{B}_{bi} \, dV \quad (4.67)$$

while the shear part of the stiffness matrix $\mathbf{K}_{S_{B3}}$ can be calculated as follows, using the previous expressions of $\mathbf{B}_{S_{sj}}$ and $\mathbf{B}_{S_{\tau i}}$

$$\mathbf{K}_{S_{B3}} = \int_V \mathbf{B}_{sj}^T \mathbf{C}_s \mathbf{B}_{si} \, dV \quad (4.68)$$

In the previous equations the matrices $\mathbf{C}_b$ and $\mathbf{C}_s$ are the bending and shear terms of the material coefficients matrix $\mathbf{C}$ (see Eq. (3.12)). These values are the following

$$\mathbf{C}_b = C_{11}$$

$$\quad (4.69)$$

$$\mathbf{C}_s = C_{44}$$

Considering Eqs. (4.63) and (4.65), the bending part of the stiffness matrix becomes

$$\mathbf{K}_{b_{B3}} = \int_V \begin{bmatrix} 0 \\ F_{sy}\mathbf{J}^{-1}N_{1,r}(r) \\ 0 \\ F_{sy}\mathbf{J}^{-1}N_{2,r}(r) \\ 0 \\ F_{sy}\mathbf{J}^{-1}N_{3,r}(r) \end{bmatrix} C_{11} \begin{bmatrix} 0 & F_{\tau y}\mathbf{J}^{-1} & 0 & F_{\tau y}\mathbf{J}^{-1} & 0 & F_{\tau y}\mathbf{J}^{-1} \\ & N_{1,r}(r) & & N_{2,r}(r) & & N_{3,r}(r) \end{bmatrix} dV \quad (4.70)$$

Considering Eq. (4.64) and (4.66), the shear part of the stiffness matrix becomes

$$
\mathbf{K}_{sB3} = \int_V
\begin{bmatrix}
F_{sx}\mathbf{J}^{-1}N_{1,r}(r) \\[2mm]
F_{sy,x}N_1(r) \\[2mm]
F_{sx}\mathbf{J}^{-1}N_{2,r}(r) \\[2mm]
F_{sy,x}N_2(r) \\[2mm]
F_{sx}\mathbf{J}^{-1}N_{3,r}(r) \\[2mm]
F_{sy,x}N_3(r)
\end{bmatrix}
C_{44}
\begin{bmatrix}
F_{\tau x}\mathbf{J}^{-1} & F_{\tau y,x} & F_{\tau x}\mathbf{J}^{-1} & F_{\tau y,x} & F_{\tau x}\mathbf{J}^{-1} & F_{\tau y,x} \\
N_{1,r}(r) & N_1(r) & N_{2,r}(r) & N_2(r) & N_{3,r}(r) & N_3(r)
\end{bmatrix} dV
$$

$$(4.71)$$

The following relations have to be taken into account

Expansion and shape functions

$$
F_{\tau x} = -1; \qquad F_{\tau y,x} = 1; \qquad F_{sx} = -1; \qquad F_{sy,x} = 1; \qquad F_{\tau y} = x
$$

$$
F_{sy} = x; \qquad F_{\tau x}F_{sx} = 1; \qquad F_{\tau y}F_{sy} = x^2; \qquad F_{\tau y,x}F_{sx} = -1
$$

$$
F_{\tau x}F_{sy,x} = -1; \qquad F_{\tau y,x}F_{sy,x} = 1
$$

$$
det\,\mathbf{J} = [\tfrac{L}{2}] \quad ; \quad \mathbf{J}^{-1} = \left[\tfrac{2}{L}\right]
$$

$$
N_1(r) = \tfrac{1}{2}r\,(r-1) \quad ; \quad N_{1,r}(r) = \frac{\partial}{\partial r}\left[\tfrac{1}{2}r\,(r-1)\right] = r - \tfrac{1}{2}
$$

$$
N_2(r) = (1+r)\,(1-r) \quad ; \quad N_{2,r}(r) = \frac{\partial}{\partial r}[(1+r)\,(1-r)] = -2r
$$

$$
N_3(r) = \tfrac{1}{2}r\,(r+1) \quad ; \quad N_{3,r}(r) = \frac{\partial}{\partial r}\left[\tfrac{1}{2}r\,(r+1)\right] = r + \tfrac{1}{2}
$$

$$
N_1(r)\mathbf{J}^{-1}N_{1,r}(r) = \tfrac{1}{2}r\,(r-1)\left[\tfrac{2}{L}\right]\left(r - \tfrac{1}{2}\right) = \frac{r^3}{L} - \frac{3}{2}\frac{r^2}{L} + \frac{1}{2}\frac{r}{L}
$$

$$
\mathbf{J}^{-1}N_{1,r}(r)\mathbf{J}^{-1}N_{1,r}(r) = \left[\tfrac{2}{L}\right]\left(r - \tfrac{1}{2}\right)\left[\tfrac{2}{L}\right]\left(r - \tfrac{1}{2}\right) = 4\frac{r^2}{L^2} - 4\frac{r}{L^2} + \frac{1}{L^2}
$$

$$
\mathbf{J}^{-1}N_{1,r}(r)\mathbf{J}^{-1}N_{2,r}(r) = \left[\tfrac{2}{L}\right]\left(r - \tfrac{1}{2}\right)\left[\tfrac{2}{L}\right](-2r) = -8\frac{r^2}{L^2} + 4\frac{1}{L^2}r
$$

$$
\mathbf{J}^{-1}N_{1,r}(r)N_2(r) = \left[\tfrac{2}{L}\right]\left(r - \tfrac{1}{2}\right)(1+r)\,(1-r) = -2\frac{r^3}{L} + \frac{r^2}{L} + 2\frac{r}{L} - \frac{1}{L}
$$

$$
\mathbf{J}^{-1}N_{1,r}(r)[\mathbf{J}]^{-1}N_{3,r}(r) = \left[\tfrac{2}{L}\right]\left(r - \tfrac{1}{2}\right)\left[\tfrac{2}{L}\right]\left(r + \tfrac{1}{2}\right) = 4\frac{r^2}{L^2} - \frac{1}{L^2}
$$

$$
\mathbf{J}^{-1}N_{1,r}(r)N_3(r) = \left[\tfrac{2}{L}\right]\left(r - \tfrac{1}{2}\right)\left(\tfrac{1}{2}r\,(r+1)\right) = \frac{r^3}{L} + \frac{1}{2}\frac{r^2}{L} - \frac{1}{2}\frac{r}{L}
$$

Expansion and shape functions

$$N_1(r)N_1(r) = \frac{1}{2}r(r-1)\,\frac{1}{2}r(r-1) = \frac{1}{4}r^4 - \frac{1}{2}r^3 + \frac{1}{4}r^2$$

$$N_1(r)\mathbf{J}^{-1}N_{2,r}(r) = \frac{1}{2}r(r-1)\left[\frac{2}{L}\right](-2r) = -2\frac{r^3}{L} + 2\frac{r^2}{L}$$

$$N_1(r)N_2(r) = \frac{1}{2}r(r-1)(1+r)(1-r) = -\frac{1}{2}r^4 + \frac{1}{2}r^3 + \frac{1}{2}r^2 - \frac{1}{2}r$$

$$N_1(r)\mathbf{J}^{-1}N_{3,r}(r) = \frac{1}{2}r(r-1)\left[\frac{2}{L}\right]\left(r+\frac{1}{2}\right) = \frac{r^3}{L} - \frac{1}{2}\frac{r^2}{L} - \frac{1}{2}\frac{r}{L}$$

$$N_1(r)N_3(r) = \frac{1}{2}r(r-1)\,\frac{1}{2}r(r+1) = \frac{1}{4}r^4 - \frac{1}{4}r^2$$

$$\mathbf{J}^{-1}N_{2,r}(r)\mathbf{J}^{-1}N_{2,r}(r) = \left[\frac{2}{L}\right](-2r)\left[\frac{2}{L}\right](-2r) = 16\frac{r^2}{L^2}$$

$$\mathbf{J}^{-1}N_{2,r}(r)N_2(r) = \left[\frac{2}{L}\right](-2r)(1+r)(1-r) = 4\frac{r^3}{L} - 4\frac{r}{L}$$

$$\mathbf{J}^{-1}N_{2,r}(r)\mathbf{J}^{-1}N_{3,r}(r) = \left[\frac{2}{L}\right](-2r)\left[\frac{2}{L}\right]\left(r+\frac{1}{2}\right) = -8\frac{r^2}{L^2} - 4\frac{r}{L^2}$$

$$\mathbf{J}^{-1}N_{2,r}(r)N_3(r) = \left[\frac{2}{L}\right](-2r)\left(\frac{1}{2}r(r+1)\right) = -2\frac{r^3}{L} - 2\frac{r^2}{L}$$

$$N_2(r)N_2(r)) = ((1+r)(1-r))((1+r)(1-r)) = r^4 - 2r^2 + 1$$

$$N_2(r)\mathbf{J}^{-1}N_{3,r}(r) = ((1+r)(1-r))\left[\frac{2}{L}\right]\left(r+\frac{1}{2}\right) = -2\frac{r^3}{L} - \frac{r^2}{L} + 2\frac{r}{L} + \frac{1}{L}$$

$$N_2(r)N_3(r) = ((1+r)(1-r))\left(\frac{1}{2}r(r+1)\right) = -\frac{1}{2}r^4 - \frac{1}{2}r^3 + \frac{1}{2}r^2 + \frac{1}{2}r$$

$$\mathbf{J}^{-1}N_{3,r}(r)\mathbf{J}^{-1}N_{3,r}(r) = \left[\frac{2}{L}\right]\left(r+\frac{1}{2}\right)\left[\frac{2}{L}\right]\left(r+\frac{1}{2}\right) = 4\frac{r^2}{L^2} + 4\frac{r}{L^2} + \frac{1}{L^2}$$

$$\mathbf{J}^{-1}N_{3,r}(r)N_3(r) = \left[\frac{2}{L}\right]\left(r+\frac{1}{2}\right)\left(\frac{1}{2}r(r+1)\right) = \frac{r^3}{L} + \frac{3}{2}\frac{r^2}{L} + \frac{1}{2}\frac{r}{L}$$

$$N_3(r)N_3(r) = \left(\frac{1}{2}r(r+1)\right)\left(\frac{1}{2}r(r+1)\right) = \frac{1}{4}r^4 + \frac{1}{2}r^3 + \frac{1}{4}r^2$$

4.4.1.1 FI Gauss Quadrature for B3 Element

The bending and the shear stiffness matrices are computed using FI Gauss Quadrature. For this purpose, the considerations made in Sects. 3.7.3 and 3.7.4 apply. From the previous relations, it is evident that the highest degree of the polynomial is $n = 4$.

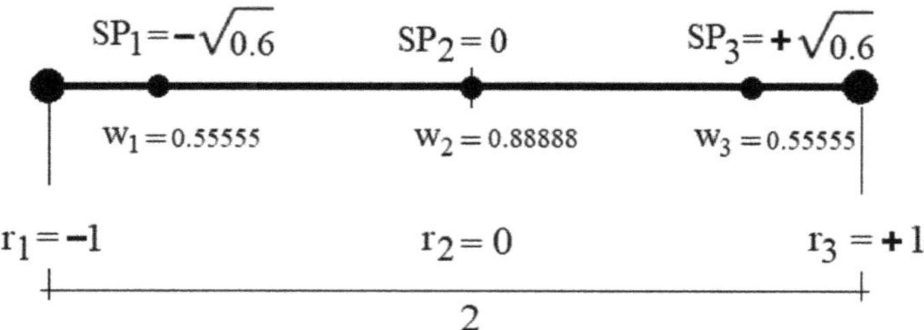

Fig. 4.4 Position of the three sampling points along the B3 beam and related weights

Recalling the relation of Eq. (3.100), the minimum number of SPs to perform a FI Gauss Quadrature is obtained.

$$n \leqslant 2SP - 1 \qquad SP \geqslant \frac{(n+1)}{2} \qquad SP \geqslant \frac{(4+1)}{2} \qquad SP \geqslant 2.5 \qquad (4.72)$$

As a result, the minimum number of SPs for the FI Gauss Quadrature is 3. Figure 4.4 shows the positions of the three SPs along the B3 beam and the related weights. The relations shown in Sect. 3.7.3 must be considered to perform the FI Gauss Quadrature. In particular, Eqs. (3.91), (3.92) and (3.93).

4.4.1.2   Bending Stiffness Matrix of the B3 Element

The stiffness matrix can be written as follows

$$\mathbf{K}_{b_{B3}} = \int_V \begin{bmatrix} k_{11b_{B3}} & k_{12b_{B3}} & k_{13b_{B3}} & k_{14b_{B3}} & k_{15b_{B3}} & k_{16b_{B3}} \\ k_{21b_{B3}} & k_{22b_{B3}} & k_{23b_{B3}} & k_{24b_{B3}} & k_{25b_{B3}} & k_{26b_{B3}} \\ k_{31b_{B3}} & k_{32b_{B3}} & k_{33b_{B3}} & k_{34b_{B3}} & k_{35b_{B3}} & k_{36b_{B3}} \\ k_{41b_{B3}} & k_{42b_{B3}} & k_{43b_{B3}} & k_{44b_{B3}} & k_{45b_{B3}} & k_{46b_{B3}} \\ k_{51b_{B3}} & k_{52b_{B3}} & k_{53b_{B3}} & k_{54b_{B3}} & k_{55b_{B3}} & k_{56b_{B3}} \\ k_{61b_{B3}} & k_{62b_{B3}} & k_{63b_{B3}} & k_{64b_{B3}} & k_{65b_{B3}} & k_{66b_{B3}} \end{bmatrix} dV \qquad (4.73)$$

The non-zero terms of the previous matrix are those placed in positions: 22, 24, 26, 42, 44, 46, 62, 64, 66. The previous relations of shape functions are recalled to compute the terms of the bending stiffness matrix. Performing the triple product of the volume integral of Eq. (4.70), the non-zero terms of Eq. (4.73) are calculated as follows

$$k_{22bB3} = \int_V \left(C_{11} F_{sy} F_{\tau y} \mathbf{J}^{-1} N_{1,r}(r) \mathbf{J}^{-1} N_{1,r}(r) \right) dV$$

$$= C_{11} \int_A F_{sy} F_{\tau y} dA \int_L \mathbf{J}^{-1} N_{1,r}(r) \mathbf{J}^{-1} N_{1,r}(r) dL$$

$$= C_{11} \int_{-a}^{a} F_{sy} F_{\tau y} dx \int_{-b}^{b} dz \int_{-1}^{1} \mathbf{J}^{-1} N_{1,r}(r) \mathbf{J}^{-1} N_{1,r}(r) det \, \mathbf{J} dr$$

$$= C_{11} \int_{-a}^{a} x^2 dx \int_{-b}^{b} dz \int_{-1}^{1} \left(4\frac{r^2}{L^2} - 4\frac{r}{L^2} + \frac{1}{L^2} \right) \frac{L}{2} dr$$

$$(4.74)$$

It should be noted that, in the previous relation, on the basis of the following relations

$$y = \frac{L}{2}(r+1); \quad dy = \frac{L}{2}dr; \quad dy = det \, \mathbf{J} dr$$

$$r = \frac{2y}{L} - 1$$

$$y = 0 \qquad r = -1 \qquad \text{and} \qquad y = L \qquad r = 1$$

$$(4.75)$$

the following equivalence was adopted

$$\int_L \mathbf{J}^{-1} N_{1,r}(r) \mathbf{J}^{-1} N_{1,r}(r) dL = \int_0^L \mathbf{J}^{-1} N_{1,r}(r) \mathbf{J}^{-1} N_{1,r}(r) dy =$$

$$= \int_{-1}^{1} \mathbf{J}^{-1} N_{1,r}(r) \mathbf{J}^{-1} N_{1,r}(r) det \, \mathbf{J} dr$$

$$(4.76)$$

From Eq. (4.76), the surface integral of Eq. (4.74) is computed using an analytical integration

$$C_{11} \int_{-a}^{a} x^2 dx \int_{-b}^{b} dz = C_{11} \frac{4a^3 b}{3} \tag{4.77}$$

and the line integral using the three SPs Gauss quadrature

$$\int_{-1}^{1} \left(4\frac{r^2}{L^2} - 4\frac{r}{L^2} + \frac{1}{L^2} \right) \frac{L}{2} dr = \left[0.55556 \left(4\frac{\left(-\sqrt{0.6}\right)^2}{L^2} - 4\frac{\left(-\sqrt{0.6}\right)}{L^2} + \frac{1}{L^2} \right) + \right.$$

$$+ 0.88889 \left(4\frac{(0)^2}{L^2} - 4\frac{(0)}{L^2} + \frac{1}{L^2} \right) +$$

$$\left. + 0.55556 \left(4\frac{\left(\sqrt{0.6}\right)^2}{L^2} - 4\frac{\left(\sqrt{0.6}\right)}{L^2} + \frac{1}{L^2} \right) \right] \frac{L}{2} = 2,3\frac{1}{L}$$

$$(4.78)$$

The term k_{22bB3} is then equal to

$$k_{22b_{B3}} = 3, \bar{1} C_{11} \frac{a^3 b}{L} \tag{4.79}$$

Subsequently, the other terms of the matrix of Eq. (4.73) are computed

$$
\begin{aligned}
k_{24b_{B3}} &= \int_V \left(C_{11} F_{sy} F_{\tau y} \mathbf{J}^{-1} N_{1,r}(r) \mathbf{J}^{-1} N_{2,r}(r) \right) dV \\
&= C_{11} \int_A F_{sy} F_{\tau y} dA \int_L \mathbf{J}^{-1} N_{1,r}(r) \mathbf{J}^{-1} N_{2,r}(r) dL \\
&= C_{11} \int_{-a}^{a} F_{sy} F_{\tau y} dx \int_{-b}^{b} dz \int_{-1}^{1} \mathbf{J}^{-1} N_{1,r}(r) \mathbf{J}^{-1} N_{2,r}(r) det\, \mathbf{J} dr \\
&= C_{11} \int_{-a}^{a} x^2 dx \int_{-b}^{b} dz \int_{-1}^{1} \left(-8\frac{r^2}{L^2} + 4\frac{1}{L^2}r \right) \frac{L}{2} dr
\end{aligned}
\tag{4.80}
$$

The surface integral of the term $k_{24b_{B3}}$ is equal to that of Eq. (4.77), and the line integral is computed as follows using the three SPs Gauss quadrature.

$$
\int_{-1}^{1} \left(-8\frac{r^2}{L^2} + 4\frac{1}{L^2}r \right) \frac{L}{2} dr = \left[0.55556 \left(-8\frac{\left(-\sqrt{0.6}\right)^2}{L^2} + 4\frac{1}{L^2}\left(-\sqrt{0.6}\right) \right) \right.
$$
$$
+ 0.88889 \left(-8\frac{(0)^2}{L^2} + 4\frac{1}{L^2}0 \right) + \tag{4.81}
$$
$$
\left. + 0.55556 \left(-8\frac{\left(\sqrt{0.6}\right)^2}{L^2} + 4\frac{1}{L^2}\sqrt{0.6} \right) \right] \frac{L}{2} = -2,\bar{6}\frac{1}{L}
$$

The term $k_{24b_{B3}}$ is then equal to

$$k_{24b_{B3}} = -3,\bar{5} C_{11} \frac{a^3 b}{L} \tag{4.82}$$

Regarding the $k_{26b_{B3}}$ term

$$
\begin{aligned}
k_{26b_{B3}} &= \int_V \left(C_{11} F_{sy} F_{\tau y} \mathbf{J}^{-1} N_{1,r}(r) \mathbf{J}^{-1} N_{3,r}(r) \right) dV \\
&= C_{11} \int_A F_{sy} F_{\tau y} dA \int_L \mathbf{J}^{-1} N_{1,r}(r) \mathbf{J}^{-1} N_{3,r}(r) dL \\
&= C_{11} \int_{-a}^{a} F_{sy} F_{\tau y} dx \int_{-b}^{b} dz \int_{-1}^{1} \mathbf{J}^{-1} N_{1,r}(r) \mathbf{J}^{-1} N_{3,r}(r) det\, \mathbf{J} dr \\
&= C_{11} \int_{-a}^{a} x^2 dx \int_{-b}^{b} dz \int_{-1}^{1} \left(4\frac{r^2}{L^2} - \frac{1}{L^2} \right) \frac{L}{2} dr
\end{aligned}
\tag{4.83}
$$

The surface integral of the term $k_{26b_{B3}}$ is equal to that of Eq. (4.77), and the line integral is computed as follows using the three SPs Gauss quadrature.

$$\int_{-1}^{1}\left(4\frac{r^2}{L^2}-\frac{1}{L^2}\right)\frac{L}{2}dr=\left[0.55556\left(4\frac{\left(-\sqrt{0.6}\right)^2}{L^2}-\frac{1}{L^2}\right)+0.88889\left(4\frac{(0)^2}{L^2}-\frac{1}{L^2}\right)+\right.$$

$$\left.+0.55556\left(4\frac{\left(\sqrt{0.6}\right)^2}{L^2}-\frac{1}{L^2}\right)\right]\frac{L}{2}=0,\overline{3}\frac{1}{L}$$

$$(4.84)$$

The term $k_{26b_{B3}}$ is then equal to

$$k_{26b_{B3}}=0,\overline{4}C_{11}\frac{a^3b}{L} \tag{4.85}$$

Regarding the $k_{42b_{B3}}$ term

$$k_{42b_{B3}}=\int_{V}\left(C_{11}F_{sy}F_{\tau y}\mathbf{J}^{-1}N_{2,r}(r)\mathbf{J}^{-1}N_{1,r}(r)\right)dV$$

$$=C_{11}\int_{A}F_{sy}F_{\tau y}dA\int_{L}\mathbf{J}^{-1}N_{2,r}(r)\mathbf{J}^{-1}N_{1,r}(r)dL$$

$$=C_{11}\int_{-a}^{a}F_{sy}F_{\tau y}dx\int_{-b}^{b}dz\int_{-1}^{1}\mathbf{J}^{-1}N_{2,r}(r)\mathbf{J}^{-1}N_{1,r}(r)det\,\mathbf{J}dr$$

$$=C_{11}\int_{-a}^{a}x^2dx\int_{-b}^{b}dz\int_{-1}^{1}\left(4\frac{1}{L^2}r-8\frac{r^2}{L^2}\right)\frac{L}{2}dr \tag{4.86}$$

The term $k_{42b_{B3}}$ is then equal to the term $k_{24b_{B3}}$ of Eq. (4.82) and it is equal to

$$k_{42b_{B3}}=-3,\overline{5}C_{11}\frac{a^3b}{L} \tag{4.87}$$

Regarding the $k_{44b_{B3}}$ term

$$k_{44b_{B3}}=\int_{V}\left(C_{11}F_{sy}F_{\tau y}\mathbf{J}^{-1}N_{2,r}(r)\mathbf{J}^{-1}N_{2,r}(r)\right)dV$$

$$=C_{11}\int_{A}F_{sy}F_{\tau y}dA\int_{L}\mathbf{J}^{-1}N_{2,r}(r)\mathbf{J}^{-1}N_{2,r}(r)dL$$

$$=C_{11}\int_{-a}^{a}F_{sy}F_{\tau y}dx\int_{-b}^{b}dz\int_{-1}^{1}\mathbf{J}^{-1}N_{2,r}(r)\mathbf{J}^{-1}N_{2,r}(r)det\,\mathbf{J}dr$$

$$=C_{11}\int_{-a}^{a}x^2dx\int_{-b}^{b}dz\int_{-1}^{1}\left(16\frac{r^2}{L^2}\right)\frac{L}{2}dr \tag{4.88}$$

The surface integral of the term $k_{44b_{B3}}$ is equal to that of Eq. (4.77), and the line integral is computed as follows using the three SPs Gauss quadrature.

$$\int_{-1}^{1}\left(16\frac{r^2}{L^2}\right)\frac{L}{2}dr = \left[0.55556\left(16\frac{\left(-\sqrt{0.6}\right)^2}{L^2}\right) + 0.88889\left(16\frac{(0)^2}{L^2}\right) + \right.$$

$$\left. +0.55556\left(16\frac{\left(\sqrt{0.6}\right)^2}{L^2}\right)\right]\frac{L}{2} = 7,\overline{1}\frac{1}{L} \tag{4.89}$$

The term $k_{44b_{B3}}$ is then equal to

$$k_{44b_{B3}} = 7,\overline{1}C_{11}\frac{a^3b}{L} \tag{4.90}$$

Regarding the $k_{46b_{B3}}$ term

$$
\begin{aligned}
k_{46b_{B3}} &= \int_V \left(C_{11}F_{sy}F_{\tau y}\mathbf{J}^{-1}N_{2,r}(r)\mathbf{J}^{-1}N_{3,r}(r)\right)dV \\
&= C_{11}\int_A F_{sy}F_{\tau y}dA\int_L \mathbf{J}^{-1}N_{2,r}(r)\mathbf{J}^{-1}N_{3,r}(r)dL \\
&= C_{11}\int_{-a}^{a}F_{sy}F_{\tau y}dx\int_{-b}^{b}dz\int_{-1}^{1}\mathbf{J}^{-1}N_{2,r}(r)\mathbf{J}^{-1}N_{3,r}(r)det\,\mathbf{J}dr \\
&= C_{11}\int_{-a}^{a}x^2dx\int_{-b}^{b}dz\int_{-1}^{1}\left(-8\frac{r^2}{L^2}-4\frac{r}{L^2}\right)\frac{L}{2}dr
\end{aligned}
\tag{4.91}
$$

The surface integral of the term $k_{46b_{B3}}$ is equal to that of Eq. (4.77), and the line integral is computed as follows using the three SPs Gauss quadrature.

$$\int_{-1}^{1}\left(-8\frac{r^2}{L^2}-4\frac{r}{L^2}\right)\frac{L}{2}dr = \left[0.55556\left(-8\frac{\left(-\sqrt{0.6}\right)^2}{L^2}-4\frac{1}{L^2}\left(-\sqrt{0.6}\right)\right) + 0.88889\left(-8\frac{(0)^2}{L^2}-4\frac{1}{L^2}0\right) + \right.$$

$$\left. +0.55556\left(-8\frac{\left(\sqrt{0.6}\right)^2}{L^2}-4\frac{1}{L^2}\sqrt{0.6}\right)\right]\frac{L}{2} = -2,\overline{6}\frac{1}{L} \tag{4.92}$$

The term $k_{46b_{B3}}$ is then equal to

$$k_{46b_{B3}} = -3,\overline{5}C_{11}\frac{a^3b}{L} \tag{4.93}$$

Regarding the $k_{62b_{B3}}$ term

$$k_{62b_{B3}} = \int_V \left(C_{11} F_{sy} F_{\tau y} \mathbf{J}^{-1} N_{3,r}(r) \mathbf{J}^{-1} N_{1,r}(r) \right) dV$$

$$= C_{11} \int_A F_{sy} F_{\tau y} dA \int_L \mathbf{J}^{-1} N_{3,r}(r) \mathbf{J}^{-1} N_{1,r}(r)(r) dL$$

$$= C_{11} \int_{-a}^{a} F_{sy} F_{\tau y} dx \int_{-b}^{b} dz \int_{-1}^{1} \mathbf{J}^{-1} N_{3,r}(r) \mathbf{J}^{-1} N_{1,r}(r) det\,\mathbf{J} dr$$

$$= C_{11} \int_{-a}^{a} x^2 dx \int_{-b}^{b} dz \int_{-1}^{1} \left(-\frac{1}{L^2} + 4\frac{r^2}{L^2} \right) \frac{L}{2} dr$$

(4.94)

The term $k_{62b_{B3}}$ is then equal to the term $k_{26b_{B3}}$ of Eq. (4.85) and it is equal to

$$k_{62b_{B3}} = 0,\overline{4} C_{11} \frac{a^3 b}{L}$$

(4.95)

Regarding the $k_{64b_{B3}}$ term

$$k_{64b_{B3}} = \int_V \left(C_{11} F_{sy} F_{\tau y} \mathbf{J}^{-1} N_{3,r}(r) \mathbf{J}^{-1} N_{2,r}(r) \right) dV$$

$$= C_{11} \int_A F_{sy} F_{\tau y} dA \int_L \mathbf{J}^{-1} N_{3,r}(r) \mathbf{J}^{-1} N_{2,r}(r)(r) dL$$

$$= C_{11} \int_{-a}^{a} F_{sy} F_{\tau y} dx \int_{-b}^{b} dz \int_{-1}^{1} \mathbf{J}^{-1} N_{3,r}(r) \mathbf{J}^{-1} N_{2,r}(r) det\,\mathbf{J} dr$$

$$= C_{11} \int_{-a}^{a} x^2 dx \int_{-b}^{b} dz \int_{-1}^{1} \left(-4\frac{r}{L^2} - 8\frac{r^2}{L^2} \right) \frac{L}{2} dr$$

(4.96)

The term $k_{64b_{B3}}$ is then equal to the term $k_{46b_{B3}}$ of Eq. (4.93) and it is equal to

$$k_{64b_{B3}} = -3,\overline{5} C_{11} \frac{a^3 b}{L}$$

(4.97)

Regarding the $k_{66b_{B3}}$ term

$$k_{66b_{B3}} = \int_V \left(C_{11} F_{sy} F_{\tau y} \mathbf{J}^{-1} N_{3,r}(r) \mathbf{J}^{-1} N_{3,r}(r) \right) dV$$

$$= C_{11} \int_A F_{sy} F_{\tau y} dA \int_L \mathbf{J}^{-1} N_{3,r}(r) \mathbf{J}^{-1} N_{3,r}(r) dL$$

$$= C_{11} \int_{-a}^{a} F_{sy} F_{\tau y} dx \int_{-b}^{b} dz \int_{-1}^{1} \mathbf{J}^{-1} N_{3,r}(r) \mathbf{J}^{-1} N_{3,r}(r) det\,\mathbf{J} dr$$

$$= C_{11} \int_{-a}^{a} x^2 dx \int_{-b}^{b} dz \int_{-1}^{1} \left(4\frac{r^2}{L^2} + 4\frac{r}{L^2} + \frac{1}{L^2} \right) \frac{L}{2} dr$$

(4.98)

The surface integral of the term $k_{66b_{B3}}$ is equal to that of Eq. (4.77), and the line integral is computed as follows using the three SPs Gauss quadrature.

$$\int_{-1}^{1}\left(4\frac{r^2}{L^2}+4\frac{r}{L^2}+\frac{1}{L^2}\right)\frac{L}{2}dr = \left[0.55556\left(4\frac{\left(-\sqrt{0.6}\right)^2}{L^2}+4\frac{\left(-\sqrt{0.6}\right)}{L^2}+\frac{1}{L^2}\right)\right.$$
$$+0.88889\left(4\frac{(0)^2}{L^2}+4\frac{(0)}{L^2}+\frac{1}{L^2}\right)+$$
$$\left.+0.55556\left(4\frac{\left(\sqrt{0.6}\right)^2}{L^2}+4\frac{\left(\sqrt{0.6}\right)}{L^2}+\frac{1}{L^2}\right)\right]\frac{L}{2}=2,\overline{3}\frac{1}{L}$$

$$(4.99)$$

The term $k_{66b_{B3}}$ is then equal to

$$k_{66b_{B3}} = 3,\overline{1}C_{11}\frac{a^3b}{L} \tag{4.100}$$

Based on the terms computed previously, the bending stiffneess matrix of the B3 beam element, obtained using a FI Gauss quadrature, is the following

$$\mathbf{K}_{b_{B3}} = \begin{bmatrix} 0 & 0 & 0 & 0 & 0 & 0 \\ 0 & 3,\overline{1}C_{11}\dfrac{a^3b}{L} & 0 & -3,\overline{5}C_{11}\dfrac{a^3b}{L} & 0 & 0,\overline{4}C_{11}\dfrac{a^3b}{L} \\ 0 & 0 & 0 & 0 & 0 & 0 \\ 0 & -3,\overline{5}C_{11}\dfrac{a^3b}{L} & 0 & 7,\overline{1}C_{11}\dfrac{a^3b}{L} & 0 & -3,\overline{5}C_{11}\dfrac{a^3b}{L} \\ 0 & 0 & 0 & 0 & 0 & 0 \\ 0 & 0,\overline{4}C_{11}\dfrac{a^3b}{L} & 0 & -3,\overline{5}C_{11}\dfrac{a^3b}{L} & 0 & 3,\overline{1}C_{11}\dfrac{a^3b}{L} \end{bmatrix} dV \tag{4.101}$$

4.4.1.3 Shear Stiffness Matrix of the B3 Element

As stated at the beginning of the Sect. 4.4.1.1, the computation of the shear stiffness matrix must be made using a three SPs FI Gauss quadrature. Figure 4.4 shows the positions of the SPs along the B3 beam and the related weights.

The shear stiffness matrix can be written, in a general form, as follows

$$
\mathbf{K}_{SB3} =
\begin{bmatrix}
k_{11s\,B3} & k_{12s\,B3} & k_{13s\,B3} & k_{14s\,B3} & k_{15s\,B3} & k_{16s\,B3} \\
k_{21s\,B3} & k_{22s\,B3} & k_{23s\,B3} & k_{24s\,B3} & k_{25s\,B3} & k_{26s\,B3} \\
k_{31s\,B3} & k_{32s\,B3} & k_{33s\,B3} & k_{34s\,B3} & k_{35s\,B3} & k_{36s\,B3} \\
k_{41s\,B3} & k_{42s\,B3} & k_{43s\,B3} & k_{44s\,B3} & k_{45s\,B3} & k_{46s\,B3} \\
k_{51s\,B3} & k_{52s\,B3} & k_{53s\,B3} & k_{54s\,B3} & k_{55s\,B3} & k_{56s\,B3} \\
k_{61s\,B3} & k_{62s\,B3} & k_{63s\,B3} & k_{64s\,B3} & k_{65s\,B3} & k_{66s\,B3}
\end{bmatrix}
\tag{4.102}
$$

Performing the triple product on the volume integral of Eq. (4.71), the terms of the matrix in Eq. (4.102) can be written as follows

$$
\begin{aligned}
k_{11s\,B3} &= \int_V C_{44} F_{sx} \mathbf{J}^{-1} N_{1,r}(r) F_{\tau x} \mathbf{J}^{-1} N_{1,r}(r) dV \\
&= C_{44} \int_A F_{sx} F_{\tau x} dA \int_L \mathbf{J}^{-1} N_{1,r}(r) \mathbf{J}^{-1} N_{1,r}(r) dL \\
&= C_{44} \int_{-a}^{a} F_{sx} F_{\tau x} dx \int_{-b}^{b} dz \int_{-1}^{1} \mathbf{J}^{-1} N_{1,r}(r) \mathbf{J}^{-1} N_{1,r}(r) \det \mathbf{J} dr \\
&= C_{44} \int_{-a}^{a} (1)\, dx \int_{-b}^{b} dz \int_{-1}^{1} \left(4\frac{r^2}{L^2} - 4\frac{r}{L^2} + \frac{1}{L^2} \right) \frac{L}{2} dr
\end{aligned}
\tag{4.103}
$$

The surface integral of the term $k_{11s\,B3}$ is computed using an analytical approach, while the line integral is computed using the three SPs Gauss quadrature. The computation of the surface integral is done considering the relations of Eq. 4.4.1.

$$
C_{44} \int_{-a}^{a} 1\, dx \int_{-b}^{b} dz = C_{44} 4ab
\tag{4.104}
$$

The computation of the line integral is done using the three SPs Gauss quadrature

$$
\begin{aligned}
\int_{-1}^{1} \left(4\frac{r^2}{L^2} - 4\frac{r}{L^2} + \frac{1}{L^2} \right) \frac{L}{2} dr &= \left[0.55556 \left(4\frac{\left(-\sqrt{0.6}\right)^2}{L^2} - 4\frac{\left(-\sqrt{0.6}\right)}{L^2} + \frac{1}{L^2} \right) \right. \\
&\quad + 0.88889 \left(4\frac{(0)^2}{L^2} - 4\frac{(0)}{L^2} + \frac{1}{L^2} \right) + \\
&\quad \left. + 0.55556 \left(4\frac{\left(\sqrt{0.6}\right)^2}{L^2} - 4\frac{\left(\sqrt{0.6}\right)}{L^2} + \frac{1}{L^2} \right) \right] \frac{L}{2} = 2{,}\overline{3}\frac{1}{L}
\end{aligned}
\tag{4.105}
$$

The term $k_{11s\,B3}$ is then equal to

$$k_{11s_{B3}} = 9,\bar{3}C_{44}\frac{ab}{L} \tag{4.106}$$

Regarding the term $k_{12s_{B3}}$

$$
\begin{aligned}
k_{12s_{B3}} &= \int_V C_{44} F_{sx} \mathbf{J}^{-1} N_{1,r}(r) F_{\tau y,x} N_1(r) dV \\
&= C_{44} \int_A F_{\tau y,x} F_{sx} dA \int_L \mathbf{J}^{-1} N_{1,r}(r) N_1(r) dL \\
&= C_{44} \int_{-a}^{a} F_{\tau y,x} F_{sx} dx \int_{-b}^{b} dz \int_{-1}^{1} \mathbf{J}^{-1} N_{1,r}(r) N_1(r) det\, \mathbf{J} dr \\
&= C_{44} \int_{-a}^{a} (-1)\, dx \int_{-b}^{b} dz \int_{-1}^{1} \left(\frac{r^3}{L} - \frac{3}{2}\frac{r^2}{L} + \frac{1}{2}\frac{r}{L}\right)\frac{L}{2} dr
\end{aligned}
\tag{4.107}
$$

The surface integral of the term $k_{12s_{B3}}$ is computed using an analytical approach, while the line integral is computed using the three SPs Gauss quadrature. The computation of the surface integral is done considering the relations of Eq. 4.4.1.

$$C_{44} \int_{-a}^{a} (-1)\, dx \int_{-b}^{b} dz = -C_{44}4ab \tag{4.108}$$

The computation of the line integral is done using the three SPs Gauss quadrature

$$
\begin{aligned}
\int_{-1}^{1} \left(\frac{r^3}{L} - \frac{3}{2}\frac{r^2}{L} + \frac{1}{2}\frac{r}{L}\right)\frac{L}{2} dr = & \left[0.55556 \left(\frac{\left(-\sqrt{0.6}\right)^3}{L} - \frac{3}{2}\frac{\left(-\sqrt{0.6}\right)^2}{L} + \frac{1}{2}\frac{\left(-\sqrt{0.6}\right)}{L}\right) + \right. \\
& + 0.88889 \left(\frac{(0)^3}{L} - \frac{3}{2}\frac{(0)^2}{L} + \frac{1}{2}\frac{(0)}{L}\right) + \\
& \left. + 0.55556 \left(\frac{\left(\sqrt{0.6}\right)^3}{L} - \frac{3}{2}\frac{\left(\sqrt{0.6}\right)^2}{L} + \frac{1}{2}\frac{\left(\sqrt{0.6}\right)}{L}\right) \right]\frac{L}{2} = -0.5
\end{aligned}
\tag{4.109}
$$

The term $k_{12s_{B3}}$ is then equal to

$$k_{12s_{B3}} = 2C_{44}ab \tag{4.110}$$

Regarding the term $k_{13s_{B3}}$

$$
k_{13sB3} = \int_V C_{44} F_{sx} \mathbf{J}^{-1} N_{1,r}(r) F_{\tau x} \mathbf{J}^{-1} N_{2,r}(r) dV
$$

$$
= C_{44} \int_A F_{\tau x} F_{sx} dA \int_L \mathbf{J}^{-1} N_{1,r}(r) \mathbf{J}^{-1} N_{2,r}(r) dL
$$

$$
= C_{44} \int_{-a}^{a} F_{\tau x} F_{sx} dx \int_{-b}^{b} dz \int_{-1}^{1} \mathbf{J}^{-1} N_{1,r}(r) \mathbf{J}^{-1} N_{2,r}(r) det \mathbf{J} dr
$$

$$
= C_{44} \int_{-a}^{a} (-1) dx \int_{-b}^{b} dz \int_{-1}^{1} \left(-8\frac{r^2}{L^2} + 4\frac{1}{L^2}r \right) \frac{L}{2} dr
$$

(4.111)

The surface integral of the term k_{13sB3} is computed using an analytical approach, while the line integral is computed using the three SPs Gauss quadrature. The computation of the surface integral is done considering the relations of Eq. 4.4.1.

$$
C_{44} \int_{-a}^{a} (1) dx \int_{-b}^{b} dz = C_{44} 4ab
$$

(4.112)

The computation of the line integral is done using the three SPs Gauss quadrature

$$
\int_{-1}^{1} \left(-8\frac{r^2}{L^2} + 4\frac{1}{L^2}r \right) \frac{L}{2} dr = \left[0.55556 \left(-8\frac{\left(-\sqrt{0.6}\right)^2}{L^2} + 4\frac{\left(-\sqrt{0.6}\right)}{L^2} \right) \right.
$$

$$
+ 0.88889 \left(-8\frac{(0)^2}{L^2} + 4\frac{(0)^2}{L^2} \right)
$$

(4.113)

$$
\left. + 0.55556 \left(-8\frac{\left(\sqrt{0.6}\right)^2}{L^2} + 4\frac{\left(\sqrt{0.6}\right)}{L^2} \right) \right] \frac{L}{2} = -2,\bar{6}\frac{1}{L}
$$

The term k_{13sB3} is then equal to

$$
k_{13sB3} = -10,\bar{6}C_{44}\frac{ab}{L}
$$

(4.114)

Regarding the term k_{14sB3}

$$
k_{14sB3} = \int_V C_{44} F_{sx} \mathbf{J}^{-1} N_{1,r}(r) F_{\tau y,x} N_2(r) dV
$$

$$
= C_{44} \int_A F_{\tau y,x} F_{sx} dA \int_L \mathbf{J}^{-1} N_{1,r}(r) N_2(r) dL
$$

$$
= C_{44} \int_{-a}^{a} F_{\tau y,x} F_{sx} dx \int_{-b}^{b} dz \int_{-1}^{1} \mathbf{J}^{-1} N_{1,r}(r) N_2(r) det \mathbf{J} dr
$$

$$
= C_{44} \int_{-a}^{a} (-1) dx \int_{-b}^{b} dz \int_{-1}^{1} \left(-2\frac{r^3}{L} + \frac{r^2}{L} + 2\frac{r}{L} - \frac{1}{L} \right) \frac{L}{2} dr
$$

(4.115)

The surface integral of the term k_{14sB3} is computed using an analytical approach, while the line integral is computed using the three SPs Gauss quadrature. The com-

putation of the surface integral is done considering the relations of Eq. 4.4.1.

$$C_{44} \int_{-a}^{a} (-1)\, dx \int_{-b}^{b} dz = -C_{44} 4ab \tag{4.116}$$

The computation of the line integral is done using the three SPs Gauss quadrature

$$\int_{-1}^{1} \left(-2\frac{r^3}{L} + \frac{r^2}{L} + 2\frac{r}{L} - \frac{1}{L} \right) \frac{L}{2} dr = \left[0.55556 \left(-2\frac{\left(-\sqrt{0.6}\right)^3}{L} + \frac{\left(-\sqrt{0.6}\right)^2}{L} + 2\frac{\left(-\sqrt{0.6}\right)}{L} - \frac{1}{L} \right) + \right.$$

$$+ 0.88889 \left(-2\frac{(0)^3}{L} + \frac{(0)^2}{L} + 2\frac{(0)}{L} - \frac{1}{L} \right) \Bigg] +$$

$$+ \left[0.55556 \left(-2\frac{\left(\sqrt{0.6}\right)^3}{L} + \frac{\left(\sqrt{0.6}\right)^2}{L} + 2\frac{\left(\sqrt{0.6}\right)}{L} - \frac{1}{L} \right) \right] \frac{L}{2}$$

$$= -0,\overline{6} \tag{4.117}$$

The term $k_{14s_{B3}}$ is then equal to

$$k_{14s_{B3}} = 2,\overline{6}\,C_{44}ab \tag{4.118}$$

Regarding the term $k_{15s_{B3}}$

$$\begin{aligned}
k_{15s_{B3}} &= \int_{V} C_{44} F_{sx} \mathbf{J}^{-1} N_{1,r}(r) F_{\tau x} \mathbf{J}^{-1} N_{3,r}(r) dV \\
&= C_{44} \int_{A} F_{\tau x} F_{sx} dA \int_{L} \mathbf{J}^{-1} N_{1,r}(r) \mathbf{J}^{-1} N_{3,r}(r) dL \\
&= C_{44} \int_{-a}^{a} F_{\tau y,x} F_{sx} dx \int_{-b}^{b} dz \int_{-1}^{1} \mathbf{J}^{-1} N_{1,r}(r) \mathbf{J}^{-1} N_{3,r}(r) \det \mathbf{J} dr \\
&= C_{44} \int_{-a}^{a} (1)\, dx \int_{-b}^{b} dz \int_{-1}^{1} \left(4\frac{r^2}{L^2} - \frac{1}{L^2} \right) \frac{L}{2} dr
\end{aligned} \tag{4.119}$$

The surface integral of the term $k_{12s_{B3}}$ is computed using an analytical approach, while the line integral is computed using the three SPs Gauss quadrature. The computation of the surface integral is done considering the relations of Eq. 4.4.1.

$$C_{44} \int_{-a}^{a} (1)\, dx \int_{-b}^{b} dz = C_{44} 4ab \tag{4.120}$$

The computation of the line integral is done using the three SPs Gauss quadrature

$$\int_{-1}^{1}\left(4\frac{r^2}{L^2}-\frac{1}{L^2}\right)\frac{L}{2}dr=\left[0.55556\left(4\frac{\left(-\sqrt{0.6}\right)^2}{L^2}-\frac{1}{L^2}\right)+0.88889\left(4\frac{(0)^2}{L^2}-\frac{1}{L^2}\right)+\right.$$

$$\left.+0.55556\left(4\frac{\left(\sqrt{0.6}\right)^2}{L^2}-\frac{1}{L^2}\right)\right]\frac{L}{2}=0,\bar{3}\frac{1}{L}$$

$$(4.121)$$

The term $k_{15s_{B3}}$ is then equal to

$$k_{15s_{B3}}=1,\bar{3}C_{44}\frac{ab}{L}\tag{4.122}$$

Regarding the term $k_{16s_{B3}}$

$$
\begin{aligned}
k_{16s_{B3}}&=\int_V C_{44}F_{sx}\mathbf{J}^{-1}N_{1,r}(r)F_{\tau y,x}N_3(r)dV\\
&=C_{44}\int_A F_{\tau y,x}F_{sx}dA\int_L \mathbf{J}^{-1}N_{1,r}(r)N_3(r)dL\\
&=C_{44}\int_{-a}^{a}F_{\tau y,x}F_{sx}dx\int_{-b}^{b}dz\int_{-1}^{1}\mathbf{J}^{-1}N_{1,r}(r)N_3(r)det\,\mathbf{J}dr\\
&=C_{44}\int_{-a}^{a}(-1)\,dx\int_{-b}^{b}dz\int_{-1}^{1}\left(\frac{r^3}{L}+\frac{1}{2}\frac{r^2}{L}-\frac{1}{2}\frac{r}{L}\right)\frac{L}{2}dr
\end{aligned}
$$

$$(4.123)$$

The surface integral of the term $k_{12s_{B3}}$ is computed using an analytical approach, while the line integral is computed using the three SPs Gauss quadrature. The computation of the surface integral is done considering the relations of Eq. 4.4.1.

$$C_{44}\int_{-a}^{a}(-1)\,dx\int_{-b}^{b}dz=-C_{44}4ab\tag{4.124}$$

The computation of the line integral is done using the three SPs Gauss quadrature

$$\int_{-1}^{1}\left(\frac{r^3}{L}+\frac{1}{2}\frac{r^2}{L}-\frac{1}{2}\frac{r}{L}\right)\frac{L}{2}dr=\left[0.55556\left(\frac{\left(-\sqrt{0.6}\right)^3}{L}+\frac{1}{2}\frac{\left(-\sqrt{0.6}\right)^2}{L}-\frac{1}{2}\frac{\left(-\sqrt{0.6}\right)}{L}\right)+\right.$$

$$+0.88889\left(\frac{(0)^3}{L}+\frac{1}{2}\frac{(0)^2}{L}-\frac{1}{2}\frac{(0)}{L}\right)+$$

$$\left.+0.55556\left(\frac{\left(\sqrt{0.6}\right)^3}{L}+\frac{1}{2}\frac{\left(\sqrt{0.6}\right)^2}{L}-\frac{1}{2}\frac{\left(\sqrt{0.6}\right)}{L}\right)\right]\frac{L}{2}=0,\bar{16}$$

$$(4.125)$$

The term $k_{16s_{B3}}$ is then equal to

$$k_{16s\,B3} = -0,\overline{6}C_{44}ab \tag{4.126}$$

Regarding the term $k_{21s\,B3}$

$$
\begin{aligned}
k_{21s\,B3} &= \int_V C_{44} F_{sy,x} N_1(r) F_{\tau x} \mathbf{J}^{-1} N_{1,r}(r)\,dV \\
&= C_{44} \int_A F_{sy,x} F_{\tau x}\,dA \int_L \mathbf{J}^{-1} N_{1,r}(r) N_1(r)\,dL \\
&= C_{44} \int_{-a}^{a} F_{\tau y,x} F_{sx}\,dx \int_{-b}^{b} dz \int_{-1}^{1} \mathbf{J}^{-1} N_{1,r}(r) N_1(r)\,det\,\mathbf{J}\,dr \\
&= C_{44} \int_{-a}^{a} (-1)\,dx \int_{-b}^{b} dz \int_{-1}^{1} \left(\frac{r^3}{L} - \frac{3}{2}\frac{r^2}{L} + \frac{1}{2}\frac{r}{L} \right) \frac{L}{2}\,dr
\end{aligned}
\tag{4.127}
$$

The term $k_{21s\,B3}$ is equal to the term $k_{12s\,B3}$ of Eq. (4.110), and it is equal to

$$k_{21s\,B3} = 2C_{44}ab \tag{4.128}$$

Regarding the term $k_{22s\,B3}$

$$
\begin{aligned}
k_{22s\,B3} &= \int_V C_{44} F_{sy,x} N_1(r) F_{\tau y,x} N_1(r)\,dV \\
&= C_{44} \int_A F_{\tau y,x} F_{sy,x}\,dA \int_L N_1(r) N_1(r)\,dL \\
&= C_{44} \int_{-a}^{a} F_{\tau y,x} F_{sy,x}\,dx \int_{-b}^{b} dz \int_{-1}^{1} N_1(r) N_1(r)\,det\,\mathbf{J}\,dr \\
&= C_{44} \int_{-a}^{a} (1)\,dx \int_{-b}^{b} dz \int_{-1}^{1} \left(\frac{1}{4}r^4 - \frac{1}{2}r^3 + \frac{1}{4}r^2 \right) \frac{L}{2}\,dr
\end{aligned}
\tag{4.129}
$$

The surface integral of the term $k_{22s\,B3}$ is computed using an analytical approach, while the line integral is computed using the three SPs Gauss quadrature. The computation of the surface integral is done considering the relations of Eq. 4.4.1.

$$C_{44} \int_{-a}^{a} (1)\,dx \int_{-b}^{b} dz = C_{44}4ab \tag{4.130}$$

The computation of the line integral is done using the three SPs Gauss quadrature

$$\int_{-1}^{1} \left(\frac{1}{4}r^4 - \frac{1}{2}r^3 + \frac{1}{4}r^2 \right) \frac{L}{2} dr = \left[0.55556 \left(\frac{1}{4} \left(-\sqrt{0.6}\right)^4 - \frac{1}{2} \left(-\sqrt{0.6}\right)^3 + \frac{1}{4} \left(-\sqrt{0.6}\right)^2 \right) + \right.$$

$$+ 0.88889 \left(\frac{1}{4}0^4 - \frac{1}{2}0^3 + \frac{1}{4}0^2 \right) +$$

$$\left. + 0.55556 \left(\frac{1}{4} \left(\sqrt{0.6}\right)^4 - \frac{1}{2} \left(\sqrt{0.6}\right)^3 + \frac{1}{4} \left(\sqrt{0.6}\right)^2 \right) \right] \frac{L}{2} = 0,1\overline{3}L$$

$$(4.131)$$

The term $k_{22s\,B3}$ is then equal to

$$k_{22s\,B3} = 0,5\overline{3}C_{44}ab \qquad (4.132)$$

Regarding the term $k_{23s\,B3}$

$$k_{23s\,B3} = \int_V C_{44} F_{sy,x} N_1(r) F_{\tau x} \mathbf{J}^{-1} N_{2,r}(r) dV$$

$$= C_{44} \int_A F_{\tau x} F_{sy,x} dA \int_L N_1(r) \mathbf{J}^{-1} N_{2,r}(r) dL$$

$$= C_{44} \int_{-a}^{a} F_{\tau x} F_{sy,x} dx \int_{-b}^{b} dz \int_{-1}^{1} N_1(r) \mathbf{J}^{-1} N_{2,r}(r) det\,\mathbf{J} dr$$

$$= C_{44} \int_{-a}^{a} (-1)\,dx \int_{-b}^{b} dz \int_{-1}^{1} \left(-2\frac{r^3}{L} + 2\frac{r^2}{L} \right) \frac{L}{2} dr$$

$$(4.133)$$

The surface integral of the term $k_{23s\,B3}$ is computed using an analytical approach, while the line integral is computed using the three SPs Gauss quadrature. The computation of the surface integral is done considering the relations of Eq. 4.4.1.

$$C_{44} \int_{-a}^{a} (-1)\,dx \int_{-b}^{b} dz = -C_{44}4ab \qquad (4.134)$$

The computation of the line integral is done using the three SPs Gauss quadrature

$$\int_{-1}^{1} \left(-2\frac{r^3}{L} + 2\frac{r^2}{L} \right) \frac{L}{2} dr = \left[0.55556 \left(-2\frac{\left(-\sqrt{0.6}\right)^3}{L} + 2\frac{\left(-\sqrt{0.6}\right)^2}{L} \right) + 0.88889 \left(-2\frac{(0)^3}{L} + 2\frac{(0)^2}{L} \right) + \right.$$

$$\left. + 0.55556 \left(-2\frac{\left(\sqrt{0.6}\right)^3}{L} + 2\frac{\left(\sqrt{0.6}\right)^2}{L} \right) \right] \frac{L}{2} = 0,\overline{6}$$

$$(4.135)$$

The term $k_{23s\,B3}$ is then equal to

$$k_{23s\,B3} = -2,\overline{6}C_{44}ab \qquad (4.136)$$

Regarding the term $k_{24s\,B3}$

$$
\begin{aligned}
k_{24sB3} &= \int_V C_{44} F_{sy,x} N_1(r) F_{\tau y,x} N_2(r) dV \\
&= C_{44} \int_A F_{\tau y,x} F_{sy,x} dA \int_L N_1(r) N_2(r) dL \\
&= C_{44} \int_{-a}^{a} F_{\tau y,x} F_{sy,x} dx \int_{-b}^{b} dz \int_{-1}^{1} N_1(r) N_2(r) det\,\mathbf{J} dr \\
&= C_{44} \int_{-a}^{a} (1)\, dx \int_{-b}^{b} dz \int_{-1}^{1} \left(-\frac{1}{2}r^4 + \frac{1}{2}r^3 + \frac{1}{2}r^2 - \frac{1}{2}r \right) \frac{L}{2} dr
\end{aligned}
\tag{4.137}
$$

The surface integral of the term k_{24sB3} is computed using an analytical approach, while the line integral is computed using the three SPs Gauss quadrature. The computation of the surface integral is done considering the relations of Eq. 4.4.1.

$$
C_{44} \int_{-a}^{a} (1)\, dx \int_{-b}^{b} dz = C_{44} 4ab
\tag{4.138}
$$

The computation of the line integral is done using the three SPs Gauss quadrature

$$
\begin{aligned}
\int_{-1}^{1} \left(-\frac{1}{2}r^4 + \frac{1}{2}r^3 + \frac{1}{2}r^2 - \frac{1}{2}r \right) \frac{L}{2} dr &= \left[0.55556 \left(-\frac{1}{2}\left(-\sqrt{0.6}\right)^4 + \frac{1}{2}\left(-\sqrt{0.6}\right)^3 + \right. \right. \\
&\quad \left. \left. +\frac{1}{2}\left(-\sqrt{0.6}\right)^2 - \frac{1}{2}\left(-\sqrt{0.6}\right) \right) + \right. \\
&\quad \left. +0.88889 \left(-\frac{1}{2}(0)^4 + \frac{1}{2}(0)^3 + \frac{1}{2}(0)^2 - \frac{1}{2}(0) \right) \right] + \\
&\quad + \left[0.55556 \left(-\frac{1}{2}\left(\sqrt{0.6}\right)^4 + \frac{1}{2}\left(\sqrt{0.6}\right)^3 + \right. \right. \\
&\quad \left. \left. +\frac{1}{2}\left(\sqrt{0.6}\right)^2 - \frac{1}{2}\left(\sqrt{0.6}\right) \right) \right] \frac{L}{2} = 0,0\overline{6}
\end{aligned}
\tag{4.139}
$$

The term k_{24sB3} is then equal to

$$
k_{24sB3} = 0,2\overline{6} C_{44} ab
\tag{4.140}
$$

Regarding the term k_{25sB3}

$$
\begin{aligned}
k_{25sB3} &= \int_V C_{44} F_{sy,x} N_1(r) F_{\tau x} \mathbf{J}^{-1} N_{3,r}(r) dV \\
&= C_{44} \int_A F_{\tau x} F_{sy,x} dA \int_L N_1(r) \mathbf{J}^{-1} N_{3,r}(r) dL \\
&= C_{44} \int_{-a}^{a} F_{\tau x} F_{sy,x} dx \int_{-b}^{b} dz \int_{-1}^{1} N_1(r) \mathbf{J}^{-1} N_{3,r}(r) det\,\mathbf{J} dr \\
&= C_{44} \int_{-a}^{a} (-1)\, dx \int_{-b}^{b} dz \int_{-1}^{1} \left(\frac{r^3}{L} - \frac{1}{2}\frac{r^2}{L} - \frac{1}{2}\frac{r}{L} \right) \frac{L}{2} dr
\end{aligned}
\tag{4.141}
$$

The surface integral of the term k_{25sB3} is computed using an analytical approach, while the line integral is computed using the three SPs Gauss quadrature. The com-

putation of the surface integral is done considering the relations of Eq. 4.4.1.

$$C_{44} \int_{-a}^{a} (-1)\, dx \int_{-b}^{b} dz = -C_{44} 4ab \tag{4.142}$$

The computation of the line integral is done using the three SPs Gauss quadrature

$$\int_{-1}^{1} \left(\frac{r^3}{L} - \frac{1}{2}\frac{r^2}{L} - \frac{1}{2}\frac{r}{L} \right) \frac{L}{2}\, dr = \left[0.55556 \left(\frac{\left(-\sqrt{0.6}\right)^3}{L} - \frac{1}{2}\frac{\left(-\sqrt{0.6}\right)^2}{L} - \frac{1}{2}\frac{\left(-\sqrt{0.6}\right)}{L} \right) \right.$$

$$+ 0.88889 \left(\frac{(0)^3}{L} - \frac{1}{2}\frac{(0)^2}{L} - \frac{1}{2}\frac{(0)}{L} \right)$$

$$\left. + 0.55556 \left(\frac{\left(\sqrt{0.6}\right)^3}{L} - \frac{1}{2}\frac{\left(\sqrt{0.6}\right)^2}{L} - \frac{1}{2}\frac{\left(\sqrt{0.6}\right)}{L} \right) \right] \frac{L}{2} = -0,1\overline{6} \tag{4.143}$$

The term $k_{25s\,B3}$ is then equal to

$$k_{25s\,B3} = 0,\overline{6} C_{44} ab \tag{4.144}$$

Regarding the term $k_{26s\,B3}$

$$
\begin{aligned}
k_{26s\,B3} &= \int_V C_{44} F_{sy,x} N_1(r) F_{\tau y,x} N_3(r)\, dV \\
&= C_{44} \int_A F_{\tau y,x} F_{sy,x}\, dA \int_L N_1(r) N_3(r)\, dL \\
&= C_{44} \int_{-a}^{a} F_{\tau y,x} F_{sy,x}\, dx \int_{-b}^{b} dz \int_{-1}^{1} N_1(r) N_3(r)\, det\, \mathbf{J}\, dr \\
&= C_{44} \int_{-a}^{a} (1)\, dx \int_{-b}^{b} dz \int_{-1}^{1} \left(\frac{1}{4} r^4 - \frac{1}{4} r^2 \right) \frac{L}{2}\, dr
\end{aligned}
\tag{4.145}
$$

The surface integral of the term $k_{26s\,B3}$ is computed using an analytical approach, while the line integral is computed using the three SPs Gauss quadrature. The computation of the surface integral is done considering the relations of Eq. 4.4.1.

$$C_{44} \int_{-a}^{a} (1)\, dx \int_{-b}^{b} dz = C_{44} 4ab \tag{4.146}$$

The computation of the line integral is done using the three SPs Gauss quadrature

$$\int_{-1}^{1} \left(\frac{1}{4} r^4 - \frac{1}{4} r^2 \right) \frac{L}{2} dr = \left[0.55556 \left(\frac{1}{4} \left(-\sqrt{0.6} \right)^4 - \frac{1}{4} \left(-\sqrt{0.6} \right)^2 \right) + 0.88889 \left(\frac{1}{4} (0)^4 - \frac{1}{4} (0)^2 \right) + \right.$$

$$\left. + 0.55556 \left(\frac{1}{4} \left(\sqrt{0.6} \right)^4 - \frac{1}{4} \left(\sqrt{0.6} \right)^2 \right) \right] \frac{L}{2} = -0,0\overline{3}$$

$$\tag{4.147}$$

The term k_{26sB3} is then equal to

$$k_{26sB3} = -0,1\overline{3} C_{44} ab \tag{4.148}$$

Regarding the term k_{31sB3}

$$\begin{aligned}
k_{31sB3} &= \int_V C_{44} F_{sx} \mathbf{J}^{-1} N_{2,r}(r) F_{\tau x} \mathbf{J}^{-1} N_{1,r}(r) dV \\
&= C_{44} \int_A F_{\tau x} F_{sx} dA \int_L \mathbf{J}^{-1} N_{2,r}(r) \mathbf{J}^{-1} N_{1,r}(r) dL \\
&= C_{44} \int_{-a}^{a} F_{\tau x} F_{sx} dx \int_{-b}^{b} dz \int_{-1}^{1} \mathbf{J}^{-1} N_{2,r}(r) \mathbf{J}^{-1} N_{1,r}(r) \det \mathbf{J} dr \\
&= C_{44} \int_{-a}^{a} (-1) dx \int_{-b}^{b} dz \int_{-1}^{1} \left(-8 \frac{r^2}{L^2} + 4 \frac{1}{L^2} r \right) \frac{L}{2} dr
\end{aligned} \tag{4.149}$$

The term k_{31sB3} is equal to the term k_{13sB3} and it is equal to

$$k_{31sB3} = -10,\overline{6} C_{44} \frac{ab}{L} \tag{4.150}$$

Regarding the term k_{32sB3}

$$\begin{aligned}
k_{32sB3} &= \int_V C_{44} F_{sx} \mathbf{J}^{-1} N_{2,r}(r) F_{\tau y,x} N_1(r) dV \\
&= C_{44} \int_A F_{\tau y,x} F_{sx} dA \int_L \mathbf{J}^{-1} N_{2,r}(r) N_1(r) dL \\
&= C_{44} \int_{-a}^{a} F_{\tau y,x} F_{sx} dx \int_{-b}^{b} dz \int_{-1}^{1} \mathbf{J}^{-1} N_{2,r}(r) N_1(r) \det \mathbf{J} dr \\
&= C_{44} \int_{-a}^{a} (-1) dx \int_{-b}^{b} dz \int_{-1}^{1} \left(-2 \frac{r^3}{L} + 2 \frac{r^2}{L} \right) \frac{L}{2} dr
\end{aligned} \tag{4.151}$$

The term k_{32sB3} is equal to the term k_{23sB3} and it is equal to

$$k_{32sB3} = -2,\overline{6} C_{44} ab \tag{4.152}$$

Regarding the term $k_{33s\,B3}$

$$
\begin{aligned}
k_{33s\,B3} &= \int_V C_{44}\,F_{sx}\mathbf{J}^{-1}N_{2,r}(r)\,F_{\tau x}\mathbf{J}^{-1}N_{2,r}(r)\,dV \\
&= C_{44}\int_A F_{\tau x}\,F_{sx}\,dA \int_L \mathbf{J}^{-1}N_{2,r}(r)\mathbf{J}^{-1}N_{2,r}(r)\,dL \\
&= C_{44}\int_{-a}^{a} F_{\tau x}\,F_{sx}\,dx \int_{-b}^{b} dz \int_{-1}^{1}\mathbf{J}^{-1}N_{2,r}(r)\mathbf{J}^{-1}N_{2,r}(r)\,det\,\mathbf{J}\,dr \\
&= C_{44}\int_{-a}^{a}(1)\,dx \int_{-b}^{b} dz \int_{-1}^{1}\left(16\frac{r^2}{L^2}\right)\frac{L}{2}\,dr
\end{aligned}
\tag{4.153}
$$

The surface integral of the term $k_{33s\,B3}$ is computed using an analytical approach, while the line integral is computed using the three SPs Gauss quadrature. The computation of the surface integral is done considering the relations of Eq. 4.4.1.

$$
C_{44}\int_{-a}^{a}(1)\,dx \int_{-b}^{b} dz = C_{44}4ab
\tag{4.154}
$$

The computation of the line integral is done using the three SPs Gauss quadrature

$$
\int_{-1}^{1}\left(16\frac{r^2}{L^2}\right)\frac{L}{2}\,dr = \left[0.55556\left(16\frac{\left(-\sqrt{0.6}\right)^2}{L^2}\right) + 0.88889\left(16\frac{(0)^2}{L^2}\right) + \right.
$$
$$
\left. +0.55556\left(16\frac{\left(\sqrt{0.6}\right)^2}{L^2}\right)\right]\frac{L}{2} = 5,\bar{3}
\tag{4.155}
$$

The term $k_{33s\,B3}$ is then equal to

$$
k_{33s\,B3} = 21,\bar{3}C_{44}ab
\tag{4.156}
$$

Regarding the term $k_{34s\,B3}$

$$
\begin{aligned}
k_{34s\,B3} &= \int_V C_{44}\,F_{sx}\mathbf{J}^{-1}N_{2,r}(r)\,F_{\tau y,x}N_2(r)\,dV \\
&= C_{44}\int_A F_{\tau y,x}\,F_{sx}\,dA \int_L \mathbf{J}^{-1}N_{2,r}(r)N_2(r)\,dL \\
&= C_{44}\int_{-a}^{a} F_{\tau y,x}\,F_{sx}\,dx \int_{-b}^{b} dz \int_{-1}^{1}\mathbf{J}^{-1}N_{2,r}(r)N_2(r)\,det\,\mathbf{J}\,dr \\
&= C_{44}\int_{-a}^{a}(-1)\,dx \int_{-b}^{b} dz \int_{-1}^{1}\left(4\frac{r^3}{L} - 4\frac{r}{L}\right)\frac{L}{2}\,dr
\end{aligned}
\tag{4.157}
$$

The surface integral of the term $k_{34s\,B3}$ is computed using an analytical approach, while the line integral is computed using the three SPs Gauss quadrature. The com-

putation of the surface integral is done considering the relations of Eq. 4.4.1.

$$C_{44} \int_{-a}^{a} (-1)\, dx \int_{-b}^{b} dz = -C_{44} 4ab \qquad (4.158)$$

The computation of the line integral is done using the three SPs Gauss quadrature

$$\int_{-b}^{b} dz \int_{-1}^{1} \left(4\frac{r^3}{L} - 4\frac{r}{L}\right) \frac{L}{2} dr = \left[0.55556\left(4\frac{\left(-\sqrt{0.6}\right)^3}{L} - 4\frac{\left(-\sqrt{0.6}\right)}{L}\right) + 0.88889\left(4\frac{(0)^3}{L} - 4\frac{(0)}{L}\right) + \right.$$
$$\left. +0.55556\left(4\frac{\left(\sqrt{0.6}\right)^3}{L} - 4\frac{\left(\sqrt{0.6}\right)}{L}\right)\right]\frac{L}{2} = 0$$

$$(4.159)$$

The term $k_{34s\,B3}$ is then equal to

$$k_{34s\,B3} = 0 \qquad (4.160)$$

Regarding the term $k_{35s\,B3}$

$$\begin{aligned}
k_{35s\,B3} &= \int_{V} C_{44} F_{sx} \mathbf{J}^{-1} N_{2,r}(r) F_{\tau x} \mathbf{J}^{-1} N_{3,r}(r) dV \\
&= C_{44} \int_{A} F_{\tau x} F_{sx} dA \int_{L} \mathbf{J}^{-1} N_{2,r}(r) \mathbf{J}^{-1} N_{3,r}(r) dL \\
&= C_{44} \int_{-a}^{a} F_{\tau x} F_{sx} dx \int_{-b}^{b} dz \int_{-1}^{1} \mathbf{J}^{-1} N_{2,r}(r) \mathbf{J}^{-1} N_{3,r}(r) det\,\mathbf{J} dr \\
&= C_{44} \int_{-a}^{a} (1)\, dx \int_{-b}^{b} dz \int_{-1}^{1} \left(-8\frac{r^2}{L^2} - 4\frac{r}{L^2}\right)\frac{L}{2} dr
\end{aligned} \qquad (4.161)$$

The surface integral of the term $k_{35s\,B3}$ is computed using an analytical approach, while the line integral is computed using the three SPs Gauss quadrature. The computation of the surface integral is done considering the relations of Eq. 4.4.1.

$$C_{44} \int_{-a}^{a} (1)\, dx \int_{-b}^{b} dz = C_{44} 4ab \qquad (4.162)$$

The computation of the line integral is done using the three SPs Gauss quadrature

$$
\int_{-1}^{1}\left(-8\frac{r^2}{L^2}-4\frac{r}{L^2}\right)\frac{L}{2}dr=\left[0.55556\left(-8\frac{\left(-\sqrt{0.6}\right)^2}{L^2}-4\frac{\left(-\sqrt{0.6}\right)}{L^2}\right)\right.
$$
$$
+0.88889\left(-8\frac{(0)^2}{L^2}-4\frac{(0)}{L^2}\right)+ \tag{4.163}
$$
$$
\left.+0.55556\left(-8\frac{\left(\sqrt{0.6}\right)^2}{L^2}-4\frac{\left(\sqrt{0.6}\right)}{L^2}\right)\right]\frac{L}{2}=-2,\overline{6}
$$

The term $k_{35s\,B3}$ is then equal to

$$
k_{35s\,B3}=-10,\overline{6}C_{44}\frac{ab}{L} \tag{4.164}
$$

Regarding the term $k_{36s\,B3}$

$$
\begin{aligned}
k_{36s\,B3}&=\int_{V}C_{44}F_{sx}\mathbf{J}^{-1}N_{2,r}(r)F_{\tau y,x}N_{3}(r)dV\\
&=C_{44}\int_{A}F_{\tau y,x}F_{sx}dA\int_{L}\mathbf{J}^{-1}N_{2,r}(r)N_{3}(r)dL\\
&=C_{44}\int_{-a}^{a}F_{\tau y,x}F_{sx}dx\int_{-b}^{b}dz\int_{-1}^{1}\mathbf{J}^{-1}N_{2,r}(r)N_{3}(r)det\,\mathbf{J}dr\\
&=C_{44}\int_{-a}^{a}(-1)\,dx\int_{-b}^{b}dz\int_{-1}^{1}\left(-2\frac{r^3}{L}-2\frac{r^2}{L}\right)\frac{L}{2}dr
\end{aligned} \tag{4.165}
$$

The surface integral of the term $k_{36s\,B3}$ is computed using an analytical approach, while the line integral is computed using the three SPs Gauss quadrature. The computation of the surface integral is done considering the relations of Eq. 4.4.1.

$$
C_{44}\int_{-a}^{a}(-1)\,dx\int_{-b}^{b}dz=-C_{44}4ab \tag{4.166}
$$

The computation of the line integral is done using the three SPs Gauss quadrature

$$
\int_{-1}^{1}\left(-2\frac{r^3}{L}-2\frac{r^2}{L}\right)\frac{L}{2}dr=\left[0.55556\left(-2\frac{\left(-\sqrt{0.6}\right)^3}{L}-2\frac{\left(-\sqrt{0.6}\right)^2}{L}\right)\right.
$$
$$
+0.88889\left(-2\frac{(0)^3}{L}-2\frac{(0)^2}{L}\right)+ \tag{4.167}
$$
$$
\left.+0.55556\left(-2\frac{\left(\sqrt{0.6}\right)^3}{L}-2\frac{\left(\sqrt{0.6}\right)^2}{L}\right)\right]\frac{L}{2}=-0,\overline{6}
$$

The term $k_{36s_{B3}}$ is then equal to

$$k_{36s_{B3}} = 2,\overline{6}C_{44}ab \tag{4.168}$$

Regarding the term $k_{41s_{B3}}$

$$
\begin{aligned}
k_{41s_{B3}} &= \int_V C_{44} F_{sy,x} N_2(r) F_{\tau x} \mathbf{J}^{-1} N_{1,r}(r) dV \\
&= C_{44} \int_A F_{\tau x} F_{sy,x} dA \int_L N_2(r) \mathbf{J}^{-1} N_{1,r}(r) dL \\
&= C_{44} \int_{-a}^{a} F_{\tau x} F_{sy,x} dx \int_{-b}^{b} dz \int_{-1}^{1} N_2(r) \mathbf{J}^{-1} N_{1,r}(r) det\,\mathbf{J} dr \\
&= C_{44} \int_{-a}^{a} (-1)\, dx \int_{-b}^{b} dz \int_{-1}^{1} \left(-2\frac{r^3}{L} + \frac{r^2}{L} + 2\frac{r}{L} - \frac{1}{L} \right) \frac{L}{2} dr
\end{aligned}
\tag{4.169}
$$

The term $k_{41s_{B3}}$ is equal to the term $k_{14s_{B3}}$ and it is equal to

$$k_{41s_{B3}} = 2,\overline{6}C_{44}ab \tag{4.170}$$

Regarding the term $k_{42s_{B3}}$

$$
\begin{aligned}
k_{42s_{B3}} &= \int_V C_{44} F_{sy,x} N_2(r) F_{\tau y,x} N_1(r) dV \\
&= C_{44} \int_A F_{\tau y,x} F_{sy,x} dA \int_L N_2(r) N_1(r) dL \\
&= C_{44} \int_{-a}^{a} F_{\tau y,x} F_{sy,x} dx \int_{-b}^{b} dz \int_{-1}^{1} N_2(r) N_1(r) det\,\mathbf{J} dr \\
&= C_{44} \int_{-a}^{a} (1)\, dx \int_{-b}^{b} dz \int_{-1}^{1} \left(-\frac{1}{2}r^4 + \frac{1}{2}r^3 + \frac{1}{2}r^2 - \frac{1}{2}r \right) \frac{L}{2} dr
\end{aligned}
\tag{4.171}
$$

The term $k_{42s_{B3}}$ is equal to the term $k_{24s_{B3}}$ and it is equal to

$$k_{42s_{B3}} = 0,2\overline{6}C_{44}ab \tag{4.172}$$

Regarding the term $k_{43s_{B3}}$

$$k_{43s\,B3} = \int_V C_{44}\,F_{sy,x}\,N_2(r)\,F_{\tau x}\,\mathbf{J}^{-1}N_{2,r}(r)dV$$

$$= C_{44}\int_A F_{\tau x}\,F_{sy,x}\,dA\int_L N_2(r)\mathbf{J}^{-1}N_{2,r}(r)dL$$

$$= C_{44}\int_{-a}^{a} F_{\tau x}\,F_{sy,x}\,dx\int_{-b}^{b}dz\int_{-1}^{1}N_2(r)\mathbf{J}^{-1}N_{2,r}(r)\det\mathbf{J}dr \tag{4.173}$$

$$= C_{44}\int_{-a}^{a}(-1)\,dx\int_{-b}^{b}dz\int_{-1}^{1}\left(4\frac{r^3}{L}-4\frac{r}{L}\right)\frac{L}{2}dr$$

The term $k_{43s\,B3}$ is equal to the term $k_{34s\,B3}$ and it is equal to

$$k_{43s\,B3} = 0 \tag{4.174}$$

Regarding the term $k_{44s\,B3}$

$$k_{44s\,B3} = \int_V C_{44}\,F_{sy,x}\,N_2(r)\,F_{\tau y,x}\,N_2(r)dV$$

$$= C_{44}\int_A F_{\tau y,x}\,F_{sy,x}\,dA\int_L N_2(r)N_2(r)dL$$

$$= C_{44}\int_{-a}^{a}F_{\tau y,x}\,F_{sy,x}\,dx\int_{-b}^{b}dz\int_{-1}^{1}N_2(r)N_2(r)\det\mathbf{J}dr \tag{4.175}$$

$$= C_{44}\int_{-a}^{a}(1)\,dx\int_{-b}^{b}dz\int_{-1}^{1}\left(r^4-2r^2+1\right)\frac{L}{2}dr$$

The surface integral of the term $k_{44s\,B3}$ is computed using an analytical approach, while the line integral is computed using the three SPs Gauss quadrature. The computation of the surface integral is done considering the relations of Eq. 4.4.1.

$$C_{44}\int_{-a}^{a}(1)\,dx\int_{-b}^{b}dz = C_{44}4ab \tag{4.176}$$

The computation of the line integral is done using the three SPs Gauss quadrature

$$\int_{-1}^{1}\left(r^4-2r^2+1\right)\frac{L}{2}dr = \left[0.55556\left(\left(-\sqrt{0.6}\right)^4-2\left(-\sqrt{0.6}\right)^2+1\right)\right.$$

$$+0.88889\left((0)^4-2\,(0)^2+1\right)+ \tag{4.177}$$

$$+0.55556\left.\left(\left(\sqrt{0.6}\right)^4-2\left(\sqrt{0.6}\right)^2+1\right)\right]\frac{L}{2}=0,5\overline{3}L$$

The term $k_{44s\,B3}$ is then equal to

$$k_{44s\,B3} = 2,1\overline{3}C_{44}ab \tag{4.178}$$

Regarding the term $k_{45s\,B3}$

$$
\begin{aligned}
k_{45s\,B3} &= \int_V C_{44}\,F_{sy,x}\,N_2(r)\,F_{\tau x}\,\mathbf{J}^{-1}\,N_{3,r}(r)\,dV \\
&= C_{44}\int_A F_{\tau x}\,F_{sy,x}\,dA \int_L N_2(r)\,\mathbf{J}^{-1}\,N_{3,r}(r)\,dL \\
&= C_{44}\int_{-a}^{a} F_{\tau x}\,F_{sy,x}\,dx \int_{-b}^{b} dz \int_{-1}^{1} N_2(r)\,\mathbf{J}^{-1}\,N_{3,r}(r)\,det\,\mathbf{J}\,dr \\
&= C_{44}\int_{-a}^{a} (-1)\,dx \int_{-b}^{b} dz \int_{-1}^{1} \left(-2\frac{r^3}{L}-\frac{r^2}{L}+2\frac{r}{L}+\frac{1}{L}\right)\frac{L}{2}\,dr
\end{aligned}
\tag{4.179}
$$

The surface integral of the term $k_{45s\,B3}$ is computed using an analytical approach, while the line integral is computed using the three SPs Gauss quadrature. The computation of the surface integral is done considering the relations of Eq. 4.4.1.

$$
C_{44}\int_{-a}^{a} (-1)\,dx \int_{-b}^{b} dz = -C_{44}4ab
\tag{4.180}
$$

The computation of the line integral is done using the three SPs Gauss quadrature

$$
\begin{aligned}
\int_{-1}^{1}\left(-2\frac{r^3}{L}-\frac{r^2}{L}+2\frac{r}{L}+\frac{1}{L}\right)\frac{L}{2}\,dr &= \Bigg[0.55556\left(-2\frac{\left(-\sqrt{0.6}\right)^3}{L}-\frac{\left(-\sqrt{0.6}\right)^2}{L}+2\frac{\left(-\sqrt{0.6}\right)}{L}+\frac{1}{L}\right)+ \\
&\quad +0.88889\left(-2\frac{(0)^3}{L}-\frac{(0)^2}{L}+2\frac{(0)}{L}+\frac{1}{L}\right)\Bigg]+ \\
&\quad +\Bigg[0.55556\left(-2\frac{\left(\sqrt{0.6}\right)^3}{L}-\frac{\left(\sqrt{0.6}\right)^2}{L}+2\frac{\left(\sqrt{0.6}\right)}{L}+\frac{1}{L}\right)\Bigg]\frac{L}{2} \\
&= 0,\overline{6}
\end{aligned}
\tag{4.181}
$$

The term $k_{45s\,B3}$ is then equal to

$$
k_{45s\,B3} = -2,\overline{6}C_{44}ab
\tag{4.182}
$$

Regarding the term $k_{46s\,B3}$

$$
\begin{aligned}
k_{46s\,B3} &= \int_V C_{44} F_{sy,x} N_2(r) F_{\tau y,x} N_3(r) dV \\
&= C_{44} \int_A F_{\tau y,x} F_{sy,x} dA \int_L N_2(r) N_3(r) dL \\
&= C_{44} \int_{-a}^{a} F_{\tau y,x} F_{sy,x} dx \int_{-b}^{b} dz \int_{-1}^{1} N_2(r) N_3(r) \det \mathbf{J} dr \\
&= C_{44} \int_{-a}^{a} (1)\, dx \int_{-b}^{b} dz \int_{-1}^{1} \left(-\frac{1}{2} r^4 - \frac{1}{2} r^3 + \frac{1}{2} r^2 + \frac{1}{2} r \right) \frac{L}{2} dr
\end{aligned}
\tag{4.183}
$$

The surface integral of the term $k_{46s\,B3}$ is computed using an analytical approach, while the line integral is computed using the three SPs Gauss quadrature. The computation of the surface integral is done considering the relations of Eq. 4.4.1.

$$
C_{44} \int_{-a}^{a} (1)\, dx \int_{-b}^{b} dz = C_{44} 4ab
\tag{4.184}
$$

The computation of the line integral is done using the three SPs Gauss quadrature

$$
\begin{aligned}
\int_{-1}^{1} \left(-\frac{1}{2} r^4 - \frac{1}{2} r^3 + \frac{1}{2} r^2 + \frac{1}{2} r \right) \frac{L}{2} dr &= \Bigg[0.55556 \left(-\frac{1}{2} \left(-\sqrt{0.6}\right)^4 - \frac{1}{2} \left(-\sqrt{0.6}\right)^3 + \right. \\
&\quad \left. + \frac{1}{2} \left(-\sqrt{0.6}\right)^2 + \frac{1}{2} \left(-\sqrt{0.6}\right) \right) + \\
&\quad + 0.88889 \left(-\frac{1}{2} (0)^4 - \frac{1}{2} (0)^3 + \frac{1}{2} (0)^2 + \frac{1}{2} (0) \right) \Bigg] + \\
&\quad + \Bigg[0.55556 \left(-\frac{1}{2} \left(\sqrt{0.6}\right)^4 - \frac{1}{2} \left(\sqrt{0.6}\right)^3 + \right. \\
&\quad \left. + \frac{1}{2} \left(\sqrt{0.6}\right)^2 + \frac{1}{2} \left(\sqrt{0.6}\right) \right) \Bigg] \frac{L}{2} = 0,0\overline{6}
\end{aligned}
\tag{4.185}
$$

The term $k_{46s\,B3}$ is then equal to

$$
k_{46s\,B3} = 0,2\overline{6} C_{44} ab
\tag{4.186}
$$

Regarding the term $k_{51s\,B3}$

$$
\begin{aligned}
k_{51s\,B3} &= \int_V C_{44} F_{sx} \mathbf{J}^{-1} N_{3,r}(r) F_{\tau x} \mathbf{J}^{-1} N_{1,r}(r) dV \\
&= C_{44} \int_A F_{\tau x} F_{sx} dA \int_L \mathbf{J}^{-1} N_{3,r}(r) \mathbf{J}^{-1} N_{1,r}(r) dL \\
&= C_{44} \int_{-a}^{a} F_{\tau y,x} F_{sx} dx \int_{-b}^{b} dz \int_{-1}^{1} \mathbf{J}^{-1} N_{3,r}(r) \mathbf{J}^{-1} N_{1,r}(r) \det \mathbf{J} dr \\
&= C_{44} \int_{-a}^{a} (1)\, dx \int_{-b}^{b} dz \int_{-1}^{1} \left(4\frac{r^2}{L^2} - \frac{1}{L^2} \right) \frac{L}{2} dr
\end{aligned}
\tag{4.187}
$$

The term $k_{51s_{B3}}$ is equal to the term $k_{15s_{B3}}$ and it is equal to

$$k_{51s_{B3}} = 1,\bar{3}C_{44}\frac{ab}{L} \tag{4.188}$$

Regarding the term $k_{52s_{B3}}$

$$
\begin{aligned}
k_{52s_{B3}} &= \int_V C_{44} F_{sx} \mathbf{J}^{-1} N_{3,r}(r) F_{\tau y,x} N_1(r) dV \\
&= C_{44} \int_A F_{\tau y,x} F_{sx} dA \int_L \mathbf{J}^{-1} N_{3,r}(r) N_1(r) dL \\
&= C_{44} \int_{-a}^{a} F_{\tau y,x} F_{sx} dx \int_{-b}^{b} dz \int_{-1}^{1} \mathbf{J}^{-1} N_{3,r}(r) N_1(r) det\,\mathbf{J} dr \\
&= C_{44} \int_{-a}^{a} (-1)\, dx \int_{-b}^{b} dz \int_{-1}^{1} \left(\frac{r^3}{L} - \frac{1}{2}\frac{r^2}{L} - \frac{1}{2}\frac{r}{L} \right) \frac{L}{2} dr
\end{aligned}
\tag{4.189}
$$

The term $k_{52s_{B3}}$ is equal to the term $k_{25s_{B3}}$ and it is equal to

$$k_{52s_{B3}} = 0,\bar{6}C_{44}ab \tag{4.190}$$

Regarding the term $k_{53s_{B3}}$

$$
\begin{aligned}
k_{53s_{B3}} &= \int_V C_{44} F_{sx} \mathbf{J}^{-1} N_{3,r}(r) F_{\tau x} \mathbf{J}^{-1} N_{2,r}(r) dV \\
&= C_{44} \int_A F_{\tau x} F_{sx} dA \int_L \mathbf{J}^{-1} N_{3,r}(r) \mathbf{J}^{-1} N_{2,r}(r) dL \\
&= C_{44} \int_{-a}^{a} F_{\tau x} F_{sx} dx \int_{-b}^{b} dz \int_{-1}^{1} \mathbf{J}^{-1} N_{3,r}(r) \mathbf{J}^{-1} N_{2,r}(r) det\,\mathbf{J} dr \\
&= C_{44} \int_{-a}^{a} (1)\, dx \int_{-b}^{b} dz \int_{-1}^{1} \left(-8\frac{r^2}{L^2} - 4\frac{r}{L^2} \right) \frac{L}{2} dr
\end{aligned}
\tag{4.191}
$$

The term $k_{53s_{B3}}$ is equal to the term $k_{35s_{B3}}$ and it is equal to

$$k_{53s_{B3}} = -10,\bar{6}C_{44}\frac{ab}{L} \tag{4.192}$$

Regarding the term $k_{54s\,B3}$

$$k_{54s\,B3} = \int_V C_{44} F_{sx} \mathbf{J}^{-1} N_{3,r}(r) F_{\tau y,x} N_2(r) dV$$

$$= C_{44} \int_A F_{\tau y,x} F_{sx} dA \int_L \mathbf{J}^{-1} N_{3,r}(r) N_2(r) dL$$

$$= C_{44} \int_{-a}^{a} F_{\tau y,x} F_{sx} dx \int_{-b}^{b} dz \int_{-1}^{1} \mathbf{J}^{-1} N_{3,r}(r) N_2(r) \det \mathbf{J} dr$$

$$= C_{44} \int_{-a}^{a} (-1) dx \int_{-b}^{b} dz \int_{-1}^{1} \left(-2\frac{r^3}{L} - \frac{r^2}{L} + 2\frac{r}{L} + \frac{1}{L} \right) \frac{L}{2} dr$$

(4.193)

The term $k_{54s\,B3}$ is equal to the term $k_{45s\,B3}$ and it is equal to

$$k_{54s\,B3} = -2,\overline{6} C_{44} ab$$

(4.194)

Regarding the term $k_{55s\,B3}$

$$k_{55s\,B3} = \int_V C_{44} F_{sx} \mathbf{J}^{-1} N_{3,r}(r) F_{\tau x} \mathbf{J}^{-1} N_{3,r}(r) dV$$

$$= C_{44} \int_A F_{\tau x} F_{sx} dA \int_L \mathbf{J}^{-1} N_{3,r}(r) \mathbf{J}^{-1} N_{3,r}(r) dL$$

$$= C_{44} \int_{-a}^{a} F_{\tau x} F_{sx} dx \int_{-b}^{b} dz \int_{-1}^{1} \mathbf{J}^{-1} N_{3,r}(r) \mathbf{J}^{-1} N_{3,r}(r) \det \mathbf{J} dr$$

$$= C_{44} \int_{-a}^{a} (1) dx \int_{-b}^{b} dz \int_{-1}^{1} \left(4\frac{r^2}{L^2} + 4\frac{r}{L^2} + \frac{1}{L^2} \right) \frac{L}{2} dr$$

(4.195)

The surface integral of the term $k_{55s\,B3}$ is computed using an analytical approach, while the line integral is computed using the three SPs Gauss quadrature. The computation of the surface integral is done considering the relations of Eq. 4.4.1.

$$C_{44} \int_{-a}^{a} 1 dx \int_{-b}^{b} dz = C_{44} 4ab$$

(4.196)

The computation of the line integral is done using the three SPs Gauss quadrature

$$\int_{-1}^{1}\left(4\frac{r^2}{L^2}+4\frac{r}{L^2}+\frac{1}{L^2}\right)\frac{L}{2}dr = \left[0.55556\left(4\frac{\left(-\sqrt{0.6}\right)^2}{L^2}+4\frac{\left(-\sqrt{0.6}\right)}{L^2}+\frac{1}{L^2}\right)\right.$$

$$+0.88889\left(4\frac{(0)^2}{L^2}+4\frac{(0)}{L^2}+\frac{1}{L^2}\right)+$$

$$\left.+0.55556\left(4\frac{\left(\sqrt{0.6}\right)^2}{L^2}+4\frac{\left(\sqrt{0.6}\right)}{L^2}+\frac{1}{L^2}\right)\right]\frac{L}{2}=2,\overline{3}\frac{1}{L}$$

$$(4.197)$$

The term $k_{55s\,B3}$ is then equal to

$$k_{55s\,B3}=9,\overline{3}C_{44}\frac{ab}{L} \tag{4.198}$$

Regarding the term $k_{56s\,B3}$

$$\begin{aligned}k_{56s\,B3}&=\int_{V}C_{44}F_{sx}\mathbf{J}^{-1}N_{3,r}(r)F_{\tau y,x}N_3(r)dV\\&=C_{44}\int_{A}F_{\tau y,x}F_{sx}dA\int_{L}\mathbf{J}^{-1}N_{3,r}(r)N_3(r)dL\\&=C_{44}\int_{-a}^{a}F_{\tau y,x}F_{sx}dx\int_{-b}^{b}dz\int_{-1}^{1}\mathbf{J}^{-1}N_{3,r}(r)N_3(r)det\,\mathbf{J}dr\\&=C_{44}\int_{-a}^{a}(-1)\,dx\int_{-b}^{b}dz\int_{-1}^{1}\left(\frac{r^3}{L}+\frac{3}{2}\frac{r^2}{L}+\frac{1}{2}\frac{r}{L}\right)\frac{L}{2}dr\end{aligned} \tag{4.199}$$

The surface integral of the term $k_{56s\,B3}$ is computed using an analytical approach, while the line integral is computed using the three SPs Gauss quadrature. The computation of the surface integral is done considering the relations of Eq. 4.4.1.

$$C_{44}\int_{-a}^{a}(-1)\,dx\int_{-b}^{b}dz=-C_{44}4ab \tag{4.200}$$

The computation of the line integral is done using the three SPs Gauss quadrature

$$\int_{-1}^{1}\left(\frac{r^3}{L}+\frac{3}{2}\frac{r^2}{L}+\frac{1}{2}\frac{r}{L}\right)\frac{L}{2}dr=\left[0.55556\left(\frac{\left(-\sqrt{0.6}\right)^3}{L}+\frac{3}{2}\frac{\left(-\sqrt{0.6}\right)^2}{L}+\frac{1}{2}\frac{\left(-\sqrt{0.6}\right)}{L}\right)\right.$$

$$+0.88889\left(\frac{(0)^3}{L}+\frac{3}{2}\frac{(0)^2}{L}+\frac{1}{2}\frac{(0)}{L}\right)+$$

$$\left.+0.55556\left(\frac{\left(\sqrt{0.6}\right)^3}{L}+\frac{3}{2}\frac{\left(\sqrt{0.6}\right)^2}{L}+\frac{1}{2}\frac{\left(\sqrt{0.6}\right)}{L}\right)\right]\frac{L}{2}=0.5$$

$$(4.201)$$

The term $k_{56s_{B3}}$ is then equal to

$$k_{56s_{B3}} = -2C_{44}ab \tag{4.202}$$

Regarding the term $k_{61s_{B3}}$

$$
\begin{aligned}
k_{61s_{B3}} &= \int_V C_{44} F_{sy,x} N_3(r) F_{\tau x} \mathbf{J}^{-1} N_{1,r}(r) dV \\
&= C_{44} \int_A F_{\tau x} F_{sy,x} dA \int_L N_3(r) \mathbf{J}^{-1} N_{1,r}(r) dL \\
&= C_{44} \int_{-a}^{a} F_{\tau x} F_{sy,x} dx \int_{-b}^{b} dz \int_{-1}^{1} N_3(r) \mathbf{J}^{-1} N_{1,r}(r) det\,\mathbf{J}\,dr \\
&= C_{44} \int_{-a}^{a} (-1) dx \int_{-b}^{b} dz \int_{-1}^{1} \left(\frac{r^3}{L} + \frac{1}{2}\frac{r^2}{L} - \frac{1}{2}\frac{r}{L} \right) \frac{L}{2} dr
\end{aligned}
\tag{4.203}
$$

The term $k_{61s_{B3}}$ is equal to the term $k_{16s_{B3}}$ and it is equal to

$$k_{61s_{B3}} = -0,\bar{6}C_{44}ab \tag{4.204}$$

Regarding the term $k_{62s_{B3}}$

$$
\begin{aligned}
k_{62s_{B3}} &= \int_V C_{44} F_{sy,x} N_3(r) F_{\tau y,x} N_1(r) dV \\
&= C_{44} \int_A F_{\tau y,x} F_{sy,x} dA \int_L N_3(r) N_1(r) dL \\
&= C_{44} \int_{-a}^{a} F_{\tau y,x} F_{sy,x} dx \int_{-b}^{b} dz \int_{-1}^{1} N_3(r) N_1(r) det\,\mathbf{J}\,dr \\
&= C_{44} \int_{-a}^{a} (1) dx \int_{-b}^{b} dz \int_{-1}^{1} \left(\frac{1}{4}r^4 - \frac{1}{4}r^2 \right) \frac{L}{2} dr
\end{aligned}
\tag{4.205}
$$

The term $k_{62s_{B3}}$ is equal to the term $k_{26s_{B3}}$ and it is equal to

$$k_{62s_{B3}} = -0,1\bar{3}C_{44}ab \tag{4.206}$$

Regarding the term $k_{63s_{B3}}$

$$
\begin{aligned}
k_{63sB3} &= \int_V C_{44} F_{sy,x} N_3(r) F_{\tau x} \mathbf{J}^{-1} N_{2,r}(r) dV \\
&= C_{44} \int_A F_{\tau x} F_{sy,x} dA \int_L N_3(r) \mathbf{J}^{-1} N_{2,r}(r) dL \\
&= C_{44} \int_{-a}^{a} F_{\tau x} F_{sy,x} dx \int_{-b}^{b} dz \int_{-1}^{1} \mathbf{J}^{-1} N_{2,r}(r) N_3(r) \det \mathbf{J} dr \\
&= C_{44} \int_{-a}^{a} (-1) dx \int_{-b}^{b} dz \int_{-1}^{1} \left(-2\frac{r^3}{L} - 2\frac{r^2}{L} \right) \frac{L}{2} dr
\end{aligned}
\tag{4.207}
$$

The term k_{63sB3} is equal to the term k_{36sB3} and it is equal to

$$
k_{63sB3} = 2,\overline{6} C_{44} ab
\tag{4.208}
$$

Regarding the term k_{64sB3}

$$
\begin{aligned}
k_{64sB3} &= \int_V C_{44} F_{sy,x} N_3(r) F_{\tau y,x} N_2(r) dV \\
&= C_{44} \int_A F_{\tau y,x} F_{sy,x} dA \int_L N_3(r) N_2(r) dL \\
&= C_{44} \int_{-a}^{a} F_{\tau y,x} F_{sy,x} dx \int_{-b}^{b} dz \int_{-1}^{1} N_3(r) N_2(r) \det \mathbf{J} dr \\
&= C_{44} \int_{-a}^{a} (1) dx \int_{-b}^{b} dz \int_{-1}^{1} \left(-\frac{1}{2} r^4 - \frac{1}{2} r^3 + \frac{1}{2} r^2 + \frac{1}{2} r \right) \frac{L}{2} dr
\end{aligned}
\tag{4.209}
$$

The term k_{64sB3} is equal to the term k_{46sB3} and it is equal to

$$
k_{64sB3} = 0,2\overline{6} C_{44} ab
\tag{4.210}
$$

Regarding the term k_{65sB3}

$$
\begin{aligned}
k_{65sB3} &= \int_V C_{44} F_{sy,x} N_3(r) F_{\tau x} \mathbf{J}^{-1} N_{3,r}(r) dV \\
&= C_{44} \int_A F_{\tau x} F_{sy,x} dA \int_L N_3(r) \mathbf{J}^{-1} N_{3,r}(r) dL \\
&= C_{44} \int_{-a}^{a} F_{\tau x} F_{sy,x} dx \int_{-b}^{b} dz \int_{-1}^{1} \mathbf{J}^{-1} N_{3,r}(r) N_3(r) \det \mathbf{J} dr \\
&= C_{44} \int_{-a}^{a} (-1) dx \int_{-b}^{b} dz \int_{-1}^{1} \left(\frac{r^3}{L} + \frac{3}{2}\frac{r^2}{L} + \frac{1}{2}\frac{r}{L} \right) \frac{L}{2} dr
\end{aligned}
\tag{4.211}
$$

The term k_{65sB3} is equal to the term k_{56sB3} and it is equal to

$$k_{65s_{B3}} = -2C_{44}ab \tag{4.212}$$

Regarding the term $k_{66s_{B3}}$

$$
\begin{aligned}
k_{66s_{B3}} &= \int_V C_{44} F_{sy,x} N_3(r) F_{\tau y,x} N_3(r) dV \\
&= C_{44} \int_A F_{\tau y,x} F_{sy,x} dA \int_L N_3(r) N_3(r) dL \\
&= C_{44} \int_{-a}^{a} F_{\tau y,x} F_{sy,x} dx \int_{-b}^{b} dz \int_{-1}^{1} N_3(r) N_3(r) \det \mathbf{J} dr \\
&= C_{44} \int_{-a}^{a} (1)\, dx \int_{-b}^{b} dz \int_{-1}^{1} \left(\frac{1}{4}r^4 + \frac{1}{2}r^3 + \frac{1}{4}r^2\right) \frac{L}{2} dr
\end{aligned}
\tag{4.213}
$$

The surface integral of the term $k_{66s_{B3}}$ is computed using an analytical approach, while the line integral is computed using the three SPs Gauss quadrature. The computation of the surface integral is done considering the relations of Eq. 4.4.1.

$$C_{44} \int_{-a}^{a} (1)\, dx \int_{-b}^{b} dz = C_{44} 4ab \tag{4.214}$$

The computation of the line integral is done using the three SPs Gauss quadrature

$$
\begin{aligned}
\int_{-1}^{1} \left(\frac{1}{4}r^4 + \frac{1}{2}r^3 + \frac{1}{4}r^2\right)\frac{L}{2} dr &= \Bigg[0.55556\left(\frac{1}{4}\left(-\sqrt{0.6}\right)^4 + \frac{1}{2}\left(-\sqrt{0.6}\right)^3 + \frac{1}{4}\left(-\sqrt{0.6}\right)^2\right) + \\
&\quad +0.88889\left(\frac{1}{4}(0)^4 + \frac{1}{2}(0)^3 + \frac{1}{4}(0)^2\right) + \\
&\quad +0.55556\left(\frac{1}{4}\left(\sqrt{0.6}\right)^4 + \frac{1}{2}\left(\sqrt{0.6}\right)^3 + \frac{1}{4}\left(\sqrt{0.6}\right)^2\right)\Bigg]\frac{L}{2} = 0,1\overline{3}
\end{aligned}
\tag{4.215}
$$

The term $k_{66s_{B3}}$ is then equal to

$$k_{66s_{B3}} = 0,5\overline{3}C_{44}ab \tag{4.216}$$

Based on the terms computed previously, the shear stiffness matrix of a B3 beam element, obtained using the FI Gauss quadrature, is the following

$$\mathbf{K}_{SB3} = \begin{bmatrix} 9,\overline{3}C_{44}\dfrac{ab}{L} & 2C_{44}ab & -10,\overline{6}C_{44}\dfrac{ab}{L} & 2,\overline{6}C_{44}ab & 1,\overline{3}C_{44}\dfrac{ab}{L} & -0,\overline{6}C_{44}ab \\[2ex] 2C_{44}ab & 0,5\overline{3}C_{44}ab & 0,2\overline{6}C_{44}ab & 0,2\overline{6}C_{44}ab & -0,1\overline{3}C_{44}ab & -0,1\overline{3}C_{44}ab \\[2ex] -10,\overline{6}C_{44}\dfrac{ab}{L} & -2,\overline{6}C_{44}ab & 21,\overline{3}C_{44}ab & 0 & -10,\overline{6}C_{44}\dfrac{ab}{L} & 2,\overline{6}C_{44}ab \\[2ex] 2,\overline{6}C_{44}ab & 0,2\overline{6}C_{44}ab & 0 & 2,1\overline{3}C_{44}ab & -2,\overline{6}C_{44}ab & 0,2\overline{6}C_{44}ab \\[2ex] 1,\overline{3}C_{44}\dfrac{ab}{L} & 0,\overline{6}C_{44}ab & -10,\overline{6}C_{44}\dfrac{ab}{L} & -2,\overline{6}C_{44}ab & 9,\overline{3}C_{44}\dfrac{ab}{L} & -2C_{44}ab \\[2ex] -0,\overline{6}C_{44}ab & -0,1\overline{3}C_{44}ab & 2,\overline{6}C_{44}ab & 0,2\overline{6}C_{44}ab & -2C_{44}ab & 0,5\overline{3}C_{44}ab \end{bmatrix} \quad (4.217)$$

The complete stiffness matrix of the B3 element is obtained summing the two previous matrices of Eqs. (4.101) and (4.217)

$$\mathbf{K}_{B3} = \begin{bmatrix} 9,\overline{3}C_{44}\dfrac{ab}{L} & 2C_{44}ab & -10,\overline{6}C_{44}\dfrac{ab}{L} & 2,\overline{6}C_{44}ab & 1,\overline{3}C_{44}\dfrac{ab}{L} & -0,\overline{6}C_{44}ab \\[2ex] 2C_{44}ab & 0,5\overline{3}C_{44}ab+3,\overline{1}C_{11}\dfrac{a^3b}{L} & 0,2\overline{6}C_{44}ab & 0,2\overline{6}C_{44}ab-3,\overline{5}C_{11}\dfrac{a^3b}{L} & -0,1\overline{3}C_{44}ab & -0,1\overline{3}C_{44}ab+0,\overline{4}C_{11}\dfrac{a^3b}{L} \\[2ex] -10,\overline{6}C_{44}\dfrac{ab}{L} & -2,\overline{6}C_{44}ab & 21,\overline{3}C_{44}ab & 0 & -10,\overline{6}C_{44}\dfrac{ab}{L} & 2,\overline{6}C_{44}ab \\[2ex] 2,\overline{6}C_{44}ab & 0,2\overline{6}C_{44}ab-3,\overline{5}C_{11}\dfrac{a^3b}{L} & 0 & 2,1\overline{3}C_{44}ab+7,\overline{1}C_{11}\dfrac{a^3b}{L} & -2,\overline{6}C_{44}ab & 0,2\overline{6}C_{44}ab-3,\overline{5}C_{11}\dfrac{a^3b}{L} \\[2ex] 1,\overline{3}C_{44}\dfrac{ab}{L} & 0,\overline{6}C_{44}ab & -10,\overline{6}C_{44}\dfrac{ab}{L} & -2,\overline{6}C_{44}ab & 9,\overline{3}C_{44}\dfrac{ab}{L} & -2C_{44}ab \\[2ex] -0,\overline{6}C_{44}ab & -0,1\overline{3}C_{44}ab+0,\overline{4}C_{11}\dfrac{a^3b}{L} & 2,\overline{6}C_{44}ab & 0,2\overline{6}C_{44}ab-3,\overline{5}C_{11}\dfrac{a^3b}{L} & -2C_{44}ab & 0,5\overline{3}C_{44}ab+3,\overline{1}C_{11}\dfrac{a^3b}{L} \end{bmatrix}$$

$$(4.218)$$

Fig. 4.5 Schematic representation of a four-node beam (B4 element)

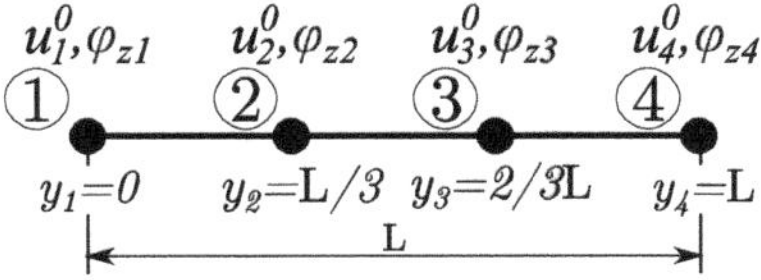

4.4.2 Cubic Element: B4 Element

Figure 4.5 shows the representation of a cubic B4 element. To obtain the stiffness matrix of a B4 beam element, the considerations made in the previous section apply. The shape functions related to a B4 element are reported below using the natural coordinates.

$$N_1(r) = -\frac{9}{16}\left(r + \frac{1}{3}\right)\left(r - \frac{1}{3}\right)(r - 1)$$

$$N_2(r) = \frac{27}{16}(r + 1)\left(r - \frac{1}{3}\right)(r - 1)$$

$$N_3(r) = -\frac{27}{16}(r + 1)\left(r + \frac{1}{3}\right)(r - 1)$$

$$N_4(r) = \frac{9}{16}\left(r + \frac{1}{3}\right)\left(r - \frac{1}{3}\right)(r + 1)$$

(4.219)

The relation between local and natural coordinates for the B4 beam element is equal to that for the B2 and B3 beam elements, and it is reported below

$$y = \frac{L}{2}(1 + r)$$

(4.220)

$$r = \frac{2y}{L} - 1$$

(4.221)

The Jacobian **J** related to the B4 beam length can be computed as follow (using the expression of Eq. (4.220))

$$\mathbf{J} = \left[\frac{\partial y}{\partial r}\right] = \left[\frac{\partial}{\partial r}\left(\left(\frac{1}{2}(1 + r)\right)L\right)\right] = \left[\frac{L}{2}\right]$$

(4.222)

The determinant of the Jacobian **J** is

$$det\,\mathbf{J} = det\left[\frac{L}{2}\right] = \frac{L}{2}$$

(4.223)

and the following relation applies

$$dy = \frac{L}{2}dr = det\,\mathbf{J}\,dr$$

(4.224)

The inverse of the Jacobian $\mathbf{J}$ is

$$\mathbf{J}^{-1} = \left[\frac{L}{2}\right]^{-1} = \left[\frac{2}{L}\right] \tag{4.225}$$

The unknowns vector of the B4 element is the following:

$$\begin{Bmatrix} u_1^0 \\ \phi_{z1} \\ u_2^0 \\ \phi_{z2} \\ u_3^0 \\ \phi_{z3} \\ u_4^0 \\ \phi_{z4} \end{Bmatrix} \tag{4.226}$$

4.4.2.1 Bending and Shear Stiffness Matrix for B4 Element

The stiffnes matrix of the B4 beam can be split into two parts, i.e. the bending and the shear parts. The strain components of interest are

$$\boldsymbol{\epsilon} = \begin{Bmatrix} \epsilon_{yy} \\ \epsilon_{xy} \end{Bmatrix} = \begin{bmatrix} 0 & F_{\tau y}\dfrac{\partial}{\partial y} \\ F_{\tau x}\dfrac{\partial}{\partial y} & F_{\tau y,x} \end{bmatrix} \begin{Bmatrix} u^0(y) \\ \phi_x(y) \end{Bmatrix} \tag{4.227}$$

With reference to the Fig. 4.5 and considering the shape functions of Eq. (4.219), the displacements $u^0(y)$ and $\phi_z(y)$ can be expressed as follows

$$u^0(y) = N_1(r)u_1^0 + N_2(r)u_2^0 + N_3(r)u_3^0 + N_4(r)u_4^0$$

$$\phi_z(y) = N_1(r)\phi_{z1} + N_2(r)\phi_{z2} + N_3(r)\phi_{z3} + N_4(r)\phi_{z4} \tag{4.228}$$

and

$$u^0(y) = \left[-\frac{9}{16}\left(r+\frac{1}{3}\right)\left(r-\frac{1}{3}\right)(r-1)\right]u_1^0 + \left[\frac{27}{16}(r+1)\left(r-\frac{1}{3}\right)(r-1)\right]u_2^0 +$$

$$+ \left[-\frac{27}{16}(r+1)\left(r+\frac{1}{3}\right)(r-1)\right]u_3^0 + \left[\frac{9}{16}\left(r+\frac{1}{3}\right)\left(r-\frac{1}{3}\right)(r+1)\right]u_4^0$$

$$\tag{4.229}$$

$$\phi_z(y) = \left[-\frac{9}{16}\left(r+\frac{1}{3}\right)\left(r-\frac{1}{3}\right)(r-1)\right]\phi_{z1} + \left[\frac{27}{16}(r+1)\left(r-\frac{1}{3}\right)(r-1)\right]\phi_{z2} +$$

$$+ \left[-\frac{27}{16}(r+1)\left(r+\frac{1}{3}\right)(r-1)\right]\phi_{z3} + \left[\frac{9}{16}\left(r+\frac{1}{3}\right)\left(r-\frac{1}{3}\right)(r+1)\right]\phi_{z4}$$

Equation (4.229), if written in matrix form, becomes

$$\left\{ \begin{array}{c} u^0(y) \\ \phi_x(y) \end{array} \right\} =$$

$$\left[\begin{array}{cccc} -\dfrac{9}{16}\left(r+\dfrac{1}{3}\right)\left(r-\dfrac{1}{3}\right)(r-1) & 0 & \dfrac{27}{16}(r+1)\left(r-\dfrac{1}{3}\right)(r-1) & 0 \\[2mm] 0 & -\dfrac{9}{16}\left(r+\dfrac{1}{3}\right)\left(r-\dfrac{1}{3}\right)(r-1) & 0 & \dfrac{27}{16}(r+1)\left(r-\dfrac{1}{3}\right)(r-1) \end{array} \right.$$

$$\left. \begin{array}{cccc} -\dfrac{27}{16}(r+1)\left(r+\dfrac{1}{3}\right)(r-1) & 0 & \dfrac{9}{16}\left(r+\dfrac{1}{3}\right)\left(r-\dfrac{1}{3}\right)(r+1) & 0 \\[2mm] 0 & -\dfrac{27}{16}(r+1)\left(r+\dfrac{1}{3}\right)(r-1) & 0 & \dfrac{9}{16}\left(r+\dfrac{1}{3}\right)\left(r-\dfrac{1}{3}\right)(r+1) \end{array} \right] \left\{ \begin{array}{c} u^0_1 \\ \phi_{z1} \\ u^0_2 \\ \phi_{z2} \\ u^0_3 \\ \phi_{z3} \\ u^0_4 \\ \phi_{z4} \end{array} \right\}$$

$$\tag{4.230}$$

By introducing Eq. (4.230) into Eq. (4.227), the following expression is obtained

$$\epsilon = \left\{ \begin{array}{c} \epsilon_{yy} \\ \epsilon_{xy} \end{array} \right\} = \left[\begin{array}{cc} 0 & F_{\tau y}\dfrac{\partial}{\partial y} \\[2mm] F_{\tau x}\dfrac{\partial}{\partial y} & F_{\tau y,x} \end{array} \right] \times$$

$$\times \left[\begin{array}{cccc} -\dfrac{9}{16}\left(r+\dfrac{1}{3}\right)\left(r-\dfrac{1}{3}\right)(r-1) & 0 & \dfrac{27}{16}(r+1)\left(r-\dfrac{1}{3}\right)(r-1) & 0 \\[2mm] 0 & -\dfrac{9}{16}\left(r+\dfrac{1}{3}\right)\left(r-\dfrac{1}{3}\right)(r-1) & 0 & \dfrac{27}{16}(r+1)\left(r-\dfrac{1}{3}\right)(r-1) \end{array} \right.$$

$$\left. \begin{array}{cccc} -\dfrac{27}{16}(r+1)\left(r+\dfrac{1}{3}\right)(r-1) & 0 & \dfrac{9}{16}\left(r+\dfrac{1}{3}\right)\left(r-\dfrac{1}{3}\right)(r+1) & 0 \\[2mm] 0 & -\dfrac{27}{16}(r+1)\left(r+\dfrac{1}{3}\right)(r-1) & 0 & \dfrac{9}{16}\left(r+\dfrac{1}{3}\right)\left(r-\dfrac{1}{3}\right)(r+1) \end{array} \right] \left\{ \begin{array}{c} u^0_1 \\ \phi_{z1} \\ u^0_2 \\ \phi_{z2} \\ u^0_3 \\ \phi_{z3} \\ u^0_4 \\ \phi_{z4} \end{array} \right\}$$

$$\tag{4.231}$$

In the previous relation, the following equivalence applies

$$\frac{\partial}{\partial y} = \mathbf{J}^{-1}\frac{\partial}{\partial r} \tag{4.232}$$

Introducing the relation in Eq. (4.232) into Eq. (4.231), it is obtained

$$\epsilon = \begin{Bmatrix} \epsilon_{yy} \\ \epsilon_{xy} \end{Bmatrix} = \begin{bmatrix} 0 & F_{\tau y} \mathbf{J}^{-1} \dfrac{\partial}{\partial r} \\ F_{\tau x} \mathbf{J}^{-1} \dfrac{\partial}{\partial r} & F_{\tau y,x} \end{bmatrix} \times$$

$$\times \begin{bmatrix} -\dfrac{9}{16}\left(r+\dfrac{1}{3}\right)\left(r-\dfrac{1}{3}\right)(r-1) & 0 & \dfrac{27}{16}(r+1)\left(r-\dfrac{1}{3}\right)(r-1) & 0 \\ 0 & -\dfrac{9}{16}\left(r+\dfrac{1}{3}\right)\left(r-\dfrac{1}{3}\right)(r-1) & 0 & \dfrac{27}{16}(r+1)\left(r-\dfrac{1}{3}\right)(r-1) \end{bmatrix}$$

$$\begin{matrix} -\dfrac{27}{16}(r+1)\left(r+\dfrac{1}{3}\right)(r-1) & 0 & \dfrac{9}{16}\left(r+\dfrac{1}{3}\right)\left(r-\dfrac{1}{3}\right)(r+1) & 0 \\ 0 & -\dfrac{27}{16}(r+1)\left(r+\dfrac{1}{3}\right)(r-1) & 0 & \dfrac{9}{16}\left(r+\dfrac{1}{3}\right)\left(r-\dfrac{1}{3}\right)(r+1) \end{matrix} \Big] \begin{Bmatrix} u_1^0 \\ \phi_{z1} \\ u_2^0 \\ \phi_{z2} \\ u_3^0 \\ \phi_{z3} \\ u_4^0 \\ \phi_{z4} \end{Bmatrix}$$

$$(4.233)$$

Generalizing the expression of Eq. (4.233), it can be written as

$$\epsilon = \begin{Bmatrix} \epsilon_{yy} \\ \epsilon_{xy} \end{Bmatrix} = \begin{bmatrix} 0 & F_{\tau y} \mathbf{J}^{-1} \dfrac{\partial}{\partial r} \\ F_{\tau x} \mathbf{J}^{-1} \dfrac{\partial}{\partial r} & F_{\tau y,x} \end{bmatrix} \begin{bmatrix} N_1(r) & 0 & N_2(r) & 0 & N_3(r) & 0 & N_4(r) & 0 \\ 0 & N_1(r) & 0 & N_2(r) & 0 & N_3(r) & 0 & N_4(r) \end{bmatrix} \begin{Bmatrix} u_1^0 \\ \phi_{z1} \\ u_2^0 \\ \phi_{z2} \\ u_3^0 \\ \phi_{z3} \\ u_4^0 \\ \phi_{z4} \end{Bmatrix} \quad (4.234)$$

and executing the derivatives

$$\epsilon = \begin{Bmatrix} \epsilon_{yy} \\ \epsilon_{xy} \end{Bmatrix} =$$

$$\begin{bmatrix} 0 & F_{\tau y} \mathbf{J}^{-1} N_{1,r}(r) & 0 & F_{\tau y} \mathbf{J}^{-1} N_{2,r}(r) \\ F_{\tau x} \mathbf{J}^{-1} N_{1,r}(r) & F_{\tau y,x} N_1(r) & F_{\tau x} \mathbf{J}^{-1} N_{2,r}(r) & F_{\tau y,x} N_2(r) \end{bmatrix}$$

$$\begin{matrix} 0 & F_{\tau y} \mathbf{J}^{-1} N_{3,r}(r) & 0 & F_{\tau y} \mathbf{J}^{-1} N_{4,r}(r) \\ F_{\tau x} \mathbf{J}^{-1} N_{3,r}(r) & F_{\tau y,x} N_3(r) & F_{\tau x} \mathbf{J}^{-1} N_{4,r}(r) & F_{\tau y,x} N_4(r) \end{matrix} \Big] \begin{Bmatrix} u_1^0 \\ \phi_{z1} \\ u_2^0 \\ \phi_{z2} \\ u_3^0 \\ \phi_{z3} \\ u_4^0 \\ \phi_{z4} \end{Bmatrix}$$

$$(4.235)$$

The previous expression can be split into two parts: one related to the longitudinal strains ϵ_{yy}, which is the bending part, and one related to the shear strains ϵ_{xy}, which is the shear part. By performing this partition, the following relation is obtained

$$\epsilon_{yy} = \begin{bmatrix} 0 & F_{\tau y}\mathbf{J}^{-1}N_{1,r}(r) & 0 & F_{\tau y}\mathbf{J}^{-1}N_{2,r}(r) & 0 & F_{\tau y}\mathbf{J}^{-1}N_{3,r}(r) & 0 & F_{\tau y}\mathbf{J}^{-1}N_{4,r}(r) \end{bmatrix} \begin{Bmatrix} u_1^0 \\ \phi_{z1} \\ u_2^0 \\ \phi_{z2} \\ u_3^0 \\ \phi_{z3} \\ u_4^0 \\ \phi_{z4} \end{Bmatrix} \quad (4.236)$$

and

$$\epsilon_{xy} =$$

$$\begin{bmatrix} F_{\tau x}\mathbf{J}^{-1}N_{1,r}(r) & F_{\tau y,x}N_1(r) & F_{\tau x}\mathbf{J}^{-1}N_{2,r}(r) & F_{\tau y,x}N_2(r) \\ & & & \\ F_{\tau x}\mathbf{J}^{-1}N_{3,r}(r) & F_{\tau y,x}N_3(r) & F_{\tau x}\mathbf{J}^{-1}N_{4,r}(r) & F_{\tau y,x}N_4(r) \end{bmatrix} \begin{Bmatrix} u_1^0 \\ \phi_{z1} \\ u_2^0 \\ \phi_{z2} \\ u_3^0 \\ \phi_{z3} \\ u_4^0 \\ \phi_{z4} \end{Bmatrix} \quad (4.237)$$

The following matrices $\mathbf{B}_{b_{\tau i}}$ (related to the bending part) and the matrix $\mathbf{B}_{s_{sj}}$ (related to the shear part) can be introduced

$$\mathbf{B}_{b_{\tau i}} = \begin{bmatrix} 0 & F_{\tau y}\mathbf{J}^{-1}N_{1,r}(r) & 0 & F_{\tau y}\mathbf{J}^{-1}N_{2,r}(r) & 0 & F_{\tau y}\mathbf{J}^{-1}N_{3,r}(r) & 0 & F_{\tau y}\mathbf{J}^{-1}N_{4,r}(r) \end{bmatrix} \quad (4.238)$$

$$\mathbf{B}_{s_{sj}} =$$
$$\begin{bmatrix} F_{\tau x}\mathbf{J}^{-1} & F_{\tau y,x} & F_{\tau x}\mathbf{J}^{-1} & F_{\tau y,x} & F_{\tau x}\mathbf{J}^{-1} & F_{\tau y,x} & F_{\tau x}\mathbf{J}^{-1} & F_{\tau y,x} \\ N_{1,r}(r) & N_1(r) & N_{2,r}(r) & N_2(r) & N_{3,r}(r) & N_3(r) & N_{4,r}(r) & N_4(r) \end{bmatrix} \quad (4.239)$$

Introducing the expansion functions F_{sx} and F_{sy} for the virtual variations

$$\mathbf{B}_{b_{sj}} = \begin{bmatrix} 0 & F_{sy}\mathbf{J}^{-1}N_{1,r}(r) & 0 & F_{sy}\mathbf{J}^{-1}N_{2,r}(r) & 0 & F_{sy}\mathbf{J}^{-1}N_{3,r}(r) & 0 & F_{sy}\mathbf{J}^{-1}N_{4,r}(r) \end{bmatrix} \quad (4.240)$$

$$\mathbf{B}_{b_{sj}} = \begin{bmatrix} F_{sx}\mathbf{J}^{-1} & F_{sy,x} & F_{sx}\mathbf{J}^{-1} & F_{sy,x} & F_{sx}\mathbf{J}^{-1} & F_{sy,x} & F_{sx}\mathbf{J}^{-1} & F_{sy,x} \\ N_{1,r}(r) & N_1(r) & N_{2,r}(r) & N_2(r) & N_{3,r}(r) & N_3(r) & N_{4,r}(r) & N_4(r) \end{bmatrix} \quad (4.241)$$

The bending part of the stiffness matrix $\mathbf{K}_{b_{B4}}$ can be calculated as follows, using the previous expressions of $\mathbf{B}_{b_{\tau i}}$ and $\mathbf{B}_{b_{sj}}$

$$\mathbf{K}_{b_{B4}} = \int_V \mathbf{B}_{b_{sj}}^T \mathbf{C}_s \mathbf{B}_{b_{\tau i}} \, dV \tag{4.242}$$

The stiffness part of the stiffness matrix $\mathbf{K}_{s_{B4}}$ can be calculated as follows, using the previous expressions of $\mathbf{B}_{s_{\tau i}}$ and $\mathbf{B}_{s_{sj}}$

$$\mathbf{K}_{s_{B4}} = \int_V \mathbf{B}_{s_{sj}}^T \mathbf{C}_s \mathbf{B}_{s_{\tau i}} \, dV \tag{4.243}$$

In the previous equations the matrices $\mathbf{C}_b$ and $\mathbf{C}_s$ are the bending and shear terms of the material coefficients matrix $\mathbf{C}$ (see Eq. (3.12)). These values are the following

$$\mathbf{C}_b = C_{11}$$

$$\tag{4.244}$$

$$\mathbf{C}_s = C_{44}$$

Considering Eqs. (4.238) and (4.240), the bending part of the stiffness matrix becomes

$$\mathbf{K}_{b_{B4}} = \int_V \begin{bmatrix} 0 \\ F_{sy}\mathbf{J}^{-1}N_{1,r}(r) \\ 0 \\ F_{sy}\mathbf{J}^{-1}N_{2,r}(r) \\ 0 \\ F_{sy}\mathbf{J}^{-1}N_{3,r}(r) \\ 0 \\ F_{sy}\mathbf{J}^{-1}N_{4,r}(r) \end{bmatrix} C_{11} \begin{bmatrix} 0 & F_{\tau y}\mathbf{J}^{-1}N_{1,r}(r) & 0 & F_{\tau y}\mathbf{J}^{-1}N_{2,r}(r) & 0 & F_{\tau y}\mathbf{J}^{-1}N_{3,r}(r) & 0 & F_{\tau y}\mathbf{J}^{-1}N_{4,r}(r) \end{bmatrix} dV$$

$$\tag{4.245}$$

Considering Eqs. (4.239) and (4.241), the shear part of the stiffness matrix becomes

$$\mathbf{K}_{S_{B4}} = \int_V \left(\begin{bmatrix} F_{sx}\mathbf{J}^{-1}N_{1,r}(r) \\[6pt] F_{sy,x}N_1(r) \\[6pt] F_{sx}\mathbf{J}^{-1}N_{2,r}(r) \\[6pt] F_{sy,x}N_2(r) \\[6pt] F_{sx}\mathbf{J}^{-1}N_{3,r}(r) \\[6pt] F_{sy,x}N_3(r) \\[6pt] F_{sx}\mathbf{J}^{-1}N_{4,r}(r) \\[6pt] F_{sy,x}N_4(r) \end{bmatrix} C_{44} \begin{bmatrix} F_{\tau x}\mathbf{J}^{-1} & F_{\tau y,x} & F_{\tau x}\mathbf{J}^{-1} & F_{\tau y,x} \\ N_{1,r}(r) & N_1(r) & N_{2,r}(r) & N_2(r) \\[6pt] F_{\tau x}\mathbf{J}^{-1} & F_{\tau y,x} & F_{\tau x}\mathbf{J}^{-1} & F_{\tau y,x} \\ N_{3,r}(r) & N_3(r) & N_{4,r}(r) & N_4(r) \end{bmatrix} \right) dV$$

$$(4.246)$$

The following relations must be considered

Expansion and shape functions

$$F_{\tau x} = -1; \quad F_{\tau y,x} = 1; \quad F_{sx} = -1; \quad F_{sy,x} = 1; \quad F_{\tau y} = x$$

$$F_{sy} = x; \quad F_{\tau x}F_{sx} = 1; \quad F_{\tau y}F_{sy} = x^2; \quad F_{\tau y,x}F_{sx} = -1$$

$$F_{\tau x}F_{sy,x} = -1; \quad F_{\tau y,x}F_{sy,x} = 1$$

$$\det \mathbf{J} = \frac{L}{2} \; ; \quad \mathbf{J}^{-1} = \left[\frac{2}{L} \right]$$

$$N_1(r) = -\frac{9}{16}\left(r + \frac{1}{3}\right)\left(r - \frac{1}{3}\right)(r - 1)$$

$$N_{1,r}(r) = \frac{\partial}{\partial r}\left[-\frac{9}{16}\left(r + \frac{1}{3}\right)\left(r - \frac{1}{3}\right)(r - 1)\right] = -\frac{27}{16}r^2 + \frac{9}{8}r + \frac{1}{16}$$

Expansion and shape functions

$$N_2(r) = \frac{27}{16}\,(r+1)\left(r-\frac{1}{3}\right)(r-1)$$

$$N_{2,r}(r) = \frac{\partial}{\partial r}\left[\frac{27}{16}\,(r+1)\left(r-\frac{1}{3}\right)(r-1)\right] = \frac{81}{16}r^2 - \frac{9}{8}r - \frac{27}{16}$$

$$N_3(r) = -\frac{27}{16}\,(r+1)\left(r+\frac{1}{3}\right)(r-1)$$

$$N_{3,r}(r) = \frac{\partial}{\partial r}\left[-\frac{27}{16}\,(r+1)\left(r+\frac{1}{3}\right)(r-1)\right] = -\frac{81}{16}r^2 - \frac{9}{8}r + \frac{27}{16}$$

$$N_4(r) = \frac{9}{16}\left(r+\frac{1}{3}\right)\left(r-\frac{1}{3}\right)(r+1)$$

$$N_{4,r}(r) = \frac{\partial}{\partial r}\left[\frac{9}{16}\left(r+\frac{1}{3}\right)\left(r-\frac{1}{3}\right)(r+1)\right] = \frac{27}{16}r^2 + \frac{9}{8}r - \frac{1}{16}$$

$$\mathbf{J}^{-1}N_{1,r}(r)\mathbf{J}^{-1}N_{1,r}(r) = \left[\frac{2}{L}\right]\left(-\frac{27}{16}r^2 + \frac{9}{8}r + \frac{1}{16}\right)\left[\frac{2}{L}\right]\left(-\frac{27}{16}r^2 + \frac{9}{8}r + \frac{1}{16}\right)$$

$$= \frac{729}{64L^2}r^4 - \frac{243}{16L^2}r^3 + \frac{135}{32L^2}r^2 + \frac{9}{16L^2}r + \frac{1}{64L^2}$$

$$\mathbf{J}^{-1}N_{1,r}(r)\mathbf{J}^{-1}N_{2,r}(r) = \left[\frac{2}{L}\right]\left(-\frac{27}{16}r^2 + \frac{9}{8}r + \frac{1}{16}\right)\left[\frac{2}{L}\right]\left(\frac{81}{16}r^2 - \frac{9}{8}r - \frac{27}{16}\right)$$

$$= -\frac{2187}{64L^2}r^4 + \frac{243}{8L^2}r^3 + \frac{243}{32L^2}r^2 - \frac{63}{8L^2}r - \frac{27}{64L^2}$$

$$\mathbf{J}^{-1}N_{1,r}(r)\mathbf{J}^{-1}N_{3,r}(r) = \left[\frac{2}{L}\right]\left(-\frac{27}{16}r^2 + \frac{9}{8}r + \frac{1}{16}\right)\left[\frac{2}{L}\right]\left(-\frac{81}{16}r^2 - \frac{9}{8}r + \frac{27}{16}\right)$$

$$= \frac{2187}{64L^2}r^4 - \frac{243}{16L^2}r^3 - \frac{567}{32L^2}r^2 + \frac{117}{16L^2}r + \frac{27}{64L^2}$$

$$\mathbf{J}^{-1}N_{1,r}(r)\mathbf{J}^{-1}N_{4,r}(r) = \left[\frac{2}{L}\right]\left(-\frac{27}{16}r^2 + \frac{9}{8}r + \frac{1}{16}\right)\left[\frac{2}{L}\right]\left(\frac{27}{16}r^2 + \frac{9}{8}r - \frac{1}{16}\right)$$

$$= -\frac{729}{64L^2}r^4 + \frac{189}{32L^2}r^2 - \frac{1}{64L^2}$$

Expansion and shape functions

$$\mathbf{J}^{-1}N_{2,r}(r)\mathbf{J}^{-1}N_{1,r}(r) = \left[\frac{2}{L}\right]\left(\frac{81}{16}r^2 - \frac{9}{8}r - \frac{27}{16}\right)\left[\frac{2}{L}\right]\left(-\frac{27}{16}r^2 + \frac{9}{8}r + \frac{1}{16}\right)$$

$$= -\frac{2187}{64L^2}r^4 + \frac{243}{8L^2}r^3 + \frac{243}{32L^2}r^2 - \frac{63}{8L^2}r - \frac{27}{64L^2}$$

$$\mathbf{J}^{-1}N_{2,r}(r)\mathbf{J}^{-1}N_{2,r}(r) = \left[\frac{2}{L}\right]\left(\frac{81}{16}r^2 - \frac{9}{8}r - \frac{27}{16}\right)\left[\frac{2}{L}\right]\left(\frac{81}{16}r^2 - \frac{9}{8}r - \frac{27}{16}\right)$$

$$= \frac{6561}{64L^2}r^4 - \frac{729}{16L^2}r^3 - \frac{2025}{32L^2}r^2 + \frac{243}{16L^2}r + \frac{729}{64L^2}$$

$$\mathbf{J}^{-1}N_{2,r}(r)\mathbf{J}^{-1}N_{3,r}(r) = \left[\frac{2}{L}\right]\left(\frac{81}{16}r^2 - \frac{9}{8}r - \frac{27}{16}\right)\left[\frac{2}{L}\right]\left(-\frac{81}{16}r^2 - \frac{9}{8}r + \frac{27}{16}\right)$$

$$= -\frac{6561}{64L^2}r^4 + \frac{2349}{32L^2}r^2 - \frac{729}{64L^2}$$

$$\mathbf{J}^{-1}N_{2,r}(r)\mathbf{J}^{-1}N_{4,r}(r) = \left[\frac{2}{L}\right]\left(\frac{81}{16}r^2 - \frac{9}{8}r - \frac{27}{16}\right)\left[\frac{2}{L}\right]\left(\frac{27}{16}r^2 + \frac{9}{8}r - \frac{1}{16}\right)$$

$$= \frac{2187}{64L^2}r^4 + \frac{243}{16L^2}r^3 - \frac{567}{32L^2}r^2 - \frac{117}{16L^2}r + \frac{27}{64L^2}$$

$$\mathbf{J}^{-1}N_{3,r}(r)\mathbf{J}^{-1}N_{1,r}(r) = \left[\frac{2}{L}\right]\left(-\frac{81}{16}r^2 - \frac{9}{8}r + \frac{27}{16}\right)\left[\frac{2}{L}\right]\left(-\frac{27}{16}r^2 + \frac{9}{8}r + \frac{1}{16}\right)$$

$$= \frac{2187}{64L^2}r^4 - \frac{243}{16L^2}r^3 - \frac{567}{32L^2}r^2 + \frac{117}{16L^2}r + \frac{27}{64L^2}$$

$$\mathbf{J}^{-1}N_{3,r}(r)\mathbf{J}^{-1}N_{2,r}(r) = \left[\frac{2}{L}\right]\left(-\frac{81}{16}r^2 - \frac{9}{8}r + \frac{27}{16}\right)\left[\frac{2}{L}\right]\left(\frac{81}{16}r^2 - \frac{9}{8}r - \frac{27}{16}\right)$$

$$= -\frac{6561}{64L^2}r^4 + \frac{2349}{32L^2}r^2 - \frac{729}{64L^2}$$

> **Expansion and shape functions**
>
> $$\mathbf{J}^{-1}N_{3,r}(r)\mathbf{J}^{-1}N_{3,r}(r)=\left[\frac{2}{L}\right]\left(-\frac{81}{16}r^2-\frac{9}{8}r+\frac{27}{16}\right)\left[\frac{2}{L}\right]\left(-\frac{81}{16}r^2-\frac{9}{8}r+\frac{27}{16}\right)$$
>
> $$=\frac{6561}{64L^2}r^4+\frac{729}{16L^2}r^3-\frac{2025}{32L^2}r^2-\frac{243}{16L^2}r+\frac{729}{64L^2}$$
>
> $$\mathbf{J}^{-1}N_{3,r}(r)\mathbf{J}^{-1}N_{4,r}(r)=\left[\frac{2}{L}\right]\left(-\frac{81}{16}r^2-\frac{9}{8}r+\frac{27}{16}\right)\left[\frac{2}{L}\right]\left(\frac{27}{16}r^2+\frac{9}{8}r-\frac{1}{16}\right)$$
>
> $$=-\frac{2187}{64L^2}r^4-\frac{243}{8L^2}r^3+\frac{243}{32L^2}r^2+\frac{63}{8L^2}r-\frac{27}{64L^2}$$
>
> $$\mathbf{J}^{-1}N_{4,r}(r)\mathbf{J}^{-1}N_{1,r}(r)=\left[\frac{2}{L}\right]\left(\frac{27}{16}r^2+\frac{9}{8}r-\frac{1}{16}\right)\left[\frac{2}{L}\right]\left(-\frac{27}{16}r^2+\frac{9}{8}r+\frac{1}{16}\right)$$
>
> $$=-\frac{729}{64L^2}r^4+\frac{189}{32L^2}r^2-\frac{1}{64L^2}$$
>
> $$\mathbf{J}^{-1}N_{4,r}(r)\mathbf{J}^{-1}N_{2,r}(r)=\left[\frac{2}{L}\right]\left(\frac{27}{16}r^2+\frac{9}{8}r-\frac{1}{16}\right)\left[\frac{2}{L}\right]\left(\frac{81}{16}r^2-\frac{9}{8}r-\frac{27}{16}\right)$$
>
> $$=\frac{2187}{64L^2}r^4+\frac{243}{16L^2}r^3-\frac{567}{32L^2}r^2-\frac{117}{16L^2}r+\frac{27}{64L^2}$$
>
> $$\mathbf{J}^{-1}N_{4,r}(r)\mathbf{J}^{-1}N_{3,r}(r)=\left[\frac{2}{L}\right]\left(\frac{27}{16}r^2+\frac{9}{8}r-\frac{1}{16}\right)\left[\frac{2}{L}\right]\left(-\frac{81}{16}r^2-\frac{9}{8}r+\frac{27}{16}\right)$$
>
> $$=-\frac{2187}{64L^2}r^4-\frac{243}{8L^2}r^3+\frac{243}{32L^2}r^2+\frac{63}{8L^2}r-\frac{27}{64L^2}$$
>
> $$\mathbf{J}^{-1}N_{4,r}(r)\mathbf{J}^{-1}N_{4,r}(r)=\left[\frac{2}{L}\right]\left(\frac{27}{16}r^2+\frac{9}{8}r-\frac{1}{16}\right)\left[\frac{2}{L}\right]\left(\frac{27}{16}r^2+\frac{9}{8}r-\frac{1}{16}\right)$$
>
> $$=\frac{729}{64L^2}r^4+\frac{243}{16L^2}r^3+\frac{135}{32L^2}r^2-\frac{9}{16L^2}r+\frac{1}{64L^2}$$

Expansion and shape functions

$$N_1(r)\mathbf{J}^{-1}N_{1,r}(r) = \left(-\frac{9}{16}\left(r+\frac{1}{3}\right)\left(r-\frac{1}{3}\right)(r-1)\right)\left[\frac{2}{L}\right]\left(-\frac{27}{16}r^2+\frac{9}{8}r+\frac{1}{16}\right)$$

$$= \frac{243}{128L}r^5 - \frac{405}{128L}r^4 + \frac{63}{64L}r^3 + \frac{27}{64L}r^2 - \frac{17}{128L}r - \frac{1}{128L}$$

$$N_1(r)N_1(r) = \left(-\frac{9}{16}\left(r+\frac{1}{3}\right)\left(r-\frac{1}{3}\right)(r-1)\right)\left(-\frac{9}{16}\left(r+\frac{1}{3}\right)\left(r-\frac{1}{3}\right)(r-1)\right)$$

$$= \frac{81}{256}r^6 - \frac{81}{128}r^5 + \frac{63}{256}r^4 + \frac{9}{64}r^3 - \frac{17}{256}r^2 - \frac{1}{128}r + \frac{1}{256}$$

$$N_1(r)\mathbf{J}^{-1}N_{2,r}(r) = \left(-\frac{9}{16}\left(r+\frac{1}{3}\right)\left(r-\frac{1}{3}\right)(r-1)\right)\left[\frac{2}{L}\right]\left(\frac{81}{16}r^2-\frac{9}{8}r-\frac{27}{16}\right)$$

$$= -\frac{729}{128L}r^5 + \frac{891}{128L}r^4 + \frac{81}{64L}r^3 - \frac{171}{64L}r^2 - \frac{9}{128L}r + \frac{27}{128L}$$

$$N_2(r)\mathbf{J}^{-1}N_{1,r}(r) = \left(\frac{27}{16}(r+1)\left(r-\frac{1}{3}\right)(r-1)\right)\left[\frac{2}{L}\right]\left(-\frac{27}{16}r^2+\frac{9}{8}r+\frac{1}{16}\right)$$

$$= -\frac{729}{128L}r^5 + \frac{729}{128L}r^4 + \frac{297}{64L}r^3 - \frac{369}{64L}r^2 + \frac{135}{128L}r + \frac{9}{128L}$$

$$N_1(r)N_2(r) = \left(-\frac{9}{16}\left(r+\frac{1}{3}\right)\left(r-\frac{1}{3}\right)(r-1)\right)\left(\frac{27}{16}(r+1)\left(r-\frac{1}{3}\right)(r-1)\right)$$

$$= -\frac{243}{256}r^6 + \frac{81}{64L}r^5 + \frac{189}{256}r^4 - \frac{45}{32}r^3 + \frac{63}{256}r^2 + \frac{9}{64}r - \frac{9}{256}$$

$$N_1(r)\mathbf{J}^{-1}N_{3,r}(r) = \left(-\frac{9}{16}\left(r+\frac{1}{3}\right)\left(r-\frac{1}{3}\right)(r-1)\right)\left[\frac{2}{L}\right]\left(-\frac{81}{16}r^2-\frac{9}{8}r+\frac{27}{16}\right)$$

$$= \frac{729}{128L}r^5 - \frac{567}{128L}r^4 - \frac{243}{64L}r^3 + \frac{153}{64L}r^2 + \frac{45}{128L}r - \frac{27}{128L}$$

$$N_3(r)\mathbf{J}^{-1}N_{1,r}(r) = \left(-\frac{27}{16}(r+1)\left(r+\frac{1}{3}\right)(r-1)\right)\left[\frac{2}{L}\right]\left(-\frac{27}{16}r^2+\frac{9}{8}r+\frac{1}{16}\right)$$

$$= \frac{729}{128L}r^5 - \frac{243}{128L}r^4 - \frac{459}{64L}r^3 + \frac{117}{64L}r^2 + \frac{189}{128L}r + \frac{9}{128L}$$

Expansion and shape functions

$$N_1(r)N_3(r) = \left(-\frac{9}{16}\left(r+\frac{1}{3}\right)\left(r-\frac{1}{3}\right)(r-1)\right)\left(-\frac{27}{16}(r+1)\left(r+\frac{1}{3}\right)(r-1)\right)$$

$$= \frac{243}{256}r^6 - \frac{81}{64L}r^5 - \frac{351}{256}r^4 + \frac{45}{64}r^3 + \frac{117}{256}r^2 - \frac{9}{128}r - \frac{9}{256}$$

$$N_1(r)\mathbf{J}^{-1}N_{4,r}(r) = \left(-\frac{9}{16}\left(r+\frac{1}{3}\right)\left(r-\frac{1}{3}\right)(r-1)\right)\left[\frac{2}{L}\right]\left(\frac{27}{16}r^2 + \frac{9}{8}r - \frac{1}{16}\right)$$

$$= -\frac{243}{128L}r^5 + \frac{81}{128L}r^4 + \frac{99}{64L}r^3 - \frac{9}{64L}r^2 - \frac{19}{128L}r + \frac{1}{128L}$$

$$N_4(r)\mathbf{J}^{-1}N_{1,r}(r) = \left(\frac{9}{16}\left(r+\frac{1}{3}\right)\left(r-\frac{1}{3}\right)(r+1)\right)\left[\frac{2}{L}\right]\left(-\frac{27}{16}r^2 + \frac{9}{8}r + \frac{1}{16}\right)$$

$$= -\frac{243}{128L}r^5 - \frac{81}{128L}r^4 + \frac{99}{64L}r^3 + \frac{9}{64L}r^2 - \frac{19}{128L}r - \frac{1}{128L}$$

$$N_1(r)N_4(r) = \left(-\frac{9}{16}\left(r+\frac{1}{3}\right)\left(r-\frac{1}{3}\right)(r-1)\right)\left(\frac{9}{16}\left(r+\frac{1}{3}\right)\left(r-\frac{1}{3}\right)(r+1)\right)$$

$$= -\frac{81}{256}r^6 + \frac{99}{256}r^4 - \frac{19}{256}r^2 + \frac{1}{256}$$

$$N_2(r)\mathbf{J}^{-1}N_{2,r}(r) = \left(\frac{27}{16}(r+1)\left(r-\frac{1}{3}\right)(r-1)\right)\left[\frac{2}{L}\right]\left(\frac{81}{16}r^2 - \frac{9}{8}r - \frac{27}{16}\right)$$

$$= \frac{2187}{128L}r^5 - \frac{1215}{128L}r^4 - \frac{1377}{64L}r^3 + \frac{729}{64L}r^2 + \frac{567}{128L}r - \frac{243}{128L}$$

$$N_2(r)N_2(r) = \left(\frac{27}{16}(r+1)\left(r-\frac{1}{3}\right)(r-1)\right)\left(\frac{27}{16}(r+1)\left(r-\frac{1}{3}\right)(r-1)\right)$$

$$= \frac{729}{256}r^6 - \frac{243}{256}r^5 - \frac{1377}{256}r^4 + \frac{243}{64}r^3 + \frac{567}{64}r^2 - \frac{243}{128}r + \frac{81}{256}$$

$$N_2(r)\mathbf{J}^{-1}N_{3,r}(r) = \left(\frac{27}{16}(r+1)\left(r-\frac{1}{3}\right)(r-1)\right)\left[\frac{2}{L}\right]\left(-\frac{81}{16}r^2 - \frac{9}{8}r + \frac{27}{16}\right)$$

$$= -\frac{2187}{128L}r^5 + \frac{243}{128L}r^4 + \frac{1539}{64L}r^3 - \frac{243}{64L}r^2 - \frac{891}{128L}r + \frac{243}{128L}$$

Expansion and shape functions

$$N_3(r)\mathbf{J}^{-1}N_{2,r}(r) = \left(-\frac{27}{16}(r+1)\left(r+\frac{1}{3}\right)(r-1)\right)\left[\frac{2}{L}\right]\left(\frac{81}{16}r^2 - \frac{9}{8}r - \frac{27}{16}\right)$$

$$= -\frac{2187}{128L}r^5 - \frac{243}{128L}r^4 + \frac{1539}{64L}r^3 + \frac{243}{64L}r^2 - \frac{891}{128L}r - \frac{243}{128L}$$

$$N_2(r)N_3(r) = \left(\frac{27}{16}(r+1)\left(r-\frac{1}{3}\right)(r-1)\right)\left(-\frac{27}{16}(r+1)\left(r+\frac{1}{3}\right)(r-1)\right)$$

$$= -\frac{79}{256}r^6 + \frac{1539}{256}r^4 - \frac{891}{256}r^2 + \frac{81}{256}$$

$$N_2(r)\mathbf{J}^{-1}N_{4,r}(r) = \left(\frac{27}{16}(r+1)\left(r-\frac{1}{3}\right)(r-1)\right)\left[\frac{2}{L}\right]\left(\frac{27}{16}r^2 + \frac{9}{8}r - \frac{1}{16}\right)$$

$$= \frac{729}{128L}r^5 + \frac{243}{128L}r^4 - \frac{459}{64L}r^3 - \frac{117}{64L}r^2 + \frac{189}{128L}r - \frac{9}{128L}$$

$$N_4(r)\mathbf{J}^{-1}N_{2,r}(r) = \left(\frac{9}{16}\left(r+\frac{1}{3}\right)\left(r-\frac{1}{3}\right)(r+1)\right)\left[\frac{2}{L}\right]\left(\frac{81}{16}r^2 - \frac{9}{8}r - \frac{27}{16}\right)$$

$$= \frac{729}{128L}r^5 + \frac{567}{128L}r^4 - \frac{243}{64L}r^3 - \frac{153}{64L}r^2 + \frac{45}{128L}r + \frac{27}{128L}$$

$$N_2(r)N_4(r) = \left(\frac{27}{16}(r+1)\left(r-\frac{1}{3}\right)(r-1)\right)\left(\frac{9}{16}\left(r+\frac{1}{3}\right)\left(r-\frac{1}{3}\right)(r+1)\right)$$

$$= \frac{243}{256}r^6 + \frac{81}{256}r^5 - \frac{351}{256}r^4 - \frac{45}{64}r^3 + \frac{117}{64}r^2 + \frac{9}{128}r - \frac{9}{256}$$

$$N_3(r)\mathbf{J}^{-1}N_{3,r}(r) = \left(-\frac{27}{16}(r+1)\left(r+\frac{1}{3}\right)(r-1)\right)\left[\frac{2}{L}\right]\left(-\frac{81}{16}r^2 - \frac{9}{8}r + \frac{27}{16}\right)$$

$$= \frac{2187}{128L}r^5 + \frac{1215}{128L}r^4 - \frac{1377}{64L}r^3 - \frac{729}{64L}r^2 + \frac{567}{128L}r + \frac{243}{128L}$$

$$N_3(r)N_3(r) = \left(-\frac{27}{16}(r+1)\left(r+\frac{1}{3}\right)(r-1)\right)\left(-\frac{27}{16}(r+1)\left(r+\frac{1}{3}\right)(r-1)\right)$$

$$= \frac{729}{256}r^6 + \frac{243}{256}r^5 - \frac{1377}{256}r^4 - \frac{243}{64}r^3 + \frac{567}{64}r^2 + \frac{243}{128}r + \frac{81}{256}$$

> **Expansion and shape functions**
>
> $$N_3(r)\mathbf{J}^{-1}N_{4,r}(r) = \left(-\frac{27}{16}(r+1)\left(r+\frac{1}{3}\right)(r-1)\right)\left[\frac{2}{L}\right]\left(\frac{27}{16}r^2 + \frac{9}{8}r - \frac{1}{16}\right)$$
>
> $$= -\frac{729}{128L}r^5 - \frac{729}{128L}r^4 + \frac{297}{64L}r^3 + \frac{369}{64L}r^2 + \frac{135}{128L}r - \frac{9}{128L}$$
>
> $$N_4(r)\mathbf{J}^{-1}N_{3,r}(r) = \left(\frac{9}{16}\left(r+\frac{1}{3}\right)\left(r-\frac{1}{3}\right)(r+1)\right)\left[\frac{2}{L}\right]\left(-\frac{81}{16}r^2 - \frac{9}{8}r + \frac{27}{16}\right)$$
>
> $$= -\frac{729}{128L}r^5 - \frac{891}{128L}r^4 + \frac{81}{64L}r^3 + \frac{171}{64L}r^2 - \frac{9}{128L}r - \frac{27}{128L}$$
>
> $$N_3(r)N_4(r) = \left(-\frac{27}{16}(r+1)\left(r+\frac{1}{3}\right)(r-1)\right)\left(\frac{9}{16}\left(r+\frac{1}{3}\right)\left(r-\frac{1}{3}\right)(r+1)\right)$$
>
> $$= -\frac{243}{256}r^6 - \frac{81}{64}r^5 + \frac{189}{256}r^4 + \frac{45}{32}r^3 + \frac{63}{256}r^2 - \frac{9}{64}r - \frac{9}{256}$$
>
> $$N_4(r)\mathbf{J}^{-1}N_{4,r}(r) = \left(\frac{9}{16}\left(r+\frac{1}{3}\right)\left(r-\frac{1}{3}\right)(r+1)\right)\left[\frac{2}{L}\right]\left(\frac{27}{16}r^2 + \frac{9}{8}r - \frac{1}{16}\right)$$
>
> $$= \frac{243}{128L}r^5 + \frac{405}{128L}r^4 + \frac{63}{64L}r^3 - \frac{27}{64L}r^2 - \frac{17}{128L}r + \frac{1}{128L}$$
>
> $$N_4(r)N_4(r) = \left(\frac{9}{16}\left(r+\frac{1}{3}\right)\left(r-\frac{1}{3}\right)(r+1)\right)\left(\frac{9}{16}\left(r+\frac{1}{3}\right)\left(r-\frac{1}{3}\right)(r+1)\right)$$
>
> $$= \frac{81}{256}r^6 + \frac{81}{128}r^5 + \frac{63}{256}r^4 - \frac{9}{64}r^3 - \frac{17}{256}r^2 + \frac{1}{128}r + \frac{1}{256}$$

4.4.2.2   FI Gauss Quadrature for B4 Element

The bending and the shear stiffness matrices are computed using FI Gauss Quadrature. For this purpose, the considerations made in Sects. 3.7.3 and 3.7.4. It is evident that the highest degree of the polynomial is $n = 6$.

Recalling the relation of Eq. (3.100), the minimum number of SPs to perform a FI Gauss quadrature is obtained.

$$n \leqslant 2SP - 1 \qquad SP \geqslant \frac{(n+1)}{2} \qquad SP \geqslant \frac{(6+1)}{2} \qquad SP \geqslant 3.5 \qquad (4.247)$$

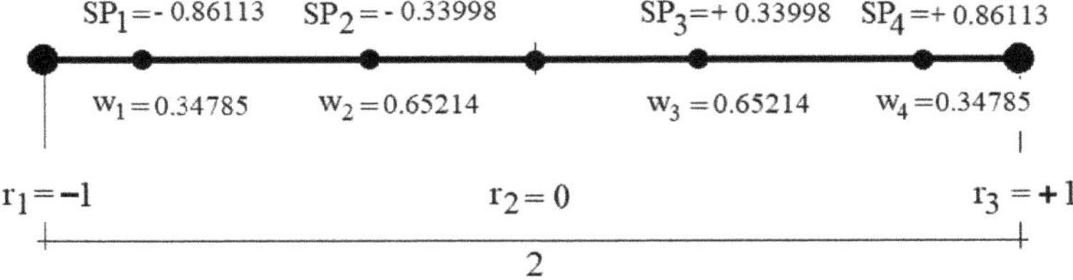

Fig. 4.6 Position of four SPs along the B4 beam and related weights

As a result, the minimum number of SPs for the FI Gauss quadrature is $SP = 4$. Figure 4.6 shows the positions of the four SPs along the B4 beam and the related weights. The relations shown in Sect. 3.7.3 must be considered to evecute the FI Gauss quadrature. In particular, Eqs. (3.91), (3.92) and (3.93).

4.4.2.3   Bending Stiffness Matrix of the B4 Element

The bending stiffness matrix can be written as follows

$$\mathbf{K}_{b_{B4}} = \begin{bmatrix} k_{11b_{B4}} & k_{12b_{B4}} & k_{13b_{B4}} & k_{14b_{B4}} & k_{15b_{B4}} & k_{16b_{B4}} & k_{17b_{B4}} & k_{18b_{B4}} \\ k_{21b_{B4}} & k_{22b_{B4}} & k_{23b_{B4}} & k_{24b_{B4}} & k_{25b_{B4}} & k_{26b_{B4}} & k_{27b_{B4}} & k_{28b_{B4}} \\ k_{31b_{B4}} & k_{32b_{B4}} & k_{33b_{B4}} & k_{34b_{B4}} & k_{35b_{B4}} & k_{36b_{B4}} & k_{37b_{B4}} & k_{38b_{B4}} \\ k_{41b_{B4}} & k_{42b_{B4}} & k_{43b_{B4}} & k_{44b_{B4}} & k_{45b_{B4}} & k_{46b_{B4}} & k_{47b_{B4}} & k_{48b_{B4}} \\ k_{51b_{B4}} & k_{52b_{B4}} & k_{53b_{B4}} & k_{54b_{B4}} & k_{55b_{B4}} & k_{56b_{B4}} & k_{57b_{B4}} & k_{58b_{B4}} \\ k_{61b_{B4}} & k_{62b_{B4}} & k_{63b_{B4}} & k_{64b_{B4}} & k_{65b_{B4}} & k_{66b_{B4}} & k_{67b_{B4}} & k_{68b_{B4}} \\ k_{71b_{B4}} & k_{72b_{B4}} & k_{73b_{B4}} & k_{74b_{B4}} & k_{75b_{B4}} & k_{76b_{B4}} & k_{77b_{B4}} & k_{78b_{B4}} \\ k_{81b_{B4}} & k_{82b_{B4}} & k_{83b_{B4}} & k_{84b_{B4}} & k_{85b_{B4}} & k_{86b_{B4}} & k_{87b_{B4}} & k_{88b_{B4}} \end{bmatrix} \tag{4.248}$$

The non-zero terms of the previous matrix are those placed in the positions: 22, 24, 26, 28, 42, 44, 46, 48, 62, 64, 66, 68, 82, 84, 86, 88. The relations of Eqs. (4.75) and (4.76) are recalled to compute the terms of the bending stiffness matrix. Furthermore, from Eq. (4.246) the following relation applies

$$F_{\tau y} F_{sy} = x^2 \tag{4.249}$$

The non-zero terms of the bending stiffness matrix part are computed

$$k_{22b_{B4}} = \int_V \left(C_{11} F_{sy} F_{\tau y} \mathbf{J}^{-1} N_{1,r}(r) \mathbf{J}^{-1} N_{1,r}(r) \right) dV$$

$$= C_{11} \int_A F_{sy} F_{\tau y} dA \int_L \mathbf{J}^{-1} N_{1,r}(r) \mathbf{J}^{-1} N_{1,r}(r) dL$$

$$= C_{11} \int_{-a}^{a} F_{sy} F_{\tau y} dx \int_{-b}^{b} dz \int_{-1}^{1} \mathbf{J}^{-1} N_{1,r}(r) \mathbf{J}^{-1} N_{1,r}(r) \det \mathbf{J} dr$$

$$= C_{11} \int_{-a}^{a} x^2 dx \int_{-b}^{b} dz \int_{-1}^{1} \left(\frac{729}{64L^2} r^4 - \frac{243}{16L^2} r^3 + \frac{135}{32L^2} r^2 + \frac{9}{16L^2} r + \frac{1}{64L^2} \right) \frac{L}{2} dr$$

$$(4.250)$$

The surface integral of the term $k_{22b_{B4}}$ is computed using an analytical integration, as follows

$$C_{11} \int_{-a}^{a} x^2 dx \int_{-b}^{b} dz = C_{11} \frac{4a^3 b}{3} \tag{4.251}$$

whereas the line integral is computed using the four SPs Gauss quadrature.

$$\int_{-1}^{1} \left(\frac{729}{64L^2} r^4 - \frac{243}{16L^2} r^3 + \frac{135}{32L^2} r^2 + \frac{9}{16L^2} r + \frac{1}{64L^2} \right) \frac{L}{2} dr =$$

$$= \left[0.34785 \left(\frac{729}{64L^2} (-0.86113)^4 - \frac{243}{16L^2} (-0.86113)^3 + \frac{135}{32L^2} (-0.86113)^2 + \frac{9}{16L^2} (-0.86113) + \frac{1}{64L^2} \right) + \right.$$

$$+ 0.65214 \left(\frac{729}{64L^2} (-0.33998)^4 - \frac{243}{16L^2} (-0.33998)^3 + \frac{135}{32L^2} (-0.33998)^2 + \frac{9}{16L^2} (-0.33998) + \frac{1}{64L^2} \right) +$$

$$+ 0.65214 \left(\frac{729}{64L^2} (0.33998)^4 - \frac{243}{16L^2} (0.33998)^3 + \frac{135}{32L^2} (0.33998)^2 + \frac{9}{16L^2} (0.33998) + \frac{1}{64L^2} \right) +$$

$$\left. + 0.34785 \left(\frac{729}{64L^2} (0.86113)^4 - \frac{243}{16L^2} (0.86113)^3 + \frac{135}{32L^2} (0.86113)^2 + \frac{9}{16L^2} (0.86113) + \frac{1}{64L^2} \right) \right] \frac{L}{2} =$$

$$= \frac{3.7}{L}$$

$$(4.252)$$

The term $k_{22b_{B4}}$ is then equal to

$$k_{22b_{B4}} = 4.933 C_{11} \frac{a^3 b}{L} \tag{4.253}$$

Regarding the $k_{24b_{B4}}$ term

$$k_{24b_{B4}} = \int_V \left(C_{11} F_{sy} \mathbf{J}^{-1} N_{1,r}(r) F_{\tau y} \mathbf{J}^{-1} N_{2,r}(r) \right) dV$$

$$= C_{11} \int_A F_{sy} F_{\tau y} dA \int_L \mathbf{J}^{-1} N_{1,r}(r) \mathbf{J}^{-1} N_{2,r}(r) dL$$

$$= C_{11} \int_{-a}^{a} F_{sy} F_{\tau y} dx \int_{-b}^{b} dz \int_{-1}^{1} \mathbf{J}^{-1} N_{1,r}(r) \mathbf{J}^{-1} N_{2,r}(r) det\, \mathbf{J} dr$$

$$= C_{11} \int_{-a}^{a} x^2 dx \int_{-b}^{b} dz \int_{-1}^{1} \left(-\frac{2187}{64L^2} r^4 + \frac{243}{8L^2} r^3 + \frac{243}{32L^2} r^2 - \frac{63}{8L^2} r - \frac{27}{64L^2} \right) \frac{L}{2} dr$$

$$\tag{4.254}$$

The surface integral of the term $k_{24b_{B4}}$ is computed using an analytical integration as follows

$$C_{11} \int_{-a}^{a} x^2 dx \int_{-b}^{b} dz = C_{11} \frac{4a^3 b}{3} \tag{4.255}$$

whereas the line integral is computed using the four SPs Gauss quadrature.

$$\int_{-1}^{1} \left(-\frac{2187}{64L^2} r^4 + \frac{243}{8L^2} r^3 + \frac{243}{32L^2} r^2 - \frac{63}{8L^2} r - \frac{27}{64L^2} \right) \frac{L}{2} dr =$$

$$= \left[0.34785 \left(\frac{-2187}{64L^2} (-0.86113)^4 + \frac{243}{8L^2} (-0.86113)^3 + \frac{243}{32L^2} (-0.86113)^2 - \frac{63}{8L^2} (-0.86113) - \frac{27}{64L^2} \right) + \right.$$

$$+ 0.65214 \left(\frac{-2187}{64L^2} (-0.33998)^4 + \frac{243}{8L^2} (-0.33998)^3 + \frac{243}{32L^2} (-0.33998)^2 - \frac{63}{8L^2} (-0.33998) - \frac{27}{64L^2} \right) +$$

$$+ 0.65214 \left(\frac{-2187}{64L^2} (0.33998)^4 + \frac{243}{8L^2} (0.33998)^3 + \frac{135}{32L^2} (0.33998)^2 - \frac{63}{8L^2} (0.33998) - \frac{27}{64L^2} \right) +$$

$$+ 0.34785 \left(\frac{-2187}{64L^2} (0.86113)^4 + \frac{243}{8L^2} (0.86113)^3 + \frac{243}{32L^2} (0.86113)^2 - \frac{63}{8L^2} (0.86113) - \frac{27}{64L^2} \right) \left. \right] \frac{L}{2} =$$

$$= \frac{-4.725}{L}$$

$$\tag{4.256}$$

The term $k_{24b_{B4}}$ is then equal to

$$k_{24b_{B4}} = -6.3 C_{11} \frac{a^3 b}{L} \tag{4.257}$$

Regarding the $k_{26b_{B4}}$ term

$$\begin{aligned}
k_{26b_{B4}} &= \int_V \left(C_{11} F_{sy} \mathbf{J}^{-1} N_{1,r}(r) F_{\tau y} \mathbf{J}^{-1} N_{3,r}(r) \right) dV \\
&= C_{11} \int_A F_{sy} F_{\tau y} dA \int_L \mathbf{J}^{-1} N_{1,r}(r) \mathbf{J}^{-1} N_{3,r}(r) dL \\
&= C_{11} \int_{-a}^{a} F_{sy} F_{\tau y} dx \int_{-b}^{b} dz \int_{-1}^{1} \mathbf{J}^{-1} N_{1,r}(r) \mathbf{J}^{-1} N_{3,r}(r) det\, \mathbf{J} dr \\
&= C_{11} \int_{-a}^{a} x^2 dx \int_{-b}^{b} dz \int_{-1}^{1} \left(\frac{2187}{64L^2} r^4 - \frac{243}{16L^2} r^3 - \frac{567}{32L^2} r^2 + \frac{117}{16L^2} r + \frac{27}{64L^2} \right) \frac{L}{2} dr
\end{aligned}$$

$$(4.258)$$

The surface integral of the term $k_{26b_{B4}}$ is computed using an analytical integration as follows

$$C_{11} \int_{-a}^{a} x^2 dx \int_{-b}^{b} dz = C_{11} \frac{4a^3 b}{3} \tag{4.259}$$

whereas the line integral is computed using the four SPs Gauss quadrature.

$$\int_{-1}^{1} \left(\frac{2187}{64L^2} r^4 - \frac{243}{16L^2} r^3 - \frac{567}{32L^2} r^2 + \frac{117}{16L^2} r + \frac{27}{64L^2} \right) \frac{L}{2} dr =$$

$$= \left[0.34785 \left(\frac{2187}{64L^2} (-0.86113)^4 - \frac{243}{16L^2} (-0.86113)^3 - \frac{567}{32L^2} (-0.86113)^2 + \frac{117}{16L^2} (-0.86113) + \frac{27}{64L^2} \right) + \right.$$

$$+ 0.65214 \left(\frac{2187}{64L^2} (-0.33998)^4 - \frac{243}{16L^2} (-0.33998)^3 - \frac{567}{32L^2} (-0.33998)^2 + \frac{117}{16L^2} (-0.33998) + \frac{27}{64L^2} \right) +$$

$$+ 0.65214 \left(\frac{2187}{64L^2} (0.33998)^4 - \frac{243}{16L^2} (0.33998)^3 - \frac{567}{32L^2} (0.33998)^2 + \frac{117}{16L^2} (0.33998) + \frac{27}{64L^2} \right) +$$

$$\left. + 0.34785 \left(\frac{2187}{64L^2} (0.86113)^4 - \frac{243}{16L^2} (0.86113)^3 - \frac{567}{32L^2} (0.86113)^2 + \frac{117}{16L^2} (0.86113) + \frac{27}{64L^2} \right) \right] \frac{L}{2} =$$

$$= \frac{1.35}{L}$$

$$(4.260)$$

The term $k_{26b_{B4}}$ is then equal to

$$k_{26b_{B4}} = 1.8 C_{11} \frac{a^3 b}{L} \tag{4.261}$$

Regarding the $k_{28b_{B4}}$ term

$$k_{28b_{B4}} = \int_V \left(C_{11} F_{sy} \mathbf{J}^{-1} N_{1,r}(r) F_{\tau y} \mathbf{J}^{-1} N_{4,r}(r) \right) dV$$

$$= C_{11} \int_A F_{sy} F_{\tau y} dA \int_L \mathbf{J}^{-1} N_{1,r}(r) \mathbf{J}^{-1} N_{4,r}(r) dL$$

$$= C_{11} \int_{-a}^a F_{sy} F_{\tau y} dx \int_{-b}^b dz \int_{-1}^1 \mathbf{J}^{-1} N_{1,r}(r) \mathbf{J}^{-1} N_{4,r}(r) det \, \mathbf{J} dr$$

$$= C_{11} \int_{-a}^a x^2 dx \int_{-b}^b dz \int_{-1}^1 \left(-\frac{729}{64L^2} r^4 + \frac{189}{32L^2} r^2 - \frac{1}{64L^2} \right) \frac{L}{2} dr$$

(4.262)

The surface integral of the term $k_{28b_{B4}}$ is computed using an analytical integration as follows

$$C_{11} \int_{-a}^a x^2 dx \int_{-b}^b dz = C_{11} \frac{4a^3 b}{3}$$

(4.263)

whereas the line integral is computed using the four SPs Gauss quadrature.

$$\int_{-1}^1 \left(-\frac{729}{64L^2} r^4 + \frac{189}{32L^2} r^2 - \frac{1}{64L^2} \right) \frac{L}{2} dr =$$

$$= \left[0.34785 \left(-\frac{729}{64L^2} (-0.86113)^4 + \frac{189}{32L^2} (-0.86113)^2 - \frac{1}{64L^2} \right) + \right.$$

$$+ 0.65214 \left(-\frac{729}{64L^2} (-0.33998)^4 + \frac{189}{32L^2} (-0.33998)^2 - \frac{1}{64L^2} \right) +$$

$$+ 0.65214 \left(-\frac{729}{64L^2} (0.33998)^4 + \frac{189}{32L^2} (0.33998)^2 - \frac{1}{64L^2} \right) +$$

$$\left. + 0.34785 \left(-\frac{729}{64L^2} (0.86113)^4 + \frac{189}{32L^2} (0.86113)^2 - \frac{1}{64L^2} \right) \right] \frac{L}{2} =$$

(4.264)

$$= \frac{-0.325}{L}$$

The term $k_{28b_{B4}}$ is then equal to

$$k_{28b_{B4}} = -0.4333 C_{11} \frac{a^3 b}{L}$$

(4.265)

Regarding the $k_{42b_{B4}}$ term

$$
\begin{aligned}
k_{42b\,B4} &= \int_V \left(C_{11} F_{sy} \mathbf{J}^{-1} N_{2,r}(r) F_{\tau y} \mathbf{J}^{-1} N_{1,r}(r) \right) dV \\
&= C_{11} \int_A F_{sy} F_{\tau y} dA \int_L \mathbf{J}^{-1} N_{2,r}(r) \mathbf{J}^{-1} N_{1,r}(r) dL \\
&= C_{11} \int_{-a}^{a} F_{sy} F_{\tau y} dx \int_{-b}^{b} dz \int_{-1}^{1} \mathbf{J}^{-1} N_{2,r}(r) \mathbf{J}^{-1} N_{1,r}(r) det\,\mathbf{J}\,dr \\
&= C_{11} \int_{-a}^{a} x^2 dx \int_{-b}^{b} dz \int_{-1}^{1} \left(-\frac{2187}{64L^2} r^4 + \frac{243}{8L^2} r^3 + \frac{243}{32L^2} r^2 - \frac{63}{8L^2} r - \frac{27}{64L^2} \right) \frac{L}{2} dr
\end{aligned}
\tag{4.266}
$$

The term $k_{42b\,B4}$ is equal to the term $k_{24b\,B4}$

$$
k_{42b\,B4} = -6.3 C_{11} \frac{a^3 b}{L}
\tag{4.267}
$$

Regarding the $k_{44b\,B4}$ term

$$
\begin{aligned}
k_{44b\,B4} &= \int_V \left(C_{11} F_{sy} \mathbf{J}^{-1} N_{2,r}(r) F_{\tau y} \mathbf{J}^{-1} N_{2,r}(r) \right) dV \\
&= C_{11} \int_A F_{sy} F_{\tau y} dA \int_L \mathbf{J}^{-1} N_{2,r}(r) \mathbf{J}^{-1} N_{2,r}(r) dL \\
&= C_{11} \int_{-a}^{a} F_{sy} F_{\tau y} dx \int_{-b}^{b} dz \int_{-1}^{1} \mathbf{J}^{-1} N_{2,r}(r) \mathbf{J}^{-1} N_{2,r}(r) det\,\mathbf{J}\,dr \\
&= C_{11} \int_{-a}^{a} x^2 dx \int_{-b}^{b} dz \int_{-1}^{1} \left(\frac{6561}{64L^2} r^4 - \frac{729}{16L^2} r^3 - \frac{2025}{32L^2} r^2 + \frac{243}{16L^2} r + \frac{729}{64L^2} \right) \frac{L}{2} dr
\end{aligned}
\tag{4.268}
$$

The surface integral of the term $k_{44b\,B4}$ is computed using an analytical integration as follows

$$
C_{11} \int_{-a}^{a} x^2 dx \int_{-b}^{b} dz = C_{11} \frac{4a^3 b}{3}
\tag{4.269}
$$

whereas the line integral is computed using the four SPs Gauss quadrature.

$$
\int_{-1}^{1} \left(\frac{6561}{64L^2} r^4 - \frac{729}{16L^2} r^3 - \frac{2025}{32L^2} r^2 + \frac{243}{16L^2} r + \frac{729}{64L^2} \right) \frac{L}{2} dr =
$$

$$
= \left[0.34785 \left(\frac{6561}{64L^2} (-0.86113)^4 - \frac{729}{16L^2} (-0.86113)^3 - \frac{2025}{32L^2} (-0.86113)^2 + \frac{243}{16L^2} (-0.86113) + \frac{729}{64L^2} \right) + \right.
$$

$$
+ 0.65214 \left(\frac{6561}{64L^2} (-0.33998)^4 - \frac{729}{16L^2} (-0.33998)^3 - \frac{2025}{32L^2} (-0.33998)^2 + \frac{243}{16L^2} (-0.33998) + \frac{729}{64L^2} \right) +
$$

$$
+ 0.65214 \left(\frac{6561}{64L^2} (0.33998)^4 - \frac{729}{16L^2} (0.33998)^3 - \frac{2025}{32L^2} (0.33998)^2 + \frac{243}{16L^2} (0.33998) + \frac{1}{64L^2} \right) +
$$

$$
\left. + 0.34785 \left(\frac{6561}{64L^2} (0.86113)^4 - \frac{729}{16L^2} (0.86113)^3 - \frac{2025}{32L^2} (0.86113)^2 + \frac{243}{16L^2} (0.86113) + \frac{729}{64L^2} \right) \right] \frac{L}{2} =
$$

$$
= \frac{10.8}{L}
\tag{4.270}
$$

The term $k_{44b_{B4}}$ is then equal to

$$k_{44b_{B4}} = 14.4 C_{11} \frac{a^3 b}{L} \tag{4.271}$$

Regarding the $k_{46b_{B4}}$ term

$$
\begin{aligned}
k_{46b_{B4}} &= \int_V \left(C_{11} F_{sy} \mathbf{J}^{-1} N_{2,r}(r) F_{\tau y} \mathbf{J}^{-1} N_{3,r}(r) \right) dV \\
&= C_{11} \int_A F_{sy} F_{\tau y} dA \int_L \mathbf{J}^{-1} N_{2,r}(r) \mathbf{J}^{-1} N_{3,r}(r) dL \\
&= C_{11} \int_{-a}^{a} F_{sy} F_{\tau y} dx \int_{-b}^{b} dz \int_{-1}^{1} \mathbf{J}^{-1} N_{2,r}(r) \mathbf{J}^{-1} N_{3,r}(r) \det \mathbf{J} dr \\
&= C_{11} \int_{-a}^{a} x^2 dx \int_{-b}^{b} dz \int_{-1}^{1} \left(-\frac{6561}{64L^2} r^4 + \frac{2349}{32L^2} r^2 - \frac{729}{64L^2} \right) \frac{L}{2} dr
\end{aligned}
\tag{4.272}
$$

The surface integral of the term $k_{46b_{B4}}$ is computed using an analytical integration as follows

$$C_{11} \int_{-a}^{a} x^2 dx \int_{-b}^{b} dz = C_{11} \frac{4a^3 b}{3} \tag{4.273}$$

whereas the line integral is computed using the four SPs Gauss quadrature.

$$
\begin{aligned}
\int_{-1}^{1} &\left(-\frac{6561}{64L^2} r^4 + \frac{2349}{32L^2} r^2 - \frac{729}{64L^2} \right) \frac{L}{2} dr = \\
&= \left[0.34785 \left(-\frac{6561}{64L^2} (-0.86113)^4 + \frac{2349}{32L^2} (-0.86113)^2 - \frac{729}{64L^2} \right) + \right. \\
&\quad + 0.65214 \left(-\frac{6561}{64L^2} (-0.33998)^4 + \frac{2349}{32L^2} (-0.33998)^2 - \frac{729}{64L^2} \right) + \\
&\quad + 0.65214 \left(-\frac{6561}{64L^2} (0.33998)^4 + \frac{2349}{32L^2} (0.33998)^2 - \frac{729}{64L^2} \right) + \\
&\quad \left. + 0.34785 \left(-\frac{6561}{64L^2} (0.86113)^4 + \frac{2349}{32L^2} (0.86113)^2 - \frac{729}{64L^2} \right) \right] \frac{L}{2} = \\
&= \frac{-7.425}{L}
\end{aligned}
\tag{4.274}
$$

The term $k_{46b_{B4}}$ is then equal to

$$k_{46b_{B4}} = -9.9 C_{11} \frac{a^3 b}{L} \tag{4.275}$$

Regarding the $k_{48b_{B4}}$ term

$$
\begin{aligned}
k_{48b_{B4}} &= \int_V \left(C_{11} F_{sy} \mathbf{J}^{-1} N_{2,r}(r) F_{\tau y} \mathbf{J}^{-1} N_{4,r}(r) \right) dV \\
&= C_{11} \int_A F_{sy} F_{\tau y} dA \int_L \mathbf{J}^{-1} N_{2,r}(r) \mathbf{J}^{-1} N_{4,r}(r) dL \\
&= C_{11} \int_{-a}^{a} F_{sy} F_{\tau y} dx \int_{-b}^{b} dz \int_{-1}^{1} \mathbf{J}^{-1} N_{2,r}(r) \mathbf{J}^{-1} N_{4,r}(r) det \, \mathbf{J} dr \\
&= C_{11} \int_{-a}^{a} x^2 dx \int_{-b}^{b} dz \int_{-1}^{1} \left(\frac{2187}{64L^2} r^4 - \frac{243}{16L^2} r^3 - \frac{567}{32L^2} r^2 + \frac{117}{16L^2} r + \frac{27}{64L^2} \right) \frac{L}{2} dr
\end{aligned}
\tag{4.276}
$$

The surface integral of the term $k_{48b_{B4}}$ is computed using an analytical integration as follows

$$C_{11} \int_{-a}^{a} x^2 dx \int_{-b}^{b} dz = C_{11} \frac{4a^3 b}{3} \tag{4.277}$$

whereas the line integral is computed using the four SPs Gauss quadrature.

$$
\begin{aligned}
&\int_{-1}^{1} \left(\frac{2187}{64L^2} r^4 + \frac{243}{16L^2} r^3 - \frac{567}{32L^2} r^2 - \frac{117}{16L^2} r + \frac{27}{64L^2} \right) \frac{L}{2} dr = \\[2mm]
&= \Bigg[0.34785 \left(\frac{2187}{64L^2} (-0.86113)^4 + \frac{243}{16L^2} (-0.86113)^3 - \frac{567}{32L^2} (-0.86113)^2 - \frac{117}{16L^2} (-0.86113) + \frac{27}{64L^2} \right) + \\[2mm]
&\quad + 0.65214 \left(\frac{2187}{64L^2} (-0.33998)^4 + \frac{243}{16L^2} (-0.33998)^3 - \frac{567}{32L^2} (-0.33998)^2 - \frac{117}{16L^2} (-0.33998) + \frac{27}{64L^2} \right) + \\[2mm]
&\quad + 0.65214 \left(\frac{2187}{64L^2} (0.33998)^4 - \frac{243}{16L^2} (0.33998)^3 - \frac{567}{32L^2} (0.33998)^2 + \frac{117}{16L^2} (0.33998) + \frac{27}{64L^2} \right) + \\[2mm]
&\quad + 0.34785 \left(\frac{2187}{64L^2} (0.86113)^4 + \frac{243}{16L^2} (0.86113)^3 - \frac{567}{32L^2} (0.86113)^2 - \frac{117}{16L^2} (0.86113) + \frac{27}{64L^2} \right) \Bigg] \frac{L}{2} = \\[2mm]
&= \frac{1.35}{L}
\end{aligned}
\tag{4.278}
$$

The term $k_{48b_{B4}}$ is then equal to

$$k_{48b_{B4}} = 1.8 C_{11} \frac{a^3 b}{L} \tag{4.279}$$

Regarding the $k_{62b_{B4}}$ term

$$k_{62bB4} = \int_V \left(C_{11} F_{sy} \mathbf{J}^{-1} N_{3,r}(r) F_{\tau y} \mathbf{J}^{-1} N_{1,r}(r) \right) dV$$

$$= C_{11} \int_A F_{sy} F_{\tau y} dA \int_L \mathbf{J}^{-1} N_{3,r}(r) \mathbf{J}^{-1} N_{1,r}(r) dL$$

$$= C_{11} \int_{-a}^{a} F_{sy} F_{\tau y} dx \int_{-b}^{b} dz \int_{-1}^{1} \mathbf{J}^{-1} N_{3,r}(r) \mathbf{J}^{-1} N_{1,r}(r) \det \mathbf{J} dr$$

$$= C_{11} \int_{-a}^{a} x^2 dx \int_{-b}^{b} dz \int_{-1}^{1} \left(\frac{2187}{64L^2} r^4 - \frac{243}{16L^2} r^3 - \frac{567}{32L^2} r^2 + \frac{117}{16L^2} r + \frac{27}{64L^2} \right) \frac{L}{2} dr$$

$$\tag{4.280}$$

The term k_{62bB4} is equal to the term k_{26bB4}

$$k_{62bB4} = 1.8 C_{11} \frac{a^3 b}{L} \tag{4.281}$$

Regarding the k_{64bB4} term

$$k_{64bB4} = \int_V \left(C_{11} F_{sy} \mathbf{J}^{-1} N_{3,r}(r) F_{\tau y} \mathbf{J}^{-1} N_{2,r}(r) \right) dV$$

$$= C_{11} \int_A F_{sy} F_{\tau y} dA \int_L \mathbf{J}^{-1} N_{3,r}(r) \mathbf{J}^{-1} N_{2,r}(r) dL$$

$$= C_{11} \int_{-a}^{a} F_{sy} F_{\tau y} dx \int_{-b}^{b} dz \int_{-1}^{1} \mathbf{J}^{-1} N_{3,r}(r) \mathbf{J}^{-1} N_{2,r}(r) \det \mathbf{J} dr$$

$$= C_{11} \int_{-a}^{a} x^2 dx \int_{-b}^{b} dz \int_{-1}^{1} \left(-\frac{6561}{64L^2} r^4 + \frac{2349}{32L^2} r^2 - \frac{729}{64L^2} \right) \frac{L}{2} dr$$

$$\tag{4.282}$$

The term k_{64bB4} is equal to the term k_{46bB4}

$$k_{64bB4} = -9.9 C_{11} \frac{a^3 b}{L} \tag{4.283}$$

Regarding the k_{66bB4} term

$$k_{66bB4} = \int_V \left(C_{11} F_{sy} \mathbf{J}^{-1} N_{3,r}(r) F_{\tau y} \mathbf{J}^{-1} N_{3,r}(r) \right) dV$$

$$= C_{11} \int_A F_{sy} F_{\tau y} dA \int_L \mathbf{J}^{-1} N_{3,r}(r) \mathbf{J}^{-1} N_{3,r}(r) dL$$

$$= C_{11} \int_{-a}^{a} F_{sy} F_{\tau y} dx \int_{-b}^{b} dz \int_{-1}^{1} \mathbf{J}^{-1} N_{3,r}(r) \mathbf{J}^{-1} N_{3,r}(r) \det \mathbf{J} dr$$

$$= C_{11} \int_{-a}^{a} x^2 dx \int_{-b}^{b} dz \int_{-1}^{1} \left(\frac{6561}{64L^2} r^4 + \frac{729}{16L^2} r^3 - \frac{2025}{32L^2} r^2 - \frac{243}{16L^2} r + \frac{729}{64L^2} \right) \frac{L}{2} dr$$

$$\tag{4.284}$$

The surface integral of the term k_{66bB4} is computed using an analytical integration as follows

$$C_{11} \int_{-a}^{a} x^2 dx \int_{-b}^{b} dz = C_{11} \frac{4a^3 b}{3} \tag{4.285}$$

whereas the line integral is computed using the four SPs Gauss quadrature.

$$\int_{-1}^{1} \left(\frac{6561}{64L^2} r^4 + \frac{729}{16L^2} r^3 - \frac{2025}{32L^2} r^2 - \frac{243}{16L^2} r + \frac{729}{64L^2} \right) \frac{L}{2} dr =$$

$$= \left[0.34785 \left(\frac{6561}{64L^2} (-0.86113)^4 + \frac{729}{16L^2} (-0.86113)^3 - \frac{2025}{32L^2} (-0.86113)^2 - \frac{243}{16L^2} (-0.86113) + \frac{729}{64L^2} \right) + \right.$$

$$+ 0.65214 \left(\frac{6561}{64L^2} (-0.33998)^4 + \frac{729}{16L^2} (-0.33998)^3 - \frac{2025}{32L^2} (-0.33998)^2 - \frac{243}{16L^2} (-0.33998) + \frac{729}{64L^2} \right) +$$

$$+ 0.65214 \left(\frac{6561}{64L^2} (0.33998)^4 + \frac{729}{16L^2} (0.33998)^3 - \frac{2025}{32L^2} (0.33998)^2 - \frac{243}{16L^2} (0.33998) + \frac{1}{64L^2} \right) +$$

$$\left. + 0.34785 \left(\frac{6561}{64L^2} (0.86113)^4 + \frac{729}{16L^2} (0.86113)^3 - \frac{2025}{32L^2} (0.86113)^2 - \frac{243}{16L^2} (0.86113) + \frac{729}{64L^2} \right) \right] \frac{L}{2} =$$

$$= \frac{10.8}{L}$$

$$(4.286)$$

The term $k_{66b_{B4}}$ is then equal to

$$k_{66b_{B4}} = 14.4 C_{11} \frac{a^3 b}{L} \tag{4.287}$$

Regarding the $k_{68b_{B4}}$ term

$$\begin{aligned} k_{68b_{B4}} &= \int_V \left(C_{11} F_{sy} \mathbf{J}^{-1} N_{3,r}(r) F_{\tau y} \mathbf{J}^{-1} N_{4,r}(r) \right) dV \\ &= C_{11} \int_A F_{sy} F_{\tau y} dA \int_L \mathbf{J}^{-1} N_{3,r}(r) \mathbf{J}^{-1} N_{4,r}(r) dL \\ &= C_{11} \int_{-a}^{a} F_{sy} F_{\tau y} dx \int_{-b}^{b} dz \int_{-1}^{1} \mathbf{J}^{-1} N_{3,r}(r) \mathbf{J}^{-1} N_{4,r}(r) \det \mathbf{J} dr \\ &= C_{11} \int_{-a}^{a} x^2 dx \int_{-b}^{b} dz \int_{-1}^{1} \left(-\frac{2187}{64L^2} r^4 - \frac{243}{8L^2} r^3 + \frac{243}{32L^2} r^2 + \frac{63}{8L^2} r - \frac{27}{64L^2} \right) \frac{L}{2} dr \end{aligned}$$

$$(4.288)$$

The surface integral of the term $k_{68b_{B4}}$ is computed using an analytical integration as follows

$$C_{11} \int_{-a}^{a} x^2 dx \int_{-b}^{b} dz = C_{11} \frac{4a^3 b}{3} \tag{4.289}$$

whereas the line integral is computed using the four SPs Gauss quadrature.

$$\int_{-1}^{1} \left(-\frac{2187}{64L^2}r^4 - \frac{243}{8L^2}r^3 + \frac{243}{32L^2}r^2 + \frac{63}{8L^2}r - \frac{27}{64L^2} \right) \frac{L}{2} dr =$$

$$= \left[0.34785 \left(-\frac{2187}{64L^2}(-0.86113)^4 - \frac{243}{8L^2}(-0.86113)^3 + \frac{243}{32L^2}(-0.86113)^2 + \frac{63}{8L^2}(-0.86113) - \frac{27}{64L^2} \right) + \right.$$

$$+ 0.65214 \left(-\frac{2187}{64L^2}(-0.33998)^4 - \frac{243}{8L^2}(-0.33998)^3 + \frac{243}{32L^2}(-0.33998)^2 + \frac{63}{8L^2}(-0.33998) - \frac{27}{64L^2} \right) +$$

$$+ 0.65214 \left(-\frac{2187}{64L^2}(0.33998)^4 - \frac{243}{8L^2}(0.33998)^3 + \frac{243}{32L^2}(0.33998)^2 + \frac{63}{8L^2}(0.33998) - \frac{27}{64L^2} \right) +$$

$$\left. + 0.34785 \left(-\frac{2187}{64L^2}(0.86113)^4 - \frac{243}{8L^2}(0.86113)^3 + \frac{243}{32L^2}(0.86113)^2 + \frac{63}{8L^2}(0.86113) - \frac{27}{64L^2} \right) \right] \frac{L}{2} =$$

$$= \frac{-4.725}{L} \tag{4.290}$$

The term $k_{68b_{B4}}$ is then equal to

$$k_{68b_{B4}} = -6.3 C_{11} \frac{a^3 b}{L} \tag{4.291}$$

Regarding the $k_{82b_{B4}}$ term

$$
\begin{aligned}
k_{82b_{B4}} &= \int_V \left(C_{11} F_{sy} \mathbf{J}^{-1} N_{4,r}(r) F_{\tau y} \mathbf{J}^{-1} N_{1,r}(r) \right) dV \\
&= C_{11} \int_A F_{sy} F_{\tau y} dA \int_L \mathbf{J}^{-1} N_{4,r}(r) \mathbf{J}^{-1} N_{1,r}(r) dL \\
&= C_{11} \int_{-a}^{a} F_{sy} F_{\tau y} dx \int_{-b}^{b} dz \int_{-1}^{1} \mathbf{J}^{-1} N_{4,r}(r) \mathbf{J}^{-1} N_{1,r}(r) det\, \mathbf{J} dr \\
&= C_{11} \int_{-a}^{a} x^2 dx \int_{-b}^{b} dz \int_{-1}^{1} \left(-\frac{729}{64L^2}r^4 + \frac{189}{32L^2}r^2 - \frac{1}{64L^2} \right) \frac{L}{2} dr
\end{aligned}
\tag{4.292}
$$

The term $k_{82b_{B4}}$ is equal to the term $k_{28b_{B4}}$

$$k_{82b_{B4}} = -0.4333 C_{11} \frac{a^3 b}{L} \tag{4.293}$$

Regarding the $k_{84b_{B4}}$ term

$$
\begin{aligned}
k_{84b_{B4}} &= \int_V \left(C_{11} F_{sy} \mathbf{J}^{-1} N_{4,r}(r) F_{\tau y} \mathbf{J}^{-1} N_{2,r}(r) \right) dV \\
&= C_{11} \int_A F_{sy} F_{\tau y} dA \int_L \mathbf{J}^{-1} N_{4,r}(r) \mathbf{J}^{-1} N_{2,r}(r) dL \\
&= C_{11} \int_{-a}^{a} F_{sy} F_{\tau y} dx \int_{-b}^{b} dz \int_{-1}^{1} \mathbf{J}^{-1} N_{4,r}(r) \mathbf{J}^{-1} N_{2,r}(r) det \mathbf{J} dr \\
&= C_{11} \int_{-a}^{a} x^2 dx \int_{-b}^{b} dz \int_{-1}^{1} \left(\frac{2187}{64L^2} r^4 + \frac{243}{16L^2} r^3 - \frac{567}{32L^2} r^2 - \frac{117}{16L^2} r + \frac{27}{64L^2} \right) \frac{L}{2} dr
\end{aligned}
$$
(4.294)

The term $k_{84b_{B4}}$ is equal to the term $k_{48b_{B4}}$

$$
k_{84b_{B4}} = 1.8 C_{11} \frac{a^3 b}{L}
$$
(4.295)

Regarding the $k_{86b_{B4}}$ term

$$
\begin{aligned}
k_{86b_{B4}} &= \int_V \left(C_{11} F_{sy} \mathbf{J}^{-1} N_{4,r}(r) F_{\tau y} \mathbf{J}^{-1} N_{3,r}(r) \right) dV \\
&= C_{11} \int_A F_{sy} F_{\tau y} dA \int_L \mathbf{J}^{-1} N_{4,r}(r) \mathbf{J}^{-1} N_{3,r}(r) dL \\
&= C_{11} \int_{-a}^{a} F_{sy} F_{\tau y} dx \int_{-b}^{b} dz \int_{-1}^{1} \mathbf{J}^{-1} N_{4,r}(r) \mathbf{J}^{-1} N_{3,r}(r) det \mathbf{J} dr \\
&= C_{11} \int_{-a}^{a} x^2 dx \int_{-b}^{b} dz \int_{-1}^{1} \left(-\frac{2187}{64L^2} r^4 - \frac{243}{8L^2} r^3 + \frac{243}{32L^2} r^2 + \frac{63}{8L^2} r - \frac{27}{64L^2} \right) \frac{L}{2} dr
\end{aligned}
$$
(4.296)

The term $k_{86b_{B4}}$ is equal to the term $k_{68b_{B4}}$

$$
k_{86b_{B4}} = -6.3 C_{11} \frac{a^3 b}{L}
$$
(4.297)

Regarding the $k_{88b_{B4}}$ term

$$
\begin{aligned}
k_{88b_{B4}} &= \int_V \left(C_{11} F_{sy} F_{\tau y} \mathbf{J}^{-1} N_{4,r}(r) \mathbf{J}^{-1} N_{4,r}(r) \right) dV \\
&= C_{11} \int_A F_{sy} F_{\tau y} dA \int_L \mathbf{J}^{-1} N_{4,r}(r) \mathbf{J}^{-1} N_{4,r}(r) dL \\
&= C_{11} \int_{-a}^{a} F_{sy} F_{\tau y} dx \int_{-b}^{b} dz \int_{-1}^{1} \mathbf{J}^{-1} N_{4,r}(r) \mathbf{J}^{-1} N_{4,r}(r) det \mathbf{J} dr \\
&= C_{11} \int_{-a}^{a} x^2 dx \int_{-b}^{b} dz \int_{-1}^{1} \left(\frac{729}{64L^2} r^4 + \frac{243}{16L^2} r^3 + \frac{135}{32L^2} r^2 - \frac{9}{16L^2} r + \frac{1}{64L^2} \right) \frac{L}{2} dr
\end{aligned}
$$
(4.298)

The surface integral of the term $k_{88b_{B4}}$ is computed using an analytical integration as follows

$$
C_{11} \int_{-a}^{a} x^2 dx \int_{-b}^{b} dz = C_{11} \frac{4a^3 b}{3}
$$
(4.299)

whereas the line integral is computed using the four SPs Gauss quadrature.

$$\int_{-1}^{1} \left(\frac{729}{64L^2}r^4 + \frac{243}{16L^2}r^3 + \frac{135}{32L^2}r^2 - \frac{9}{16L^2}r + \frac{1}{64L^2} \right) \frac{L}{2} dr =$$

$$= \left[0.34785 \left(\frac{729}{64L^2}(-0.86113)^4 + \frac{243}{16L^2}(-0.86113)^3 + \frac{135}{32L^2}(-0.86113)^2 - \frac{9}{16L^2}(-0.86113) + \frac{1}{64L^2} \right) + \right.$$

$$+ 0.65214 \left(\frac{729}{64L^2}(-0.33998)^4 + \frac{243}{16L^2}(-0.33998)^3 + \frac{135}{32L^2}(-0.33998)^2 - \frac{9}{16L^2}(-0.33998) + \frac{1}{64L^2} \right) +$$

$$+ 0.65214 \left(\frac{729}{64L^2}(0.33998)^4 - \frac{243}{16L^2}(0.33998)^3 + \frac{135}{32L^2}(0.33998)^2 + \frac{9}{16L^2}(0.33998) + \frac{1}{64L^2} \right) +$$

$$\left. + 0.34785 \left(\frac{729}{64L^2}(0.86113)^4 + \frac{243}{16L^2}(0.86113)^3 + \frac{135}{32L^2}(0.86113)^2 - \frac{9}{16L^2}(0.86113) + \frac{1}{64L^2} \right) \right] \frac{L}{2} =$$

$$= \frac{3.7}{L}$$

$$(4.300)$$

The term $k_{88b_{B4}}$ is then equal to

$$k_{88b_{B4}} = 4.933 C_{11} \frac{a^3 b}{L} \tag{4.301}$$

Based on the terms computed previously, the bending stiffneess matrix of the B4 beam element, obtained using a FI Gauss quadrature, is the following

$$\mathbf{K}_{b_{B4}} = \begin{bmatrix} 0 & 0 & 0 & 0 & 0 & 0 & 0 & 0 \\ 0 & 4.933C_{11}\frac{a^3b}{L} & 0 & -6.3C_{11}\frac{a^3b}{L} & 0 & 1.8C_{11}\frac{a^3b}{L} & 0 & -0.4333C_{11}\frac{a^3b}{L} \\ 0 & 0 & 0 & 0 & 0 & 0 & 0 & 0 \\ 0 & -6.3C_{11}\frac{a^3b}{L} & 0 & 14.4C_{11}\frac{a^3b}{L} & 0 & -9.9C_{11}\frac{a^3b}{L} & 0 & 1.8C_{11}\frac{a^3b}{L} \\ 0 & 0 & 0 & 0 & 0 & 0 & 0 & 0 \\ 0 & 1.8C_{11}\frac{a^3b}{L} & 0 & -9.9C_{11}\frac{a^3b}{L} & 0 & 14.4C_{11}\frac{a^3b}{L} & 0 & -6.3C_{11}\frac{a^3b}{L} \\ 0 & 0 & 0 & 0 & 0 & 0 & 0 & 0 \\ 0 & -0.4333C_{11}\frac{a^3b}{L} & 0 & k_{84b_{B4}}=1.8C_{11}\frac{a^3b}{L} & 0 & -6.3C_{11}\frac{a^3b}{L} & 0 & 4.933C_{11}\frac{a^3b}{L} \end{bmatrix}$$

$$(4.302)$$

4.4.2.4 Shear Stiffness Matrix Part for B4 Element

As stated at the beginning of the Sect. 4.4.2.2, the computation of the shear stiffness matrix must be made using a four SPs FI Gauss quadrature. The shear stiffness matrix can be written, in a general form, as follows

$$
\mathbf{K_{S_{B4}}} =
\begin{bmatrix}
k_{11s_{B4}} & k_{12s_{B4}} & k_{13s_{B4}} & k_{14s_{B4}} & k_{15s_{B4}} & k_{16s_{B4}} & k_{17s_{B4}} & k_{18s_{B4}} \\
k_{21s_{B4}} & k_{22s_{B4}} & k_{23s_{B4}} & k_{24s_{B4}} & k_{25s_{B4}} & k_{26s_{B4}} & k_{27s_{B4}} & k_{28s_{B4}} \\
k_{31s_{B4}} & k_{32s_{B4}} & k_{33s_{B4}} & k_{34s_{B4}} & k_{35s_{B4}} & k_{36s_{B4}} & k_{37s_{B4}} & k_{38s_{B4}} \\
k_{41s_{B4}} & k_{42s_{B4}} & k_{43s_{B4}} & k_{44s_{B4}} & k_{45s_{B4}} & k_{46s_{B4}} & k_{47s_{B4}} & k_{48s_{B4}} \\
k_{51s_{B4}} & k_{52s_{B4}} & k_{53s_{B4}} & k_{54s_{B4}} & k_{55s_{B4}} & k_{56s_{B4}} & k_{57s_{B4}} & k_{58s_{B4}} \\
k_{61s_{B4}} & k_{62s_{B4}} & k_{63s_{B4}} & k_{64s_{B4}} & k_{65s_{B4}} & k_{66s_{B4}} & k_{67s_{B4}} & k_{68s_{B4}} \\
k_{71s_{B4}} & k_{72s_{B4}} & k_{73s_{B4}} & k_{74s_{B4}} & k_{75s_{B4}} & k_{76s_{B4}} & k_{77s_{B4}} & k_{78s_{B4}} \\
k_{81s_{B4}} & k_{82s_{B4}} & k_{83s_{B4}} & k_{84s_{B4}} & k_{85s_{B4}} & k_{86s_{B4}} & k_{87s_{B4}} & k_{88s_{B4}}
\end{bmatrix}
\tag{4.303}
$$

Performing the triple product on the volume integral of Eq. (4.246), the terms of the matrix in Eq. (4.303) can be written as follows

$$
\begin{aligned}
k_{11s_{B4}} &= \int_V \left(C_{44} F_{sy,x} N_2(r) F_{\tau x} \mathbf{J}^{-1} N_{2,r}(r) \right) dV \\
&= C_{44} \int_A F_{sy,x} F_{\tau x} dA \int_L N_2(r) \mathbf{J}^{-1} N_{2,r}(r) dL \\
&= C_{44} \int_{-a}^{a} F_{sy,x} F_{\tau x} dx \int_{-b}^{b} dz \int_{-1}^{1} N_2(r) \mathbf{J}^{-1} N_{2,r}(r) det\,\mathbf{J} dr \\
&= C_{44} \int_{-a}^{a} 1 dx \int_{-b}^{b} dz \int_{-1}^{1} \left(\frac{729}{64L^2} r^4 - \frac{243}{16L^2} r^3 + \frac{135}{32L^2} r^2 + \frac{9}{16L^2} r + \frac{1}{64L^2} \right) \frac{L}{2} dr
\end{aligned}
\tag{4.304}
$$

The surface integral of the term $k_{11s_{B4}}$ is computed, using an analytical integration, as follows

$$
C_{44} \int_{-a}^{a} 1 dx \int_{-b}^{b} dz = C_{44} 4ab
\tag{4.305}
$$

while the line integral, computed using four SPs Gauss quadrature, is equal to that of Eqs. (4.250) and (4.252)

$$
\int_{-1}^{1} \left(\frac{729}{64L^2} r^4 - \frac{243}{16L^2} r^3 + \frac{135}{32L^2} r^2 + \frac{9}{16L^2} r + \frac{1}{64L^2} \right) \frac{L}{2} dr = \frac{3.7}{L}
\tag{4.306}
$$

The term $k_{11s\,B4}$ is then equal to

$$k_{11s\,B4} = 14.8 C_{44} \frac{ab}{L} \tag{4.307}$$

Regarding the $k_{12s\,B4}$ term

$$
\begin{aligned}
k_{12s\,B4} &= \int_V \left(C_{44} F_{sx} \mathbf{J}^{-1} N_{1,r}(r) F_{\tau y,x} N_1(r) \right) dV \\
&= C_{44} \int_A F_{sx} F_{\tau y,x} dA \int_L \mathbf{J}^{-1} N_{1,r}(r) N_1(r) dL \\
&= C_{44} \int_{-a}^{a} F_{sx} F_{\tau y,x} dx \int_{-b}^{b} dz \int_{-1}^{1} \mathbf{J}^{-1} N_{1,r}(r) N_1(r) det \, \mathbf{J} dr \\
&= C_{44} \int_{-a}^{a} (-1) dx \int_{-b}^{b} dz \int_{-1}^{1} \left(\frac{243}{128L} r^5 - \frac{405}{128L} r^4 + \frac{63}{64L} r^3 + \frac{27}{64L} r^2 - \frac{17}{128L} r - \frac{1}{128L} \right) \frac{L}{2} dr
\end{aligned}
\tag{4.308}
$$

The surface integral of the term $k_{12s\,B4}$ is computed, using an analytical integration

$$C_{44} \int_{-a}^{a} (-1) dx \int_{-b}^{b} dz = -C_{44} 4ab \tag{4.309}$$

whereas the line integral is computed using the four SPs Gauss quadrature.

$$
\begin{aligned}
&\int_{-1}^{1} \left(\frac{243}{128L} r^5 - \frac{405}{128L} r^4 + \frac{63}{64L} r^3 + \frac{27}{64L} r^2 - \frac{17}{128L} r - \frac{1}{128L} \right) \frac{L}{2} dr = \\
&= \left[0.34785 \left(\frac{243}{128L} (-0.86113)^5 - \frac{405}{128L} (-0.86113)^4 + \right. \right. \\
&\quad \left. + \frac{63}{64L} (-0.86113)^3 + \frac{27}{64L} (-0.86113)^2 - \frac{17}{128L} (-0.86113) - \frac{1}{128L} \right) + \\
&\quad + 0.65214 \left(\frac{243}{128L} (-0.33998)^5 - \frac{405}{128L} (-0.33998)^4 + \right. \\
&\quad \left. + \frac{63}{64L} (-0.33998)^3 + \frac{27}{64L} (-0.33998)^2 - \frac{17}{128L} (-0.33998) - \frac{1}{128L} \right) \\
&\quad + 0.65214 \left(\frac{243}{128L} (0.33998)^5 - \frac{405}{128L} (0.33998)^4 + \right. \\
&\quad \left. + \frac{63}{64L} (0.33998)^3 + \frac{27}{64L} (0.33998)^2 - \frac{17}{128L} (0.33998) - \frac{1}{128L} \right) \\
&\quad + 0.34785 \left(\frac{243}{128L} (0.86113)^5 - \frac{405}{128L} (0.86113)^4 + \right. \\
&\quad \left. \left. + \frac{63}{64L} (0.86113)^3 + \frac{27}{64L} (0.86113)^2 - \frac{17}{128L} (0.86113) - \frac{1}{128L} \right) \right] \frac{L}{2} = \\
&= -0.5
\end{aligned}
\tag{4.310}
$$

The term $k_{12s\,B4}$ is then equal to

$$k_{12s\,B4} = 2 C_{44} ab \tag{4.311}$$

Regarding the $k_{13s_{B4}}$ term

$$k_{13s_{B4}} = \int_V \left(C_{44} F_{sx} \mathbf{J}^{-1} N_{1,r}(r) F_{\tau x} \mathbf{J}^{-1} N_{2,r}(r) \right) dV$$

$$= C_{44} \int_A F_{sx} F_{\tau x} \, dA \int_L \mathbf{J}^{-1} N_{1,r}(r \mathbf{J}^{-1} N_{2,r}(r) dL$$

$$= C_{44} \int_{-a}^{a} F_{sx} F_{\tau x} \, dx \int_{-b}^{b} dz \int_{-1}^{1} \mathbf{J}^{-1} N_{1,r}(r \mathbf{J}^{-1} N_{2,r}(r) \det \mathbf{J} \, dr$$

$$= C_{44} \int_{-a}^{a} 1 \, dx \int_{-b}^{b} dz \int_{-1}^{1} \left(-\frac{2187}{64L^2} r^4 + \frac{243}{8L^2} r^3 + \frac{243}{32L^2} r^2 - \frac{63}{8L^2} r - \frac{27}{64L^2} \right) \frac{L}{2} dr$$

$$\tag{4.312}$$

The surface integral of the term $k_{13s_{B4}}$ is computed, using an analytical integration

$$C_{44} \int_{-a}^{a} 1 \, dx \int_{-b}^{b} dz = C_{44} 4ab \tag{4.313}$$

while the line integral, computed using the four SPs Gauss quadrature, is equal to that of Eqs. (4.254), (4.256) and (4.257) and reported below:

$$\int_{-1}^{1} \left(-\frac{2187}{64L^2} r^4 + \frac{243}{8L^2} r^3 + \frac{243}{32L^2} r^2 - \frac{63}{8L^2} r - \frac{27}{64L^2} \right) \frac{L}{2} dr = \frac{-4.725}{L} \tag{4.314}$$

The term $k_{13s_{B4}}$ is then equal to

$$k_{13s_{B4}} = -18.9 C_{44} \frac{ab}{L} \tag{4.315}$$

Regarding the $k_{14s_{B4}}$ term

$$k_{14s_{B4}} = \int_V \left(C_{44} F_{sx} \mathbf{J}^{-1} N_{1,r}(r) F_{\tau y,x} N_2(r) \right) dV$$

$$= C_{44} \int_A F_{sx} F_{\tau y,x} \, dA \int_L \mathbf{J}^{-1} N_{1,r}(r) N_2(r) dL$$

$$= C_{44} \int_{-a}^{a} F_{sx} F_{\tau y,x} \, dx \int_{-b}^{b} dz \int_{-1}^{1} \mathbf{J}^{-1} N_{1,r}(r) N_2(r) \det \mathbf{J} \, dr$$

$$= C_{44} \int_{-a}^{a} (-1) \, dx \int_{-b}^{b} dz \int_{-1}^{1} \left(-\frac{729}{128L} r^5 + \frac{729}{128L} r^4 + \frac{297}{64L} r^3 - \frac{369}{64L} r^2 + \frac{135}{128L} r + \frac{9}{128L} \right) \frac{L}{2} dr$$

$$\tag{4.316}$$

The surface integral of the term $k_{14s_{B4}}$ is computed, using an analytical integratiom

$$C_{44} \int_{-a}^{a} (-1) \, dx \int_{-b}^{b} dz = -C_{44} 4ab \tag{4.317}$$

whereas the line integral is computed using the four SPs Gauss quadrature.

$$\int_{-1}^{1}\left(-\frac{729}{128L}r^5+\frac{729}{128L}r^4+\frac{297}{64L}r^3-\frac{369}{64L}r^2+\frac{135}{128L}r+\frac{9}{128L}\right)\frac{L}{2}dr=$$

$$=\left[0.34785\left(-\frac{729}{128L}(-0.86113)^5+\frac{729}{128L}(-0.86113)^4+\right.\right.$$

$$+\frac{297}{64L}(-0.86113)^3-\frac{369}{64L}(-0.86113)^2+\frac{135}{128L}(-0.86113)+\frac{9}{128L}\bigg)+$$

$$+0.65214\left(-\frac{729}{128L}(-0.33998)^5+\frac{729}{128L}(-0.33998)^4+\right.$$

$$+\frac{297}{64L}(-0.33998)^3-\frac{369}{64L}(-0.33998)^2+\frac{135}{128L}(-0.33998)+\frac{9}{128L}\bigg)$$

$$+0.65214\left(-\frac{729}{128L}(0.33998)^5+\frac{729}{128L}(0.33998)^4+\right.$$

$$+\frac{297}{64L}(0.33998)^3-\frac{369}{64L}(0.33998)^2+\frac{135}{128L}(0.33998)+\frac{9}{128L}\bigg)$$

$$+0.34785\left(-\frac{729}{128L}(0.86113)^5+\frac{729}{128L}(0.86113)^4+\right.$$

$$+\frac{297}{64L}(0.86113)^3-\frac{369}{64L}(0.86113)^2+\frac{135}{128L}(0.86113)+\frac{9}{128L}\bigg)\bigg]\frac{L}{2}=$$

$$=-0.7125 \tag{4.318}$$

The term $k_{14s\,B4}$ is then equal to

$$k_{14s\,B4}=2.85C_{44}ab \tag{4.319}$$

Regarding the $k_{15s\,B4}$ term

$$k_{15s\,B4}=\int_V\left(C_{44}F_{sx}\mathbf{J}^{-1}N_{1,r}(r)F_{\tau x}\mathbf{J}^{-1}N_{3,r}(r)\right)dV$$

$$=C_{44}\int_A F_{sx}F_{\tau x}dA\int_L \mathbf{J}^{-1}N_{1,r}(r\mathbf{J}^{-1}N_{3,r}(r)dL$$

$$=C_{44}\int_{-a}^{a}F_{sx}F_{\tau x}dx\int_{-b}^{b}dz\int_{-1}^{1}\mathbf{J}^{-1}N_{1,r}(r\mathbf{J}^{-1}N_{3,r}(r)det\,\mathbf{J}dr$$

$$=C_{44}\int_{-a}^{a}1dx\int_{-b}^{b}dz\int_{-1}^{1}\left(\frac{2187}{64L^2}r^4-\frac{243}{16L^2}r^3-\frac{567}{32L^2}r^2+\frac{117}{16L^2}r+\frac{27}{64L^2}\right)\frac{L}{2}dr \tag{4.320}$$

The surface integral of the term $k_{15s\,B4}$ is computed, using an analytical integration

$$C_{44}\int_{-a}^{a}1dx\int_{-b}^{b}dz=C_{44}4ab \tag{4.321}$$

while the line integral,computed using the four SPs Gauss quadrature, is equal to
that of Eqs. (4.258), (4.260) and (4.261)

$$\int_{-1}^{1} \left(\frac{2187}{64L^2} r^4 - \frac{243}{16L^2} r^3 - \frac{567}{32L^2} r^2 + \frac{117}{16L^2} r + \frac{27}{64L^2} \right) \frac{L}{2} dr = \frac{1.35}{L} \quad (4.322)$$

The term $k_{15s\,B4}$ is then equal to

$$k_{15s\,B4} = 5.4 C_{44} \frac{ab}{L} \quad (4.323)$$

Regarding the $k_{16s\,B4}$ term

$$
\begin{aligned}
k_{16s\,B4} &= \int_V \left(C_{44} F_{sx} \mathbf{J}^{-1} N_{1,r}(r) F_{\tau y,x} N_3(r) \right) dV \\
&= C_{44} \int_A F_{sx} F_{\tau y,x} dA \int_L \mathbf{J}^{-1} N_{1,r}(r) N_3(r) dL \\
&= C_{44} \int_{-a}^{a} F_{sx} F_{\tau y,x} dx \int_{-b}^{b} dz \int_{-1}^{1} \mathbf{J}^{-1} N_{1,r}(r) N_3(r) \det \mathbf{J} dr \\
&= C_{44} \int_{-a}^{a} (-1) dx \int_{-b}^{b} dz \int_{-1}^{1} \left(\frac{729}{128L} r^5 - \frac{243}{128L} r^4 - \frac{459}{64L} r^3 + \frac{117}{64L} r^2 + \frac{189}{128L} r + \frac{9}{128L} \right) \frac{L}{2} dr
\end{aligned}
$$
$$(4.324)$$

The surface integral of the term $k_{16s\,B4}$ is computed, using an analytical integration

$$C_{44} \int_{-a}^{a} (-1) dx \int_{-b}^{b} dz = -C_{44} 4ab \quad (4.325)$$

whereas the line integral is computed using the four SPs Gauss quadrature.

$$
\begin{aligned}
&\int_{-1}^{1} \left(\frac{729}{128L} r^5 - \frac{243}{128L} r^4 - \frac{459}{64L} r^3 + \frac{117}{64L} r^2 + \frac{189}{128L} r + \frac{9}{128L} \right) \frac{L}{2} dr = \\
&= \left[0.34785 \left(\frac{729}{128L} (-0.86113)^5 - \frac{243}{128L} (-0.86113)^4 - \right. \right. \\
&\quad \left. - \frac{459}{64L} (-0.86113)^3 + \frac{117}{64L} (-0.86113)^2 + \frac{189}{128L} (-0.86113) + \frac{9}{128L} \right) + \\
&\quad + 0.65214 \left(\frac{729}{128L} (-0.33998)^5 - \frac{243}{128L} (-0.33998)^4 - \right. \\
&\quad \left. - \frac{459}{64L} (-0.33998)^3 + \frac{117}{64L} (-0.33998)^2 + \frac{189}{128L} (-0.33998) + \frac{9}{128L} \right) \\
&\quad + 0.65214 \left(\frac{729}{128L} (0.33998)^5 - \frac{243}{128L} (0.33998)^4 - \right. \\
&\quad \left. - \frac{459}{64L} (0.33998)^3 + \frac{117}{64L} (0.33998)^2 + \frac{189}{128L} (0.33998) + \frac{9}{128L} \right) \\
&\quad + 0.34785 \left(\frac{729}{128L} (0.86113)^5 - \frac{243}{128L} (0.86113)^4 - \right. \\
&\quad \left. \left. - \frac{459}{64L} (0.86113)^3 + \frac{117}{64L} (0.86113)^2 + \frac{189}{128L} (0.86113) + \frac{9}{128L} \right) \right] \frac{L}{2} = \\
&= 0.3
\end{aligned}
$$
$$(4.326)$$

The term $k_{16s\,B4}$ is then equal to

$$k_{16s\,B4} = -1.2 C_{44} ab \tag{4.327}$$

Regarding the $k_{17s\,B4}$ term

$$
\begin{aligned}
k_{17s\,B4} &= \int_V \left(C_{44} F_{sx} \mathbf{J}^{-1} N_{1,r}(r) F_{\tau x} \mathbf{J}^{-1} N_{4,r}(r) \right) dV \\
&= C_{44} \int_A F_{sx} F_{\tau x} dA \int_L \mathbf{J}^{-1} N_{1,r}(r \mathbf{J}^{-1} N_{4,r}(r) dL \\
&= C_{44} \int_{-a}^{a} F_{sx} F_{\tau x} dx \int_{-b}^{b} dz \int_{-1}^{1} \mathbf{J}^{-1} N_{1,r}(r \mathbf{J}^{-1} N_{4,r}(r) \det \mathbf{J} dr \\
&= C_{44} \int_{-a}^{a} 1 dx \int_{-b}^{b} dz \int_{-1}^{1} \left(-\frac{729}{64L^2} r^4 + \frac{189}{32L^2} r^2 - \frac{1}{64L^2} \right) \frac{L}{2} dr
\end{aligned}
\tag{4.328}
$$

The surface integral of the term $k_{17s\,B4}$ is computed, using an analytical integration

$$C_{44} \int_{-a}^{a} 1 dx \int_{-b}^{b} dz = C_{44} 4ab \tag{4.329}$$

while the line integral, computed using the four SPs Gauss quadrature, is equal to that of Eqs. (4.262), (4.264) and (4.265)

$$\int_{-1}^{1} \left(-\frac{729}{64L^2} r^4 + \frac{189}{32L^2} r^2 - \frac{1}{64L^2} \right) \frac{L}{2} dr = \frac{-0325}{L} \tag{4.330}$$

The term $k_{17s\,B4}$ is then equal to

$$k_{17s\,B4} = -1.3 C_{44} \frac{ab}{L} \tag{4.331}$$

Regarding the $k_{18s\,B4}$ term

$$
\begin{aligned}
k_{18s\,B4} &= \int_V \left(C_{44} F_{sx} \mathbf{J}^{-1} N_{1,r}(r) F_{\tau y,x} N_4(r) \right) dV \\
&= C_{44} \int_A F_{sx} F_{\tau y,x} dA \int_L \mathbf{J}^{-1} N_{1,r}(r) N_4(r) dL \\
&= C_{44} \int_{-a}^{a} F_{sx} F_{\tau y,x} dx \int_{-b}^{b} dz \int_{-1}^{1} \mathbf{J}^{-1} N_{1,r}(r) N_4(r) \det \mathbf{J} dr \\
&= C_{44} \int_{-a}^{a} (-1) dx \int_{-b}^{b} dz \int_{-1}^{1} \left(-\frac{243}{128L} r^5 - \frac{81}{128L} r^4 + \frac{99}{64L} r^3 + \frac{9}{64L} r^2 - \frac{19}{128L} r - \frac{1}{128L} \right) \frac{L}{2} dr
\end{aligned}
\tag{4.332}
$$

The surface integral of the term $k_{18s\,B4}$ is computed, using an analytical integration

$$C_{44} \int_{-a}^{a} (-1) dx \int_{-b}^{b} dz = -C_{44} 4ab \tag{4.333}$$

whereas the line integral is computed using the four SPs Gauss quadrature.

$$\int_{-1}^{1} \left(-\frac{243}{128L}r^5 - \frac{81}{128L}r^4 + \frac{99}{64L}r^3 + \frac{9}{64L}r^2 - \frac{19}{128L}r - \frac{1}{128L} \right) \frac{L}{2} dr =$$

$$= \left[0.34785 \left(-\frac{243}{128L}(-0.86113)^5 - \frac{81}{128L}(-0.86113)^4 + \frac{99}{64L}(-0.86113)^3 + \right.\right.$$

$$+ \frac{9}{64L}(-0.86113)^2 - \frac{19}{128L}(-0.86113) - \frac{1}{128L} \right) +$$

$$+ 0.65214 \left(-\frac{243}{128L}(-0.33998)^5 - \frac{81}{128L}(-0.33998)^4 + \frac{99}{64L}(-0.33998)^3 + \right.$$

$$+ \frac{9}{64L}(-0.33998)^2 - \frac{19}{128L}(-0.33998) - \frac{1}{128L} \right)$$

$$+ 0.65214 \left(-\frac{243}{128L}(0.33998)^5 - \frac{81}{128L}(0.33998)^4 + \frac{99}{64L}(0.33998)^3 + \right.$$

$$+ \frac{9}{64L}(0.33998)^2 - \frac{19}{128L}(0.33998) - \frac{1}{128L} \right)$$

$$+ 0.34785 \left(-\frac{243}{128L}(0.86113)^5 - \frac{81}{128L}(0.86113)^4 + \frac{99}{64L}(0.86113)^3 + \right.$$

$$+ \frac{9}{64L}(0.86113)^2 - \frac{19}{128L}(0.86113) - \frac{1}{128L} \right) \bigg] \frac{L}{2} =$$

$$= -0.0875 \tag{4.334}$$

The term $k_{18s_{B4}}$ is then equal to

$$k_{18s_{B4}} = 0.35 C_{44} ab \tag{4.335}$$

Regarding the $k_{21s_{B4}}$ term

$$k_{21s_{B4}} = \int_V \left(C_{44} F_{sy,x} N_1(r) F_{\tau x} \mathbf{J}^{-1} N_{1,r}(r) \right) dV$$

$$= C_{44} \int_A F_{sy,x} F_{\tau x} dA \int_L N_1(r) \mathbf{J}^{-1} N_{1,r}(r) dL$$

$$= C_{44} \int_{-a}^{a} F_{sy,x} F_{\tau x} dx \int_{-b}^{b} dz \int_{-1}^{1} N_1(r) \mathbf{J}^{-1} N_{1,r}(r) \det \mathbf{J} dr$$

$$= C_{44} \int_{-a}^{a} (-1) dx \int_{-b}^{b} dz \int_{-1}^{1} \left(\frac{243}{128L}r^5 - \frac{405}{128L}r^4 + \frac{63}{64L}r^3 + \frac{27}{64L}r^2 - \frac{17}{128L}r - \frac{1}{128L} \right) \frac{L}{2} dr \tag{4.336}$$

The term $k_{21s_{B4}}$ is equal to the term $k_{12s_{B4}}$, see Eqs. (4.308) and (4.311)

$$k_{21s_{B4}} = 2 C_{44} ab \tag{4.337}$$

Regarding the $k_{22s_{B4}}$ term

$$
\begin{aligned}
k_{22s\,B4} &= \int_V \left(C_{44}\, F_{sy,x}\, N_1(r)\, F_{\tau y,x}\, N_1(r) \right) dV \\
&= C_{44} \int_A F_{sy,x}\, F_{\tau y,x}\, dA \int_L N_1(r) N_1(r)\, dL \\
&= C_{44} \int_{-a}^{a} F_{sy,x}\, F_{\tau y,x}\, dx \int_{-b}^{b} dz \int_{-1}^{1} N_1(r) N_1(r)\, \det \mathbf{J}\, dr \\
&= C_{44} \int_{-a}^{a} (1)\, dx \int_{-b}^{b} dz \int_{-1}^{1} \left(\frac{81}{256} r^6 - \frac{81}{128} r^5 + \frac{63}{256} r^4 + \frac{9}{64} r^3 - \frac{17}{256} r^2 - \frac{1}{128} r + \frac{1}{256} \right) \frac{L}{2}\, dr
\end{aligned}
\tag{4.338}
$$

The surface integral of the term $k_{22s\,B4}$ is computed, using an analytical integration

$$
C_{44} \int_{-a}^{a} (1)\, dx \int_{-b}^{b} dz = C_{44} 4ab
\tag{4.339}
$$

whereas the line integral is computed using the four SPs Gauss quadrature.

$$
\begin{aligned}
\int_{-1}^{1} &\left(\frac{81}{256} r^6 - \frac{81}{128L} r^5 + \frac{63}{256} r^4 + \frac{9}{64} r^3 - \frac{17}{256} r^2 - \frac{1}{128} r + \frac{1}{256} \right) \frac{L}{2}\, dr = \\
&= \Bigg[0.34785 \left(\frac{81}{256} (-0.86113)^6 - \frac{81}{128L} (-0.86113)^5 + \frac{63}{256} (-0.86113)^4 \right. \\
&\qquad\left. + \frac{9}{64} (-0.86113)^3 - \frac{17}{256} (-0.86113)^2 - \frac{1}{128} (-0.86113) + \frac{1}{256} \right) + \\
&\quad + 0.65214 \left(\frac{81}{256} (-0.33998)^6 - \frac{81}{128L} (-0.33998)^5 + \frac{63}{256} (-0.33998)^4 \right. \\
&\qquad\left. + \frac{9}{64} (-0.33998)^3 - \frac{17}{256} (-0.33998)^2 - \frac{1}{128} (-0.33998) + \frac{1}{256} \right) \\
&\quad + 0.65214 \left(\frac{81}{256} (0.33998)^6 - \frac{81}{128L} (0.33998)^5 + \frac{63}{256} (0.33998)^4 \right. \\
&\qquad\left. + \frac{9}{64} (0.33998)^3 - \frac{17}{256} (0.33998)^2 - \frac{1}{128} (0.33998) + \frac{1}{256} \right) \\
&\quad + 0.34785 \left(\frac{81}{256} (0.86113)^6 - \frac{81}{128L} (0.86113)^5 + \frac{63}{256} (0.86113)^4 \right. \\
&\qquad\left. + \frac{9}{64} (0.86113)^3 - \frac{17}{256} (0.86113)^2 - \frac{1}{128} (0.86113) + \frac{1}{256} \right) \Bigg] \frac{L}{2} = \\
&= 0.07619L
\end{aligned}
\tag{4.340}
$$

The term $k_{22s\,B4}$ is then equal to

$$
k_{22s\,B4} = 0.3047 C_{44} ab L
\tag{4.341}
$$

Regarding the $k_{23s\,B4}$ term

$$
\begin{aligned}
k_{23s\,B4} &= \int_V \left(C_{44}\, F_{sy,x}\, N_1(r)\, F_{\tau x}\, \mathbf{J}^{-1}\, N_{2,r}(r) \right) dV \\[2mm]
&= C_{44} \int_A F_{sy,x}\, F_{\tau x}\, dA \int_L N_1(r)\, \mathbf{J}^{-1}\, N_{2,r}(r)\, dL \\[2mm]
&= C_{44} \int_{-a}^{a} F_{sy,x}\, F_{\tau x}\, dx \int_{-b}^{b} dz \int_{-1}^{1} N_1(r)\, \mathbf{J}^{-1}\, N_{2,r}(r)\, \det \mathbf{J}\, dr \\[2mm]
&= C_{44} \int_{-a}^{a} (-1)\, dx \int_{-b}^{b} dz \int_{-1}^{1} \left(-\frac{729}{128L} r^5 + \frac{891}{128L} r^4 + \frac{81}{64L} r^3 - \frac{171}{64L} r^2 - \frac{9}{128L} r + \frac{27}{128L} \right) \frac{L}{2}\, dr
\end{aligned}
$$
$$(4.342)$$

The surface integral of the term $k_{23s\,B4}$ is computed, using an analytical integration

$$
C_{44} \int_{-a}^{a} (-1)\, dx \int_{-b}^{b} dz = -C_{44} 4ab \tag{4.343}
$$

whereas the line integral is computed using the four SPs Gauss quadrature.

$$
\begin{aligned}
\int_{-1}^{1} &\left(-\frac{729}{128L} r^5 + \frac{891}{128L} r^4 + \frac{81}{64L} r^3 - \frac{171}{64L} r^2 - \frac{9}{128L} r + \frac{27}{128L} \right) \frac{L}{2}\, dr = \\[2mm]
&= \Bigg[0.34785 \left(-\frac{729}{128L} (-0.86113)^5 + \frac{891}{128L} (-0.86113)^4 \right. \\[2mm]
&\quad \left. + \frac{81}{64L} (-0.86113)^3 - \frac{171}{64L} (-0.86113)^2 - \frac{9}{128L} (-0.86113) + \frac{27}{128L} \right) + \\[2mm]
&\quad + 0.65214 \left(-\frac{729}{128L} (-0.33998)^5 + \frac{891}{128L} (-0.33998)^4 \right. \\[2mm]
&\quad \left. + \frac{81}{64L} (-0.33998)^3 - \frac{171}{64L} (-0.33998)^2 - \frac{9}{128L} (-0.33998) + \frac{27}{128L} \right) \\[2mm]
&\quad + 0.65214 \left(-\frac{729}{128L} (0.33998)^5 + \frac{891}{128L} (0.33998)^4 \right. \\[2mm]
&\quad \left. + \frac{81}{64L} (0.33998)^3 - \frac{171}{64L} (0.33998)^2 - \frac{9}{128L} (0.33998) + \frac{27}{128L} \right) \\[2mm]
&\quad + 0.34785 \left(-\frac{729}{128L} (0.86113)^5 + \frac{891}{128L} (0.86113)^4 \right. \\[2mm]
&\quad \left. + \frac{81}{64L} (0.86113)^3 - \frac{171}{64L} (0.86113)^2 - \frac{9}{128L} (0.86113) + \frac{27}{128L} \right) \Bigg] \frac{L}{2} = \\[2mm]
&= 0.7125
\end{aligned}
$$
$$(4.344)$$

The term $k_{23s\,B4}$ is then equal to

$$
k_{23s\,B4} = -2.85\, C_{44}\, ab \tag{4.345}
$$

Regarding the $k_{24s\,B4}$ term

$$k_{24s\,B4} = \int_V (C_{44}\, F_{sy,x}\, N_1(r)\, F_{\tau y,x}\, N_2(r))\, dV$$

$$= C_{44} \int_A F_{sy,x}\, F_{\tau y,x}\, dA \int_L N_1(r) N_2(r)\, dL$$

$$= C_{44} \int_{-a}^{a} F_{sy,x}\, F_{\tau y,x}\, dx \int_{-b}^{b} dz \int_{-1}^{1} N_1(r) N_2(r)\, det\, \mathbf{J}\, dr$$

$$= C_{44} \int_{-a}^{a} (1)\, dx \int_{-b}^{b} dz \int_{-1}^{1} \left(-\frac{243}{256} r^6 + \frac{81}{64L} r^5 + \frac{189}{256} r^4 - \frac{45}{32} r^3 + \frac{63}{256} r^2 + \frac{9}{64} r - \frac{9}{256} \right) \frac{L}{2}\, dr$$

$$\tag{4.346}$$

The surface integral of the term $k_{24s\,B4}$ is computed, using an analytical integration

$$C_{44} \int_{-a}^{a} (1)\, dx \int_{-b}^{b} dz = C_{44} 4ab \tag{4.347}$$

whereas the line integral is computed using the four SPs Gauss quadrature.

$$\int_{-1}^{1} \left(-\frac{243}{256} r^6 + \frac{81}{64L} r^5 + \frac{189}{256} r^4 - \frac{45}{32} r^3 + \frac{63}{256} r^2 + \frac{9}{64} r - \frac{9}{256} \right) \frac{L}{2}\, dr =$$

$$= \Bigg[0.34785 \left(-\frac{243}{256}(-0.86113)^6 + \frac{81}{64L}(-0.86113)^5 + \frac{189}{256}(-0.86113)^4 \right.$$

$$\left. -\frac{45}{32}(-0.86113)^3 + \frac{63}{256}(-0.86113)^2 + \frac{9}{64}(-0.86113) - \frac{9}{256} \right) +$$

$$+ 0.65214 \left(-\frac{243}{256}(-0.33998)^6 + \frac{81}{64L}(-0.33998)^5 + \frac{189}{256}(-0.33998)^4 - \right.$$

$$\left. -\frac{45}{32}(-0.33998)^3 + \frac{63}{256}(-0.33998)^2 + \frac{9}{64}(-0.33998) - \frac{9}{256} \right)$$

$$+ 0.65214 \left(-\frac{243}{256}(0.33998)^6 + \frac{81}{64L}(0.33998)^5 + \frac{189}{256}(0.33998)^4 - \right.$$

$$\left. -\frac{45}{32}(0.33998)^3 + \frac{63}{256}(0.33998)^2 + \frac{9}{64}(0.33998) - \frac{9}{256} \right)$$

$$+ 0.34785 \left(-\frac{243}{256}(0.86113)^6 + \frac{81}{64L}(0.86113)^5 + \frac{189}{256}(0.86113)^4 - \right.$$

$$\left. -\frac{45}{32}(0.86113)^3 + \frac{63}{256}(0.86113)^2 + \frac{9}{64}(0.86113) - \frac{9}{256} \right) \Bigg] \frac{L}{2} =$$

$$= 0.05893L \tag{4.348}$$

The term $k_{24s\,B4}$ is then equal to

$$k_{24s\,B4} = 0.2357 C_{44} abL \tag{4.349}$$

Regarding the $k_{25s\,B4}$ term

$$k_{25s_{B4}} = \int_V \left(C_{44} F_{sy,x} N_1(r) F_{\tau x} \mathbf{J}^{-1} N_{3,r}(r) \right) dV$$

$$= C_{44} \int_A F_{sy,x} F_{\tau x} dA \int_L N_1(r) \mathbf{J}^{-1} N_{3,r}(r) dL$$

$$= C_{44} \int_{-a}^a F_{sy,x} F_{\tau x} dx \int_{-b}^b dz \int_{-1}^1 N_1(r) \mathbf{J}^{-1} N_{3,r}(r) det\, \mathbf{J} dr$$

$$= C_{44} \int_{-a}^a (-1)\, dx \int_{-b}^b dz \int_{-1}^1 \left(\frac{729}{128L} r^5 - \frac{567}{128L} r^4 - \frac{243}{64L} r^3 + \frac{153}{64L} r^2 + \frac{45}{128L} r - \frac{27}{128L} \right) \frac{L}{2} dr$$

$$(4.350)$$

The surface integral of the term $k_{25s_{B4}}$ is computed, using an analytical integration

$$C_{44} \int_{-a}^a (-1)\, dx \int_{-b}^b dz = -C_{44} 4ab \tag{4.351}$$

whereas the line integral is computed using the four SPs Gauss quadrature.

$$\int_{-1}^1 \left(\frac{729}{128L} r^5 - \frac{567}{128L} r^4 - \frac{243}{64L} r^3 + \frac{153}{64L} r^2 + \frac{45}{128L} r - \frac{27}{128L} \right) \frac{L}{2} dr =$$

$$= \left[0.34785 \left(\frac{729}{128L} (-0.86113)^5 - \frac{567}{128L} (-0.86113)^4 \right.\right.$$

$$\left. - \frac{243}{64L} (-0.86113)^3 + \frac{153}{64L} (-0.86113)^2 + \frac{45}{128L} (-0.86113) - \frac{27}{128L} \right) +$$

$$+ 0.65214 \left(\frac{729}{128L} (-0.33998)^5 - \frac{567}{128L} (-0.33998)^4 \right.$$

$$\left. - \frac{243}{64L} (-0.33998)^3 + \frac{153}{64L} (-0.33998)^2 + \frac{45}{128L} (-0.33998) - \frac{27}{128L} \right)$$

$$+ 0.65214 \left(\frac{729}{128L} (0.33998)^5 - \frac{567}{128L} (0.33998)^4 \right.$$

$$\left. - \frac{243}{64L} (0.33998)^3 + \frac{153}{64L} (0.33998)^2 + \frac{45}{128L} (0.33998) - \frac{27}{128L} \right)$$

$$+ 0.34785 \left(\frac{729}{128L} (0.86113)^5 - \frac{567}{128L} (0.86113)^4 \right.$$

$$\left. - \frac{243}{64L} (0.86113)^3 + \frac{153}{64L} (0.86113)^2 + \frac{45}{128L} (0.86113) - \frac{27}{128L} \right) \right] \frac{L}{2} =$$

$$= -0.3 \tag{4.352}$$

The term $k_{25s_{B4}}$ is then equal to

$$k_{25s_{B4}} = 1.2 C_{44} ab \tag{4.353}$$

Regarding the $k_{26s_{B4}}$ term

$$
\begin{aligned}
k_{26sB4} &= \int_V \left(C_{44} F_{sy,x} N_1(r) F_{\tau y,x} N_3(r) \right) dV \\
&= C_{44} \int_A F_{sy,x} F_{\tau y,x} dA \int_L N_1(r) N_3(r) dL \\
&= C_{44} \int_{-a}^{a} F_{sy,x} F_{\tau y,x} dx \int_{-b}^{b} dz \int_{-1}^{1} N_1(r) N_3(r) \det \mathbf{J} dr \\
&= C_{44} \int_{-a}^{a} (1) dx \int_{-b}^{b} dz \int_{-1}^{1} \left(\frac{243}{256} r^6 - \frac{81}{64L} r^5 - \frac{351}{256} r^4 + \frac{45}{64} r^3 + \frac{117}{256} r^2 - \frac{9}{128} r - \frac{9}{256} \right) \frac{L}{2} dr
\end{aligned}
$$

$$(4.354)$$

The surface integral of the term k_{26sB4} is computed, using an analytical integration

$$
C_{44} \int_{-a}^{a} (1)\, dx \int_{-b}^{b} dz = C_{44} 4ab \tag{4.355}
$$

whereas the line integral is computed using the four SPs Gauss quadrature.

$$
\begin{aligned}
\int_{-1}^{1} &\left(\frac{243}{256} r^6 - \frac{81}{64L} r^5 - \frac{351}{256} r^4 + \frac{45}{64} r^3 + \frac{117}{256} r^2 - \frac{9}{128} r - \frac{9}{256} \right) \frac{L}{2} dr = \\
&= \Bigg[0.34785 \left(\frac{243}{256} (-0.86113)^6 - \frac{81}{64L} (-0.86113)^5 - \frac{351}{256} (-0.86113)^4 \right. \\
&\qquad \left. + \frac{45}{64} (-0.86113)^3 + \frac{117}{256} (-0.86113)^2 - \frac{9}{128} (-0.86113) - \frac{9}{256} \right) + \\
&\quad + 0.65214 \left(\frac{243}{256} (-0.33998)^6 - \frac{81}{64L} (-0.33998)^5 - \frac{351}{256} (-0.33998)^4 \right. \\
&\qquad \left. + \frac{45}{64} (-0.33998)^3 + \frac{117}{256} (-0.33998)^2 - \frac{9}{128} (-0.33998) - \frac{9}{256} \right) \\
&\quad + 0.65214 \left(\frac{243}{256} (0.33998)^6 - \frac{81}{64L} (0.33998)^5 - \frac{351}{256} (0.33998)^4 \right. \\
&\qquad \left. + \frac{45}{64} (0.33998)^3 + \frac{117}{256} (0.33998)^2 - \frac{9}{128} (0.33998) - \frac{9}{256} \right) \\
&\quad + 0.34785 \left(\frac{243}{256} (0.86113)^6 - \frac{81}{64L} (0.86113)^5 - \frac{351}{256} (0.86113)^4 \right. \\
&\qquad \left. + \frac{45}{64} (0.86113)^3 + \frac{117}{256} (0.86113)^2 - \frac{9}{128} (0.86113) - \frac{9}{256} \right) \Bigg] \frac{L}{2} = \\
&= -0.02143 L
\end{aligned}
$$

$$(4.356)$$

The term k_{26sB4} is then equal to

$$
k_{26sB4} = -0.08571 C_{44} ab L \tag{4.357}
$$

Regarding the k_{27sB4} term

$$
\begin{aligned}
k_{27s\,B4} &= \int_V \left(C_{44}\, F_{sy,x}\, N_1(r)\, F_{\tau x}\, \mathbf{J}^{-1} N_{4,r}(r) \right) dV \\
&= C_{44} \int_A F_{sy,x}\, F_{\tau x}\, dA \int_L N_1(r)\, \mathbf{J}^{-1} N_{4,r}(r)\, dL \\
&= C_{44} \int_{-a}^{a} F_{sy,x}\, F_{\tau x}\, dx \int_{-b}^{b} dz \int_{-1}^{1} N_1(r)\, \mathbf{J}^{-1} N_{4,r}(r)\, det\, \mathbf{J}\, dr \\
&= C_{44} \int_{-a}^{a} (-1)\, dx \int_{-b}^{b} dz \int_{-1}^{1} \left(-\frac{243}{128L} r^5 + \frac{81}{128L} r^4 + \frac{99}{64L} r^3 - \frac{9}{64L} r^2 - \frac{19}{128L} r + \frac{1}{128L} \right) \frac{L}{2} dr
\end{aligned}
$$

$$\tag{4.358}$$

The surface integral of the term $k_{27s\,B4}$ is computed, using an analytical integration

$$
C_{44} \int_{-a}^{a} (-1)\, dx \int_{-b}^{b} dz = -C_{44} 4ab
\tag{4.359}
$$

whereas the line integral is computed using the four SPs Gauss quadrature.

$$
\begin{aligned}
\int_{-1}^{1} &\left(-\frac{243}{128L} r^5 + \frac{81}{128L} r^4 + \frac{99}{64L} r^3 - \frac{9}{64L} r^2 - \frac{19}{128L} r + \frac{1}{128L} \right) \frac{L}{2} dr = \\
&= \Bigg[0.34785 \left(-\frac{243}{128L} (-0.86113)^5 + \frac{81}{128L} (-0.86113)^4 + \right. \\
&\quad \left. + \frac{99}{64L} (-0.86113)^3 - \frac{9}{64L} (-0.86113)^2 - \frac{19}{128L} (-0.86113) + \frac{1}{128L} \right) + \\
&\quad + 0.65214 \left(-\frac{243}{128L} (-0.33998)^5 + \frac{81}{128L} (-0.0.33998)^4 + \right. \\
&\quad \left. + \frac{99}{64L} (-0.0.33998)^3 - \frac{9}{64L} (-0.0.33998)^2 - \frac{19}{128L} (-0.33998) + \frac{1}{128L} \right) \\
&\quad + 0.65214 \left(-\frac{243}{128L} (0.33998)^5 + \frac{81}{128L} (0.0.33998)^4 + \right. \\
&\quad \left. + \frac{99}{64L} (0.0.33998)^3 - \frac{9}{64L} (0.0.33998)^2 - \frac{19}{128L} (0.33998) + \frac{1}{128L} \right) \\
&\quad + 0.34785 \left(-\frac{243}{128L} (0.86113)^5 + \frac{81}{128L} (0.86113)^4 + \right. \\
&\quad \left. + \frac{99}{64L} (0.86113)^3 - \frac{9}{64L} (0.86113)^2 - \frac{19}{128L} (0.86113) + \frac{1}{128L} \right) \Bigg] \frac{L}{2} = \\
&= 0.0875
\end{aligned}
$$

$$\tag{4.360}$$

The term $k_{27s\,B4}$ is then equal to

$$
k_{27s\,B4} = -0.35\, C_{44} ab
\tag{4.361}
$$

Regarding the $k_{28s\,B4}$ term

$$
\begin{aligned}
k_{28sB4} &= \int_V \left(C_{44} F_{sy,x} N_1(r) F_{\tau y,x} N_4(r) \right) dV \\
&= C_{44} \int_A F_{sy,x} F_{\tau y,x} dA \int_L N_1(r) N_4(r) dL \\
&= C_{44} \int_{-a}^{a} F_{sy,x} F_{\tau y,x} dx \int_{-b}^{b} dz \int_{-1}^{1} N_1(r) N_4(r) \det \mathbf{J} dr \\
&= C_{44} \int_{-a}^{a} (1)\, dx \int_{-b}^{b} dz \int_{-1}^{1} \left(-\frac{81}{256} r^6 + \frac{99}{256} r^4 - \frac{19}{256} r^2 + \frac{1}{256} \right) \frac{L}{2} dr
\end{aligned}
\tag{4.362}
$$

The surface integral of the term k_{28sB4} is computed, using an analytical integration

$$
C_{44} \int_{-a}^{a} (1)\, dx \int_{-b}^{b} dz = C_{44} 4ab
\tag{4.363}
$$

whereas the line integral is computed using the four SPs Gauss quadrature.

$$
\int_{-1}^{1} \left(-\frac{81}{256} r^6 + \frac{99}{256} r^4 - \frac{19}{256} r^2 + \frac{1}{256} \right) \frac{L}{2} dr =
$$

$$
= \left[0.34785 \left(-\frac{81}{256} (-0.86113)^6 + \frac{99}{256} (-0.86113)^4 - \frac{19}{256} (-0.86113)^2 - \frac{9}{128} (-0.86113) + \frac{1}{256} \right) + \right.
$$

$$
+ 0.65214 \left(-\frac{81}{256} (-0.33998)^6 + \frac{99}{256} (-0.33998)^4 - \frac{19}{256} (-0.33998)^2 - \frac{9}{128} (-0.33998) + \frac{1}{256} \right)
$$

$$
+ 0.65214 \left(-\frac{81}{256} (0.33998)^6 + \frac{99}{256} (0.33998)^4 - \frac{19}{256} (0.33998)^2 - \frac{9}{128} (0.33998) + \frac{1}{256} \right)
$$

$$
\left. + 0.34785 \left(-\frac{81}{256} (0.86113)^6 + \frac{99}{256} (0.86113)^4 - \frac{19}{256} (0.86113)^2 - \frac{9}{128} (0.86113) + \frac{1}{256} \right) \right] \frac{L}{2} =
$$

$$
= 0.0113 L 1
\tag{4.364}
$$

The term k_{28sB4} is then equal to

$$
k_{28sB4} = 0.04524 C_{44} ab L
\tag{4.365}
$$

Regarding the k_{31sB4} term

$$
\begin{aligned}
k_{31sB4} &= \int_V \left(C_{44} F_{sx} \mathbf{J}^{-1} N_{2,r}(r) F_{\tau x} \mathbf{J}^{-1} N_{1,r}(r) \right) dV \\
&= C_{44} \int_A F_{sx} F_{\tau x} dA \int_L \mathbf{J}^{-1} N_{2,r}(r) \mathbf{J}^{-1} N_{1,r}(r) dL \\
&= C_{44} \int_{-a}^{a} F_{sx} F_{\tau x} dx \int_{-b}^{b} dz \int_{-1}^{1} \mathbf{J}^{-1} N_{2,r}(r) \mathbf{J}^{-1} N_{1,r}(r) \det \mathbf{J} dr \\
&= C_{44} \int_{-a}^{a} (1)\, dx \int_{-b}^{b} dz \int_{-1}^{1} \left(-\frac{2187}{64L^2} r^4 + \frac{243}{8L^2} r^3 + \frac{243}{32L^2} r^2 - \frac{63}{8L^2} r - \frac{27}{64L^2} \right) \frac{L}{2} dr
\end{aligned}
\tag{4.366}
$$

The term k_{31sB4} is equal to the term k_{13sB4}, see Eqs. (4.312) and (4.315)

$$
k_{31sB4} = -18.9 C_{44} \frac{ab}{L}
\tag{4.367}
$$

Regarding the k_{32sB4} term

$$
\begin{aligned}
k_{32sB4} &= \int_V \left(C_{44} F_{sx} \mathbf{J}^{-1} N_{2,r}(r) F_{\tau y,x} N_1(r) \right) dV \\
&= C_{44} \int_A F_{sx} F_{\tau y,x} dA \int_L \mathbf{J}^{-1} N_{2,r}(r) N_1(r) dL \\
&= C_{44} \int_{-a}^{a} F_{sx} F_{\tau y,x} dx \int_{-b}^{b} dz \int_{-1}^{1} N_1(r) \mathbf{J}^{-1} N_{1,r}(r) \det \mathbf{J} dr \\
&= C_{44} \int_{-a}^{a} (-1)\, dx \int_{-b}^{b} dz \int_{-1}^{1} \left(-\frac{729}{128L} r^5 + \frac{891}{128L} r^4 + \frac{81}{64L} r^3 - \frac{171}{64L} r^2 - \frac{9}{128L} r + \frac{27}{128L} \right) \frac{L}{2} dr
\end{aligned}
\tag{4.368}
$$

The term k_{32sB4} is equal to the term k_{23sB4}, see Eqs. (4.342) and (4.345).

$$
k_{32sB4} = -2.85 C_{44} ab
\tag{4.369}
$$

Regarding the k_{33sB4} term

$$
\begin{aligned}
k_{33sB4} &= \int_V \left(C_{44} F_{sx} \mathbf{J}^{-1} N_{2,r}(r) F_{\tau x} \mathbf{J}^{-1} N_{2,r}(r) \right) dV \\
&= C_{44} \int_A F_{sx} F_{\tau x} dA \int_L \mathbf{J}^{-1} N_{2,r}(r) \mathbf{J}^{-1} N_{2,r}(r) dL \\
&= C_{44} \int_{-a}^{a} F_{sx} F_{\tau x} dx \int_{-b}^{b} dz \int_{-1}^{1} \mathbf{J}^{-1} N_{2,r}(r) \mathbf{J}^{-1} N_{2,r}(r)) \det \mathbf{J} dr \\
&= C_{44} \int_{-a}^{a} 1\, dx \int_{-b}^{b} dz \int_{-1}^{1} \left(\frac{6561}{64L^2} r^4 - \frac{729}{16L^2} r^3 - \frac{2025}{32L^2} r^2 + \frac{243}{16L^2} r + \frac{729}{64L^2} \right) \frac{L}{2} dr
\end{aligned}
\tag{4.370}
$$

The surface integral of the term k_{33sB4} is computed, using an analytical integration

$$
C_{44} \int_{-a}^{a} 1\, dx \int_{-b}^{b} dz = C_{44} 4ab
\tag{4.371}
$$

while the line integral, computed using four SPs Gauss quadrature, is equal to that of Eqs. (4.268) and (4.270)

$$
\int_{-1}^{1} \left(\frac{6561}{64L^2} r^4 - \frac{729}{16L^2} r^3 - \frac{2025}{32L^2} r^2 + \frac{243}{16L^2} r + \frac{729}{64L^2} \right) \frac{L}{2} dr = \frac{10.8}{L}
\tag{4.372}
$$

The term $k_{33s\,B4}$ is then equal to

$$k_{33s\,B4} = 43.2\,C_{44}\frac{ab}{L} \tag{4.373}$$

Regarding the $k_{34s\,B4}$ term

$$
\begin{aligned}
k_{34s\,B4} &= \int_V \left(C_{44}\,F_{sx}\,\mathbf{J}^{-1}\,N_{2,r}(r)\,F_{\tau y,x}\,N_2(r) \right) dV \\[4pt]
&= C_{44} \int_A F_{sx}\,F_{\tau y,x}\,dA \int_L \mathbf{J}^{-1}\,N_{2,r}(r)\,N_2(r)\,dL \\[4pt]
&= C_{44} \int_{-a}^{a} F_{sx}\,F_{\tau y,x}\,dx \int_{-b}^{b} dz \int_{-1}^{1} \mathbf{J}^{-1}\,N_{2,r}(r)\,N_2(r)\,det\,\mathbf{J}\,dr \\[4pt]
&= C_{44} \int_{-a}^{a} (-1)\,dx \int_{-b}^{b} dz \int_{-1}^{1} \left(\frac{2187}{128L}r^5 - \frac{1215}{128L}r^4 - \frac{1377}{64L}r^3 + \frac{729}{64L}r^2 + \frac{567}{128L}r - \frac{243}{128L} \right)\frac{L}{2}\,dr
\end{aligned}
\tag{4.374}
$$

The surface integral of the term $k_{34s\,B4}$ is computed, using an analytical integration

$$C_{44} \int_{-a}^{a} (-1)\,dx \int_{-b}^{b} dz = -C_{44}4ab \tag{4.375}$$

whereas the line integral is computed using the four SPs Gauss quadrature.

$$
\begin{aligned}
\int_{-1}^{1} &\left(\frac{2187}{128L}r^5 - \frac{1215}{128L}r^4 - \frac{1377}{64L}r^3 + \frac{729}{64L}r^2 + \frac{567}{128L}r - \frac{243}{128L} \right)\frac{L}{2}\,dr = \\[4pt]
=\Bigg[&0.34785\left(\frac{2187}{128L}(-0.86113)^5 - \frac{1215}{128L}(-0.86113)^4 - \right. \\[4pt]
&\left. -\frac{1377}{64L}(-0.86113)^3 + \frac{729}{64L}(-0.86113)^2 + \frac{567}{128L}(-0.86113) - \frac{243}{128L} \right) + \\[4pt]
&+ 0.65214\left(\frac{2187}{128L}(-0.33998)^5 - \frac{1215}{128L}(-0.33998)^4 - \right. \\[4pt]
&\left. -\frac{1377}{64L}(-0.33998)^3 + \frac{729}{64L}(-0.33998)^2 + \frac{567}{128L}(-0.33998) - \frac{243}{128L} \right) \\[4pt]
&+ 0.65214\left(\frac{2187}{128L}(0.33998)^5 - \frac{1215}{128L}(0.33998)^4 - \right. \\[4pt]
&\left. -\frac{1377}{64L}(0.33998)^3 + \frac{729}{64L}(0.33998)^2 + \frac{567}{128L}(0.33998) - \frac{243}{128L} \right) \\[4pt]
&+ 0.34785\left(\frac{2187}{128L}(0.86113)^5 - \frac{1215}{128L}(0.86113)^4 - \right. \\[4pt]
&\left. -\frac{1377}{64L}(0.86113)^3 + \frac{729}{64L}(0.86113)^2 + \frac{567}{128L}(0.86113) - \frac{243}{128L} \right) \Bigg]\frac{L}{2} = \\[4pt]
&= 0.0
\end{aligned}
\tag{4.376}
$$

The term $k_{34s\,B4}$ is then equal to

$$k_{34s\,B4} = 0.0 \tag{4.377}$$

Regarding the $k_{35s_{B4}}$ term

$$k_{35s_{B4}} = \int_V \left(C_{44} F_{sx} \mathbf{J}^{-1} N_{2,r}(r) F_{\tau x} \mathbf{J}^{-1} N_{3,r}(r) \right) dV$$

$$= C_{44} \int_A F_{sx} F_{\tau x} dA \int_L \mathbf{J}^{-1} N_{2,r}(r) \mathbf{J}^{-1} N_{3,r}(r) dL$$

$$= C_{44} \int_{-a}^{a} F_{sx} F_{\tau x} dx \int_{-b}^{b} dz \int_{-1}^{1} \mathbf{J}^{-1} N_{2,r}(r) \mathbf{J}^{-1} N_{3,r}(r)) \det \mathbf{J} dr$$

$$= C_{44} \int_{-a}^{a} 1 dx \int_{-b}^{b} dz \int_{-1}^{1} \left(-\frac{6561}{64L^2} r^4 + \frac{2349}{32L^2} r^2 - \frac{729}{64L^2} \right) \frac{L}{2} dr \tag{4.378}$$

The surface integral of the term $k_{35s_{B4}}$ is computed, using an analytical integration

$$C_{44} \int_{-a}^{a} 1 dx \int_{-b}^{b} dz = C_{44} 4ab \tag{4.379}$$

while the line integral,computed using four SPs Gauss quadrature, is equal to that of Eqs. (4.272) and (4.274)

$$\int_{-1}^{1} \left(-\frac{6561}{64L^2} r^4 + \frac{2349}{32L^2} r^2 - \frac{729}{64L^2} \right) \frac{L}{2} dr = \frac{-7.425}{L} \tag{4.380}$$

The term $k_{35s_{B4}}$ is then equal to

$$k_{35s_{B4}} = -29.7 C_{44} \frac{ab}{L} \tag{4.381}$$

Regarding the $k_{36s_{B4}}$ term

$$k_{36s_{B4}} = \int_V \left(C_{44} F_{sx} \mathbf{J}^{-1} N_{2,r}(r) F_{\tau y,x} N_3(r) \right) dV$$

$$= C_{44} \int_A F_{sx} F_{\tau y,x} dA \int_L \mathbf{J}^{-1} N_{2,r}(r) N_3(r) dL$$

$$= C_{44} \int_{-a}^{a} F_{sx} F_{\tau y,x} dx \int_{-b}^{b} dz \int_{-1}^{1} \mathbf{J}^{-1} N_{2,r}(r) N_3(r) \det \mathbf{J} dr$$

$$= C_{44} \int_{-a}^{a} (-1) dx \int_{-b}^{b} dz \int_{-1}^{1} \left(-\frac{2187}{128L} r^5 - \frac{243}{128L} r^4 + \frac{1539}{64L} r^3 + \frac{243}{64L} r^2 - \frac{891}{128L} r - \frac{243}{128L} \right) \frac{L}{2} dr \tag{4.382}$$

The surface integral of the term $k_{36s_{B4}}$ is computed, using an analytical integration

$$C_{44} \int_{-a}^{a} (-1) dx \int_{-b}^{b} dz = -C_{44} 4ab \tag{4.383}$$

whereas the line integral is computed using the four SPs Gauss quadrature.

$$\int_{-1}^{1} \left(-\frac{2187}{128L}r^5 - \frac{243}{128L}r^4 + \frac{1539}{64L}r^3 + \frac{243}{64L}r^2 - \frac{891}{128L}r - \frac{243}{128L} \right) \frac{L}{2} dr =$$

$$= \left[0.34785 \left(-\frac{2187}{128L}(-0.86113)^5 - \frac{243}{128L}(-0.86113)^4 + \frac{1539}{64L}(-0.86113)^3 + \frac{243}{64L}(-0.86113)^2 - \frac{891}{128L}(-0.86113) - \frac{243}{128L} \right) + \right.$$

$$+ 0.65214 \left(-\frac{2187}{128L}(-0.33998)^5 - \frac{243}{128L}(-0.33998)^4 + \frac{1539}{64L}(-0.33998)^3 + \frac{243}{64L}(-0.33998)^2 - \frac{891}{128L}(-0.33998) - \frac{243}{128L} \right)$$

$$+ 0.65214 \left(-\frac{2187}{128L}(0.33998)^5 - \frac{243}{128L}(0.33998)^4 + \frac{1539}{64L}(0.33998)^3 + \frac{243}{64L}(0.33998)^2 - \frac{891}{128L}(0.33998) - \frac{243}{128L} \right)$$

$$+ 0.34785 \left. \left(-\frac{2187}{128L}(0.86113)^5 - \frac{243}{128L}(0.86113)^4 + \frac{1539}{64L}(0.86113)^3 + \frac{243}{64L}(0.86113)^2 - \frac{891}{128L}(0.86113) - \frac{243}{128L} \right) \right] \frac{L}{2} =$$

$$= -1.012$$

$$(4.384)$$

The term $k_{36s\,B4}$ is then equal to

$$k_{36s\,B4} = 4.05 C_{44} ab \tag{4.385}$$

Regarding the $k_{37s\,B4}$ term

$$k_{37s\,B4} = \int_V \left(C_{44} F_{sx} \mathbf{J}^{-1} N_{2,r}(r) F_{\tau x} \mathbf{J}^{-1} N_{4,r}(r) \right) dV$$

$$= C_{44} \int_A F_{sx} F_{\tau x} dA \int_L \mathbf{J}^{-1} N_{2,r}(r) \mathbf{J}^{-1} N_{4,r}(r) dL$$

$$= C_{44} \int_{-a}^{a} F_{sx} F_{\tau x} dx \int_{-b}^{b} dz \int_{-1}^{1} \mathbf{J}^{-1} N_{2,r}(r) \mathbf{J}^{-1} N_{4,r}(r)) \det \mathbf{J} dr$$

$$= C_{44} \int_{-a}^{a} 1 dx \int_{-b}^{b} dz \int_{-1}^{1} \left(\frac{2187}{64L^2}r^4 + \frac{243}{16L^2}r^3 - \frac{567}{32L^2}r^2 - \frac{117}{16L^2}r + \frac{27}{64L^2} \right) \frac{L}{2} dr$$

$$(4.386)$$

The surface integral of the term $k_{37s\,B4}$ is computed, using an analytical integration

$$C_{44} \int_{-a}^{a} 1 dx \int_{-b}^{b} dz = C_{44} 4ab \tag{4.387}$$

while the line integral, computed using four SPs Gauss quadrature, is equal to that of Eqs. (4.276) and (4.278)

$$\int_{-1}^{1} \left(\frac{2187}{64L^2}r^4 + \frac{243}{16L^2}r^3 - \frac{567}{32L^2}r^2 - \frac{117}{16L^2}r + \frac{27}{64L^2} \right) \frac{L}{2} dr = \frac{1.35}{L} \tag{4.388}$$

The term $k_{37s\,B4}$ is then equal to

$$k_{37s\,B4} = 5.4 C_{44} \frac{ab}{L} \tag{4.389}$$

Regarding the $k_{38s\,B4}$ term

$$
\begin{aligned}
k_{38s\,B4} &= \int_V \left(C_{44}\, F_{sx}\, \mathbf{J}^{-1} N_{2,r}(r)\, F_{\tau y,x} N_4(r) \right) dV \\
&= C_{44} \int_A F_{sx}\, F_{\tau y,x}\, dA \int_L \mathbf{J}^{-1} N_{2,r}(r) N_4(r)\, dL \\
&= C_{44} \int_{-a}^a F_{sx}\, F_{\tau y,x}\, dx \int_{-b}^b dz \int_{-1}^1 \mathbf{J}^{-1} N_{2,r}(r) N_4(r)\, det\, \mathbf{J} dr \\
&= C_{44} \int_{-a}^a (-1)\, dx \int_{-b}^b dz \int_{-1}^1 \left(\frac{729}{128L} r^5 + \frac{243}{128L} r^4 - \frac{459}{64L} r^3 - \frac{117}{64L} r^2 + \frac{189}{128L} r - \frac{9}{128L} \right) \frac{L}{2} dr
\end{aligned}
$$

$$(4.390)$$

The surface integral of the term $k_{38s\,B4}$ is computed, using an analytical integration

$$
C_{44} \int_{-a}^a (-1)\, dx \int_{-b}^b dz = -C_{44} 4ab
\tag{4.391}
$$

whereas the line integral is computed using the four SPs Gauss quadrature.

$$
\begin{aligned}
&\int_{-1}^1 \left(\frac{729}{128L} r^5 + \frac{243}{128L} r^4 - \frac{459}{64L} r^3 - \frac{117}{64L} r^2 + \frac{189}{128L} r - \frac{9}{128L} \right) \frac{L}{2} dr = \\
&= \left[0.34785 \left(\frac{729}{128L} (-0.86113)^5 + \frac{243}{128L} (-0.86113)^4 - \right. \right. \\
&\qquad \left. - \frac{459}{64L} (-0.86113)^3 - \frac{117}{64L} (-0.86113)^2 + \frac{189}{128L} (-0.86113) - \frac{9}{128L} \right) + \\
&\quad + 0.65214 \left(\frac{729}{128L} (-0.33998)^5 + \frac{243}{128L} (-0.33998)^4 - \right. \\
&\qquad \left. - \frac{459}{64L} (-0.33998)^3 - \frac{117}{64L} (-0.33998)^2 + \frac{189}{128L} (-0.33998) - \frac{9}{128L} \right) \\
&\quad + 0.65214 \left(\frac{729}{128L} (0.33998)^5 + \frac{243}{128L} (0.33998)^4 - \right. \\
&\qquad \left. - \frac{459}{64L} (0.33998)^3 - \frac{117}{64L} (0.33998)^2 + \frac{189}{128L} (0.33998) - \frac{9}{128L} \right) \\
&\quad + 0.34785 \left(\frac{729}{128L} (0.86113)^5 + \frac{243}{128L} (0.86113)^4 - \right. \\
&\qquad \left. \left. - \frac{459}{64L} (0.86113)^3 - \frac{117}{64L} (0.86113)^2 + \frac{189}{128L} (0.86113) - \frac{9}{128L} \right) \right] \frac{L}{2} = \\
&= 0.3
\end{aligned}
$$

$$(4.392)$$

The term $k_{38s\,B4}$ is then equal to

$$
k_{38s\,B4} = -1.2 C_{44} ab
\tag{4.393}
$$

Regarding the $k_{41s\,B4}$ term

$$k_{41s\,B4} = \int_V \left(C_{44}\, F_{sy,x}\, N_2(r)\, F_{\tau x}\, \mathbf{J}^{-1} N_{1,r}(r) \right) dV$$

$$= C_{44} \int_A F_{sy,x}\, F_{\tau x}\, dA \int_L N_2(r)\, \mathbf{J}^{-1} N_{1,r}(r)\, dL$$

$$= C_{44} \int_{-a}^{a} F_{sy,x}\, F_{\tau x}\, dx \int_{-b}^{b} dz \int_{-1}^{1} N_2(r)\, \mathbf{J}^{-1} N_{1,r}(r)\, det\, \mathbf{J}\, dr$$

$$= C_{44} \int_{-a}^{a} (1)\, dx \int_{-b}^{b} dz \int_{-1}^{1} \left(-\frac{729}{128L} r^5 + \frac{729}{128L} r^4 + \frac{297}{64L} r^3 - \frac{369}{64L} r^2 + \frac{135}{128L} r + \frac{9}{128L} \right) \frac{L}{2}\, dr$$

$$(4.394)$$

The term $k_{41s\,B4}$ is equal to the term $k_{14s\,B4}$, see Eqs. (4.316) and (4.319)

$$k_{41s\,B4} = 2.85 C_{44} ab \tag{4.395}$$

Regarding the $k_{42s\,B4}$ term

$$k_{42s\,B4} = \int_V \left(C_{44}\, F_{sy,x}\, N_2(r)\, F_{\tau y,x}\, N_2(r) \right) dV$$

$$= C_{44} \int_A F_{sy,x}\, F_{\tau y,x}\, dA \int_L N_2(r)\, N_2(r)\, dL$$

$$= C_{44} \int_{-a}^{a} F_{sy,x}\, F_{\tau y,x}\, dx \int_{-b}^{b} dz \int_{-1}^{1} N_2(r)\, N_2(r)\, det\, \mathbf{J}\, dr$$

$$= C_{44} \int_{-a}^{a} (1)\, dx \int_{-b}^{b} dz \int_{-1}^{1} \left(-\frac{243}{256} r^6 + \frac{81}{64L} r^5 + \frac{189}{256} r^4 - \frac{45}{32} r^3 + \frac{63}{256} r^2 + \frac{9}{64} r - \frac{9}{256} \right) \frac{L}{2}\, dr$$

$$(4.396)$$

The term $k_{42s\,B4}$ is equal to the term $k_{24s\,B4}$, see Eqs. (4.346) and (4.349)

$$k_{42s\,B4} = 0.2357 C_{44} abL \tag{4.397}$$

Regarding the $k_{43s\,B4}$ term

$$k_{43s\,B4} = \int_V \left(C_{44}\, F_{sy,x}\, N_2(r)\, F_{\tau x}\, \mathbf{J}^{-1} N_{2,r}(r) \right) dV$$

$$= C_{44} \int_A F_{sy,x}\, F_{\tau y,x}\, dA \int_L N_2(r)\, \mathbf{J}^{-1} N_{2,r}(r)\, dL$$

$$= C_{44} \int_{-a}^{a} F_{sy,x}\, F_{\tau y,x}\, dx \int_{-b}^{b} dz \int_{-1}^{1} N_2(r)\, \mathbf{J}^{-1} N_{2,r}(r)\, det\, \mathbf{J}\, dr$$

$$= C_{44} \int_{-a}^{a} (1)\, dx \int_{-b}^{b} dz \int_{-1}^{1} \left(\frac{2187}{128L} r^5 - \frac{1215}{128L} r^4 - \frac{1377}{64L} r^3 + \frac{729}{64L} r^2 + \frac{567}{128L} r - \frac{243}{128L} \right) \frac{L}{2}\, dr$$

$$(4.398)$$

The term $k_{43s\,B4}$ is equal to the term $k_{34s\,B4}$, see (4.374) and (4.377)

$$k_{43s\,B4} = 0.0 \tag{4.399}$$

Regarding the $k_{44s\,B4}$ term

$$
\begin{aligned}
k_{44s\,B4} &= \int_V \left(C_{44} F_{sy,x} N_2(r) F_{\tau y,x} N_2(r) \right) dV \\
&= C_{44} \int_A F_{sy,x} F_{\tau y,x}\, dA \int_L N_2(r) N_2(r)\, dL \\
&= C_{44} \int_{-a}^{a} F_{sy,x} F_{\tau y,x}\, dx \int_{-b}^{b} dz \int_{-1}^{1} N_2(r) N_2(r)\, det\,\mathbf{J} dr \\
&= C_{44} \int_{-a}^{a} (1)\, dx \int_{-b}^{b} dz \int_{-1}^{1} \left(\frac{729}{256} r^6 - \frac{243}{256} r^5 - \frac{1377}{256} r^4 + \frac{243}{64} r^3 + \frac{567}{64} r^2 - \frac{243}{128} r + \frac{81}{256} \right) \frac{L}{2} dr
\end{aligned}
$$

$$(4.400)$$

The surface integral of the term $k_{44s\,B4}$ is computed, using an analytical integration

$$
C_{44} \int_{-a}^{a} (1)\, dx \int_{-b}^{b} dz = C_{44} 4ab
$$

$$(4.401)$$

whereas the line integral is computed using the four SPs Gauss quadrature.

$$
\begin{aligned}
\int_{-1}^{1} &\left(\frac{729}{256} r^6 - \frac{243}{256} r^5 - \frac{1377}{256} r^4 + \frac{243}{64} r^3 + \frac{567}{64} r^2 - \frac{243}{128} r + \frac{81}{256} \right) \frac{L}{2} dr = \\
= \Bigg[&0.34785 \left(\frac{729}{256} (-0.86113)^6 - \frac{243}{256} (-0.86113)^5 - \frac{1377}{256} (-0.86113)^4 \right. \\
&\left. + \frac{243}{64} (-0.86113)^3 + \frac{567}{64} (-0.86113)^2 - \frac{243}{128} (-0.86113) + \frac{81}{256} \right) + \\
&+ 0.65214 \left(\frac{729}{256} (-0.33998)^6 - \frac{243}{256} (-0.33998)^5 - \frac{1377}{256} (-0.33998)^4 \right. \\
&\left. + \frac{243}{64} (-0.33998)^3 + \frac{567}{64} (-0.33998)^2 - \frac{243}{128} (-0.33998) + \frac{81}{256} \right) \\
&+ 0.65214 \left(\frac{729}{256} (0.33998)^6 - \frac{243}{256} (0.33998)^5 - \frac{1377}{256} (0.33998)^4 \right. \\
&\left. + \frac{243}{64} (0.33998)^3 + \frac{567}{64} (0.33998)^2 - \frac{243}{128} (0.33998) + \frac{81}{256} \right) \\
&+ 0.34785 \left(\frac{729}{256} (0.86113)^6 - \frac{243}{256} (0.86113)^5 - \frac{1377}{256} (0.86113)^4 \right. \\
&\left. + \frac{243}{64} (0.86113)^3 + \frac{567}{64} (0.86113)^2 - \frac{243}{128} (0.86113) + \frac{81}{256} \right) \Bigg] \frac{L}{2} = \\
= &\, 0.3857 L
\end{aligned}
$$

$$(4.402)$$

The term $k_{44s\,B4}$ is then equal to

$$
k_{44s\,B4} = 1.543 C_{44} ab L
$$

$$(4.403)$$

Regarding the $k_{45s\,B4}$ term

$$
\begin{aligned}
k_{45s\,B4} &= \int_V \left(C_{44}\,F_{sy,x}\,N_2(r)\,F_{\tau x}\,\mathbf{J}^{-1}N_{3,r}(r) \right) dV \\[2mm]
&= C_{44} \int_A F_{sy,x}\,F_{\tau x}\,dA \int_L N_2(r)\mathbf{J}^{-1}N_{3,r}(r)\,dL \\[2mm]
&= C_{44} \int_{-a}^{a} F_{sy,x}\,F_{\tau x}\,dx \int_{-b}^{b} dz \int_{-1}^{1} N_2(r)\mathbf{J}^{-1}N_{3,r}(r)\,det\,\mathbf{J}dr \\[2mm]
&= C_{44} \int_{-a}^{a} (-1)\,dx \int_{-b}^{b} dz \int_{-1}^{1} \left(-\frac{2187}{128L}r^5 + \frac{243}{128L}r^4 + \frac{1539}{64L}r^3 - \frac{243}{64L}r^2 - \frac{891}{128L}r + \frac{243}{128L} \right) \frac{L}{2}\,dr
\end{aligned}
$$

$$(4.404)$$

The surface integral of the term $k_{45s\,B4}$ is computed, using an analytical integration

$$
C_{44} \int_{-a}^{a} (-1)\,dx \int_{-b}^{b} dz = -C_{44}4ab
\tag{4.405}
$$

whereas the line integral is computed using the four SPs Gauss quadrature.

$$
\begin{aligned}
\int_{-1}^{1} &\left(-\frac{2187}{128L}r^5 + \frac{243}{128L}r^4 + \frac{1539}{64L}r^3 - \frac{243}{64L}r^2 - \frac{891}{128L}r + \frac{243}{128L} \right) \frac{L}{2}\,dr = \\[2mm]
&= \Bigg[0.34785 \left(-\frac{2187}{128L}(-0.86113)^5 + \frac{243}{128L}(-0.86113)^4 + \right. \\[2mm]
&\quad \left. +\frac{1539}{64L}(-0.86113)^3 - \frac{243}{64L}(-0.86113)^2 - \frac{891}{128L}(-0.86113) + \frac{243}{128L} \right) + \\[2mm]
&\quad + 0.65214 \left(-\frac{2187}{128L}(-0.33998)^5 + \frac{243}{128L}(-0.33998)^4 + \right. \\[2mm]
&\quad \left. +\frac{1539}{64L}(-0.33998)^3 - \frac{243}{64L}(-0.33998)^2 - \frac{891}{128L}(-0.33998) + \frac{243}{128L} \right) \\[2mm]
&\quad + 0.65214 \left(-\frac{2187}{128L}(0.33998)^5 + \frac{243}{128L}(0.33998)^4 + \right. \\[2mm]
&\quad \left. +\frac{1539}{64L}(0.33998)^3 - \frac{243}{64L}(0.33998)^2 - \frac{891}{128L}(0.33998) + \frac{243}{128L} \right) \\[2mm]
&\quad + 0.34785 \left(-\frac{2187}{128L}(0.86113)^5 + \frac{243}{128L}(0.86113)^4 + \right. \\[2mm]
&\quad \left. +\frac{1539}{64L}(0.86113)^3 - \frac{243}{64L}(0.86113)^2 - \frac{891}{128L}(0.86113) + \frac{243}{128L} \right) \Bigg] \frac{L}{2} = \\[2mm]
&= 1.012
\end{aligned}
$$

$$(4.406)$$

The term $k_{45s\,B4}$ is then equal to

$$
k_{45s\,B4} = -4.05 C_{44}ab
\tag{4.407}
$$

Regarding the $k_{46s\,B4}$ term

$$
\begin{aligned}
k_{46s\,B4} &= \int_V \left(C_{44} F_{sy,x} N_2(r) F_{\tau y,x} N_3(r) \right) dV \\
&= C_{44} \int_A F_{sy,x} F_{\tau y,x}\, dA \int_L N_2(r) N_3(r)\, dL \\
&= C_{44} \int_{-a}^{a} F_{sy,x} F_{\tau y,x}\, dx \int_{-b}^{b} dz \int_{-1}^{1} N_2(r) N_3(r)\, det\,\mathbf{J}\, dr \\
&= C_{44} \int_{-a}^{a} (1)\, dx \int_{-b}^{b} dz \int_{-1}^{1} \left(-\frac{79}{256} r^6 + \frac{1539}{256} r^4 - \frac{891}{256} r^2 + \frac{81}{256} \right) \frac{L}{2}\, dr
\end{aligned}
\tag{4.408}
$$

The surface integral of the term $k_{46s\,B4}$ is computed, using an analytical integration

$$
C_{44} \int_{-a}^{a} (1)\, dx \int_{-b}^{b} dz = C_{44}\, 4ab
\tag{4.409}
$$

whereas the line integral is computed using the four SPs Gauss quadrature.

$$
\begin{aligned}
\int_{-1}^{1} &\left(-\frac{79}{256} r^6 + \frac{1539}{256} r^4 - \frac{891}{256} r^2 + \frac{81}{256} \right) \frac{L}{2}\, dr = \\[2mm]
&= \Bigg[0.34785 \left(-\frac{79}{256} (-0.86113)^6 + \frac{1539}{256} (-0.86113)^4 - \frac{891}{256} (-0.86113)^2 + \frac{81}{256} \right) + \\[2mm]
&\quad + 0.65214 \left(-\frac{79}{256} (-0.33998)^6 + \frac{1539}{256} (-0.33998)^4 - \frac{891}{256} (-0.33998)^2 + \frac{81}{256} \right) \\[2mm]
&\quad + 0.65214 \left(-\frac{79}{256} (0.33998)^6 + \frac{1539}{256} (0.33998)^4 - \frac{891}{256} (0.33998)^2 + \frac{81}{256} \right) \\[2mm]
&\quad + 0.34785 \left(-\frac{79}{256} (0.86113)^6 + \frac{1539}{256} (0.86113)^4 - \frac{891}{256} (0.86113)^2 + \frac{81}{256} \right) \Bigg] \frac{L}{2} = \\[2mm]
&= -0.04822 L
\end{aligned}
\tag{4.410}
$$

The term $k_{46s\,B4}$ is then equal to

$$
k_{46s\,B4} = -0.1929 C_{44}\, abL
\tag{4.411}
$$

Regarding the $k_{47s\,B4}$ term

$$
\begin{aligned}
k_{47s\,B4} &= \int_V \left(C_{44}\, F_{sy,x}\, N_2(r)\, F_{\tau x}\, \mathbf{J}^{-1}\, N_{4,r}(r) \right) dV \\[6pt]
&= C_{44} \int_A F_{sy,x}\, F_{\tau x}\, dA \int_L N_2(r)\, \mathbf{J}^{-1}\, N_{4,r}(r)\, dL \\[6pt]
&= C_{44} \int_{-a}^{a} F_{sy,x}\, F_{\tau x}\, dx \int_{-b}^{b} dz \int_{-1}^{1} N_2(r)\, \mathbf{J}^{-1}\, N_{4,r}(r)\, det\, \mathbf{J}\, dr \\[6pt]
&= C_{44} \int_{-a}^{a} (-1)\, dx \int_{-b}^{b} dz \int_{-1}^{1} \left(\frac{729}{128L} r^5 + \frac{243}{128L} r^4 - \frac{459}{64L} r^3 - \frac{117}{64L} r^2 + \frac{189}{128L} r - \frac{9}{128L} \right) \frac{L}{2} dr
\end{aligned}
$$

$$(4.412)$$

The surface integral of the term $k_{47s\,B4}$ is computed, using an analytical integration

$$
C_{44} \int_{-a}^{a} (-1)\, dx \int_{-b}^{b} dz = -C_{44} 4ab \tag{4.413}
$$

whereas the line integral is computed using the four SPs Gauss quadrature.

$$
\begin{aligned}
\int_{-1}^{1} &\left(\frac{729}{128L} r^5 + \frac{243}{128L} r^4 - \frac{459}{64L} r^3 - \frac{117}{64L} r^2 + \frac{189}{128L} r - \frac{9}{128L} \right) \frac{L}{2} dr = \\[6pt]
&= \left[0.34785 \left(\frac{729}{128L} (-0.86113)^5 + \frac{243}{128L} (-0.86113)^4 - \right. \right. \\[6pt]
&\quad \left. - \frac{459}{64L} (-0.86113)^3 - \frac{117}{64L} (-0.86113)^2 + \frac{189}{128L} (-0.86113) - \frac{9}{128L} \right) + \\[6pt]
&\quad + 0.65214 \left(\frac{729}{128L} (-0.33998)^5 + \frac{243}{128L} (-0.33998)^4 - \right. \\[6pt]
&\quad \left. - \frac{459}{64L} (-0.33998)^3 - \frac{117}{64L} (-0.33998)^2 + \frac{189}{128L} (-0.33998) - \frac{9}{128L} \right) \\[6pt]
&\quad + 0.65214 \left(\frac{729}{128L} (0.33998)^5 + \frac{243}{128L} (0.33998)^4 \right. \\[6pt]
&\quad \left. - \frac{459}{64L} (0.33998)^3 - \frac{117}{64L} (0.33998)^2 + \frac{189}{128L} (0.33998) - \frac{9}{128L} \right) \\[6pt]
&\quad + 0.34785 \left(\frac{729}{128L} (0.86113)^5 + \frac{243}{128L} (0.86113)^4 - \right. \\[6pt]
&\quad \left. \left. - \frac{459}{64L} (0.86113)^3 - \frac{117}{64L} (0.86113)^2 + \frac{189}{128L} (0.86113) - \frac{9}{128L} \right) \right] \frac{L}{2} = \\[6pt]
&= -0.3
\end{aligned}
$$

$$(4.414)$$

The term $k_{47s\,B4}$ is then equal to

$$
k_{47s\,B4} = 1.2\, C_{44}\, ab \tag{4.415}
$$

Regarding the $k_{48s\,B4}$ term

$$k_{48s\,B4} = \int_V \left(C_{44}\,F_{sy,x}\,N_2(r)\,F_{\tau y,x}\,N_4(r)\right)dV$$

$$= C_{44}\int_A F_{sy,x}\,F_{\tau y,x}\,dA \int_L N_2(r)N_4(r)\,dL$$

$$= C_{44}\int_{-a}^{a} F_{sy,x}\,F_{\tau y,x}\,dx \int_{-b}^{b} dz \int_{-1}^{1} N_2(r)N_4(r)\det \mathbf{J}\,dr$$

$$= C_{44}\int_{-a}^{a}(1)\,dx \int_{-b}^{b} dz \int_{-1}^{1}\left(\frac{243}{256}r^6 + \frac{81}{256}r^5 - \frac{351}{256}r^4 - \frac{45}{64}r^3 + \frac{117}{64}r^2 + \frac{9}{128}r - \frac{9}{256}\right)\frac{L}{2}\,dr$$

$$\tag{4.416}$$

The surface integral of the term $k_{48s\,B4}$ is computed, using an analytical integration

$$C_{44}\int_{-a}^{a}(1)\,dx \int_{-b}^{b} dz = C_{44}4ab \tag{4.417}$$

whereas the line integral is computed using the four SPs Gauss quadrature.

$$\int_{-1}^{1}\left(\frac{243}{256}r^6 + \frac{81}{256}r^5 - \frac{351}{256}r^4 - \frac{45}{64}r^3 + \frac{117}{64}r^2 + \frac{9}{128}r - \frac{9}{256}\right)\frac{L}{2}\,dr =$$

$$= \Bigg[0.34785\left(\frac{243}{256}(-0.86113)^6 + \frac{81}{256}(-0.86113)^5 - \frac{351}{256}(-0.86113)^4 - \right.$$

$$\left. -\frac{45}{64}(-0.86113)^3 + \frac{117}{64}(-0.86113)^2 + \frac{9}{128}(-0.86113) - \frac{9}{256}\right) +$$

$$+ 0.65214\left(\frac{243}{256}(-0.33998)^6 + \frac{81}{256}(-0.33998)^5 - \frac{351}{256}(-0.33998)^4 - \right.$$

$$\left. -\frac{45}{64}(-0.33998)^3 + \frac{117}{64}(-0.33998)^2 + \frac{9}{128}(-0.33998) - \frac{9}{256}\right)$$

$$+ 0.65214\left(\frac{243}{256}(0.33998)^6 + \frac{81}{256}(0.33998)^5 - \frac{351}{256}(0.33998)^4 - \right.$$

$$\left. -\frac{45}{64}(0.33998)^3 + \frac{117}{64}(0.33998)^2 + \frac{9}{128}(0.33998) - \frac{9}{256}\right)$$

$$+ 0.34785\left(\frac{243}{256}(0.86113)^6 + \frac{81}{256}(0.86113)^5 - \frac{351}{256}(0.86113)^4 - \right.$$

$$\left. -\frac{45}{64}(0.86113)^3 + \frac{117}{64}(0.86113)^2 + \frac{9}{128}(0.86113) - \frac{9}{256}\right)\Bigg]\frac{L}{2} =$$

$$= -0.02143L \tag{4.418}$$

The term $k_{48s\,B4}$ is then equal to

$$k_{48s\,B4} = -0.08571\,C_{44}abL \tag{4.419}$$

Regarding the $k_{51s\,B4}$ term

$$
\begin{aligned}
k_{51s\,B4} &= \int_V \left(C_{44} F_{sx} \mathbf{J}^{-1} N_{3,r}(r) F_{\tau x} \mathbf{J}^{-1} N_{1,r}(r) \right) dV \\
&= C_{44} \int_A F_{sx} F_{\tau x}\, dA \int_L \mathbf{J}^{-1} N_{3,r}(r) \mathbf{J}^{-1} N_{1,r}(r)\, dL \\
&= C_{44} \int_{-a}^{a} F_{sx} F_{\tau x}\, dx \int_{-b}^{b} dz \int_{-1}^{1} \mathbf{J}^{-1} N_{3,r}(r) \mathbf{J}^{-1} N_{1,r}(r)\, det\, \mathbf{J}\, dr \\
&= C_{44} \int_{-a}^{a} (1)\, dx \int_{-b}^{b} dz \int_{-1}^{1} \left(\frac{2187}{64L^2} r^4 - \frac{243}{16L^2} r^3 - \frac{567}{32L^2} r^2 + \frac{117}{16L^2} r + \frac{27}{64L^2} \right) \frac{L}{2}\, dr
\end{aligned}
\tag{4.420}
$$

The term $k_{51s\,B4}$ is equal to the term $k_{15s\,B4}$, see Eqs. (4.320) and (4.323)

$$
k_{51s\,B4} = 5.4 C_{44} \frac{ab}{L}
\tag{4.421}
$$

Regarding the $k_{52s\,B4}$ term

$$
\begin{aligned}
k_{52s\,B4} &= \int_V \left(C_{44} F_{sx} \mathbf{J}^{-1} N_{3,r}(r) F_{\tau y,x} N_1(r) \right) dV \\
&= C_{44} \int_A F_{sx} F_{\tau y,x}\, dA \int_L \mathbf{J}^{-1} N_{3,r}(r) N_1(r)\, dL \\
&= C_{44} \int_{-a}^{a} F_{sx} F_{\tau y,x}\, dx \int_{-b}^{b} dz \int_{-1}^{1} \mathbf{J}^{-1} N_{3,r}(r) N_1(r)\, det\, \mathbf{J}\, dr \\
&= C_{44} \int_{-a}^{a} (1)\, dx \int_{-b}^{b} dz \int_{-1}^{1} \left(\frac{729}{128L} r^5 - \frac{567}{128L} r^4 - \frac{243}{64L} r^3 + \frac{153}{64L} r^2 + \frac{45}{128L} r - \frac{27}{128L} \right) \frac{L}{2}\, dr
\end{aligned}
\tag{4.422}
$$

The term $k_{52s\,B4}$ is equal to the term $k_{25s\,B4}$, see Eqs. (4.350) and (4.353)

$$
k_{52s\,B4} = 1.2 C_{44} ab
\tag{4.423}
$$

Regarding the $k_{53s\,B4}$ term

$$
\begin{aligned}
k_{53s\,B4} &= \int_V \left(C_{44} F_{sx} \mathbf{J}^{-1} N_{3,r}(r) F_{\tau x} \mathbf{J}^{-1} N_{2,r}(r) \right) dV \\
&= C_{44} \int_A F_{sx} F_{\tau x}\, dA \int_L \mathbf{J}^{-1} N_{3,r}(r) \mathbf{J}^{-1} N_{2,r}(r)\, dL \\
&= C_{44} \int_{-a}^{a} F_{sx} F_{\tau x}\, dx \int_{-b}^{b} dz \int_{-1}^{1} \mathbf{J}^{-1} N_{3,r}(r) \mathbf{J}^{-1} N_{2,r}(r)\, det\, \mathbf{J}\, dr \\
&= C_{44} \int_{-a}^{a} (1)\, dx \int_{-b}^{b} dz \int_{-1}^{1} \left(-\frac{6561}{64L^2} r^4 + \frac{2349}{32L^2} r^2 - \frac{729}{64L^2} \right) \frac{L}{2}\, dr
\end{aligned}
\tag{4.424}
$$

The term $k_{53s\,B4}$ is equal to the term $k_{35s\,B4}$, see Eqs. (4.378) and (4.381)

$$
k_{53s\,B4} = -29.7 C_{44} \frac{ab}{L}
\tag{4.425}
$$

Regarding the $k_{54s\,B4}$ term

$$k_{54sB4} = \int_V \left(C_{44} F_{sx} \mathbf{J}^{-1} N_{3,r}(r) F_{\tau y,x} N_2(r) \right) dV$$

$$= C_{44} \int_A F_{sx} F_{\tau y,x} dA \int_L \mathbf{J}^{-1} N_{3,r}(r) N_2(r) dL$$

$$= C_{44} \int_{-a}^{a} F_{sx} F_{\tau y,x} dx \int_{-b}^{b} dz \int_{-1}^{1} \mathbf{J}^{-1} N_{3,r}(r) N_2(r) det\,\mathbf{J} dr$$

$$= C_{44} \int_{-a}^{a} (1)\, dx \int_{-b}^{b} dz \int_{-1}^{1} \left(-\frac{2187}{128L} r^5 + \frac{243}{128L} r^4 + \frac{1539}{64L} r^3 - \frac{243}{64L} r^2 - \frac{891}{128L} r + \frac{243}{128L} \right) \frac{L}{2} dr$$

$$(4.426)$$

The term k_{54sB4} is equal to the term k_{45sB4} see Eqs. (4.404) and (4.407)

$$k_{54sB4} = -4.05 C_{44} ab \tag{4.427}$$

Regarding the k_{55sB4} term

$$k_{55sB4} = \int_V \left(C_{44} F_{sx} \mathbf{J}^{-1} N_{3,r}(r) F_{\tau x} \mathbf{J}^{-1} N_{3,r}(r) \right) dV$$

$$= C_{44} \int_A F_{sx} F_{\tau x} dA \int_L \mathbf{J}^{-1} N_{3,r}(r) \mathbf{J}^{-1} N_{3,r}(r) dL$$

$$= C_{44} \int_{-a}^{a} F_{sx} F_{\tau x} dx \int_{-b}^{b} dz \int_{-1}^{1} \mathbf{J}^{-1} N_{3,r}(r) \mathbf{J}^{-1} N_{3,r}(r)) det\,\mathbf{J} dr$$

$$= C_{44} \int_{-a}^{a} 1\, dx \int_{-b}^{b} dz \int_{-1}^{1} \left(\frac{6561}{64L^2} r^4 + \frac{729}{16L^2} r^3 - \frac{2025}{32L^2} r^2 - \frac{243}{16L^2} r + \frac{729}{64L^2} \right) \frac{L}{2} dr$$

$$(4.428)$$

The surface integral of the term k_{55sB4} is computed, using an analytical integration

$$C_{44} \int_{-a}^{a} 1\, dx \int_{-b}^{b} dz = C_{44} 4ab \tag{4.429}$$

while the line integral, computed using four SPs Gauss quadrature, is equal to that of Eqs. (4.284) and (4.286)

$$\int_{-1}^{1} \left(\frac{6561}{64L^2} r^4 - \frac{729}{16L^2} r^3 - \frac{2025}{32L^2} r^2 + \frac{243}{16L^2} r + \frac{729}{64L^2} \right) \frac{L}{2} dr = \frac{10.8}{L} \tag{4.430}$$

The term k_{55sB4} is then equal to

$$k_{55sB4} = 43.2 C_{44} \frac{ab}{L} \tag{4.431}$$

Regarding the k_{56sB4} term

$$k_{56sB4} = \int_V \left(C_{44} F_{sx} \mathbf{J}^{-1} N_{3,r}(r) F_{\tau y,x} N_3(r) \right) dV$$

$$= C_{44} \int_A F_{sx} F_{\tau y,x} dA \int_L \mathbf{J}^{-1} N_{3,r}(r) N_3(r) dL$$

$$= C_{44} \int_{-a}^{a} F_{sx} F_{\tau y,x} dx \int_{-b}^{b} dz \int_{-1}^{1} \mathbf{J}^{-1} N_{3,r}(r) N_3(r) det\,\mathbf{J} dr$$

$$= C_{44} \int_{-a}^{a} (-1)\, dx \int_{-b}^{b} dz \int_{-1}^{1} \left(\frac{2187}{128L} r^5 + \frac{1215}{128L} r^4 - \frac{1377}{64L} r^3 - \frac{729}{64L} r^2 + \frac{567}{128L} r + \frac{243}{128L} \right) \frac{L}{2} dr$$

$$(4.432)$$

The surface integral of the term $k_{56s\,B4}$ is computed, using an analytical integration

$$C_{44} \int_{-a}^{a} (-1)\, dx \int_{-b}^{b} dz = -C_{44} 4ab \tag{4.433}$$

whereas the line integral is computed using the four SPs Gauss quadrature.

$$\int_{-1}^{1} \left(\frac{2187}{128L} r^5 + \frac{1215}{128L} r^4 - \frac{1377}{64L} r^3 - \frac{729}{64L} r^2 + \frac{567}{128L} r + \frac{243}{128L} \right) \frac{L}{2} dr =$$

$$= \left[0.34785 \left(\frac{2187}{128L} (-0.86113)^5 + \frac{1215}{128L} (-0.86113)^4 - \right.\right.$$

$$- \frac{1377}{64L} (-0.86113)^3 - \frac{729}{64L} (-0.86113)^2 + \frac{567}{128L} (-0.86113) + \frac{243}{128L} \right) +$$

$$+ 0.65214 \left(\frac{2187}{128L} (-0.33998)^5 + \frac{1215}{128L} (-0.33998)^4 - \right.$$

$$- \frac{1377}{64L} (-0.33998)^3 - \frac{729}{64L} (-0.33998)^2 + \frac{567}{128L} (-0.33998) + \frac{243}{128L} \right)$$

$$+ 0.65214 \left(\frac{2187}{128L} (0.33998)^5 + \frac{1215}{128L} (0.33998)^4 - \right.$$

$$- \frac{1377}{64L} (0.33998)^3 - \frac{729}{64L} (0.33998)^2 + \frac{567}{128L} (0.33998) + \frac{243}{128L} \right)$$

$$+ 0.34785 \left(\frac{2187}{128L} (0.86113)^5 + \frac{1215}{128L} (0.86113)^4 - \right.$$

$$\left.\left. - \frac{1377}{64L} (0.86113)^3 - \frac{729}{64L} (0.86113)^2 + \frac{567}{128L} (0.86113) + \frac{243}{128L} \right) \right] \frac{L}{2} =$$

$$= 0.0 \tag{4.434}$$

The term $k_{56s\,B4}$ is then equal to

$$k_{56s\,B4} = 0.0 \tag{4.435}$$

Regarding the $k_{57s\,B4}$ term

$$k_{57s\,B4} = \int_{V} \left(C_{44} F_{sx} \mathbf{J}^{-1} N_{3,r}(r) F_{\tau x} \mathbf{J}^{-1} N_{4,r}(r) \right) dV$$

$$= C_{44} \int_{A} F_{sx} F_{\tau x}\, dA \int_{L} \mathbf{J}^{-1} N_{3,r}(r) \mathbf{J}^{-1} N_{4,r}(r)\, dL$$

$$= C_{44} \int_{-a}^{a} F_{sx} F_{\tau x}\, dx \int_{-b}^{b} dz \int_{-1}^{1} \mathbf{J}^{-1} N_{3,r}(r) \mathbf{J}^{-1} N_{4,r}(r))\, det\, \mathbf{J}\, dr \tag{4.436}$$

$$= C_{44} \int_{-a}^{a} 1\, dx \int_{-b}^{b} dz \int_{-1}^{1} \left(-\frac{2187}{64L^2} r^4 - \frac{243}{8L^2} r^3 + \frac{243}{32L^2} r^2 + \frac{63}{8L^2} r - \frac{27}{64L^2} \right) \frac{L}{2} dr$$

The surface integral of the term $k_{57s\,B4}$ is computed, using an analytical integration

$$C_{44} \int_{-a}^{a} 1\, dx \int_{-b}^{b} dz = C_{44} 4ab \tag{4.437}$$

while the line integral, computed using four SPs Gauss quadrature, is equal to that of
Eqs. (4.288) and (4.290)

$$\int_{-1}^{1} \left(-\frac{2187}{64L^2} r^4 - \frac{243}{8L^2} r^3 + \frac{243}{32L^2} r^2 + \frac{63}{8L^2} r - \frac{27}{64L^2} \right) \frac{L}{2} dr = -\frac{4.725}{L} \quad (4.438)$$

The term $k_{57 s B4}$ is then equal to

$$k_{57 s B4} = -18.9 C_{44} \frac{ab}{L} \quad (4.439)$$

Regarding the $k_{58 s B4}$ term

$$k_{58 s B4} = \int_{V} \left(C_{44} F_{sx} \mathbf{J}^{-1} N_{3,r}(r) F_{\tau y,x} N_4(r) \right) dV$$

$$= C_{44} \int_{A} F_{sx} F_{\tau y,x} dA \int_{L} \mathbf{J}^{-1} N_{3,r}(r) N_4(r) dL$$

$$= C_{44} \int_{-a}^{a} F_{sx} F_{\tau y,x} dx \int_{-b}^{b} dz \int_{-1}^{1} \mathbf{J}^{-1} N_{3,r}(r) N_4(r) \det \mathbf{J} dr$$

$$= C_{44} \int_{-a}^{a} (-1) dx \int_{-b}^{b} dz \int_{-1}^{1} \left(-\frac{729}{128L} r^5 - \frac{891}{128L} r^4 + \frac{81}{64L} r^3 + \frac{171}{64L} r^2 - \frac{9}{128L} r - \frac{27}{128L} \right) \frac{L}{2} dr \quad (4.440)$$

The surface integral of the term $k_{58 s B4}$ is computed, using an analytical integration

$$C_{44} \int_{-a}^{a} (-1) dx \int_{-b}^{b} dz = -C_{44} 4ab \quad (4.441)$$

whereas the line integral is computed using the four SPs Gauss quadrature.

$$\int_{-1}^{1} \left(-\frac{729}{128L} r^5 - \frac{891}{128L} r^4 + \frac{81}{64L} r^3 + \frac{171}{64L} r^2 - \frac{9}{128L} r - \frac{27}{128L} \right) \frac{L}{2} dr =$$

$$= \left[0.34785 \left(-\frac{729}{128L} (-0.86113)^5 - \frac{891}{128L} (-0.86113)^4 + \right. \right.$$

$$\left. + \frac{81}{64L} (-0.86113)^3 + \frac{171}{64L} (-0.86113)^2 - \frac{9}{128L} (-0.86113) - \frac{27}{128L} \right) +$$

$$+ 0.65214 \left(-\frac{729}{128L} (-0.33998)^5 - \frac{891}{128L} (-0.33998)^4 + \right.$$

$$\left. + \frac{81}{64L} (-0.33998)^3 + \frac{171}{64L} (-0.33998)^2 - \frac{9}{128L} (-0.33998) - \frac{27}{128L} \right)$$

$$+ 0.65214 \left(-\frac{729}{128L} (0.33998)^5 - \frac{891}{128L} (0.33998)^4 + \right.$$

$$\left. + \frac{81}{64L} (0.33998)^3 + \frac{171}{64L} (0.33998)^2 - \frac{9}{128L} (0.33998) - \frac{27}{128L} \right)$$

$$+ 0.34785 \left(-\frac{729}{128L} (0.86113)^5 - \frac{891}{128L} (0.86113)^4 + \right.$$

$$\left. \left. + \frac{81}{64L} (0.86113)^3 + \frac{171}{64L} (0.86113)^2 - \frac{9}{128L} (0.86113) - \frac{27}{128L} \right) \right] \frac{L}{2} =$$

$$= -0.7125 \quad (4.442)$$

The term k_{58sB4} is then equal to

$$k_{58sB4} = 2.85 C_{44} ab \tag{4.443}$$

Regarding the k_{61sB4} term

$$
\begin{aligned}
k_{61sB4} &= \int_V \left(C_{44} F_{sy,x} N_3(r) F_{\tau x} \mathbf{J}^{-1} N_{1,r}(r) \right) dV \\
&= C_{44} \int_A F_{sy,x} F_{\tau x} dA \int_L N_3(r\mathbf{J}^{-1} N_{1,r}(r) dL \\
&= C_{44} \int_{-a}^{a} F_{sy,x} F_{\tau x} dx \int_{-b}^{b} dz \int_{-1}^{1} N_3(r\mathbf{J}^{-1} N_{1,r}(r) det\,\mathbf{J} dr \\
&= C_{44} \int_{-a}^{a} (-1)\,dx \int_{-b}^{b} dz \int_{-1}^{1} \left(\frac{729}{128L} r^5 - \frac{243}{128L} r^4 - \frac{459}{64L} r^3 + \frac{117}{64L} r^2 + \frac{189}{128L} r + \frac{9}{128L} \right) \frac{L}{2} dr
\end{aligned}
\tag{4.444}
$$

The term k_{61sB4} is equal to the term k_{16sB4}, see Eqs. (4.324) and (4.327)

$$k_{61sB4} = -1.2 C_{44} ab \tag{4.445}$$

Regarding the k_{62sB4} term

$$
\begin{aligned}
k_{62sB4} &= \int_V \left(C_{44} F_{sy,x} N_3(r) F_{\tau y,x} N_1(r) \right) dV \\
&= C_{44} \int_A F_{sy,x} F_{\tau y,x} dA \int_L N_3(r) N_1(r) dL \\
&= C_{44} \int_{-a}^{a} F_{sy,x} F_{\tau y,x} dx \int_{-b}^{b} dz \int_{-1}^{1} N_3(r) N_1(r) det\,\mathbf{J} dr \\
&= C_{44} \int_{-a}^{a} (1)\,dx \int_{-b}^{b} dz \int_{-1}^{1} \left(\frac{243}{256} r^6 - \frac{81}{64L} r^5 - \frac{351}{256} r^4 + \frac{45}{64} r^3 + \frac{117}{256} r^2 - \frac{9}{128} r - \frac{9}{256} \right) \frac{L}{2} dr
\end{aligned}
\tag{4.446}
$$

The term k_{62sB4} is equal to the term k_{26sB4}, see Eqs. (4.354) and (4.357)

$$k_{62sB4} = -0.08571 C_{44} abL \tag{4.447}$$

Regarding the k_{63sB4} term

$$
\begin{aligned}
k_{63sB4} &= \int_V \left(C_{44} F_{sy,x} N_3(r) F_{\tau x} \mathbf{J}^{-1} N_{2,r}(r) \right) dV \\
&= C_{44} \int_A F_{sy,x} F_{\tau x} dA \int_L N_3(r) \mathbf{J}^{-1} N_{2,r}(r) dL \\
&= C_{44} \int_{-a}^{a} F_{sy,x} F_{\tau x} dx \int_{-b}^{b} dz \int_{-1}^{1} N_3(r) \mathbf{J}^{-1} N_{2,r}(r) det\,\mathbf{J} dr \\
&= C_{44} \int_{-a}^{a} (-1)\,dx \int_{-b}^{b} dz \int_{-1}^{1} \left(-\frac{2187}{128L} r^5 - \frac{243}{128L} r^4 + \frac{1539}{64L} r^3 + \frac{243}{64L} r^2 - \frac{891}{128L} r - \frac{243}{128L} \right) \frac{L}{2} dr
\end{aligned}
\tag{4.448}
$$

The term k_{63sB4} is equal to the term k_{36sB4}, see Eqs. (4.382) and (4.385)

$$k_{63sB4} = 4.05 C_{44} ab \tag{4.449}$$

Regarding the k_{64sB4} term

$$
\begin{aligned}
k_{64s\,B4} &= \int_V \left(C_{44} F_{sy,x} N_3(r) F_{\tau y,x} N_2(r) \right) dV \\
&= C_{44} \int_A F_{sy,x} F_{\tau y,x} dA \int_L N_3(r) N_2(r) dL \\
&= C_{44} \int_{-a}^{a} F_{sy,x} F_{\tau y,x} dx \int_{-b}^{b} dz \int_{-1}^{1} N_3(r) N_2(r) \det \mathbf{J} dr \\
&= C_{44} \int_{-a}^{a} (1)\, dx \int_{-b}^{b} dz \int_{-1}^{1} \left(-\frac{79}{256} r^6 + \frac{1539}{256} r^4 - \frac{891}{256} r^2 + \frac{81}{256} \right) \frac{L}{2} dr
\end{aligned}
\tag{4.450}
$$

The term $k_{64s\,B4}$ is equal to the term $k_{46s\,B4}$, see Eqs. (4.408) and (4.411)

$$
k_{64s\,B4} = -0.1929\, C_{44} abL
\tag{4.451}
$$

Regarding the $k_{65s\,B4}$ term

$$
\begin{aligned}
k_{65s\,B4} &= \int_V \left(C_{44} F_{sy,x} N_3(r) F_{\tau x} \mathbf{J}^{-1} N_{3,r}(r) \right) dV \\
&= C_{44} \int_A F_{sy,x} F_{\tau x} dA \int_L N_3(r) \mathbf{J}^{-1} N_{3,r}(r) dL \\
&= C_{44} \int_{-a}^{a} F_{sy,x} F_{\tau x} dx \int_{-b}^{b} dz \int_{-1}^{1} N_3(r) \mathbf{J}^{-1} N_{3,r}(r) \det \mathbf{J} dr \\
&= C_{44} \int_{-a}^{a} (1)\, dx \int_{-b}^{b} dz \int_{-1}^{1} \left(\frac{2187}{128L} r^5 + \frac{1215}{128L} r^4 - \frac{1377}{64L} r^3 - \frac{729}{64L} r^2 + \frac{567}{128L} r + \frac{243}{128L} \right) \frac{L}{2} dr
\end{aligned}
\tag{4.452}
$$

The term $k_{65s\,B4}$ is equal to the term $k_{56s\,B4}$, see Eqs. (4.432) and (4.435)

$$
k_{65s\,B4} = 0.0
\tag{4.453}
$$

Regarding the $k_{66s\,B4}$ term

$$
\begin{aligned}
k_{66s\,B4} &= \int_V \left(C_{44} F_{sy,x} N_3(r) F_{\tau y,x} N_3(r) \right) dV \\
&= C_{44} \int_A F_{sy,x} F_{\tau y,x} dA \int_L N_3(r) N_3(r) dL \\
&= C_{44} \int_{-a}^{a} F_{sy,x} F_{\tau y,x} dx \int_{-b}^{b} dz \int_{-1}^{1} N_3(r) N_3(r) \det \mathbf{J} dr \\
&= C_{44} \int_{-a}^{a} (1)\, dx \int_{-b}^{b} dz \int_{-1}^{1} \left(\frac{729}{256} r^6 + \frac{243}{256} r^5 - \frac{1377}{256} r^4 - \frac{243}{64} r^3 + \frac{567}{64} r^2 + \frac{243}{128} r + \frac{81}{256} \right) \frac{L}{2} dr
\end{aligned}
\tag{4.454}
$$

The surface integral of the term $k_{66s\,B4}$ is computed, using an analytical integration

$$
C_{44} \int_{-a}^{a} (1)\, dx \int_{-b}^{b} dz = C_{44}\, 4ab
\tag{4.455}
$$

whereas the line integral is computed using the four SPs Gauss quadrature.

$$\int_{-1}^{1} \left(\frac{729}{256} r^6 + \frac{243}{256} r^5 - \frac{1377}{256} r^4 - \frac{243}{64} r^3 + \frac{567}{64} r^2 + \frac{243}{128} r + \frac{81}{256} \right) \frac{L}{2} dr =$$

$$= \left[0.34785 \left(\frac{729}{256} (-0.86113)^6 + \frac{243}{256} (-0.86113)^5 - \frac{1377}{256} (-0.86113)^4 \right. \right.$$

$$\left. - \frac{243}{64} (-0.86113)^3 + \frac{567}{64} (-0.86113)^2 + \frac{243}{128} (-0.86113) + \frac{81}{256} \right) +$$

$$+ 0.65214 \left(\frac{729}{256} (-0.33998)^6 + \frac{243}{256} (-0.33998)^5 - \frac{1377}{256} (-0.33998)^4 \right.$$

$$\left. - \frac{243}{64} (-0.33998)^3 + \frac{567}{64} (-0.33998)^2 + \frac{243}{128} (-0.33998) + \frac{81}{256} \right)$$

$$+ 0.65214 \left(\frac{729}{256} (0.33998)^6 + \frac{243}{256} (0.33998)^5 - \frac{1377}{256} (0.33998)^4 \right.$$

$$\left. - \frac{243}{64} (0.33998)^3 + \frac{567}{64} (0.33998)^2 + \frac{243}{128} (0.33998) + \frac{81}{256} \right)$$

$$+ 0.34785 \left(\frac{729}{256} (0.86113)^6 + \frac{243}{256} (0.86113)^5 - \frac{1377}{256} (0.86113)^4 \right.$$

$$\left. \left. - \frac{243}{64} (0.86113)^3 + \frac{567}{64} (0.86113)^2 + \frac{243}{128} (0.86113) + \frac{81}{256} \right) \right] \frac{L}{2} =$$

$$= 0.3857L \tag{4.456}$$

The term $k_{66s\,B4}$ is then equal to

$$k_{66s\,B4} = 1.543 C_{44} abL \tag{4.457}$$

Regarding the $k_{67s\,B4}$ term

$$k_{67s\,B4} = \int_{V} \left(C_{44} F_{sy,x} N_3(r) F_{\tau x} \mathbf{J}^{-1} N_{4,r}(r) \right) dV$$

$$= C_{44} \int_{A} F_{sy,x} F_{\tau x} dA \int_{L} N_3(r) \mathbf{J}^{-1} N_{4,r}(r) dL$$

$$= C_{44} \int_{-a}^{a} F_{sy,x} F_{\tau x} dx \int_{-b}^{b} dz \int_{-1}^{1} N_3(r) \mathbf{J}^{-1} N_{4,r}(r) \det \mathbf{J} dr$$

$$= C_{44} \int_{-a}^{a} (-1) dx \int_{-b}^{b} dz \int_{-1}^{1} \left(-\frac{729}{128L} r^5 - \frac{729}{128L} r^4 + \frac{297}{64L} r^3 + \frac{369}{64L} r^2 + \frac{135}{128L} r - \frac{9}{128L} \right) \frac{L}{2} dr \tag{4.458}$$

The surface integral of the term $k_{67s\,B4}$ is computed, using an analytical integration

$$C_{44} \int_{-a}^{a} (-1) dx \int_{-b}^{b} dz = -C_{44} 4ab \tag{4.459}$$

whereas the line integral is computed using the four SPs Gauss quadrature.

$$\int_{-1}^{1} \left(-\frac{729}{128L}r^5 - \frac{729}{128L}r^4 + \frac{297}{64L}r^3 + \frac{369}{64L}r^2 + \frac{135}{128L}r - \frac{9}{128L} \right) \frac{L}{2} dr =$$

$$= \left[0.34785 \left(-\frac{729}{128L}(-0.86113)^5 - \frac{729}{128L}(-0.86113)^4 + \right. \right.$$

$$+ \frac{297}{64L}(-0.86113)^3 + \frac{369}{64L}(-0.86113)^2 + \frac{135}{128L}r(-0.86113) - \frac{9}{128L} \left. \right) +$$

$$+ 0.65214 \left(-\frac{729}{128L}(-0.33998)^5 - \frac{729}{128L}(-0.33998)^4 + \right.$$

$$+ \frac{297}{64L}(-0.33998)^3 + \frac{369}{64L}(-0.33998)^2 + \frac{135}{128L}r(-0.33998) - \frac{9}{128L} \left. \right)$$

$$+ 0.65214 \left(-\frac{729}{128L}(0.33998)^5 - \frac{729}{128L}(0.33998)^4 + \right.$$

$$+ \frac{297}{64L}(0.33998)^3 + \frac{369}{64L}(0.33998)^2 + \frac{135}{128L}r(0.33998) - \frac{9}{128L} \left. \right)$$

$$+ 0.34785 \left(-\frac{729}{128L}(0.86113)^5 - \frac{729}{128L}(0.86113)^4 + \right.$$

$$+ \frac{297}{64L}(0.86113)^3 + \frac{369}{64L}(0.86113)^2 + \frac{135}{128L}r(0.86113) - \frac{9}{128L} \left. \right) \left. \right] \frac{L}{2} =$$

$$= 0.7125 \tag{4.460}$$

The term $k_{67s_{B4}}$ is then equal to

$$k_{67s_{B4}} = -2.85C_{44}ab \tag{4.461}$$

Regarding the $k_{68s_{B4}}$ term

$$k_{68s_{B4}} = \int_V \left(C_{44} F_{sy,x} N_3(r) F_{\tau y,x} N_4(r) \right) dV$$

$$= C_{44} \int_A F_{sy,x} F_{\tau y,x} dA \int_L N_3(r) N_4(r) dL$$

$$= C_{44} \int_{-a}^{a} F_{sy,x} F_{\tau y,x} dx \int_{-b}^{b} dz \int_{-1}^{1} N_3(r) N_4(r) det\, \mathbf{J} dr$$

$$= C_{44} \int_{-a}^{a} (1)\, dx \int_{-b}^{b} dz \int_{-1}^{1} \left(-\frac{243}{256}r^6 - \frac{81}{64}r^5 + \frac{189}{256}r^4 + \frac{45}{32}r^3 + \frac{63}{256}r^2 - \frac{9}{64}r - \frac{9}{256} \right) \frac{L}{2} dr \tag{4.462}$$

The surface integral of the term $k_{68s_{B4}}$ is computed, using an analytical integration

$$C_{44} \int_{-a}^{a} (1)\, dx \int_{-b}^{b} dz = C_{44}4ab \tag{4.463}$$

whereas the line integral is computed using the four SPs Gauss quadrature.

$$\int_{-1}^{1} \left(-\frac{243}{256}r^6 - \frac{81}{64}r^5 + \frac{189}{256}r^4 + \frac{45}{32}r^3 + \frac{63}{256}r^2 - \frac{9}{64}r - \frac{9}{256} \right) \frac{L}{2} dr =$$

$$= \left[0.34785 \left(-\frac{243}{256} (-0.86113)^6 - \frac{81}{64} (-0.86113)^5 + \frac{189}{256} (-0.86113)^4 \right. \right.$$

$$\left. +\frac{45}{32} (-0.86113)^3 + \frac{63}{256} (-0.86113)^2 - \frac{9}{64} (-0.86113) - \frac{9}{256} \right) +$$

$$+ 0.65214 \left(-\frac{243}{256} (-0.33998)^6 - \frac{81}{64} (-0.33998)^5 + \frac{189}{256} (-0.33998)^4 \right.$$

$$\left. +\frac{45}{32} (-0.33998)^3 + \frac{63}{256} (-0.33998)^2 - \frac{9}{64} (-0.33998) - \frac{9}{256} \right)$$

$$+ 0.65214 \left(-\frac{243}{256} (0.33998)^6 - \frac{81}{64} (0.33998)^5 + \frac{189}{256} (0.33998)^4 \right.$$

$$\left. +\frac{45}{32} (0.33998)^3 + \frac{63}{256} (0.33998)^2 - \frac{9}{64} (0.33998) - \frac{9}{256} \right)$$

$$+ 0.34785 \left(-\frac{243}{256} (0.86113)^6 - \frac{81}{64} (0.86113)^5 + \frac{189}{256} (0.86113)^4 \right.$$

$$\left. \left. +\frac{45}{32} (0.86113)^3 + \frac{63}{256} (0.86113)^2 - \frac{9}{64} (0.86113) - \frac{9}{256} \right) \right] \frac{L}{2} =$$

$$= 0.05893L \tag{4.464}$$

The term $k_{68s\,B4}$ is then equal to

$$k_{68s\,B4} = 0.23571 C_{44} abL \tag{4.465}$$

Regarding the $k_{71s\,B4}$ term

$$k_{71s\,B4} = \int_V \left(C_{44} F_{sx} \mathbf{J}^{-1} N_{4,r}(r) F_{\tau x} \mathbf{J}^{-1} N_{1,r}(r) \right) dV$$

$$= C_{44} \int_A F_{sx} F_{\tau x} dA \int_L \mathbf{J}^{-1} N_{4,r}(r) \mathbf{J}^{-1} N_{1,r}(r) dL \tag{4.466}$$

$$= C_{44} \int_{-a}^{a} F_{sx} F_{\tau x} dx \int_{-b}^{b} dz \int_{-1}^{1} \mathbf{J}^{-1} N_{4,r}(r) \mathbf{J}^{-1} N_{1,r}(r) \det \mathbf{J} dr$$

$$= C_{44} \int_{-a}^{a} (1) dx \int_{-b}^{b} dz \int_{-1}^{1} \left(-\frac{729}{64L^2}r^4 + \frac{189}{32L^2}r^2 - \frac{1}{64L^2} \right) \frac{L}{2} dr$$

The term $k_{71s\,B4}$ is equal to the term $k_{17s\,B4}$, see Eqs. (4.328) and (4.331)

$$k_{71s\,B4} = -1.3 C_{44} \frac{ab}{L} \tag{4.467}$$

Regarding the $k_{72s\,B4}$ term

$$
\begin{aligned}
k_{72s\,B4} &= \int_V \left(C_{44} F_{sx} \mathbf{J}^{-1} N_{4,r}(r) F_{\tau y,x} N_1(r) \right) dV \\
&= C_{44} \int_A F_{sx} F_{\tau y,x}\, dA \int_L \mathbf{J}^{-1} N_{4,r}(r) N_1(r)\, dL \\
&= C_{44} \int_{-a}^{a} F_{sx} F_{\tau y,x}\, dx \int_{-b}^{b} dz \int_{-1}^{1} \mathbf{J}^{-1} N_{4,r}(r) N_1(r)\, det\, \mathbf{J}\, dr \\
&= C_{44} \int_{-a}^{a} (1)\, dx \int_{-b}^{b} dz \int_{-1}^{1} \left(-\frac{243}{128L} r^5 + \frac{81}{128L} r^4 + \frac{99}{64L} r^3 - \frac{9}{64L} r^2 - \frac{19}{128L} r + \frac{1}{128L} \right) \frac{L}{2}\, dr
\end{aligned}
$$

$$(4.468)$$

The term $k_{72s\,B4}$ is equal to the term $k_{27s\,B4}$, see Eqs. (4.358) and (4.361)

$$
k_{72s\,B4} = -0.35 C_{44} ab \tag{4.469}
$$

Regarding the $k_{73s\,B4}$ term

$$
\begin{aligned}
k_{73s\,B4} &= \int_V \left(C_{44} F_{sx} \mathbf{J}^{-1} N_{4,r}(r) F_{\tau x} \mathbf{J}^{-1} N_{2,r}(r) \right) dV \\
&= C_{44} \int_A F_{sx} F_{\tau x}\, dA \int_L \mathbf{J}^{-1} N_{4,r}(r) \mathbf{J}^{-1} N_{2,r}(r)\, dL \\
&= C_{44} \int_{-a}^{a} F_{sx} F_{\tau x}\, dx \int_{-b}^{b} dz \int_{-1}^{1} \mathbf{J}^{-1} N_{4,r}(r) \mathbf{J}^{-1} N_{2,r}(r)\, det\, \mathbf{J}\, dr \\
&= C_{44} \int_{-a}^{a} (-1)\, dx \int_{-b}^{b} dz \int_{-1}^{1} \left(\frac{2187}{64L^2} r^4 + \frac{243}{16L^2} r^3 - \frac{567}{32L^2} r^2 - \frac{117}{16L^2} r + \frac{27}{64L^2} \right) \frac{L}{2}\, dr
\end{aligned}
$$

$$(4.470)$$

The term $k_{73s\,B4}$ is equal to the term $k_{37s\,B4}$, see Eqs. (4.386) and (4.389)

$$
k_{73s\,B4} = 5.4 C_{44} \frac{ab}{L} \tag{4.471}
$$

Regarding the $k_{74s\,B4}$ term

$$
\begin{aligned}
k_{74s\,B4} &= \int_V \left(C_{44} F_{sx} \mathbf{J}^{-1} N_{4,r}(r) F_{\tau y,x} N_2(r) \right) dV \\
&= C_{44} \int_A F_{sx} F_{\tau y,x}\, dA \int_L \mathbf{J}^{-1} N_{4,r}(r) N_2(r)\, dL \\
&= C_{44} \int_{-a}^{a} F_{sx} F_{\tau y,x}\, dx \int_{-b}^{b} dz \int_{-1}^{1} \mathbf{J}^{-1} N_{4,r}(r) N_2(r)\, det\, \mathbf{J}\, dr \\
&= C_{44} \int_{-a}^{a} (-1)\, dx \int_{-b}^{b} dz \int_{-1}^{1} \left(\frac{729}{128L} r^5 + \frac{243}{128L} r^4 - \frac{459}{64L} r^3 - \frac{117}{64L} r^2 + \frac{189}{128L} r - \frac{9}{128L} \right) \frac{L}{2}\, dr
\end{aligned}
$$

$$(4.472)$$

The term $k_{74s\,B4}$ is equal to the term $k_{47s\,B4}$, see Eqs. (4.412) and (4.415)

$$
k_{74s\,B4} = 1.2 C_{44} ab \tag{4.473}
$$

Regarding the $k_{75s\,B4}$ term

$$
\begin{aligned}
k_{75s\,B4} &= \int_V \left(C_{44}\, F_{sx}\mathbf{J}^{-1}N_{4,r}(r)\,F_{\tau x}\mathbf{J}^{-1}N_{3,r}(r) \right) dV \\
&= C_{44} \int_A F_{sx}\, F_{\tau x}\, dA \int_L \mathbf{J}^{-1}N_{4,r}(r)\mathbf{J}^{-1}N_{3,r}(r)\, dL \\
&= C_{44} \int_{-a}^{a} F_{sx}\, F_{\tau x}\, dx \int_{-b}^{b} dz \int_{-1}^{1} \mathbf{J}^{-1}N_{4,r}(r)\mathbf{J}^{-1}N_{3,r}(r)\, det\,\mathbf{J}\, dr \\
&= C_{44} \int_{-a}^{a} (1)\, dx \int_{-b}^{b} dz \int_{-1}^{1} \left(-\frac{2187}{64L^2}r^4 - \frac{243}{8L^2}r^3 + \frac{243}{32L^2}r^2 + \frac{63}{8L^2}r - \frac{27}{64L^2} \right) \frac{L}{2}\, dr
\end{aligned}
\tag{4.474}
$$

The term $k_{75s\,B4}$ is equal to the term $k_{57s\,B4}$, see Eqs. (4.436) and (4.439)

$$
k_{75s\,B4} = -18.9 C_{44} \frac{ab}{L}
\tag{4.475}
$$

Regarding the $k_{76s\,B4}$ term

$$
\begin{aligned}
k_{76s\,B4} &= \int_V \left(C_{44}\, F_{sx}\mathbf{J}^{-1}N_{4,r}(r)\,F_{\tau y,x}N_{3}(r) \right) dV \\
&= C_{44} \int_A F_{sx}\, F_{\tau y,x}\, dA \int_L \mathbf{J}^{-1}N_{4,r}(r)N_{3}(r)\, dL \\
&= C_{44} \int_{-a}^{a} F_{sx}\, F_{\tau y,x}\, dx \int_{-b}^{b} dz \int_{-1}^{1} \mathbf{J}^{-1}N_{4,r}(r)N_{3}(r)\, det\,\mathbf{J}\, dr \\
&= C_{44} \int_{-a}^{a} (-1)\, dx \int_{-b}^{b} dz \int_{-1}^{1} \left(-\frac{729}{128L}r^5 - \frac{729}{128L}r^4 + \frac{297}{64L}r^3 + \frac{369}{64L}r^2 + \frac{135}{128L}r - \frac{9}{128L} \right) \frac{L}{2}\, dr
\end{aligned}
$$
$$
\tag{4.476}
$$

The term $k_{76s\,B4}$ is equal to the term $k_{67s\,B4}$, see Eqs. (4.458) and (4.461)

$$
k_{76s\,B4} = -2.85 C_{44} ab
\tag{4.477}
$$

Regarding the $k_{77s\,B4}$ term

$$
\begin{aligned}
k_{77s\,B4} &= \int_V \left(C_{44}\, F_{sx}\mathbf{J}^{-1}N_{4,r}(r)\,F_{\tau x}\mathbf{J}^{-1}N_{4,r}(r) \right) dV \\
&= C_{44} \int_A F_{sx}\, F_{\tau x}\, dA \int_L \mathbf{J}^{-1}N_{4,r}(r)\mathbf{J}^{-1}N_{4,r}(r)\, dL \\
&= C_{44} \int_{-a}^{a} F_{sx}\, F_{\tau x}\, dx \int_{-b}^{b} dz \int_{-1}^{1} \mathbf{J}^{-1}N_{4,r}(r)\mathbf{J}^{-1}N_{4,r}(r))\, det\,\mathbf{J}\, dr \\
&= C_{44} \int_{-a}^{a} 1\, dx \int_{-b}^{b} dz \int_{-1}^{1} \left(\frac{729}{64L^2}r^4 + \frac{243}{16L^2}r^3 + \frac{135}{32L^2}r^2 - \frac{9}{16L^2}r + \frac{1}{64L^2} \right) \frac{L}{2}\, dr
\end{aligned}
\tag{4.478}
$$

The surface integral of the term $k_{77s\,B4}$ is computed, using an analytical integration

$$
C_{44} \int_{-a}^{a} 1\, dx \int_{-b}^{b} dz = C_{44}\, 4ab
\tag{4.479}
$$

while the line integral,computed using four SPs Gauss quadrature, is equal to that of Eqs. (4.298) and (4.300)

$$\int_{-1}^{1} \left(\frac{729}{64L^2}r^4 + \frac{243}{16L^2}r^3 + \frac{135}{32L^2}r^2 - \frac{9}{16L^2}r + \frac{1}{64L^2} \right) \frac{L}{2} dr = \frac{3.7}{L} \quad (4.480)$$

The term $k_{77s_{B4}}$ is then equal to

$$k_{77s_{B4}} = 14.8 C_{44} \frac{ab}{L} \quad (4.481)$$

Regarding the $k_{78s_{B4}}$ term

$$\begin{aligned}
k_{78s_{B4}} &= \int_V \left(C_{44} F_{sx} \mathbf{J}^{-1} N_{4,r}(r) F_{\tau y,x} N_4(r) \right) dV \\
&= C_{44} \int_A F_{sx} F_{\tau y,x} dA \int_L \mathbf{J}^{-1} N_{4,r}(r) N_4(r) dL \\
&= C_{44} \int_{-a}^{a} F_{sx} F_{\tau y,x} dx \int_{-b}^{b} dz \int_{-1}^{1} \mathbf{J}^{-1} N_{4,r}(r) N_4(r) det\,\mathbf{J} dr \\
&= C_{44} \int_{-a}^{a} (-1)\, dx \int_{-b}^{b} dz \int_{-1}^{1} \left(\frac{243}{128L}r^5 + \frac{405}{128L}r^4 + \frac{63}{64L}r^3 - \frac{27}{64L}r^2 - \frac{17}{128L}r + \frac{1}{128L} \right) \frac{L}{2} dr
\end{aligned}$$
$$(4.482)$$

The surface integral of the term $k_{78s_{B4}}$ is computed, using an analytical integration

$$C_{44} \int_{-a}^{a} (-1)\, dx \int_{-b}^{b} dz = -C_{44} 4ab \quad (4.483)$$

whereas the line integral is computed using the four SPs Gauss quadrature.

$$\begin{aligned}
&\int_{-1}^{1} \left(\frac{243}{128L}r^5 + \frac{405}{128L}r^4 + \frac{63}{64L}r^3 - \frac{27}{64L}r^2 - \frac{17}{128L}r + \frac{1}{128L} \right) \frac{L}{2} dr = \\
&= \Bigg[0.34785 \left(\frac{243}{128L}(-0.86113)^5 + \frac{405}{128L}(-0.86113)^4 \right. \\
&\quad + \frac{63}{64L}(-0.86113)^3 - \frac{27}{64L}(-0.86113)^2 - \frac{17}{128L}(-0.86113) + \frac{1}{128L} \Bigg) + \\
&\quad + 0.652140.34785 \left(\frac{243}{128L}(-0.33998)^5 + \frac{405}{128L}(-0.33998)^4 \right. \\
&\quad + \frac{63}{64L}(-0.33998)^3 - \frac{27}{64L}(-0.33998)^2 - \frac{17}{128L}(-0.33998) + \frac{1}{128L} \Bigg) \\
&\quad + 0.652140.34785 \left(\frac{243}{128L}(0.33998)^5 + \frac{405}{128L}(0.33998)^4 \right. \\
&\quad + \frac{63}{64L}(0.33998)^3 - \frac{27}{64L}(0.33998)^2 - \frac{17}{128L}(0.33998) + \frac{1}{128L} \Bigg) \\
&\quad + 0.34785 \left(\frac{243}{128L}(0.86113)^5 + \frac{405}{128L}(0.86113)^4 \right. \\
&\quad + \frac{63}{64L}(0.86113)^3 - \frac{27}{64L}(0.86113)^2 - \frac{17}{128L}(0.86113) + \frac{1}{128L} \Bigg) \Bigg] \frac{L}{2} = \\
&= 0.5
\end{aligned}$$
$$(4.484)$$

The term $k_{78s_{B4}}$ is then equal to

$$k_{78s\,B4} = -2.0C_{44}ab \tag{4.485}$$

Regarding the $k_{81s\,B4}$ term

$$
\begin{aligned}
k_{81s\,B4} &= \int_V \left(C_{44}F_{sy,x}N_4(r)F_{\tau x}\mathbf{J}^{-1}N_{1,r}(r) \right) dV \\
&= C_{44}\int_A F_{sy,x}F_{\tau x}dA \int_L N_4(r\mathbf{J}^{-1}N_{1,r}(r)dL \\
&= C_{44}\int_{-a}^{a} F_{sy,x}F_{\tau x}dx \int_{-b}^{b} dz \int_{-1}^{1} N_4(r\mathbf{J}^{-1}N_{1,r}(r)det\,\mathbf{J}dr \\
&= C_{44}\int_{-a}^{a} (-1)dx \int_{-b}^{b} dz \int_{-1}^{1} \left(-\frac{243}{128L}r^5 - \frac{81}{128L}r^4 + \frac{99}{64L}r^3 + \frac{9}{64L}r^2 - \frac{19}{128L}r - \frac{1}{128L} \right) \frac{L}{2}dr
\end{aligned}
\tag{4.486}
$$

The term $k_{81s\,B4}$ is equal to the term $k_{18s\,B4}$, see Eqs. (4.332) and (4.335)

$$k_{81s\,B4} = 0.35C_{44}ab \tag{4.487}$$

Regarding the $k_{82s\,B4}$ term

$$
\begin{aligned}
k_{82s\,B4} &= \int_V \left(C_{44}F_{sy,x}N_4(r)F_{\tau y,x}N_1(r) \right) dV \\
&= C_{44}\int_A F_{sy,x}F_{\tau y,x}dA \int_L N_4(rN_1(r)dL \\
&= C_{44}\int_{-a}^{a} F_{sy,x}F_{\tau y,x}dx \int_{-b}^{b} dz \int_{-1}^{1} N_4(rN_1(r)det\,\mathbf{J}dr \\
&= C_{44}\int_{-a}^{a} (1)dx \int_{-b}^{b} dz \int_{-1}^{1} \left(-\frac{81}{256}r^6 + \frac{99}{256}r^4 - \frac{19}{256}r^2 + \frac{1}{256} \right) \frac{L}{2}dr
\end{aligned}
\tag{4.488}
$$

The term $k_{82s\,B4}$ is equal to the term $k_{28s\,B4}$, see Eqs. (4.362) and (4.365)

$$k_{82s\,B4} = 0.04524C_{44}abL \tag{4.489}$$

Regarding the $k_{12s\,B4}$ term

$$
\begin{aligned}
k_{83s\,B4} &= \int_V \left(C_{44}F_{sy,x}N_4(r)F_{\tau x}[J_b]^{-1}N_{2,r}(r) \right) dV \\
&= C_{44}\int_A F_{sy,x}F_{\tau x}dA \int_L N_4(r)[J_b]^{-1}N_{2,r}(r)dL \\
&= C_{44}\int_{-a}^{a} F_{sy,x}F_{\tau x}dx \int_{-b}^{b} dz \int_{-1}^{1} N_4(r)[J_b]^{-1}N_{2,r}(r)det[J_b]dr \\
&= C_{44}\int_{-a}^{a} (-1)dx \int_{-b}^{b} dz \int_{-1}^{1} \left(\frac{729}{128L}r^5 + \frac{567}{128L}r^4 - \frac{243}{64L}r^3 - \frac{153}{64L}r^2 + \frac{45}{128L}r + \frac{27}{128L} \right) \frac{L}{2}dr
\end{aligned}
\tag{4.490}
$$

The term $k_{83s\,B4}$ is equal to the term $k_{38s\,B4}$, see Eqs. (4.390) and (4.393)

$$k_{83s\,B4} = -1.2C_{44}ab \tag{4.491}$$

Regarding the $k_{84s\,B4}$ term

$$
\begin{aligned}
k_{84s\,B4} &= \int_V \left(C_{44} F_{sy,x} N_4(r) F_{\tau y,x} N_2(r) \right) dV \\
&= C_{44} \int_A F_{sy,x} F_{\tau y,x} dA \int_L N_4(r) N_2(r) dL \\
&= C_{44} \int_{-a}^{a} F_{sy,x} F_{\tau y,x} dx \int_{-b}^{b} dz \int_{-1}^{1} N_4(r) N_2(r) det[J_b] dr \\
&= C_{44} \int_{-a}^{a} (1)\, dx \int_{-b}^{b} dz \int_{-1}^{1} \left(\frac{243}{256} r^6 + \frac{81}{256} r^5 - \frac{351}{256} r^4 - \frac{45}{64} r^3 + \frac{117}{64} r^2 + \frac{9}{128} r - \frac{9}{256} \right) \frac{L}{2} dr
\end{aligned}
$$
$$(4.492)$$

The term $k_{84s\,B4}$ is equal to the term $k_{48s\,B4}$, see Eqs. (4.416) and (4.419)

$$
k_{84s\,B4} = -0.08571 C_{44} abL \tag{4.493}
$$

Regarding the $k_{85s\,B4}$ term

$$
\begin{aligned}
k_{85s\,B4} &= \int_V \left(C_{44} F_{sy,x} N_4(r) F_{\tau x} [J_b]^{-1} N_{3,r}(r) \right) dV \\
&= C_{44} \int_A F_{sy,x} F_{\tau x} dA \int_L N_4(r)[J_b]^{-1} N_{3,r}(r) dL \\
&= C_{44} \int_{-a}^{a} F_{sy,x} F_{\tau x} dx \int_{-b}^{b} dz \int_{-1}^{1} N_4(r)[J_b]^{-1} N_{3,r}(r) det[J_b] dr \\
&= C_{44} \int_{-a}^{a} (-1)\, dx \int_{-b}^{b} dz \int_{-1}^{1} \left(-\frac{729}{128L} r^5 - \frac{891}{128L} r^4 + \frac{81}{64L} r^3 + \frac{171}{64L} r^2 - \frac{9}{128L} r - \frac{27}{128L} \right) \frac{L}{2} dr
\end{aligned}
$$
$$(4.494)$$

The term $k_{85s\,B4}$ is equal to the term $k_{58s\,B4}$, see Eqs. (4.440) and (4.443)

$$
k_{85s\,B4} = 2.85 C_{44} ab \tag{4.495}
$$

Regarding the $k_{86s\,B4}$ term

$$
\begin{aligned}
k_{86s\,B4} &= \int_V \left(C_{44} F_{sy,x} N_4(r) F_{\tau y,x} N_3(r) \right) dV \\
&= C_{44} \int_A F_{sy,x} F_{\tau y,x} dA \int_L N_4(r) N_3(r) dL \\
&= C_{44} \int_{-a}^{a} F_{sy,x} F_{\tau y,x} dx \int_{-b}^{b} dz \int_{-1}^{1} N_4(r) N_3(r) det[J_b] dr \\
&= C_{44} \int_{-a}^{a} (1)\, dx \int_{-b}^{b} dz \int_{-1}^{1} \left(-\frac{243}{256} r^6 - \frac{81}{64} r^5 + \frac{189}{256} r^4 + \frac{45}{32} r^3 + \frac{63}{256} r^2 - \frac{9}{64} r - \frac{9}{256} \right) \frac{L}{2} dr
\end{aligned}
$$
$$(4.496)$$

The term $k_{86s\,B4}$ is equal to the term $k_{68s\,B4}$, see Eqs. (4.462) and (4.465)

$$k_{86s\,B4} = 0.23571 C_{44} abL \tag{4.497}$$

Regarding the $k_{87s\,B4}$ term

$$
\begin{aligned}
k_{87s\,B4} &= \int_V \left(C_{44} F_{sy,x} N_4(r) F_{\tau x} [J_b]^{-1} N_{4,r}(r) \right) dV \\
&= C_{44} \int_A F_{sy,x} F_{\tau x} dA \int_L N_4(r)[J_b]^{-1} N_{4,r}(r) dL \\
&= C_{44} \int_{-a}^{a} F_{sy,x} F_{\tau x} dx \int_{-b}^{b} dz \int_{-1}^{1} N_4(r)[J_b]^{-1} N_{4,r}(r) det[J_b] dr \\
&= C_{44} \int_{-a}^{a} (-1) dx \int_{-b}^{b} dz \int_{-1}^{1} \left(\frac{243}{128L} r^5 + \frac{405}{128L} r^4 + \frac{63}{64L} r^3 - \frac{27}{64L} r^2 - \frac{17}{128L} r + \frac{1}{128L} \right) \frac{L}{2} dr
\end{aligned}
\tag{4.498}
$$

The term $k_{87s\,B4}$ is equal to the term $k_{78s\,B4}$, see Eqs. (4.482) and (4.485)

$$k_{87s\,B4} = -2.0 C_{44} ab \tag{4.499}$$

Regarding the $k_{88s\,B4}$ term

$$
\begin{aligned}
k_{88s\,B4} &= \int_V \left(C_{44} F_{sy,x} N_4(r) F_{\tau y,x} N_4(r) \right) dV \\
&= C_{44} \int_A F_{sy,x} F_{\tau y,x} dA \int_L N_4(r) N_4(r) dL \\
&= C_{44} \int_{-a}^{a} F_{sy,x} F_{\tau y,x} dx \int_{-b}^{b} dz \int_{-1}^{1} N_4(r) N_4(r) det[J_b] dr \\
&= C_{44} \int_{-a}^{a} (1) dx \int_{-b}^{b} dz \int_{-1}^{1} \left(\frac{81}{256} r^6 + \frac{81}{128} r^5 + \frac{63}{256} r^4 - \frac{9}{64} r^3 - \frac{17}{256} r^2 + \frac{1}{128} r + \frac{1}{256} \right) \frac{L}{2} dr
\end{aligned}
\tag{4.500}
$$

The surface integral of the term $k_{88s\,B4}$ is computed, using an analytical integration

$$C_{44} \int_{-a}^{a} (1) dx \int_{-b}^{b} dz = C_{44} 4ab \tag{4.501}$$

whereas the line integral is computed using the four SPs Gauss quadrature.

$$\int_{-1}^{1} \left(\frac{81}{256} r^6 + \frac{81}{128} r^5 + \frac{63}{256} r^4 - \frac{9}{64} r^3 - \frac{17}{256} r^2 + \frac{1}{128} r + \frac{1}{256} \right) \frac{L}{2} dr =$$

$$= \left[0.34785 \left(\frac{81}{256} (-0.86113)^6 + \frac{81}{128} (-0.86113)^5 + \frac{63}{256} (-0.86113)^4 - \right.\right.$$

$$\left. - \frac{9}{64} (-0.86113)^3 - \frac{17}{256} (-0.86113)^2 + \frac{1}{128} (-0.86113) + \frac{1}{256} \right) +$$

$$+ 0.65214 \left(\frac{81}{256} (-0.33998)^6 + \frac{81}{128} (-0.33998)^5 + \frac{63}{256} (-0.33998)^4 - \right.$$

$$\left. - \frac{9}{64} (-0.33998)^3 - \frac{17}{256} (-0.33998)^2 + \frac{1}{128} (-0.33998) + \frac{1}{256} \right)$$

$$+ 0.65214 \left(\frac{81}{256} (0.33998)^6 + \frac{81}{128} (0.33998)^5 + \frac{63}{256} (0.33998)^4 - \right.$$

$$\left. - \frac{9}{64} (0.33998)^3 - \frac{17}{256} (0.33998)^2 + \frac{1}{128} (0.33998) + \frac{1}{256} \right)$$

$$+ 0.34785 \left(\frac{81}{256} (0.86113)^6 + \frac{81}{128} (0.86113)^5 + \frac{63}{256} (0.86113)^4 - \right.$$

$$\left.\left. - \frac{9}{64} (0.86113)^3 - \frac{17}{256} (0.86113)^2 + \frac{1}{128} (0.86113) + \frac{1}{256} \right) \right] \frac{L}{2} =$$

$$= 0.07619L \tag{4.502}$$

The term $k_{88s_{B4}}$ is then equal to

$$k_{88s_{B4}} = 0.3047 C_{44} abL \tag{4.503}$$

Based on the terms computed previously, the shear stiffneess matrix of a B4 beam element, obtained using a FI Gauss quadrature, is the following

$$
\mathbf{K}_{S_{B4}} =
\begin{bmatrix}
14.8C_{44}\dfrac{ab}{L} & 2C_{44}ab & -18.9C_{44}\dfrac{ab}{L} & 2.85C_{44}ab & 5.4C_{44}\dfrac{ab}{L} & -1.2C_{44}ab & -1.3C_{44}\dfrac{ab}{L} & 0.35C_{44}ab \\[2ex]
2C_{44}ab & 0.3047C_{44}abL & -2.85C_{44}ab & 0.2357C_{44}abL & 1.2C_{44}ab & -0.08571C_{44}abL & -0.35C_{44}ab & 0.04524C_{44}abL \\[2ex]
-18.9C_{44}\dfrac{ab}{L} & -2.85C_{44}ab & 43.2C_{44}\dfrac{ab}{L} & 0.0 & -29.7C_{44}\dfrac{ab}{L} & 4.05C_{44}ab & 5.4C_{44}\dfrac{ab}{L} & -1.2C_{44}ab \\[2ex]
2.85C_{44}ab & 0.2357C_{44}abL & 0.0 & 1.543C_{44}abL & -4.05C_{44}ab & -0.1929C_{44}abL & 1.2C_{44}ab & -0.08571C_{44}abL \\[2ex]
5.4C_{44}\dfrac{ab}{L} & 1.2C_{44}ab & -29.7C_{44}\dfrac{ab}{L} & -4.05C_{44}ab & 43.2C_{44}\dfrac{ab}{L} & 0.0 & -18.9C_{44}\dfrac{ab}{L} & 2.85C_{44}ab \\[2ex]
-1.2C_{44}ab & -0.08571C_{44}abL & 4.05C_{44}ab & -0.1929C_{44}abL & 0.0 & 1.543C_{44}abL & -2.85C_{44}ab & 0.23571C_{44}abL \\[2ex]
-1.3C_{44}\dfrac{ab}{L} & -0.35C_{44}ab & 5.4C_{44}\dfrac{ab}{L} & 1.2C_{44}ab & -18.9C_{44}\dfrac{ab}{L} & -2.85C_{44}ab & 14.8C_{44}\dfrac{ab}{L} & -2.0C_{44}ab \\[2ex]
0.35C_{44}ab & 0.04524C_{44}abL & -1.2C_{44}ab & -0.08571C_{44}abL & 2.85C_{44}ab & 0.23571C_{44}abL & -2.0C_{44}ab & 0.3047C_{44}abL
\end{bmatrix}
$$

The complete stiffness matrix of the B4 element is obtained summing the two matrices of Eqs. (4.302) and (4.4.2.4)

$$
\mathbf{K}_{B4} =
\begin{bmatrix}
14.8C_{44}\dfrac{ab}{L} & 2C_{44}ab & -18.9C_{44}\dfrac{ab}{L} & 2.85C_{44}ab & 5.4C_{44}\dfrac{ab}{L} & -1.2C_{44}ab & -1.3C_{44}\dfrac{ab}{L} & 0.35C_{44}ab \\[2ex]
2C_{44}ab & \begin{array}{l}0.3047C_{44}abL+\\ +4.933C_{11}\dfrac{a^{3}b}{L}\end{array} & -2.85C_{44}ab & \begin{array}{l}0.2357C_{44}abL+\\ +-6.3C_{11}\dfrac{a^{3}b}{L}\end{array} & 1.2C_{44}ab & \begin{array}{l}-0.08571C_{44}abL+\\ +1.8C_{11}\dfrac{a^{3}b}{L}\end{array} & -0.35C_{44}ab & \begin{array}{l}0.04524C_{44}abL+\\ +-0.4333C_{11}\dfrac{a^{3}b}{L}\end{array} \\[3ex]
-18.9C_{44}\dfrac{ab}{L} & -2.85C_{44}ab & 43.2C_{44}\dfrac{ab}{L} & 0.0 & -29.7C_{44}\dfrac{ab}{L} & 4.05C_{44}ab & 5.4C_{44}\dfrac{ab}{L} & -1.2C_{44}ab \\[2ex]
2.85C_{44}ab & \begin{array}{l}0.2357C_{44}abL+\\ +-6.3C_{11}\dfrac{a^{3}b}{L}\end{array} & 0.0 & \begin{array}{l}1.543C_{44}abL+\\ +14.4C_{11}\dfrac{a^{3}b}{L}\end{array} & -4.05C_{44}ab & \begin{array}{l}-0.1929C_{44}abL+\\ +-9.9C_{11}\dfrac{a^{3}b}{L}\end{array} & 1.2C_{44}ab & \begin{array}{l}-0.08571C_{44}abL+\\ +1.8C_{11}\dfrac{a^{3}b}{L}\end{array} \\[3ex]
5.4C_{44}\dfrac{ab}{L} & 1.2C_{44}ab & -29.7C_{44}\dfrac{ab}{L} & -4.05C_{44}ab & 43.2C_{44}\dfrac{ab}{L} & 0.0 & -18.9C_{44}\dfrac{ab}{L} & 2.85C_{44}ab \\[2ex]
-1.2C_{44}ab & \begin{array}{l}-0.08571C_{44}abL+\\ +1.8C_{11}\dfrac{a^{3}b}{L}\end{array} & 4.05C_{44}ab & \begin{array}{l}-0.1929C_{44}abL+\\ +-9.9C_{11}\dfrac{a^{3}b}{L}\end{array} & 0.0 & \begin{array}{l}1.543C_{44}abL+\\ +14.4C_{11}\dfrac{a^{3}b}{L}\end{array} & -2.85C_{44}ab & \begin{array}{l}0.23571C_{44}abL+\\ +-6.3C_{11}\dfrac{a^{3}b}{L}\end{array} \\[3ex]
-1.3C_{44}\dfrac{ab}{L} & -0.35C_{44}ab & 5.4C_{44}\dfrac{ab}{L} & 1.2C_{44}ab & -18.9C_{44}\dfrac{ab}{L} & -2.85C_{44}ab & 14.8C_{44}\dfrac{ab}{L} & -2.0C_{44}ab \\[2ex]
0.35C_{44}ab & \begin{array}{l}0.04524C_{44}abL+\\ +-0.4333C_{11}\dfrac{a^{3}b}{L}\end{array} & -1.2C_{44}ab & \begin{array}{l}-0.08571C_{44}abL+\\ +k_{84b_{B4}}=1.8C_{11}\dfrac{a^{3}b}{L}\end{array} & 2.85C_{44}ab & \begin{array}{l}0.23571C_{44}abL+\\ +-6.3C_{11}\dfrac{a^{3}b}{L}\end{array} & -2.0C_{44}ab & \begin{array}{l}0.3047C_{44}abL+\\ +4.933C_{11}\dfrac{a^{3}b}{L}\end{array}
\end{bmatrix}
\tag{4.504}
$$

4.5 Numerical Examples of Comparison Between FI and URI Matrices of B2 Element

4.5.1 Comparison Between FI and URI Matrices of B2 Element

To highlight the numerical effects of shear locking, examples are provided showing that, despite a relatively small discrepancy in only four terms between the two matrices (reaching $\pm 33.\overline{3}\%$ as indicated in matrices in Eq. (4.24)), the resulting outcomes can differ significantly in terms of percentage variation. Below are the stiffness matrices of a B2 element computed using FI Gauss quadrature and URI:

$$
\mathbf{K}_{FI} =
\begin{bmatrix}
\dfrac{4C_{44}ab}{L} & 2C_{44}ab & -\dfrac{4C_{44}ab}{L} & 2C_{44}ab \\[2ex]
2C_{44}ab & 0,\overline{3}(4C_{44}abL)+\dfrac{4C_{11}a^3b}{3L} & -2C_{44}ab & 0,1\overline{6}(4C_{44}abL)+\dfrac{4C_{11}a^3b}{3L} \\[2ex]
-\dfrac{4C_{44}ab}{L} & -2C_{44}ab & \dfrac{4C_{44}ab}{L} & -2C_{44}ab \\[2ex]
2C_{44}ab & 0,1\overline{6}(4C_{44}abL)-\dfrac{4C_{11}a^3b}{3L} & -2C_{44}ab & 0,\overline{3}(4C_{44}abL)+\dfrac{4C_{11}a^3b}{3L}
\end{bmatrix}
\tag{4.505}
$$

$$
\mathbf{K}_{URI} =
\begin{bmatrix}
\dfrac{4C_{44}ab}{L} & 2C_{44}ab & -\dfrac{4C_{44}ab}{L} & 2C_{44}ab \\[2ex]
2C_{44}ab & 0,25(4C_{44}abL)+\dfrac{4C_{11}a^3b}{3L} & -2C_{44}ab & 0,25(4C_{44}abL)-\dfrac{4C_{11}a^3b}{3L} \\[2ex]
-\dfrac{4C_{44}ab}{L} & -2C_{44}ab & \dfrac{4C_{44}ab}{L} & -2C_{44}ab \\[2ex]
2C_{44}ab & 0,25(4C_{44}abL)-\dfrac{4C_{11}a^3b}{3L} & -2C_{44}ab & 0,25(4C_{44}abL)+\dfrac{4C_{11}a^3b}{3L}
\end{bmatrix}
\tag{4.506}
$$

For the subsequent analyses, the following geometric and material parameters are considered

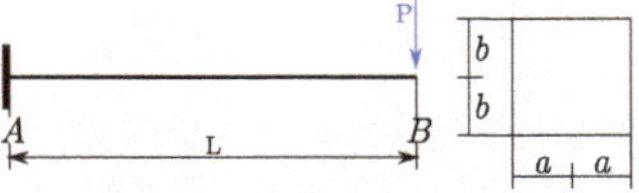

Geometric and material properties

- Geometric properties
  $a = 25$ mm, $b = 25$ mm, $L = 200$ mm

- Material properties
  $\nu = 0,\ C_{11} = 210000\ \dfrac{\text{N}}{\text{mm}^2},\ C_{44} = 105000\ \dfrac{\text{N}}{\text{mm}^2}$

Introducing the values on Eqs (4.505) and (4.506), the numerical matrices are the following.

$$\mathbf{K}_{FI} = \begin{bmatrix} 1312500 & 131250000 & -1312500 & 131250000 \\[2mm] 131250000 & \begin{matrix}0,\overline{3}(5.25e+10)+\\+(546875000)\end{matrix} & -131250000 & \begin{matrix}0,\overline{16}(5.25e+10)+\\-(546875000)\end{matrix} \\[2mm] -1312500 & -131250000 & 1312500 & -131250000 \\[2mm] 131250000 & \begin{matrix}0,\overline{16}(5.25e+10)+\\-(546875000)\end{matrix} & -131250000 & \begin{matrix}0,\overline{3}(5.25e+10)+\\+(546875000)\end{matrix} \end{bmatrix} \tag{4.507}$$

$$\mathbf{K}_{URI} = \begin{bmatrix} 1312500 & 131250000 & -1312500 & 131250000 \\[2mm] 131250000 & \begin{matrix}0,25(5.25e+10)+\\+(546875000)\end{matrix} & -131250000 & \begin{matrix}0,25(5.25e+10)+\\-(546875000)\end{matrix} \\[2mm] -1312500 & -131250000 & 1312500 & -131250000 \\[2mm] 131250000 & \begin{matrix}0,25(5.25e+10)+\\-(546875000)\end{matrix} & -131250000 & \begin{matrix}0,25(5.25e+10)+\\+(546875000)\end{matrix} \end{bmatrix} \tag{4.508}$$

Evidently, the difference between matrices $\mathbf{K}_{FI}$ and $\mathbf{K}_{URI}$ is limited to the cyan and blue colored coefficients, ranging from $+33,\overline{3}\%$ to $-33,\overline{3}\%$. **Although this numerical discrepancy is not large, the behavior of the two matrices, particularly regarding the estimation of bending stiffness and, consequently, displacement, differs significantly. As a result, the outcomes derived from the two matrices can exhibit much larger percentage deviations.**

To illustrate these numerical differences, consider a beam modeled by a single B2 element, clamped at the left end and subjected to a load at the right end. The solution, in terms of the displacement at point B in the figure, can be obtained by solving the following equation

$$\mathbf{K} \left\{ \begin{array}{c} u_1^0 \\ \phi_{z1} \\ u_2^0 \\ \phi_{z2} \end{array} \right\} = \left\{ \begin{array}{c} P_1 \\ M_1 \\ P_2 \\ M_2 \end{array} \right\} \tag{4.509}$$

where the force vector consists of forces (P_1 and P_2) and moments (M_1 and M_2). In order to solve Eq. (4.509), the matrix $\mathbf{K}$ must be penalized for the displacement components u_1^0 and ϕ_{z1}, which are those of the clamped node. Additionally, the force vector must be formulated so that only the force acting along the u_2^0 component is non-zero.

$$\left\{ \begin{array}{c} 0 \\ 0 \\ P \\ 0 \end{array} \right\} \tag{4.510}$$

The matrices $\mathbf{K}_{FI}$ and $\mathbf{K}_{URI}$ can be written as follows

$$\mathbf{K}_{FI} = \begin{bmatrix} 1312500 & 131250000 & -1312500 & 131250000 \\ 131250000 & 18046875000 & -131250000 & 8203125000 \\ -1312500 & -131250000 & 1312500 & -131250000 \\ 131250000 & 8203125000 & -131250000 & 18046875000 \end{bmatrix} \tag{4.511}$$

$$\mathbf{K}_{URI} = \begin{bmatrix} 1312500 & 131250000 & -1312500 & 131250000 \\ 131250000 & 13671875000 & -131250000 & 12578125000 \\ -1312500 & -131250000 & 1312500 & -131250000 \\ 131250000 & 12578125000 & -131250000 & 13671875000 \end{bmatrix} \tag{4.512}$$

A penalty technique is recalled for the resolution of the static problem of Eq. (4.509). The penalty values for the $\mathbf{K}_{FI}$ and $\mathbf{K}_{URI}$ matrices is equal to the maximum value on the diagonal terms $\times\ 10^5$. These two values are the following.

$$\mathbf{K}_{FI} \quad \text{Penalty value} = (18046875000 \cdot 100000) = 1.8046875 \cdot 10^{15} \tag{4.513}$$

$$\mathbf{K}_{URI} \quad \text{Penalty value} = (13671875000 \cdot 100000) = 1.3671875 \cdot 10^{15}$$

The penalized matrices are, then, the following

$$\mathbf{K}_{FI} = \begin{bmatrix} 1.8046875 \cdot 10^{15} & 131250000 & -1312500 & 131250000 \\ 131250000 & 1.8046875 \cdot 10^{15} & -131250000 & 8203125000 \\ -1312500 & -131250000 & 1312500 & -131250000 \\ 131250000 & 8203125000 & -131250000 & 18046875000 \end{bmatrix}$$

(4.514)

$$\mathbf{K}_{URI} = \begin{bmatrix} 1.3671875 \cdot 10^{15} & 131250000 & -1312500 & 131250000 \\ 131250000 & 1.3671875 \cdot 10^{15} & -131250000 & 12578125000 \\ -1312500 & -131250000 & 1312500 & -131250000 \\ 131250000 & 12578125000 & -131250000 & 13671875000 \end{bmatrix}$$

(4.515)

Considering the following force vector

$$\begin{Bmatrix} 0 \\ 0 \\ 1 \\ 0 \end{Bmatrix}$$

(4.516)

the solutions can be obtained using the below relations

$$\begin{Bmatrix} u_1^0 \\ \phi_{z1} \\ u_2^0 \\ \phi_{z2} \end{Bmatrix} = \left[\mathbf{K}_{FI_{PENALIZED}} \right]^{-1} \begin{Bmatrix} 0 \\ 0 \\ 1 \\ 0 \end{Bmatrix}$$

(4.517)

and

$$\begin{Bmatrix} u_1^0 \\ \phi_{z1} \\ u_2^0 \\ \phi_{z2} \end{Bmatrix}_C = \left[\mathbf{K}_{URI_{PENALIZED}} \right]^{-1} \begin{Bmatrix} 0 \\ 0 \\ 1 \\ 0 \end{Bmatrix}_C$$

(4.518)

The previous Eqs. (4.517) and (4.518) are equal to the following

$$
\begin{Bmatrix} u_1^0 \\ \phi_{z1} \\ u_2^0 \\ \phi_{z2} \end{Bmatrix} = \begin{bmatrix} 5.541 \cdot 10^{-16} & 0 & 5.541 \cdot 10^{-16} & 0 \\ 0 & 5.541 \cdot 10^{-16} & 1.108 \cdot 10^{-13} & 5.541 \cdot 10^{-16} \\ 5.541 \cdot 10^{-16} & 1.108 \cdot 10^{-13} & 2.794 \cdot 10^{-6} & 2.032 \cdot 10^{-8} \\ 0 & 5.541 \cdot 10^{-16} & 2.032 \cdot 10^{-8} & 2.032 \cdot 10^{-10} \end{bmatrix} \begin{Bmatrix} 0 \\ 0 \\ 1 \\ 0 \end{Bmatrix} =
$$

$$
= \begin{Bmatrix} 0 \\ 0 \\ 2.79 \cdot 10^{-6} \\ 2.03 \cdot 10^{-8} \end{Bmatrix}
$$

$$(4.519)$$

and

$$
\begin{Bmatrix} u_1^0 \\ \phi_{z1} \\ u_2^0 \\ \phi_{z2} \end{Bmatrix} = \begin{bmatrix} 7.314 \cdot 10^{-16} & 0 & 7.314 \cdot 10^{-16} & 0 \\ 0 & 7.314 \cdot 10^{-16} & 1.463 \cdot 10^{-13} & 7.314 \cdot 10^{-16} \\ 7.314 \cdot 10^{-16} & 1.463 \cdot 10^{-13} & 1.905 \cdot 10^{-5} & 1.829 \cdot 10^{-7} \\ 0 & 7.314 \cdot 10^{-16} & 1.829 \cdot 10^{-7} & 1.829 \cdot 10^{-9} \end{bmatrix} \begin{Bmatrix} 0 \\ 0 \\ 1 \\ 0 \end{Bmatrix} =
$$

$$
= \begin{Bmatrix} 0 \\ 0 \\ 1.900 \cdot 10^{-5} \\ 1.830 \cdot 10^{-7} \end{Bmatrix}
$$

$$(4.520)$$

As far as the displacement u_2^0 is concerned, from the previous equation, it has

$$
u_{2_{FI}}^0 = 2.79 \cdot 10^{-6} \text{mm} \tag{4.521}
$$

and

$$
u_{2_{URI}}^0 = 1.90 \cdot 10^{-5} \text{mm} \tag{4.522}
$$

The percentage difference between the two previous values is

$$
\left(\frac{u_{2_{URI}}^0 - u_{2_{FI}}^0}{u_{2_{FI}}^0} \right) \cdot 100 = \left(\frac{1.900 \cdot 10^{-5} - 2.79 \cdot 10^{-6}}{2.79 \cdot 10^{-6}} \right) \cdot 100 = 581\% \tag{4.523}
$$

Table 4.1 Percentage difference of the U_2^0 displacement component

Beam length	$U_{2_{FI}}^0$	$U_{2_{URI}}^0$	Percentage difference (%)
100	$1.14 \cdot 10^{-6}$	$2.67 \cdot 10^{-6}$	133
200	$2.79 \cdot 10^{-6}$	$1.90 \cdot 10^{-5}$	581
300	$4.39 \cdot 10^{-6}$	$6.29 \cdot 10^{-5}$	1332
400	$5.96 \cdot 10^{-6}$	$1.48 \cdot 10^{-4}$	2381
500	$7.51 \cdot 10^{-6}$	$2.88 \cdot 10^{-4}$	3731
600	$9.05 \cdot 10^{-6}$	$4.96 \cdot 10^{-4}$	5381
700	$1.06 \cdot 10^{-5}$	$7.87 \cdot 10^{-4}$	7331
800	$1.21 \cdot 10^{-5}$	$1.17 \cdot 10^{-3}$	9581
900	$1.36 \cdot 10^{-5}$	$1.67 \cdot 10^{-3}$	12131
1000	$1.52 \cdot 10^{-5}$	$2.29 \cdot 10^{-3}$	14981

As shown by the previously calculated percentage differences, the matrix derived using FI displays a numerical behavior that diverges significantly from that of the matrix obtained via URI. Despite a relatively small discrepancy of $\pm 33.3\%$ between the terms (see Eqs. (4.507) and (4.508)), the resulting solutions can deviate by as much as 581%, underscoring the notably different numerical performances of the two matrices. This discrepancy arises from the shear locking phenomenon.

The beam dimensions plays also a role in the shear locking effect. As indicated in Table 4.1, increasing the beam length leads to a more slender structure with greater bending flexibility, causing the FI matrix to underestimate the beam's bending stiffness compared to the URI matrix.

The numerical behavior of the shear locking effect also depends on the number of elements used to represent the structure. To investigate this effect, the beam previously considered is approximated with a varying number of B2 elements, from 1 to 5, and the u^0 displacement component at point B is computed. For the case of 2 elements, the assembled stiffness matrices for the FI and URI approaches are shown in Eqs. (4.524) and (4.525).

$$\mathbf{K}_{FI} =
\begin{bmatrix}
\dfrac{4C_{44}ab}{L_1} & 2C_{44}ab & -\dfrac{4C_{44}ab}{L_1} & 2C_{44}ab & 0 & 0 \\[2ex]
2C_{44}ab & 0,\overline{3}(4C_{44}abL_1)+\dfrac{4C_{11}a^3b}{3L_1} & -2C_{44}ab & 0,1\overline{6}(4C_{44}abL_1)-\dfrac{4C_{11}a^3b}{3L_1} & 0 & 0 \\[2ex]
-\dfrac{4C_{44}ab}{L_1} & -2C_{44}ab & \left[\dfrac{4C_{44}ab}{L_1}+\dfrac{4C_{44}ab}{L_2}\right] & [-2C_{44}ab+2C_{44}ab] & -\dfrac{4C_{44}ab}{L_2} & 2C_{44}ab \\[2ex]
2C_{44}ab & 0,1\overline{6}(4C_{44}abL_1)-\dfrac{4C_{11}a^3b}{3L_1} & [-2C_{44}ab+2C_{44}ab] & [0,\overline{3}(4C_{44}abL_1)-\dfrac{4C_{11}a^3b}{3L_1}]+[0,\overline{3}(4C_{44}abL_2)+\dfrac{4C_{11}a^3b}{3L_2}] & -2C_{44}ab & 0,1\overline{6}(4C_{44}abL_2)-\dfrac{4C_{11}a^3b}{3L_2} \\[2ex]
0 & 0 & -\dfrac{4C_{44}ab}{L_2} & -2C_{44}ab & \dfrac{4C_{44}ab}{L_2} & -2C_{44}ab \\[2ex]
0 & 0 & 2C_{44}ab & 0,1\overline{6}(4C_{44}abL_2)-\dfrac{4C_{11}a^3b}{3L_2} & -2C_{44}ab & 0,\overline{3}(4C_{44}abL_2)++\dfrac{4C_{11}a^3b}{3L_2}
\end{bmatrix}
\tag{4.524}$$

$$\mathbf{K}_{URI} =
\begin{bmatrix}
\dfrac{4C_{44}ab}{L_1} & 2C_{44}ab & -\dfrac{4C_{44}ab}{L_1} & 2C_{44}ab & 0 & 0 \\[2ex]
2C_{44}ab & 0,25(4C_{44}abL_1)+\dfrac{4C_{11}a^3b}{3L_1} & -2C_{44}ab & 0,25(4C_{44}abL_1)-\dfrac{4C_{11}a^3b}{3L_1} & 0 & 0 \\[2ex]
-\dfrac{4C_{44}ab}{L_1} & -2C_{44}ab & \left[\dfrac{4C_{44}ab}{L_1}+\dfrac{4C_{44}ab}{L_2}\right] & [-2C_{44}ab+2C_{44}ab] & -\dfrac{4C_{44}ab}{L_2} & 2C_{44}ab \\[2ex]
2C_{44}ab & 0,25(4C_{44}abL_1)-\dfrac{4C_{11}a^3b}{3L_1} & [-2C_{44}ab+2C_{44}ab] & [0,25(4C_{44}abL_1)-\dfrac{4C_{11}a^3b}{3L_1}]+[0,25(4C_{44}abL_2)+\dfrac{4C_{11}a^3b}{3L_2}] & -2C_{44}ab & 0,25(4C_{44}abL_2)-\dfrac{4C_{11}a^3b}{3L_2} \\[2ex]
0 & 0 & -\dfrac{4C_{44}ab}{L_2} & -2C_{44}ab & \dfrac{4C_{44}ab}{L_2} & -2C_{44}ab \\[2ex]
0 & 0 & 2C_{44}ab & 0,25(4C_{44}abL_2)-\dfrac{4C_{11}a^3b}{3L_2} & -2C_{44}ab & 0,25(4C_{44}abL_2)++\dfrac{4C_{11}a^3b}{3L_2}
\end{bmatrix}
\tag{4.525}$$

In the previous equations, the terms placed inside the green brackets are summed in order to account for the double stiffness values of node 2, see the picture reported below where also the geometric and material properties are reported.

Geometric and material properties

- Geometric properties
  a = 25 mm, b = 25 mm, L_1 = 300 mm, L_2 = 300 mm

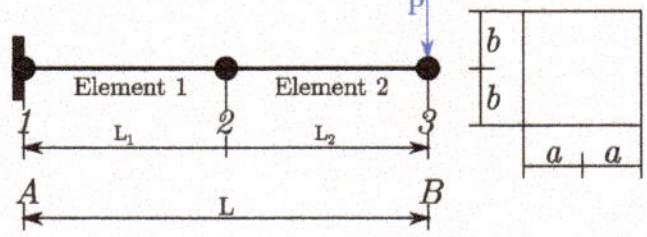

- Material properties
  $\nu = 0$, $C_{11} = 210000 \ \dfrac{\text{N}}{\text{mm}^2}$, $C_{44} = 105000 \ \dfrac{\text{N}}{\text{mm}^2}$

The related numerical FI and URI matrices are the following

$$
\mathbf{K}_{FI} =
\begin{bmatrix}
8.75 \cdot 10^5 & 1.31 \cdot 10^8 & -8.75 \cdot 10^5 & 1.31 \cdot 10^8 & 0 & 0 \\
1.31 \cdot 10^8 & 2.66 \cdot 10^{10} & -1.31 \cdot 10^8 & 1.28 \cdot 10^{10} & 0 & 0 \\
-8.75 \cdot 10^5 & -1.31 \cdot 10^8 & 1.75 \cdot 10^6 & 0 & -8.75 \cdot 10^5 & 1.31 \cdot 10^8 \\
1.31 \cdot 10^8 & 1.28 \cdot 10^{10} & 0 & 5.32 \cdot 10^{10} & -1.31 \cdot 10^8 & 1.28 \cdot 10^{10} \\
0 & 0 & -8.75 \cdot 10^5 & -1.31 \cdot 10^8 & 8.75 \cdot 10^5 & -1.31 \cdot 10^8 \\
0 & 0 & 1.31 \cdot 10^8 & 1.28 \cdot 10^{10} & -1.31 \cdot 10^8 & 2.66 \cdot 10^{10}
\end{bmatrix}
\tag{4.526}
$$

$$
\mathbf{K}_{URI} =
\begin{bmatrix}
8.75 \cdot 10^5 & 1.31 \cdot 10^8 & -8.75 \cdot 10^5 & 1.31 \cdot 10^8 & 0 & 0 \\
1.31 \cdot 10^8 & 2.01 \cdot 10^{10} & -1.31 \cdot 10^8 & 1.93 \cdot 10^{10} & 0 & 0 \\
-8.75 \cdot 10^5 & -1.31 \cdot 10^8 & 1.75 \cdot 10^6 & 0 & -8.75 \cdot 10^5 & 1.31 \cdot 10^8 \\
1.31 \cdot 10^8 & 1.93 \cdot 10^{10} & 0 & 4.01 \cdot 10^{10} & -1.31 \cdot 10^8 & 1.93 \cdot 10^{10} \\
0 & 0 & -8.75 \cdot 10^5 & -1.31 \cdot 10^8 & 8.75 \cdot 10^5 & -1.31 \cdot 10^8 \\
0 & 0 & 1.31 \cdot 10^8 & 1.93 \cdot 10^{10} & -1.31 \cdot 10^8 & 2.01 \cdot 10^{10}
\end{bmatrix}
\tag{4.527}
$$

The two matrices of Eqs. (4.526) and (4.527) can be penalized multiplying the maximum diagonal term by 10^5. The penalization must be done for the two displacement components of the clamped node 1, i.e. for u_1^0 and ϕ_{z1}. The penalty values to be used are the following

$$
\mathbf{K}_{FI} \quad \text{Penalty} \quad \text{value} = (5.32 \cdot 10^{10} \cdot 100000) = 5.32 \cdot 10^{15}
\tag{4.528}
$$

$$
\mathbf{K}_{URI} \quad \text{Penalty} \quad \text{value} = (4.01 \cdot 100000) = 4.01 \cdot 10^{15}
$$

The penalized matrices are reported below

$$
\mathbf{K}_{FI} =
\begin{bmatrix}
5.32 \cdot 10^{15} & 1.31 \cdot 10^8 & -8.75 \cdot 10^5 & 1.31 \cdot 10^8 & 0 & 0 \\
1.31 \cdot 10^8 & 5.32 \cdot 10^{15} & -1.31 \cdot 10^8 & 1.28 \cdot 10^{10} & 0 & 0 \\
-8.75 \cdot 10^5 & -1.31 \cdot 10^8 & 1.75 \cdot 10^6 & 0 & -8.75 \cdot 10^5 & 1.31 \cdot 10^8 \\
1.31 \cdot 10^8 & 1.28 \cdot 10^{10} & 0 & 5.32 \cdot 10^{10} & -1.31 \cdot 10^8 & 1.28 \cdot 10^{10} \\
0 & 0 & -8.75 \cdot 10^5 & -1.31 \cdot 10^8 & 8.75 \cdot 10^5 & -1.31 \cdot 10^8 \\
0 & 0 & 1.31 \cdot 10^8 & 1.28 \cdot 10^{10} & -1.31 \cdot 10^8 & 2.66 \cdot 10^{10}
\end{bmatrix}
\tag{4.529}
$$

$$\mathbf{K}_{URI} = \begin{bmatrix} 4.01 \cdot 10^{15} & 1.31 \cdot 10^8 & -8.75 \cdot 10^5 & 1.31 \cdot 10^8 & 0 & 0 \\ 1.31 \cdot 10^8 & 4.01 \cdot 10^{15} & -1.31 \cdot 10^8 & 1.93 \cdot 10^{10} & 0 & 0 \\ -8.75 \cdot 10^5 & -1.31 \cdot 10^8 & 1.75 \cdot 10^6 & 0 & -8.75 \cdot 10^5 & 1.31 \cdot 10^8 \\ 1.31 \cdot 10^8 & 1.93 \cdot 10^{10} & 0 & 4.01 \cdot 10^{10} & -1.31 \cdot 10^8 & 1.93 \cdot 10^{10} \\ 0 & 0 & -8.75 \cdot 10^5 & -1.31 \cdot 10^8 & 8.75 \cdot 10^5 & -1.31 \cdot 10^8 \\ 0 & 0 & 1.31 \cdot 10^8 & 1.93 \cdot 10^{10} & -1.31 \cdot 10^8 & 2.01 \cdot 10^{10} \end{bmatrix} \tag{4.530}$$

In order to solve this problem, the unknown displacements vector and the force vector are written as follows

$$\begin{Bmatrix} u_1^0 \\ \phi_{z1} \\ u_2^0 \\ \phi_{z2} \\ u_3^0 \\ \phi_{z3} \end{Bmatrix} \tag{4.531}$$

$$\begin{Bmatrix} P_1 \\ M_1 \\ P_2 \\ M_2 \\ P_3 \\ M_3 \end{Bmatrix} = \begin{Bmatrix} 0 \\ 0 \\ 0 \\ 0 \\ 1 \\ 0 \end{Bmatrix} \tag{4.532}$$

Below it is the system to be considered for solving the problem using the FI matrix:

$$\begin{Bmatrix} u_{1\,FI}^0 \\ \phi_{z1\,FI} \\ u_{2\,FI}^0 \\ \phi_{z2\,FI} \\ u_{3\,FI}^0 \\ \phi_{z3\,FI} \end{Bmatrix} = \begin{bmatrix} 5.32 \cdot 10^{15} & 1.31 \cdot 10^8 & -8.75 \cdot 10^5 & 1.31 \cdot 10^8 & 0 & 0 \\ 1.31 \cdot 10^8 & 5.32 \cdot 10^{15} & -1.31 \cdot 10^8 & 1.28 \cdot 10^{10} & 0 & 0 \\ -8.75 \cdot 10^5 & -1.31 \cdot 10^8 & 1.75 \cdot 10^6 & 0 & -8.75 \cdot 10^5 & 1.31 \cdot 10^8 \\ 1.31 \cdot 10^8 & 1.28 \cdot 10^{10} & 0 & 5.32 \cdot 10^{10} & -1.31 \cdot 10^8 & 1.28 \cdot 10^{10} \\ 0 & 0 & -8.75 \cdot 10^5 & -1.31 \cdot 10^8 & 8.75 \cdot 10^5 & -1.31 \cdot 10^8 \\ 0 & 0 & 1.31 \cdot 10^8 & 1.28 \cdot 10^{10} & -1.31 \cdot 10^8 & 2.66 \cdot 10^{10} \end{bmatrix}^{-1} \begin{Bmatrix} 0 \\ 0 \\ 0 \\ 0 \\ 1 \\ 0 \end{Bmatrix} \tag{4.533}$$

and the system to be used to solve the problem for URI matrix

$$
\begin{Bmatrix} u^0_{1_{URI}} \\ \phi_{z1_{URI}} \\ u^0_{2_{URI}} \\ \phi_{z2_{URI}} \\ u^0_{3_{URI}} \\ \phi_{z3_{URI}} \end{Bmatrix} =
\begin{bmatrix}
4.01 \cdot 10^{15} & 1.31 \cdot 10^{8} & -8.75 \cdot 10^{5} & 1.31 \cdot 10^{8} & 0 & 0 \\
1.31 \cdot 10^{8} & 4.01 \cdot 10^{15} & -1.31 \cdot 10^{8} & 1.93 \cdot 10^{10} & 0 & 0 \\
-8.75 \cdot 10^{5} & -1.31 \cdot 10^{8} & 1.75 \cdot 10^{6} & 0 & -8.75 \cdot 10^{5} & 1.31 \cdot 10^{8} \\
1.31 \cdot 10^{8} & 1.93 \cdot 10^{10} & 0 & 4.01 \cdot 10^{10} & -1.31 \cdot 10^{8} & 1.93 \cdot 10^{10} \\
0 & 0 & -8.75 \cdot 10^{5} & -1.31 \cdot 10^{8} & 8.75 \cdot 10^{5} & -1.31 \cdot 10^{8} \\
0 & 0 & 1.31 \cdot 10^{8} & 1.93 \cdot 10^{10} & -1.31 \cdot 10^{8} & 2.01 \cdot 10^{10}
\end{bmatrix}^{-1}
\begin{Bmatrix} 0 \\ 0 \\ 0 \\ 0 \\ 1 \\ 0 \end{Bmatrix}
\tag{4.534}
$$

The obtained results are the following

$$
\begin{Bmatrix} u^0_{1_{FI}} \\ \phi_{z1_{FI}} \\ u^0_{2_{FI}} \\ \phi_{z2_{FI}} \\ u^0_{3_{FI}} \\ \phi_{z3_{FI}} \end{Bmatrix} =
\begin{Bmatrix} 0 \\ 0 \\ 1.09 \cdot 10^{-5} \\ 6.50 \cdot 10^{-8} \\ \mathbf{3.48 \cdot 10^{-5}} \\ 8.66 \cdot 10^{-8} \end{Bmatrix}
\tag{4.535}
$$

$$
\begin{Bmatrix} u^0_{1_{URI}} \\ \phi_{z1_{URI}} \\ u^0_{2_{URI}} \\ \phi_{z2_{URI}} \\ u^0_{3_{URI}} \\ \phi_{z3_{URI}} \end{Bmatrix} =
\begin{Bmatrix} 0 \\ 0 \\ 1.86 \cdot 10^{-4} \\ 1.23 \cdot 10^{-6} \\ \mathbf{6.19 \cdot 10^{-4}} \\ 1.65 \cdot 10^{-6} \end{Bmatrix}
\tag{4.536}
$$

The percentage difference between the displacement numeric components $u^0_{1_{FI}}$ and $u^0_{1_{URI}}$ (see blue values in Eqs. (4.535) and (4.536)) is equal to

Table 4.2 Percentage difference of the U_2^0 displacement component using different number of elements

Element number	Element length	u_2^0 FI	u_2^0 URI	Percentage difference (%)
1	600	$9.05 \cdot 10^{-6}$	$4.96 \cdot 10^{-4}$	5381
2	300	$3.48 \cdot 10^{-5}$	$6.19 \cdot 10^{-4}$	1687
3	200	$7.34 \cdot 10^{-5}$	$6.42 \cdot 10^{-4}$	775
4	150	$1.2 \cdot 10^{-4}$	$6.50 \cdot 10^{-4}$	442
5	120	$1.70 \cdot 10^{-4}$	$6.54 \cdot 10^{-4}$	285
6	100	$2.20 \cdot 10^{-4}$	$6.56 \cdot 10^{-4}$	198
7	85.71	$2.67 \cdot 10^{-4}$	$6.57 \cdot 10^{-4}$	146
8	75	$3.11 \cdot 10^{-4}$	$6.58 \cdot 10^{-4}$	112
9	66.67	$3.50 \cdot 10^{-4}$	$6.59 \cdot 10^{-4}$	88
10	60	$3.84 \cdot 10^{-4}$	$6.59 \cdot 10^{-4}$	72

$$\left(\frac{u_{3_{URI}}^0 - u_{3_{FI}}^0}{u_{3_{FI}}^0}\right) \cdot 100 = \left(\frac{6.19 \cdot 10^{-4} - 3.48 \cdot 10^{-5}}{3.48 \cdot 10^{-5}}\right) \cdot 100 = 1681.658\%$$

$$(4.537)$$

Table 4.2 the results evaluating the impact of using different numbers of elements on the numerical behavior of FI and URI matrices.

The numerical values align with those in the previous example. Clearly, Table 4.2 shows that increasing the number of elements results in a mitigation of the effect of shear locking.

4.5.2   Comparison Between the Full and Selective Reduced Integrated Matrices of B2 Element

In order to compare the results of the B2 element based on the martix obtained using SRI and those obtained using FI, the following numerical examples are carried out. A cantilever beam is considered with the same geometry as the previous example, but with longer length equals 600 mm. Three cases are considered

$$\text{Case 1}: \ 3 \ \text{elements, each of length} \ = \ 200 \ mm$$
$$\text{Case 2}: \ 4 \ \text{elements, each of length} \ = \ 150 \ mm \qquad (4.538)$$
$$\text{Case 3}: \ 5 \ \text{elements, each of length} \ = \ 120 \ mm$$

The displacement at the tip, with reference to the Timoshenko Beam Theory, is evaluated as

$$f = \frac{PL^3}{3EI_z} + \kappa\frac{PL}{GA} = \frac{PL^3}{3E\frac{4}{3}a^3b} + 1.2\frac{PL}{G4ab}$$

$$f = \frac{10000 \cdot 600^3}{210000 \cdot 4 \cdot 25^3 \cdot 25} + 1.2 \cdot \frac{10000 \cdot 600}{105000 \cdot 4 \cdot 25 \cdot 25} = 6.610\ mm \tag{4.539}$$

where k is the shear correction factor, which in the case of a square cross-section assumes the value of 1.2. This value of Eq. (4.539) is taken as reference result. The numerical matrices, with reference to a B2 element and the Case 1, are the following.

$$\mathbf{K}_{S_{FI}} = \begin{bmatrix} 1312500 & 131250000 & -1312500 & 131250000 \\ 131250000 & 17500000000 & -131250000 & 8750000000 \\ -1312500 & -131250000 & 1312500 & -131250000 \\ 131250000 & 8750000000 & -131250000 & 17500000000 \end{bmatrix} \tag{4.540}$$

The bending part of the stiffness matrix, computed previously is reported below

$$\mathbf{K}_{b_{FI}} = \begin{bmatrix} 0 & 0 & 0 & 0 \\ 0 & 546875000 & 0 & -546875000 \\ 0 & 0 & 0 & 0 \\ 0 & -546875000 & 0 & 546875000 \end{bmatrix} \tag{4.541}$$

The complete stiffness matrix for a B2 Element computed using the FI, can be obtained summing the two previous matrices

$$\mathbf{K}_{FI} = \begin{bmatrix} 1312500 & 131250000 & -1312500 & 131250000 \\ 131250000 & 18046875000 & -131250000 & 8203125000 \\ -1312500 & -131250000 & 1312500 & -131250000 \\ 131250000 & 8203125000 & -131250000 & 18046875000 \end{bmatrix} \tag{4.542}$$

The numerical matrices for SRI is the following.

$$\mathbf{K}_{SSRI} = \begin{bmatrix} 1312500 & 131250000 & -1312500 & 131250000 \\ 131250000 & 13125000000 & -131250000 & 13125000000 \\ -1312500 & -131250000 & 1312500 & -131250000 \\ 131250000 & 13125000000 & -131250000 & 13125000000 \end{bmatrix} \quad (4.543)$$

The bending part of the stiffness matrix, computed previously is reported below

$$\mathbf{K}_{bSRI} = \begin{bmatrix} 0 & 0 & 0 & 0 \\ 0 & 546875000 & 0 & -546875000 \\ 0 & 0 & 0 & 0 \\ 0 & -546875000 & 0 & 546875000 \end{bmatrix} \quad (4.544)$$

The complete stiffness matrix for a B2 Element computed using SRI, can be obtained summing the two previous matrices

$$\mathbf{K}_{SRI} = \begin{bmatrix} 1312500 & 131250000 & -1312500 & 131250000 \\ 131250000 & 13671875000 & -131250000 & 12578125000 \\ -1312500 & -131250000 & 1312500 & -131250000 \\ 131250000 & 12578125000 & -131250000 & 13671875000 \end{bmatrix} \quad (4.545)$$

The difference between the numerical matrices obtained using FI and SRI is highlighted by the red coloured terms in the matrix of Eq. (4.542) for FI and the blue colored terms of the matrix of the Eq. (4.545) for SRI.

4.5.2.1 Results of Numerical Examples

The results, in term of maximum displacement of the tip end of the cantilever beam, obtained using FI, URI, SRI and MITC matrices, for B2, B3 and B4 elements and for the three cases mentioned in Eq. (4.538), are reported in Tables 4.3, 4.4 and 4.5. For a comparison purpose, the reference value obtained in the Eq. (4.539) is also reported.

Table 4.3 Free end displacement of a three elements cantilever beam (see above Case 1 [4.538])

Element	FI	URI	SRI	MITC	Reference
B2	0.73	6.42	6.42	6.42	
B3	6.49	6.61	6.61	6.61	6.61
B4	6.61	6.61	6.61	6.61	

Table 4.4 Free end displacement of a four elements cantilever beam (see above Case 2 [4.538])

Element	FI	URI	SRI	MITC	Reference
B2	1.20	6.50	6.50	6.50	
B3	6.56	6.61	6.61	6.61	6.61
B4	6.61	6.61	6.61	6.61	

Table 4.5 Free end displacement of a five elements cantilever beam (see above Case 3 [4.538])

Element	FI	URI	SRI	MITC	Reference
B2	1.70	6.54	6.54	6.54	
B3	6.58	6.61	6.61	6.61	6.61
B4	6.61	6.61	6.61	6.61	

4.5.2.2 Examples to Highlight the Shear Locking Phenomenon

A couple of examples are reported hereafter to demonstrate the effect of the shear locking considering the bending and the shear energies and when a rotation unknown vector is considered.

4.5.2.3 Example 1: Comparison Between Bending and Shear Strain Energies

The first example focuses on comparing the bending and shear contributions to the strain energy. This comparison shows that the stiffness matrix derived using FI over-estimates the shear stiffness while underestimating the bending stiffness. In contrast, matrices obtained through other techniques produce accurate estimates for both components. Under FI, the shear strain energy is significantly larger than the bending strain energy, whereas the strain energies determined via the alternative methods indicate that bending strain energy is the dominant and physically consistent component. These observations relate to the cantilever beam introduced in Sect. 4.538, specifically Case 1. To compute the bending and shear parts of the strain energy, the stiffness matrix is split into its bending and shear stiffness submatrices.

Table 4.6 Strain energy of the B2 element

Elements number	FI		URI	
	Bending	Shear	Bending	Shear
3	395 (11%)	3275 (89%)	32000 (99.6%)	114 (0.4%)
4	1071 (18%)	4934 (82%)	32400 (99.6%)	114 (0.4%)
5	2164 (25%)	6348 (75%)	32585 (99.6%)	114 (0.4%)

Elements number	SRI		MITC	
	Bending	Shear	Bending	Shear
3	32000 (99.6%)	114 (0.4%)	32000 (99.6%)	114 (0.4%)
4	32400 (99.6%)	114 (0.4%)	32000 (99.6%)	114 (0.4%)
5	32585 (99.6%)	114 (0.4%)	32585 (99.6%)	114 (0.4%)

$$\text{Strain energy} = \frac{1}{2} \times \mathbf{u}^T \times \mathbf{K} \times \mathbf{u} = \frac{1}{2} \times \mathbf{u}^T \times (\mathbf{K}_b + \mathbf{K}_s) \times \mathbf{u}$$
$$= \frac{1}{2} \times \mathbf{u}^T \times \mathbf{K}_b \times \mathbf{u} + \frac{1}{2} \times \mathbf{u}^T \times \mathbf{K}_s \times \mathbf{u} \tag{4.546}$$

Recalling the displacement vector $\mathbf{u}$ from case 1 (see Eq. (4.538)) the corresponding strain energies are presented in Table 4.6. The percentages represent each energy contribution to the total strain energy under its respective integration scheme.

From the previous table, it is evident that, under FI quadrature, the shear strain energy is overestimated relative to the bending strain energy. By contrast, in the other approaches, the opposite situation occurs.

4.5.2.4   Example 2: Use of a Rotations Only Displacement Field

The second example considers a displacement vector consisting only of rotations at the end node of a single B2 element. The displacement vector used in this scenario is shown below.

$$\mathbf{u}_{rot} = \begin{Bmatrix} 0 \\ \alpha \\ 0 \\ -\alpha \end{Bmatrix} \tag{4.547}$$

Since the displacement field involves only rotations, it engages only the bending stiffness of the B2 element.

4.5.2.5 Application of a Rotations Only Displacement Field to the FI Matrix

The FI matrix used is presented in Eq. (3.124). The following relation holds:

$$\mathbf{K}_{FI} = \mathbf{K}_{bFI} + \mathbf{K}_{sFI} \tag{4.548}$$

If matrices in Eq. (4.548) are multiplied by the displacement field consisting only of rotations of Eq. (4.547), the following results are obtained.

$$\mathbf{K}_{FI} \times \mathbf{u}_{rot} = \mathbf{K}_{bFULL} \times \mathbf{u}_{rot} + \mathbf{K}_{sFULL} \times \mathbf{u}_{rot} \tag{4.549}$$

Considering the matrices in Eqs. (3.120) and (3.122), the following expressions are obtained.

$$
\begin{bmatrix}
0 & 0 & 0 & 0 \\
0 & \dfrac{4C_{11}a^3 b}{3L} & 0 & -\dfrac{4C_{11}a^3 b}{3L} \\
0 & 0 & 0 & 0 \\
0 & -\dfrac{4C_{11}a^3 b}{3L} & 0 & \dfrac{4C_{11}a^3 b}{3L}
\end{bmatrix}
*
\begin{Bmatrix}
0 \\ \alpha \\ 0 \\ -\alpha
\end{Bmatrix}
+
\begin{bmatrix}
\dfrac{4C_{44}ab}{L} & 2C_{44}ab & -\dfrac{4C_{44}ab}{L} & 2C_{44}ab \\
2C_{44}ab & \dfrac{4C_{44}abL}{3} & -2C_{44}ab & \dfrac{2C_{44}abL}{3} \\
-\dfrac{4C_{44}ab}{L} & -2C_{44}ab & \dfrac{4C_{44}ab}{L} & -2C_{44}ab \\
2C_{44}ab & \dfrac{2C_{44}abL}{3} & -2C_{44}ab & \dfrac{4C_{44}abL}{3}
\end{bmatrix}
*
\begin{Bmatrix}
0 \\ \alpha \\ 0 \\ -\alpha
\end{Bmatrix}
\tag{4.550}
$$

The final result is the following.

$$
\begin{Bmatrix}
0 \\ \dfrac{8C_{11}a^3 b\alpha}{3L} \\ 0 \\ -\dfrac{8C_{11}a^3 b\alpha}{3L}
\end{Bmatrix}
+
\begin{Bmatrix}
0 \\ \dfrac{2C_{44}abL\alpha}{3} \\ 0 \\ -\dfrac{2C_{44}abL\alpha}{3}
\end{Bmatrix}
\tag{4.551}
$$

The previously derived result comprises two contributions: the bending stiffness (represented in red) and an erroneous shear stiffness contribution (represented in blue). The inaccuracies in the shear stiffness component stem exclusively from shear locking.

4.5.2.6 Application of a Rotations Only Displacement Field to the SELECTIVE REDUCED INTEGRATION Matrix

The SRI matrix, reported in Eq. (4.28), is equal to the sum of the matrices of Eq. (4.26) (bending part of the stiffness matrix) and (4.27) (shear part of the stiffness matrix). The following relation applies

$$\mathbf{K}_{SRI} = \mathbf{K}_{bSRI} + \mathbf{K}_{sSRI} \tag{4.552}$$

If these matrices of Eq. (4.552) are multiplied by the displacement field of Eq. (4.547), the following results are obtained

$$\mathbf{K}_{SRI} \times \mathbf{u}_{rot} = \mathbf{K}_{bSRI} \times \mathbf{u}_{rot} + \mathbf{K}_{sSRI} \times \mathbf{u}_{rot} \tag{4.553}$$

Using the matrices in Eqs. (4.26) and (4.27), the following expressions are obtained

$$
\begin{bmatrix}
0 & 0 & 0 & 0 \\
0 & \dfrac{4C_{11}a^3 b}{3L} & 0 & -\dfrac{4C_{11}a^3 b}{3L} \\
0 & 0 & 0 & 0 \\
0 & -\dfrac{4C_{11}a^3 b}{3L} & 0 & \dfrac{4C_{11}a^3 b}{3L}
\end{bmatrix}
* \left\{ \begin{matrix} 0 \\ \alpha \\ 0 \\ -\alpha \end{matrix} \right\}
+
\begin{bmatrix}
\dfrac{4C_{44}ab}{L} & 2C_{44}ab & -\dfrac{4C_{44}ab}{L} & 2C_{44}ab \\
2C_{44}ab & C_{44}Lab & -2C_{44}ab & C_{44}Lab \\
-\dfrac{4C_{44}ab}{L} & -2C_{44}ab & \dfrac{4C_{44}ab}{L} & -2C_{44}ab \\
2C_{44}ab & C_{44}Lab & -2C_{44}ab & C_{44}Lab
\end{bmatrix}
* \left\{ \begin{matrix} 0 \\ \alpha \\ 0 \\ -\alpha \end{matrix} \right\} \tag{4.554}
$$

The final result is the following.

$$
\left\{ \begin{matrix} 0 \\ \dfrac{8C_{11}a^3 b\alpha}{3L} \\ 0 \\ -\dfrac{8C_{11}a^3 b\alpha}{3L} \end{matrix} \right\}
+
\left\{ \begin{matrix} 0 \\ 0 \\ 0 \\ 0 \end{matrix} \right\} \tag{4.555}
$$

In the previous result, only the bending stiffness contribution (shown in red) remains, while the shear stiffness contribution (shown in blue) is zero. This outcome aligns with expectations, indicating that a matrix derived through SRI effectively eliminates the shear locking effects.

References

1. Hughes, T.J.R., Taylor, R.L., Kanoknukulchai, W.: A Simple and Efficient Finite Element for Plate Bending. Int. J. Numer. Meth. Eng. **11**, 1529–1543 (1977)
2. Bathe, K.J., Dvorkin, E.N.: A four-node plate bending element based on Mindlin/Reissner plate theory and a mixed interpolation. Int. J. Numer. Meth. Eng. **21**, 367–383 (1985)
3. Bathe, K.J., Dvorkin, E.N.: A formulation of general shell elements–the use of mixed interpolation of tensorial components. Int. J. Numer. Meth. Eng. **22**, 697–722 (1986)

Chapter 5
Stiffness Matrix of a Beam Bending Around x Axis

Graphical Abstract

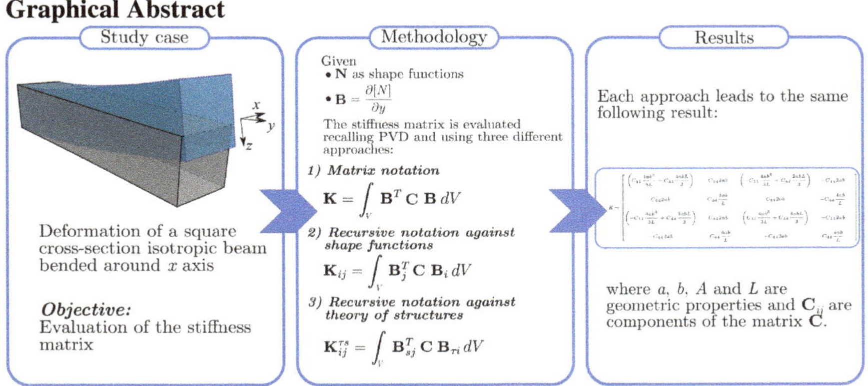

Introduction

In this chapter, the stiffness matrix of a two-node beam element is derived. The beam is a structural element with axial, torsional and bending stiffness, but only the bending stiffness is considered in this derivation. The methodology behind this derivation is presented in Sect. 5.1. Section 5.2 details the assumptions underlying the finite element beam definition and specifies the associated geometric data. The stiffness matrix derivation utilizes the PVD approach, as discussed in the preceding chapter. The same techniques are employed, including:

- Matrix notation (Sect. 5.3)
- Recursive notation against shape functions (Sect. 5.4)
- Recursive notation against Theory Of Structures (TOS) (Sect. 5.5).

© The Author(s), under exclusive license to Springer Nature Switzerland AG 2026

E. Carrera et al., *Implementation of Beam-Type Finite Elements Based on Carrera Unified Formulation*, https://doi.org/10.1007/978-3-031-95856-4_5

5.1   Structural Model of a Beam Bended Around x Axis

Figure 5.1 shows the deformed shape of a beam bending around the x axis. Three displacement components are involved in the deformation of the beam: the displacement u along the direction x, the displacement v along the direction y, the displacement w along the direction z.

The bending of a beam around x axis produces displacements in the zy plane. Then, the displacement u along the x axis can be assumed zero. Figure 5.2 shows a two-dimensional representation of the deformed beam. In this figure, $w^0(y)$ refers to the displacement along direction z axis as a function of the beam axis y, and $\phi_x(y)$ is the rotation angle around the x axis.

Given a generic point P, identified in the figure at a distance z from the midpoint of the beam cross-section, it has a displacement v_P along the longitudinal y axis. Assuming small rotations (i.e. $tan(\phi_x(y)) \simeq \phi_x(y)$), this displacement is given by $-z\phi_x(y)$.

The displacement field is the following:

$$u(x, y, z) = 0$$
$$v(x, y, z) = -z\phi_x(y) \tag{5.1}$$
$$w(x, y, z) = w^0(y)$$

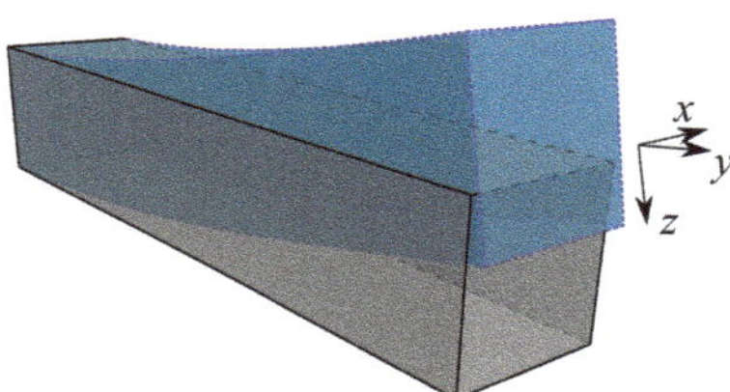

Fig. 5.1 Bending around x axis

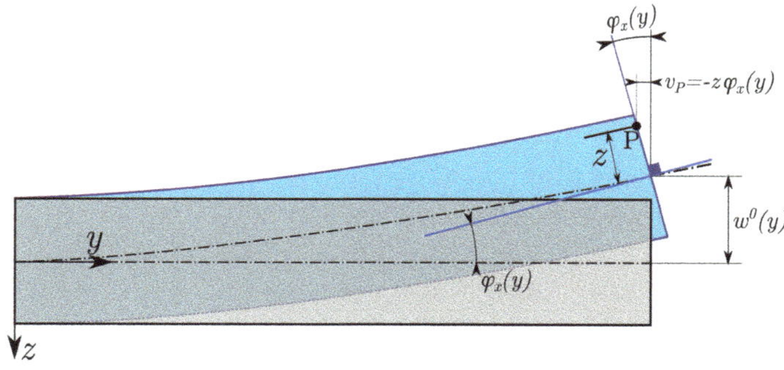

Fig. 5.2 2D assumption of bending around x axis

For the case under study, the strain components of interest are ϵ_{yy} and ϵ_{yz}. The geometric relations between these strain components and the displacement components are:

$$\epsilon_{yy} = \frac{\partial v}{\partial y}, \quad \epsilon_{yz} = \frac{\partial w}{\partial y} + \frac{\partial v}{\partial z} \tag{5.2}$$

Given the displacement components of Eq. (5.1), the following expressions for the strains are obtained

$$\epsilon_{yy} = -z\frac{\partial \phi_x(y)}{\partial y}, \quad \epsilon_{yz} = -\frac{\partial w^0(y)}{\partial y} + \phi_x \tag{5.3}$$

In a matrix form, Eq. (5.3) reads as:

$$\boldsymbol{\epsilon} = \left\{ \begin{array}{c} \epsilon_{yy} \\ \epsilon_{yz} \end{array} \right\} = \left[\begin{array}{cc} -z\dfrac{\partial}{\partial y} & 0 \\ -1 & \dfrac{\partial}{\partial y} \end{array} \right] \left\{ \begin{array}{c} \phi_x(y) \\ w^0(y) \end{array} \right\} \tag{5.4}$$

Introducing the differential operator matrix $\mathbf{b}$, Eq. (5.4) can be expressed as:

$$\mathbf{b} = \left[\begin{array}{cc} -z\dfrac{\partial}{\partial y} & 0 \\ -1 & \dfrac{\partial}{\partial y} \end{array} \right], \quad \boldsymbol{\epsilon} = \mathbf{b} \left\{ \begin{array}{c} \phi_x(y) \\ w^0(y) \end{array} \right\} = \mathbf{b} \left\{ \mathbf{s}^0 \right\} \tag{5.5}$$

where $\mathbf{s}^0$ equals to

$$\mathbf{s}^0 = \left\{ \begin{array}{c} \phi_x(y) \\ w^0(y) \end{array} \right\} \tag{5.6}$$

The constitutive relations are based on the material coefficient and the Hooke's law. The constitutive relations of interest are the following

$$\sigma_{yy} = C_{11}\epsilon_{yy}$$

$$\tag{5.7}$$

$$\sigma_{yz} = C_{44}\epsilon_{yz}$$

The expression of C_{11} is given in the Eqs. (2.6) and (2.7), while the expression of C_{44} is equal to

$$C_{44} = G \tag{5.8}$$

Eq. (5.7) can be written in matrix form, reading as

$$\sigma = \mathbf{C}\epsilon \tag{5.9}$$

where

$$\sigma = \left\{ \begin{array}{c} \sigma_{yy} \\ \sigma_{yz} \end{array} \right\} \tag{5.10}$$

$$\epsilon = \left\{ \begin{array}{c} \epsilon_{yy} \\ \epsilon_{yz} \end{array} \right\} \tag{5.11}$$

$$\mathbf{C} = \begin{bmatrix} C_{11} & 0 \\ 0 & C_{44} \end{bmatrix} \tag{5.12}$$

5.2 Finite Element Approximation

The finite element approximations of the beam under bending around x axis is considered here. This example refers to a two-node beam element as shown in Fig. 5.3, where subscripts 1 and 2 refer to node 1 and node 2 of the beam.

The coordinate system of the beam is also indicated in Fig. 5.3: the y axis is along the element length, and as a consequence the x and z axes define the cross-section of the beam. w_1^0 and w_2^0 are the displacements along the coordinates z, while the symbols ϕ_{x1} and ϕ_{x2} refer to the rotations of the cross-section of the beam around the x axis at nodes 1 and 2. The geometric characteristics and the dimensions of the beam are the same as those of the bar reported in the Sect. 2.2.

5.3 Matrix Notation

In this section the stiffness matrix of the beam is calculated by means of PVD, as described in Sect. 3.3. The variation, along the beam length, of the displacement $w^0(y)$ can be expressed as a function of the displacements of the beam nodes 1 and

Fig. 5.3 Schematic representation of a beam with 2 nodes

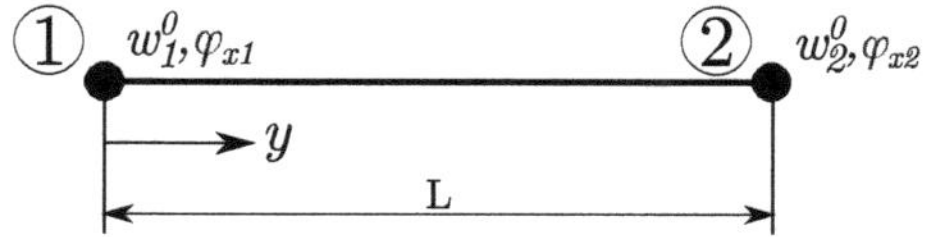

2, w_1^0 and w_2^0 respectively, and the shape functions. In the same manner, the variation of the rotation angle ϕ_x, which defines the $v^0(y)$ displacement, can be described using the rotations ϕ_{x1} and ϕ_{x2} of the beam nodes 1 and 2 respectively and the same shape functions. The shape functions used for this purpose are based on Lagrange polynomials and are the following

$$N_1(y) = 1 - \frac{y}{L}$$
$$N_2(y) = \frac{y}{L}$$

(5.13)

The expressions that describe the variation, along the beam, of the displacement $w^0(y)$ and the rotation ϕ_x, are the following

$$w^0(y) = N_1(y)w_1^0 + N_2(y)w_2^0$$

$$\phi_x(y) = N_1(y)\phi_{x1} + N_2(y)\phi_{x2}$$

(5.14)

The matrix form of Eq. (5.14) is

$$\mathbf{s}^0 = \begin{bmatrix} N_1 & 0 & N_2 & 0 \\ 0 & N_1 & 0 & N_2 \end{bmatrix} \begin{bmatrix} w_1^0 \\ \phi_{x1} \\ w_2^0 \\ \phi_{x2} \end{bmatrix} = \begin{bmatrix} N_1 & 0 & N_2 & 0 \\ 0 & N_1 & 0 & N_2 \end{bmatrix} \mathbf{S}^0$$

(5.15)

In Eq. (5.15), $\mathbf{S}^0$ is introduced to express the unknowns of the problems. Introducing the shape functions matrix $\mathbf{N}$, it is

$$\mathbf{N} = \begin{bmatrix} N_1 & 0 & N_2 & 0 \\ 0 & N_1 & 0 & N_2 \end{bmatrix}$$

(5.16)

Introducing the *displacement differentiation matrix* $\mathbf{B}$, it is obtained by deriving the shape functions used to describe the beam displacements. It is expressed in the following

$$\mathbf{B} = \mathbf{bN} = \begin{bmatrix} -z\dfrac{\partial}{\partial y} & 0 \\ -1 & \dfrac{\partial}{\partial y} \end{bmatrix} \begin{bmatrix} N_1 & 0 & N_2 & 0 \\ 0 & N_1 & 0 & N_2 \end{bmatrix} =$$

$$= \begin{bmatrix} -zN_{1,y} & 0 & -zN_{2,y} & 0 \\ -N_1 & N_{1,y} & N_2 & -N_{2,y} \end{bmatrix} \tag{5.17}$$

where

$$\begin{aligned} N_{1,y} &= \frac{\partial N_1(y)}{\partial y} \\ N_{2,y} &= \frac{\partial N_2(y)}{\partial y} \end{aligned} \tag{5.18}$$

Recalling the variation of the internal work

$$\delta L_{\text{int}} = \int_V \delta \mathbf{S}^0 \mathbf{B}^T \mathbf{C} \, \mathbf{B} \, \mathbf{S}^0 \, dV \tag{5.19}$$

the stiffness matrix $\mathbf{K}$ of the beam is expressed as

$$\mathbf{K} = \int_V \mathbf{B}^T \mathbf{C} \mathbf{B} \tag{5.20}$$

Using the previous definitions expressed in Eqs. (5.20) and (5.12), the explicit form of the stiffness matrix $\mathbf{K}$ becomes

$$\mathbf{K} = \int_V \begin{bmatrix} -zN_{1,y} & 0 & -zN_{2,y} & 0 \\ -N_1 & N_{1,y} & N_2 & -N_{2,y} \end{bmatrix}^T \begin{bmatrix} C_{11} & 0 \\ 0 & C_{44} \end{bmatrix} \begin{bmatrix} -zN_{1,y} & 0 & -zN_{2,y} & 0 \\ -N_1 & N_{1,y} & N_2 & -N_{2,y} \end{bmatrix} dV \tag{5.21}$$

$$\mathbf{K} = \int_V \begin{bmatrix} -zN_{1,y} & -N_1 \\ 0 & N_{1,y} \\ -zN_{2,y} & N_2 \\ 0 & -N_{2,y} \end{bmatrix} \begin{bmatrix} -C_{11}zN_{1,y} & 0 & -C_{11}zN_{2,y} & 0 \\ -C_{44}N_1 & C_{44}N_{1,y} & C_{44}N_2 & -C_{44}N_{2,y} \end{bmatrix} dV \tag{5.22}$$

By performing the double matrix product of Eq. 5.22, the stiffness matrix becomes

$$\mathbf{K} = \int_V \begin{bmatrix} C_{11}z^2 N_{1,y}N_{1,y} + C_{44}N_1 N_1 & -C_{44}N_1 N_{1,y} & C_{11}z^2 N_{1,y}N_{2,y} - C_{44}N_1 N_2 & C_{44}N_1 N_{2,y} \\ -C_{44}N_1 N_{1,y} & C_{44}N_{1,y}N_{1,y} & C_{44}N_{1,y}N_2 & -C_{44}N_{1,y}N_{2,y} \\ C_{11}z^2 N_{1,y}N_{2,y} + C_{44}N_1 N_2 & C_{44}N_2 N_{1,y} & C_{11}z^2 N_{2,y}N_{2,y} + C_{44}N_2 N_2 & -C_{44}N_2 N_{2,y} \\ C_{44}N_1 N_{2,y} & -C_{44}N_{1,y}N_{2,y} & -C_{44}N_2 N_{2,y} & C_{44}N_{2,y}N_{2,y} \end{bmatrix} dV$$

$$= \begin{bmatrix} k_{11} & k_{12} & k_{13} & k_{14} \\ k_{21} & k_{22} & k_{23} & k_{24} \\ k_{31} & k_{32} & k_{33} & k_{34} \\ k_{41} & k_{42} & k_{43} & k_{44} \end{bmatrix}$$

$$(5.23)$$

Subsequently, the volume integral of each term of matrix $\mathbf{K}$ of Eq. 5.23 can be evaluated separately considering the same geometrical relations and the shape functions presented in Sect. 3.3.

$$k_{11} = \int_V \left(C_{11}z^2 N_{1,y}N_{1,y} + C_{44}N_1 N_1 \right) dV$$
$$= C_{11}\int_{-a}^{a} z^2 dx \int_{-b}^{b} dz \int_0^L N_{1,y}N_{1,y}dy + C_{44}\int_{-a}^{a} dx \int_{-b}^{b} dz \int_0^L N_1 N_1 dy$$
$$= C_{11}\int_{-a}^{a} z^2 dx \int_{-b}^{b} dz \int_0^L \left[\left(-\frac{1}{L}\right)\left(-\frac{1}{L}\right)\right] dy +$$
$$+ C_{44}\int_{-a}^{a} dx \int_{-b}^{b} dz \int_0^L \left(1 - \frac{y}{L}\right)\left(1 - \frac{y}{L}\right) dy$$
$$= C_{11}\int_{-a}^{a} z^2 dx \int_{-b}^{b} dz \int_0^L \left(\frac{1}{L^2}\right) dy + C_{44}\int_{-a}^{a} dx \int_{-b}^{b} dz \int_0^L \left(1 + \frac{y^2}{L^2} - 2\frac{y}{L}\right) dy$$
$$= C_{11}\frac{I_x}{L} + C_{44}\frac{AL}{3} = C_{11}\frac{4ab^3}{3L} + C_{44}\frac{4abL}{3}$$

$$k_{12} = \int_V (-C_{44}N_1 N_{1,y})dV = -C_{44}\int_{-a}^{a} dx \int_{-b}^{b} dz \int_0^L N_1 N_{1,y}dy$$
$$= -C_{44}\int_{-a}^{a} dx \int_{-b}^{b} dz \int_0^L \left[\left(1 - \frac{y}{L}\right)\left(-\frac{1}{L}\right)\right] dy$$
$$= -C_{44}\int_{-a}^{a} dx \int_{-b}^{b} dz \int_0^L \left(-\frac{1}{L} + \frac{y}{L^2}\right) dy$$
$$= C_{44}\frac{A}{2} = C_{44}\frac{4ab}{2} = C_{44}2ab$$

$$k_{13} = \int_V \left(C_{11}z^2 N_{1,y}N_{2,y} + C_{44}N_1 N_2 \right) dV$$
$$= C_{11}\int_{-a}^{a} z^2 dx \int_{-b}^{b} dz \int_0^L N_{1,y}N_{2,y}dy + C_{44}\int_{-a}^{a} dx \int_{-b}^{b} dz \int_0^L N_1 N_2 dy$$
$$= C_{11}\int_{-a}^{a} z^2 dx \int_{-b}^{b} dz \int_0^L \left[\left(-\frac{1}{L}\right)\left(\frac{1}{L}\right)\right] dy +$$
$$+ C_{44}\int_{-a}^{a} dx \int_{-b}^{b} dz \int_0^L \left(1 - \frac{y}{L}\right)\left(\frac{y}{L}\right) dy$$
$$= C_{11}\int_{-a}^{a} z^2 dx \int_{-b}^{b} dz \int_0^L \left(-\frac{1}{L^2}\right) dy + C_{44}\int_{-a}^{a} dx \int_{-b}^{b} dz \int_0^L \left(\frac{y}{L} - \frac{y^2}{L^2}\right) dy$$
$$= -C_{11}\frac{I_x}{L} + C_{44}\frac{AL}{6} = -C_{11}\frac{4ab^3}{3L} + C_{44}\frac{4abL}{6} = -C_{11}\frac{4a^3 b}{3L} + C_{44}\frac{2abL}{3}$$

$$k_{14} = \int_V (-C_{44} N_1 N_{2,y}) dV = C_{44} \int_{-a}^{a} dx \int_{-b}^{b} dz \int_0^L N_1 N_{2,y} dy$$

$$= -C_{44} \int_{-a}^{a} dx \int_{-b}^{b} dz \int_0^L \left[\left(1 - \frac{y}{L}\right) \left(\frac{1}{L}\right) \right] dy =$$

$$= -C_{44} \int_{-a}^{a} dx \int_{-b}^{b} dz \int_0^L \left(\frac{1}{L} - \frac{y}{L^2} \right) dy$$

$$= -C_{44} \frac{A}{2} = -C_{44} \frac{4ab}{2} = -C_{44} 2ab$$

$$k_{22} = \int_V (C_{44} N_{1,y} N_{1,y}) dV = C_{44} \int_{-a}^{a} dx \int_{-b}^{b} dz \int_0^L N_{1,y} N_{1,y} dy$$

$$= C_{44} \int_{-a}^{a} dx \int_{-b}^{b} dz \int_0^L \left[\left(-\frac{1}{L}\right) \left(-\frac{1}{L}\right) \right] dy = C_{44} \int_{-a}^{a} dx \int_{-b}^{b} dz \int_0^L \frac{1}{L^2} dy$$

$$= C_{44} \frac{A}{L} = C_{44} \frac{4ab}{L}$$

$$k_{23} = \int_V (-C_{44} N_2 N_{1,y}) dV = -C_{44} \int_{-a}^{a} dx \int_{-b}^{b} dz \int_0^L N_2 N_{1,y} dy$$

$$= -C_{44} \int_{-a}^{a} dx \int_{-b}^{b} dz \int_0^L \left[\left(\frac{y}{L}\right) \left(-\frac{1}{L}\right) \right] dy =$$

$$= -C_{44} \int_{-a}^{a} dx \int_{-b}^{b} dz \int_0^L \left(-\frac{y}{L^2} \right) dy$$

$$= C_{44} \frac{A}{2} = C_{44} \frac{4ab}{2} = C_{44} 2ab$$

$$k_{24} = \int_V (C_{44} N_{1,y} N_{2,y}) dV = C_{44} \int_{-a}^{a} dx \int_{-b}^{b} dz \int_0^L N_{1,y} N_{2,y} dy$$

$$= C_{44} \int_{-a}^{a} dx \int_{-b}^{b} dz \int_0^L \left[\left(-\frac{1}{L}\right) \left(\frac{1}{L}\right) \right] dy = C_{44} \int_{-a}^{a} dx \int_{-b}^{b} dz \int_0^L -\frac{1}{L^2} dy$$

$$= -C_{44} \frac{A}{L} = -C_{44} \frac{4ab}{L}$$

$$k_{33} = \int_V \left(C_{11} z^2 N_{2,y} N_{2,y} + C_{44} N_2 N_2 \right) dV$$

$$= C_{11} \int_{-a}^{a} z^2 dx \int_{-b}^{b} dz \int_0^L N_{2,y} N_{2,y} dy + C_{44} \int_{-a}^{a} dx \int_{-b}^{b} dz \int_0^L N_2 N_2 dy$$

$$= C_{11} \int_{-a}^{a} z^2 dx \int_{-b}^{b} dz \int_0^L \left[\left(\frac{1}{L}\right) \left(\frac{1}{L}\right) \right] dy + C_{44} \int_{-a}^{a} dx \int_{-b}^{b} dz \int_0^L \left(\frac{y}{L}\right) \left(\frac{y}{L}\right) dy$$

$$= C_{11} \int_{-a}^{a} z^2 dx \int_{-b}^{b} dz \int_0^L \left(\frac{1}{L^2} \right) dy + C_{44} \int_{-a}^{a} dx \int_{-b}^{b} dz \int_0^L \left(\frac{y^2}{L^2} \right) dy$$

$$= C_{11} \frac{I_x}{L} + C_{44} \frac{AL}{3} = C_{11} \frac{4ab^3}{3L} + C_{44} \frac{4abL}{3}$$

$$k_{34} = \int_V (-C_{44} N_2 N_{2,y}) dV = -C_{44} \int_{-a}^{a} dx \int_{-b}^{b} dz \int_0^L N_2 N_{1,y} dy$$

$$= -C_{44} \int_{-a}^{a} dx \int_{-b}^{b} dz \int_0^L \left[\left(\frac{y}{L}\right) \left(-\frac{1}{L}\right) \right] dy =$$

$$= -C_{44} \int_{-a}^{a} dx \int_{-b}^{b} dz \int_0^L \left(-\frac{y}{L} + \frac{y}{L^2} \right) dy$$

$$= -C_{44} \frac{A}{2} = -C_{44} \frac{4ab}{2} = -C_{44} 2ab$$

$$k_{44} = \int_V (C_{44} N_{2,y} N_{2,y}) dV = C_{44} \int_{-a}^{a} dx \int_{-b}^{b} dz \int_0^L N_{2,y} N_{2,y} dy$$

$$= C_{44} \int_{-a}^{a} dx \int_{-b}^{b} dz \int_0^L \left[\left(\frac{1}{L}\right)\left(\frac{1}{L}\right) \right] dy = C_{44} \int_{-a}^{a} dx \int_{-b}^{b} dz \int_0^L \frac{1}{L^2} dy$$

$$= C_{44} \frac{A}{L} = C_{44} \frac{4ab}{L}$$

Considering that the matrix $\mathbf{K}$ is symmetric, i.e. $K_{21} = K_{21}$, $K_{31} = K_{13}$, $K_{32} = K_{23}$, $K_{41} = K_{14}$, $K_{42} = K_{24}$ and $K_{43} = K_{34}$, its complete form results in the following

$$\mathbf{K} = \begin{bmatrix} \left(C_{11}\dfrac{4ab^3}{3L} + C_{44}\dfrac{4abL}{3}\right) & C_{44}2ab & \left(-C_{11}\dfrac{4ab^3}{3L} + C_{44}\dfrac{2abL}{3}\right) & -C_{44}2ab \\[2ex] C_{44}2ab & C_{44}\dfrac{4ab}{L} & C_{44}2ab & -C_{44}\dfrac{4ab}{L} \\[2ex] \left(-C_{11}\dfrac{4ab^3}{3L} + C_{44}\dfrac{2abL}{3}\right) & C_{44}2ab & \left(C_{11}\dfrac{4ab^3}{3L} + C_{44}\dfrac{4abL}{3}\right) & -C_{44}2ab \\[2ex] -C_{44}2ab & -C_{44}\dfrac{4ab}{L} & -C_{44}2ab & C_{44}\dfrac{4ab}{L} \end{bmatrix} \tag{5.24}$$

Introducing the numerical data, the matrix of a beam is:

$$\mathbf{K} = \begin{bmatrix} 7005600000 & 21000000 & 3494400000 & -21000000 \\ 21000000 & 84000 & 21000000 & -84000 \\ 3494400000 & 21000000 & 7005600000 & -21000000 \\ -21000000 & -84000 & -21000000 & 84000 \end{bmatrix} \tag{5.25}$$

The comprehensive procedure for deriving the matrix presented in Eq. (5.25) is detailed in Appendix D.1, and the step-by-step implementation is provided as a MATLAB script.

5.4 Recursive Notation Against Shape Functions

In this section the stiffness matrix of the beam is calculated using a recursive notation against the shape functions. The displacements field of Eq. (5.14) can be written as follows.

$$w^0(y) = N_i(y)w_i^0$$

$$\phi_x(y) = N_i(y)\phi_{xi}$$

(5.26)

where i is either 1 or 2, according to which node of the beam is considered. **B** matrix of Eq. (5.17) can be expressed as

$$\mathbf{B} = \begin{bmatrix} B_1 & B_2 \end{bmatrix}$$

(5.27)

where

$$\mathbf{B}_1 = \begin{bmatrix} -zN_{1,y} & 0 \\ -N_1 & N_{1,y} \end{bmatrix}$$

(5.28)

and

$$\mathbf{B}_2 = \begin{bmatrix} -zN_{2,y} & 0 \\ -N_2 & N_{2,y} \end{bmatrix}$$

(5.29)

and using the recursive notation **B** matrix becomes

$$\mathbf{B}_i = \begin{bmatrix} -zN_{i,y} & 0 \\ -N_i & N_{i,y} \end{bmatrix}$$

(5.30)

In order to distinguish the variables from their virtual variation, the second index j is introduced

$$\mathbf{B}_j = \begin{bmatrix} -zN_{j,y} & 0 \\ -N_j & N_{j,y} \end{bmatrix}$$

(5.31)

The subscripts i and j are either 1 to 2. The stiffness matrix can be written as follows

$$\mathbf{K}_{ij} = \int_V \mathbf{B}_j^T \mathbf{C} \mathbf{B}_i dV = \int_V \begin{bmatrix} -zN_{j,y} & -N_j \\ 0 & N_{j,y} \end{bmatrix} \begin{bmatrix} C_{11} & 0 \\ 0 & C_{44} \end{bmatrix} \begin{bmatrix} -zN_{i,y} & 0 \\ -N_i & N_{i,y} \end{bmatrix}$$

(5.32)

By computing the triple product, the previous matrix becomes

$$\mathbf{K}_{ij} = \int_V \begin{bmatrix} \left(C_{11}z^2 N_{i,y}N_{j,y} + C_{44}N_i N_j\right) & -C_{44}N_j N_{i,y} \\ -C_{44}N_i N_{j,y} & C_{44}N_{i,y}N_{j,y} \end{bmatrix} dV$$

(5.33)

Solving the volume integral of Eq. (5.33), the following matrices are obtained

$$
\mathbf{K}_{ij} =
\begin{bmatrix}
\left(C_{11} \int_A z^2 \, dA \int_L N_{i,y} N_{j,y} \, dL + C_{44} \int_A dA \int_L N_i N_j \, dL \right) & -C_{44} \int_A dA \int_L N_j N_{i,y} \, dL \\[4ex]
-C_{44} \int_A dA \int_L N_i N_{j,y} \, dL & C_{44} \int_A dA \int_L N_{i,y} N_{j,y} \, dL
\end{bmatrix}
\tag{5.34}
$$

$$
\mathbf{K}_{ij} =
\begin{bmatrix}
\left(C_{11} \int_{-a}^{a} z^2 \, dx \int_{-b}^{b} dz \int_0^L N_{i,y} N_{j,y} \, dy + C_{44} \int_{-a}^{a} dx \int_{-b}^{b} dz \int_0^L N_i N_j \, dy \right) & -C_{44} \int_{-a}^{a} dx \int_{-b}^{b} dz \int_0^L N_j N_{i,y} \, dy \\[4ex]
-C_{44} \int_{-a}^{a} dx \int_{-b}^{b} dz \int_0^L N_i N_{j,y} \, dy & C_{44} \int_{-a}^{a} dx \int_{-b}^{b} dz \int_0^L N_{i,y} N_{j,y} \, dy
\end{bmatrix}
\tag{5.35}
$$

By looping the indexes i and j from 1 to 2 the following submatrices of the global element of the beam under consideration can be obtained:

$$
\mathbf{K}_{11} =
\begin{bmatrix}
\left(C_{11} \int_{-a}^{a} z^2 \, dx \int_{-b}^{b} dz \int_0^L N_{1,y} N_{1,y} \, dy + C_{44} \int_{-a}^{a} dx \int_{-b}^{b} dz \int_0^L N_1 N_1 \, dy \right) & -C_{44} \int_{-a}^{a} dx \int_{-b}^{b} dz \int_0^L N_1 N_{1,y} \, dy \\[4ex]
-C_{44} \int_{-a}^{a} dx \int_{-b}^{b} dz \int_0^L N_1 N_{1,y} \, dy & C_{44} \int_{-a}^{a} dx \int_{-b}^{b} dz \int_0^L N_{1,y} N_{1,y} \, dy
\end{bmatrix}
\tag{5.36}
$$

$$
\mathbf{K}_{12} =
\begin{bmatrix}
\left(C_{11} \int_{-a}^{a} z^2 \, dx \int_{-b}^{b} dz \int_0^L N_{1,y} N_{2,y} \, dy + C_{44} \int_{-a}^{a} dx \int_{-b}^{b} dz \int_0^L N_1 N_2 \, dy \right) & -C_{44} \int_{-a}^{a} dx \int_{-b}^{b} dz \int_0^L N_2 N_{1,y} \, dy \\[4ex]
-C_{44} \int_{-a}^{a} dx \int_{-b}^{b} dz \int_0^L N_1 N_{2,y} \, dy & C_{44} \int_{-a}^{a} dx \int_{-b}^{b} dz \int_0^L N_{1,y} N_{2,y} \, dy
\end{bmatrix}
\tag{5.37}
$$

$$
\mathbf{K}_{21} =
\begin{bmatrix}
\left(C_{11} \int_{-a}^{a} z^2 \, dx \int_{-b}^{b} dz \int_0^L N_{2,y} N_{1,y} \, dy + C_{44} \int_{-a}^{a} dx \int_{-b}^{b} dz \int_0^L N_2 N_1 \, dy \right) & -C_{44} \int_{-a}^{a} dx \int_{-b}^{b} dz \int_0^L N_2 N_{1,y} \, dy \\[4ex]
-C_{44} \int_{-a}^{a} dx \int_{-b}^{b} dz \int_0^L N_2 N_{1,y} \, dy & C_{44} \int_{-a}^{a} dx \int_{-b}^{b} dz \int_0^L N_{2,y} N_{1,y} \, dy
\end{bmatrix}
\tag{5.38}
$$

$$\mathbf{K}_{22} = \begin{bmatrix} \left(C_{11}\int_{-a}^{a} z^2 dx \int_{-b}^{b} dz \int_{0}^{L} N_{2,y}N_{2,y}dy + \right. & \\ \left. +C_{44}\int_{-a}^{a} dx \int_{-b}^{b} dz \int_{0}^{L} N_2 N_2 dy\right) & -C_{44}\int_{-a}^{a} dx \int_{-b}^{b} dz \int_{0}^{L} N_2 N_{2,y}dy \\ & \\ -C_{44}\int_{-a}^{a} dx \int_{-b}^{b} dz \int_{0}^{L} N_2 N_{2,y}dy & C_{44}\int_{-a}^{a} dx \int_{-b}^{b} dz \int_{0}^{L} N_{2,y}N_{2,y}dy \end{bmatrix} \tag{5.39}$$

The matrix $\mathbf{K}$ can be written as follows

$$\mathbf{K} = \begin{bmatrix} \mathbf{K}_{11} & \mathbf{K}_{12} \\ \mathbf{K}_{21} & \mathbf{K}_{22} \end{bmatrix} \tag{5.40}$$

Each component expressed from Eq. (5.36) to Eq. (5.39) can be written as follows

$$\mathbf{K}_{11} = \begin{bmatrix} k_{11} & k_{12} \\ k_{21} & k_{22} \end{bmatrix} \tag{5.41}$$

$$\mathbf{K}_{12} = \begin{bmatrix} k_{13} & k_{14} \\ k_{23} & k_{24} \end{bmatrix} \tag{5.42}$$

$$\mathbf{K}_{21} = \begin{bmatrix} k_{31} & k_{32} \\ k_{41} & k_{42} \end{bmatrix} \tag{5.43}$$

$$\mathbf{K}_{22} = \begin{bmatrix} k_{33} & k_{34} \\ k_{43} & k_{44} \end{bmatrix} \tag{5.44}$$

Eq. (5.40), considering Eqs. (5.41) to (5.44), results equal to Eq. (5.24).
The resultant stiffness matrix is then

$$
\mathbf{K} =
\begin{bmatrix}
\left(C_{11}\dfrac{4ab^3}{3L} + C_{44}\dfrac{4abL}{3} \right) & C_{44}2ab & \left(-C_{11}\dfrac{4ab^3}{3L} + C_{44}\dfrac{2abL}{3} \right) & -C_{44}2ab \\[2.5em]
C_{44}2ab & C_{44}\dfrac{4ab}{L} & C_{44}2ab & -C_{44}\dfrac{4ab}{L} \\[2.5em]
\left(-C_{11}\dfrac{4ab^3}{3L} + C_{44}\dfrac{2abL}{3} \right) & C_{44}2ab & \left(C_{11}\dfrac{4ab^3}{3L} + C_{44}\dfrac{4abL}{3} \right) & -C_{44}2ab \\[2.5em]
-C_{44}2ab & -C_{44}\dfrac{4ab}{L} & -C_{44}2ab & C_{44}\dfrac{4ab}{L}
\end{bmatrix}
\tag{5.45}
$$

Introducing the numerical data, the matrix of the beam, reads as

$$
\mathbf{K} =
\begin{bmatrix}
7005600000 & 21000000 & 3494400000 & -21000000 \\
21000000 & 84000 & 21000000 & -84000 \\
3494400000 & 21000000 & 7005600000 & -21000000 \\
-21000000 & -84000 & -21000000 & 84000
\end{bmatrix}
\tag{5.46}
$$

The comprehensive procedure for deriving the matrix presented in Eq. (5.46) is detailed in Appendix D.2, and the step-by-step implementation is provided as a MATLAB script.

5.5 Recursive Notation Against the Theory of Structure

In this section the stiffness matrix of the bar is calculated using a recursive notation against the TOS. Introducing $s_n = s_x, s_y, s_z$, Eq. (5.14) becomes

$$
\begin{aligned}
s_x &= u(x, y, z) = 0 \\
s_y &= v(x, y, z) = -z\phi_x(y) \\
s_z &= w(x, y, z) = w^0(y)
\end{aligned}
\tag{5.47}
$$

Then, the expansion functions F are introduced, so that

$$
\begin{aligned}
s_x &= F_{1_x} \times 0 \\
s_y &= F_{1_y} \times \phi_x(y) \\
s_z &= F_{1_z} \times w^0(y)
\end{aligned}
\tag{5.48}
$$

In this case

$$
F_{1_x} = 0 \qquad F_{1_y} = -z, \qquad F_{1_z} = 1
\tag{5.49}
$$

Equation (5.49) can be expressed in a generic form introducing the index τ which ranges from 1 to the number of the terms in the cross-sectional expansion. This case can then expressed as

$$s_n = F_{\tau_n} s_{\tau_n} \tag{5.50}$$

where

$$s_{1_x} = 0 \qquad s_{1_y} = \phi_x(y), \qquad s_{1_z} = w^0(y) \tag{5.51}$$

with the generic index n introduced for the directions x, y and z. The same procedure can be used for the virtual variation, by introducing the index s for the virtual variation.

$$\delta s_n = F_{s_n} s_{s_n} \tag{5.52}$$

By applying the recursive notation against TOS, $\mathbf{B}_{\tau i}$ and $\mathbf{B}_{sj}$ can be written as follows

$$\mathbf{B}_{\tau i} = \begin{bmatrix} F_{\tau y} N_{i,y} & 0 \\ F_{\tau y,z} N_i & F_{\tau z} N_{i,y} \end{bmatrix} \tag{5.53}$$

$$\mathbf{B}_{sj} = \begin{bmatrix} F_{sy} N_{j,y} & 0 \\ F_{sy,z} N_j & F_{sz} N_{j,y} \end{bmatrix} \tag{5.54}$$

The stiffness matrix can be expressed as follows

$$\mathbf{K}_{ij}^{\tau s} = \int_V \mathbf{B}_{sj}^T \mathbf{C} \mathbf{B}_{\tau i} dV =$$
$$= \int_V \begin{bmatrix} F_{sy} N_{j,y} & F_{sy,z} N_j \\ 0 & F_{sz} N_{j,y} \end{bmatrix} \begin{bmatrix} C_{11} & 0 \\ 0 & C_{44} \end{bmatrix} \begin{bmatrix} F_{\tau y} N_{i,y} & 0 \\ F_{\tau y,z} N_i & F_{\tau z} N_{i,y} \end{bmatrix} \tag{5.55}$$

By computing the matrix triple product, the previous matrix becomes

$$\mathbf{K}_{ij}^{\tau s} = \int_V \begin{bmatrix} C_{11} F_{sy} F_{\tau y} N_{i,y} N_{j,y} + \\ + C_{44} F_{sy,z} F_{\tau y,z} N_i N_j \end{bmatrix} \quad C_{44} F_{\tau z} F_{sy,z} N_j N_{i,y} \\ C_{44} F_{sz} F_{\tau y,z} N_i N_{j,y} \quad C_{44} F_{sz} F_{\tau z} N_{i,y} N_{j,y} \end{bmatrix} dV \tag{5.56}$$

Solving the volume integral of the Eq. (5.56), the following matrices are obtained

$$\mathbf{K}_{ij}^{\tau s} = \begin{bmatrix} C_{11} \int_A F_{sy} F_{\tau y} dA \int_L N_{i,y} N_{j,y} dL + & & \\ + C_{44} \int_A F_{sy,z} F_{\tau y,z} dA \int_L N_i N_j dL & C_{44} \int_A F_{\tau z} F_{sy,z} dA \int_L N_j N_{i,y} dL \\ & & \\ C_{44} \int_A F_{sz} F_{\tau y,z} dA \int_L N_i N_{j,y} dL & C_{44} \int_A F_{sz} F_{\tau z} dA \int_L N_{i,y} N_{j,y} dL \end{bmatrix}$$

$$(5.57)$$

Considering the values of the products between the expansion functions and their derivatives

Expansion functions

$$F_{sz} F_{\tau z} = (1)(1) = 1$$

$$F_{sz} F_{\tau y,z} = (1)(-1) = -1$$

$$F_{sy,z} F_{\tau z} = (-1)(1) = -1$$

$$F_{sy} F_{\tau y} = (-z)(-z) = z^2$$

$$F_{sy,z} F_{\tau y,z} = (-1)(-1) = 1$$

the stiffness matrix of Eq. (5.57) becomes the following

$$\mathbf{K}_{ij}^{\tau s} = \begin{bmatrix} C_{11} \int_{-a}^{a} z^2 dx \int_{-b}^{b} dz \int_0^L N_{i,y} N_{j,y} dy + & & \\ + C_{44} \int_{-a}^{a} 1 dx \int_{-b}^{b} dz \int_0^L N_i N_j dy & C_{44} \int_{-a}^{a} -1 dx \int_{-b}^{b} dz \int_0^L N_j N_{i,y} dy \\ & & \\ C_{44} \int_{-a}^{a} -1 dx \int_{-b}^{b} dz \int_0^L N_i N_{j,y} dy & C_{44} \int_{-a}^{a} 1 dx \int_{-b}^{b} dz \int_0^L N_{i,y} N_{j,y} dy \end{bmatrix}$$

$$(5.58)$$

The stiffness matrix of the beam then results the same as the matrix of Eq. (5.45) and it is reported below

$$\mathbf{K} == \begin{bmatrix} \left(C_{11}\dfrac{4ab^3}{3L} + C_{44}\dfrac{4abL}{3}\right) & C_{44}2ab & \left(-C_{11}\dfrac{4ab^3}{3L} + C_{44}\dfrac{2abL}{3}\right) & -C_{44}2ab \\[2ex] C_{44}2ab & C_{44}\dfrac{4ab}{L} & C_{44}2ab & -C_{44}\dfrac{4ab}{L} \\[2ex] \left(-C_{11}\dfrac{4ab^3}{3L} + C_{44}\dfrac{2abL}{3}\right) & C_{44}2ab & \left(C_{11}\dfrac{4ab^3}{3L} + C_{44}\dfrac{4abL}{3}\right) & -C_{44}2ab \\[2ex] -C_{44}2ab & -C_{44}\dfrac{4ab}{L} & -C_{44}2ab & C_{44}\dfrac{4ab}{L} \end{bmatrix} \tag{5.59}$$

The numerical matrix of the beam, calculated by using the recursive notation against TOS, is:

$$\mathbf{K} = \begin{bmatrix} 7005600000 & 21000000 & 3494400000 & -21000000 \\ 21000000 & 84000 & 21000000 & -84000 \\ 3494400000 & 21000000 & 7005600000 & -21000000 \\ -21000000 & -84000 & -21000000 & 84000 \end{bmatrix} \tag{5.60}$$

The comprehensive procedure for deriving the matrix presented in Eq. (5.60) is detailed in Appendix D.3, and the step-by-step implementation is provided as a MATLAB script.

Chapter 6
Stiffness Matrix of a Beam Subjected by a Torsion

Graphical Abstract

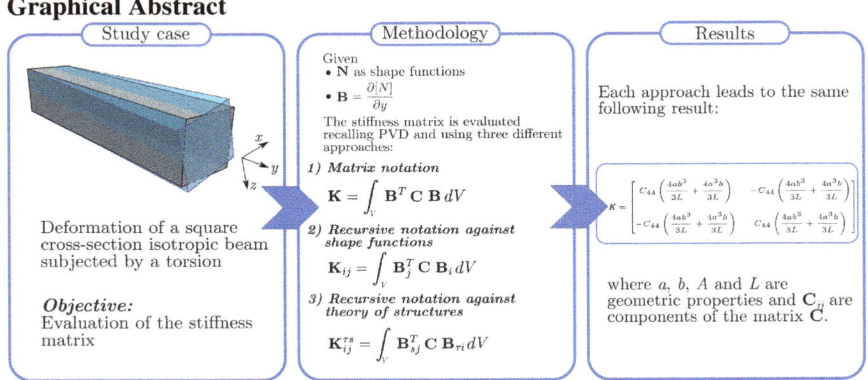

Introduction

In this chapter, the stiffness matrix of a two-node beam element is derived. The beam is a structural element with axial, torsional and bending stiffness, but only the torsional stiffness is considered in this derivation. The methodology behind this derivation is presented in Sect. 6.1. Section 6.2, details the assumptions underlying the finite element bar definition and specifies the associated geometric data. The stiffness matrix derivation utilizes the PVD approach, as discussed in the preceding chapter. The same techniques are employed, including:

- Matrix notation (Sect. 6.3)
- Recursive notation against shape functions (Sect. 6.4)
- Recursive notation against Theory Of Structures (TOS) (Sect. 6.5).

E. Carrera et al., *Implementation of Beam-Type Finite Elements Based on Carrera Unified Formulation*, https://doi.org/10.1007/978-3-031-95856-4_6

6.1   Structural Model of a Beam Subjected by a Torsion

Figure 6.1 shows the deformed shape of a beam under torsion around the y axis. Three displacement components are involved in the deformation of the beam: the displacement u which relates the direction x, the displacement v which relates the direction y, the displacement w which relates the direction z.

Figure 6.2 shows a two-dimensional representation of the deformation of the cross-section of the beam. Given a generic point P, placed at a distance x_P and z_P from the center point of the cross-section, u_P refers to the displacement along direction z, w_P refers to the displacement along direction x, and $\phi_y(y)$ is the rotation angle around the y axis. The displacement components of the point P are calculated as follows

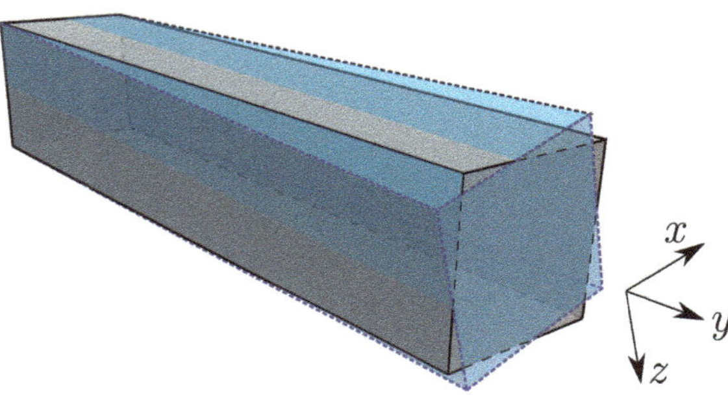

Fig. 6.1 Beam under torsion around y axis

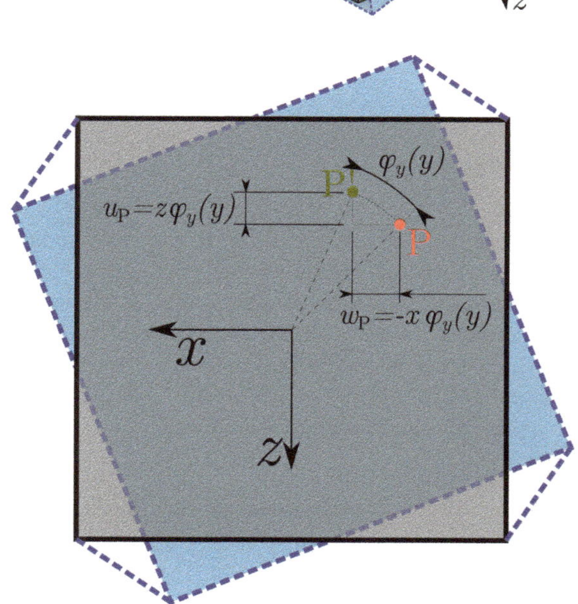

Fig. 6.2 Two-dimensional assumptions of torsion around beam axis

$$u_P = z_P \phi_y(y)$$
$$w_P = -x_P \phi_y(y) \tag{6.1}$$

Thus, the displacement field is the following:

$$u(x, y, z) = z\phi_y(y)$$
$$v(x, y, z) = 0$$
$$w(x, y, z) = -x\phi_y(y) \tag{6.2}$$

For the case under study, the strain components of interest are ϵ_{xy} and ϵ_{yz} and. The geometric relations between the strain and displacement components are the following

$$\epsilon_{xy} = \frac{\partial u}{\partial y} + \frac{\partial v}{\partial x} = \frac{\partial u}{\partial y}, \qquad \epsilon_{yz} = \frac{\partial w}{\partial y} + \frac{\partial v}{\partial z} = \frac{\partial w}{\partial y} \tag{6.3}$$

Given the displacement components of Eq. (6.2), the following expressions for the strains are obtained

$$\epsilon_{xy} = z\frac{\partial \phi_y(y)}{\partial y}, \qquad \epsilon_{yz} = -x\frac{\partial \phi_y(y)}{\partial y} \tag{6.4}$$

In the matrix form, Eq. (6.4) reads as

$$\epsilon = \begin{Bmatrix} \epsilon_{xy} \\ \epsilon_{yz} \end{Bmatrix} = \begin{bmatrix} z\dfrac{\partial}{\partial y} \\ -x\dfrac{\partial}{\partial y} \end{bmatrix} \{\phi_y(y)\} \tag{6.5}$$

Introducing the differential operator matrix $\mathbf{b}$, Eq. (6.5) can be expressed as:

$$\mathbf{b} = \begin{bmatrix} z\dfrac{\partial}{\partial y} \\ -x\dfrac{\partial}{\partial y} \end{bmatrix}, \quad \epsilon = \mathbf{b}\{\phi_y(y)\} = \mathbf{b}\mathbf{s}^0 \tag{6.6}$$

where

$$\mathbf{s}^0 = \{\phi_y(y)\} \tag{6.7}$$

The constitutive relations, in the presence of a torsion, are those between the strains ϵ_{xy} and ϵ_{yz} and the stresses σ_{xy} and σ_{yz}. These relations are based on the material coefficient and the Hooke's law. The constitutive relations of interest are the following

$$\sigma_{xy} = C_{44}\epsilon_{xy}$$

$$\sigma_{yz} = C_{44}\epsilon_{yz}$$

(6.8)

The expression of C_{44} is equal to

$$C_{44} = G \tag{6.9}$$

Eq. (6.8) can be written in matrix form, reading as

$$\boldsymbol{\sigma} = \mathbf{C}\boldsymbol{\epsilon} \tag{6.10}$$

where

$$\boldsymbol{\sigma} = \left\{ \begin{array}{c} \sigma_{xy} \\ \sigma_{yz} \end{array} \right\} \tag{6.11}$$

$$\boldsymbol{\epsilon} = \left\{ \begin{array}{c} \epsilon_{xy} \\ \epsilon_{yz} \end{array} \right\} \tag{6.12}$$

$$\mathbf{C} = \left[\begin{array}{cc} C_{44} & 0 \\ 0 & C_{44} \end{array} \right] \tag{6.13}$$

6.2   Finite Element Approximation

The finite element approximations of the beam under torsion around y axis is considered here. This example refers to a two-node beam element as shown in Fig. 6.3, where subscripts 1 and 2 refer to node 1 and node 2 of the beam.

The coordinate system of the beam is also indicated in Fig. 6.3: the y axis is along the elements length, and as a consequence the x and z axes define the cross-section of the beam. ϕ_{y1} and ϕ_{y2} refer to the rotation of the cross-section of the beam at nodes 1 and 2 respectively around the y axis. The geometric characteristics and the dimensions of the beam are the same as those in the previous chapters.

Fig. 6.3 Schematic representation of a beam with 2 nodes

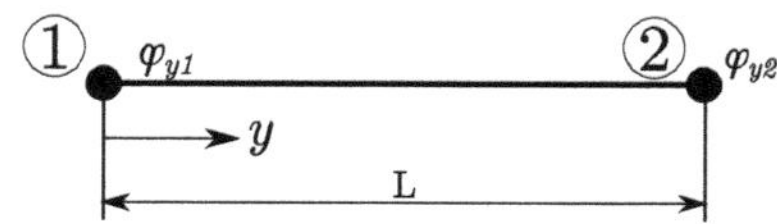

6.3 Matrix Notation

In this section the stiffness matrix of the beam under torsion is calculated by means of PVD. The displacement of interest are $u(x, y, z)$ along the x axis and $w(x, y, z)$ along the z axis, whereas the displacement $v(x, y, z)$ along the y axis is zero. These displacements depend on the rotation ϕ_y that varies over the beam length y. This variation can be expressed as a function of the rotations of the beam nodes 1 and 2, i.e. $\phi_y 1$ and $\phi_y 2$ respectively, and the shape functions. The shape functions used for this purpose are based on Lagrange polynomials and are the following

$$
\begin{aligned}
N_1(y) &= 1 - \frac{y}{L} \\
N_2(y) &= \frac{y}{L}
\end{aligned}
\tag{6.14}
$$

The expressions that describe the variation, along the beam, of the rotation ϕ_y is the following

$$
\phi_y(y) = N_1(y)\phi_{y1} + N_2(y)\phi_{y2}
\tag{6.15}
$$

The matrix form of Eq. (6.15) is

$$
\mathbf{s}^0 = \begin{bmatrix} N_1 & N_2 \end{bmatrix} \begin{bmatrix} \phi_{y1} \\ \phi_{y2} \end{bmatrix} = \begin{bmatrix} N_1 & N_2 \end{bmatrix} \mathbf{S}^0
\tag{6.16}
$$

In Eq. (6.16), $\mathbf{S}^0$ is introduced to express the unknowns of the problems. Introducing the shape functions $\mathbf{N}$ equal to

$$
\mathbf{N} = \begin{bmatrix} N_1 & N_2 \end{bmatrix}
\tag{6.17}
$$

The displacement differentiation matrix $\mathbf{B}$ is obtained by deriving the shape functions used to describe the beam displacements. It is expressed in the following

$$
\mathbf{B} = \mathbf{b}\mathbf{N} = \begin{bmatrix} z\dfrac{\partial}{\partial y} \\ -x\dfrac{\partial}{\partial y} \end{bmatrix} \begin{bmatrix} N_1 & N_2 \end{bmatrix} = \begin{bmatrix} zN_{1,y} & zN_{2,y} \\ -xN_{1,y} & -xN_{2,y} \end{bmatrix}
\tag{6.18}
$$

where

$$
\begin{aligned}
N_{1,y} &= \frac{\partial N_1(y)}{\partial y} = \frac{\partial}{\partial y}\left(1 - \frac{y}{L}\right) = -\frac{1}{L} \\[2mm]
N_{2,y} &= \frac{\partial N_2(y)}{\partial y} = \frac{\partial}{\partial y}\left(\frac{y}{L}\right) = \frac{1}{L}
\end{aligned}
\tag{6.19}
$$

Applying PVD for the case under study, as for the previous chapters, the expression of the stiffness matrix of the beam can be derived as follows

$$\mathbf{K} = \int_V \mathbf{B}^T \mathbf{C} \mathbf{B} dV \tag{6.20}$$

Using the previous definitions expressed in Eqs. (6.13) and (6.18), the explicit form of the stiffness matrix $\mathbf{K}$ becomes

$$\mathbf{K} = \int_V \begin{bmatrix} zN_{1,y} & zN_{2,y} \\ -xN_{1,y} & -xN_{2,y} \end{bmatrix}^T \begin{bmatrix} C_{44} & 0 \\ 0 & C_{44} \end{bmatrix} \begin{bmatrix} zN_{1,y} & zN_{2,y} \\ -xN_{1,y} & -xN_{2,y} \end{bmatrix} dV \tag{6.21}$$

$$\mathbf{K} = \int_V \begin{bmatrix} zN_{1,y} & -xN_{1,y} \\ zN_{2,y} & -xN_{2,y} \end{bmatrix} \begin{bmatrix} C_{44}zN_{1,y} & +C_{44}zN_{2,y} \\ -C_{44}xN_{1,y} & -C_{44}xN_{2,y} \end{bmatrix} dV \tag{6.22}$$

By performing the double matrix product of the Eq. 6.22, the stiffness matrix becomes

$$\mathbf{K} = \int_V \begin{bmatrix} C_{44}z^2 N_{1,y}N_{1,y}+ & C_{44}z^2 N_{1,y}N_{2,y}+ \\ +C_{44}x^2 N_{1,y}N_{1,y} & +C_{44}x^2 N_{1,y}N_{2,y} \\ \\ C_{44}z^2 N_{1,y}N_{2,y}+ & C_{44}z^2 N_{2,y}N_{2,y}+ \\ +C_{44}x^2 N_{1,y}N_{2,y} & +C_{44}x^2 N_{2,y}N_{2,y} \end{bmatrix} dV$$

$$= \begin{bmatrix} k_{11} & k_{12} & k_{13} & k_{14} \\ k_{21} & k_{22} & k_{23} & k_{24} \\ k_{31} & k_{32} & k_{33} & k_{34} \\ k_{41} & k_{42} & k_{43} & k_{44} \end{bmatrix} \tag{6.23}$$

Introducing the expressions of the shape functions, the stiffness matrix for a beam around the y axis becomes as follows

$$\mathbf{K} = \begin{bmatrix} C_{44}\left(\dfrac{4ab^3}{3L} + \dfrac{4a^3 b}{3L}\right) & -C_{44}\left(\dfrac{4ab^3}{3L} + \dfrac{4a^3 b}{3L}\right) \\ \\ -C_{44}\left(\dfrac{4ab^3}{3L} + \dfrac{4a^3 b}{3L}\right) & C_{44}\left(\dfrac{4ab^3}{3L} + \dfrac{4a^3 b}{3L}\right) \end{bmatrix} \tag{6.24}$$

Considering the numerical data, the matrix if the beam is

$$\mathbf{K} = \begin{bmatrix} 5600000 & -5600000 \\ -5600000 & 5600000 \end{bmatrix} \tag{6.25}$$

The comprehensive procedure for deriving the matrix presented in Eq. (6.25) is detailed in Appendix E.1, and the step-by-step implementation is provided as a MATLAB script.

6.4 Recursive Notation Against Shape Functions

In this section the stiffness matrix of the beam is calculated using a recursive notation against the shape functions. The displacements field of Eq. (6.15) can be written as follows.

$$\phi_y(y) = N_i(y)\phi_{yi} \tag{6.26}$$

where i is either 1 or 2, according to which node of the beam is considered. $\mathbf{B}$ matrix of Eq. (6.18) can be expressed as

$$\mathbf{B} = \begin{bmatrix} B_1 & B_2 \end{bmatrix} \tag{6.27}$$

where

$$\mathbf{B}_1 = \begin{bmatrix} zN_{1,y} \\ -xN_{1,y} \end{bmatrix} \tag{6.28}$$

and

$$\mathbf{B}_2 = \begin{bmatrix} zN_{2,y} \\ -xN_{2,y} \end{bmatrix} \tag{6.29}$$

Using the recursive notation, $\mathbf{B}$ matrix can be written as follows

$$\mathbf{B}_i = \begin{bmatrix} zN_{i,y} \\ -xN_{i,y} \end{bmatrix} \tag{6.30}$$

In order to distinguish the variables from their virtual variations, the second index j is introduced

$$\mathbf{B}_j = \begin{bmatrix} zN_{j,y} \\ -xN_{j,y} \end{bmatrix} \tag{6.31}$$

In the previous matrices, the indexes i and j can vary from 1 to 2.

By using the recursive notation against shape functions, the stiffness matrix can be written as follows

$$\mathbf{K}_{ij} = \int_V \mathbf{B}_j^T \mathbf{C} \mathbf{B}_i dV = \int_V \begin{bmatrix} zN_{j,y} & -xN_{j,y} \end{bmatrix} \begin{bmatrix} C_{44} & 0 \\ 0 & C_{44} \end{bmatrix} \begin{bmatrix} zN_{i,y} \\ -xN_{i,y} \end{bmatrix} \tag{6.32}$$

By performing the triple product, the previous matrix becomes

$$\mathbf{K}_{ij} = \int_V \left[\left(C_{44}z^2 N_{i,y} N_{j,y} + C_{44}x^2 N_{i,y} N_{j,y} \right) \right] dV \tag{6.33}$$

Looping the indexes i and j from 1 to 2 in the previous matrix of Eq. 6.33 as follows

$$\mathbf{K} = \begin{bmatrix} k_{11} & k_{12} \\ k_{21} & k_{22} \end{bmatrix} = \begin{bmatrix} \int_V (C_{44}z^2 N_{1,y} N_{1,y} + & \int_V (C_{44}z^2 N_{1,y} N_{2,y} + \\ + C_{44}x^2 N_{1,y} N_{1,y}) \, dV & + C_{44}x^2 N_{1,y} N_{2,y}) \, dV \\ \int_V (C_{44}z^2 N_{1,y} N_{2,y} + & \int_V (C_{44}z^2 N_{2,y} N_{2,y} + \\ + C_{44}x^2 N_{1,y} N_{2,y}) \, dV & + C_{44}x^2 N_{2,y} N_{2,y}) \, dV \end{bmatrix} \tag{6.34}$$

The resultant stiffness matrix is then

$$\mathbf{K} = \begin{bmatrix} C_{44}\left(\dfrac{4ab^3}{3L} + \dfrac{4a^3b}{3L} \right) & -C_{44}\left(\dfrac{4ab^3}{3L} + \dfrac{4a^3b}{3L} \right) \\ -C_{44}\left(\dfrac{4ab^3}{3L} + \dfrac{4a^3b}{3L} \right) & C_{44}\left(\dfrac{4ab^3}{3L} + \dfrac{4a^3b}{3L} \right) \end{bmatrix} \tag{6.35}$$

Introducing the numerical data, the matrix if the beam is

$$\mathbf{K} = \begin{bmatrix} 5600000 & -5600000 \\ -5600000 & 5600000 \end{bmatrix} \tag{6.36}$$

The comprehensive procedure for deriving the matrix presented in Eq. (6.36) is detailed in Appendix E.2, and the step-by-step implementation is provided as a MATLAB script.

6.5 Recursive Notation Against the Theory of Structure

In this section the stiffness matrix of the beam is calculated using a recursive notation against TOS. Introducing $s_n = s_x, s_y, s_z$, Eq. (6.2) becomes

$$
\begin{aligned}
s_x &= u(x, y, z) = z\phi_y(y) \\
s_y &= v(x, y, z) = 0 \\
s_z &= w(x, y, z) = -x\phi_y(y)
\end{aligned}
\tag{6.37}
$$

Then, the expansion functions F are introduced, so that

$$
\begin{aligned}
s_x &= F_{1_x} \times \phi_y(y) \\
s_y &= F_{1_y} \times 0 \\
s_z &= F_{1_z} \times \phi_y(y)
\end{aligned}
\tag{6.38}
$$

In this case

$$
F_{1_x} = z \qquad F_{1_y} = 0, \qquad F_{1_z} = -x
\tag{6.39}
$$

Equation (6.39) can be expressed in a generic form introducing the index τ which ranges from 1 to the number of the terms in the cross-sectional expansion. This case can then expressed as

$$
s_n = F_{\tau_n} s_{\tau_n}
\tag{6.40}
$$

where

$$
s_{1_x} = \phi_y(y) \qquad s_{1_y} = 0, \qquad s_{1_z} = \phi_y(y)
\tag{6.41}
$$

with the generic index n introduced for the directions x, y and z. The same procedure can be used for the virtual variation, by introducing the index s for the virtual variation.

$$
\delta s_n = F_{s_n} s_{s_n}
\tag{6.42}
$$

Using the $F_{\tau x}$ and the $F_{\tau z}$ previously defined, the matrices $\mathbf{B}_{\tau i}$ and $\mathbf{B}_{sj}$ of Eqs. (6.30) and (6.31) can be written as follows

$$
\mathbf{B}_{\tau i} =
\begin{bmatrix}
F_{\tau x} N_{i,y} \\
\\
F_{\tau z} N_{i,y}
\end{bmatrix}
\tag{6.43}
$$

$$\mathbf{B}_{sj} = \begin{bmatrix} F_{sx}N_{j,y} \\ F_{sz}N_{j,y} \end{bmatrix} \tag{6.44}$$

The stiffness matrix can be expressed as follows

$$\mathbf{K}_{ij}^{\tau s} = \int_V \mathbf{B}_{sj}^T \mathbf{C} \mathbf{B}_{\tau i} dV = \int_V \begin{bmatrix} F_{sx}N_{j,y} & F_{sz}N_{j,y} \end{bmatrix} \begin{bmatrix} C_{44} & 0 \\ 0 & C_{44} \end{bmatrix} \begin{bmatrix} F_{\tau x}N_{i,y} \\ F_{\tau z}N_{i,y} \end{bmatrix} \tag{6.45}$$

By computing the matrix triple product, the previous matrix becomes

$$\mathbf{K}_{ij}^{\tau s} = \int_V \left(C_{44} F_{sx} F_{\tau x} N_{i,y} N_{j,y} + C_{44} F_{sz} F_{\tau z} N_j N_{i,y} \right) dV \tag{6.46}$$

and consequently

Considering the values of products between the expansion functions and their derivatives

> **Expansion functions**
>
> $$F_{\tau x} = F_{1x} = z$$
>
> $$F_{sx} = F_{1x} = z$$
>
> $$F_{\tau z} = F_{1z} = -x$$
>
> $$F_{sz} = F_{1z} = -x$$
>
> $$F_{sx} F_{\tau x} = F_{1x} F_{1x} = (z)(z) = z^2$$
>
> $$F_{sz} F_{\tau z} = F_{1z} F_{1z} = (-x)(-x) = x^2$$

the stiffness matrix becomes as follows

$$\mathbf{K} = \begin{bmatrix} C_{44}\left(\dfrac{4ab^3}{3L} + \dfrac{4a^3b}{3L}\right) & -C_{44}\left(\dfrac{4ab^3}{3L} + \dfrac{4a^3b}{3L}\right) \\ -C_{44}\left(\dfrac{4ab^3}{3L} + \dfrac{4a^3b}{3L}\right) & C_{44}\left(\dfrac{4ab^3}{3L} + \dfrac{4a^3b}{3L}\right) \end{bmatrix} \tag{6.47}$$

Considering the numerical data of the problem, the numerical matrix is the following

$$\mathbf{K} = \begin{bmatrix} 5600000 & -5600000 \\ -5600000 & 5600000 \end{bmatrix} \tag{6.48}$$

The comprehensive procedure for deriving the matrix presented in Eq. (6.48) is detailed in Appendix E.3, and the step-by-step implementation is provided as a MATLAB script.

Chapter 7
Stiffness Matrix of a Beam Under Combined Load

Graphical Abstract

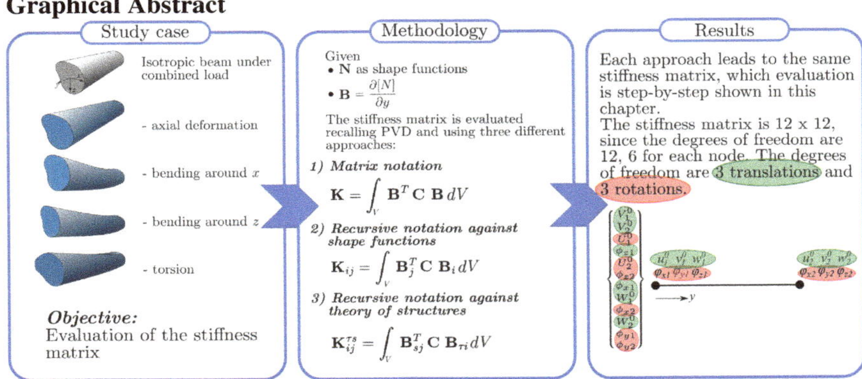

Introduction

In this chapter, the stiffness matrix of a two-node beam element is derived. The beam is a structural element with axial, torsional and bending stiffness. In this case, each stiffness is considered in this derivation. The methodology behind this derivation is presented in Sect. 7.1. Section 7.2 details the assumptions underlying the finite element beam definition and specifies the associated geometric data. The derivation of the stiffness matrix utilizes the PVD approach, as discussed in the preceding chapters. The same techniques as in the previous chapters are employed for the derivation of the stiffness matrix, including:

- Matrix notation (Sect. 7.3)
- Recursive notation against shape functions (Sect. 7.4)
- Recursive notation against Theory Of Structures (TOS) (Sect. 7.5).

E. Carrera et al., *Implementation of Beam-Type Finite Elements Based on Carrera Unified Formulation*, https://doi.org/10.1007/978-3-031-95856-4_7

7.1   Structural Model of a Complete Beam

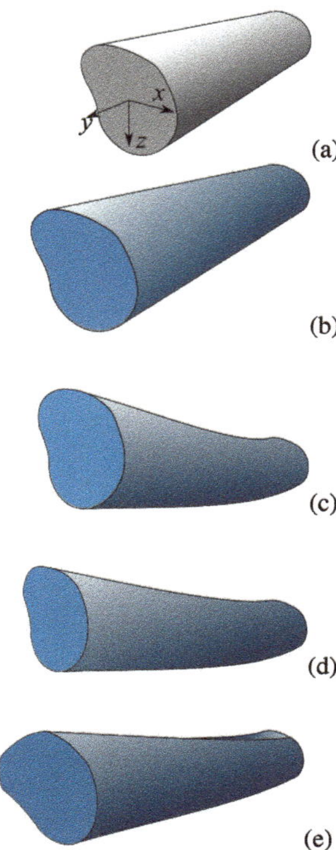

The figure (a) on the left shows the undeformed shape of the beam under study, with a generic cross-section geometry. The deformed shapes are reported as weel, undergoing axial displacement along the y axis (b), bending around the x axis (c), bending around the z axis (d), torsion around the beam axis y (e).

Three displacement components and three rotations are involved in the deformation of the beam: the displacement $u(x, y, z)$ along the direction x, the displacement $v(x, y, z)$ which relates the direction y, the displacement $w(x, y, z)$ along the direction z and the rotation $\phi_x(y)$ around the x axis, the rotation $\phi_y(y)$ around the y axis and the rotation $\phi_z(y)$ around the z axis. As a results, the displacements model for the complete beam can be written as the following:

$$
\begin{aligned}
u(x, y, z) &= -u^0(y) + z\phi_y(y) \\
v(x, y, z) &= v^0(y) \quad - z\phi_x(y) + x\phi_z(y) \\
w(x, y, z) &= w^0(y) \quad - x\phi_y(y)
\end{aligned}
\tag{7.1}
$$

In a matrix form, Eq. (7.1) can be expressed as

$$
\mathbf{s} = \left\{ \begin{array}{c} u \\ v \\ w \end{array} \right\} = \begin{bmatrix} -1 & 0 & 0 & 0 & z & 0 \\ 0 & 1 & 0 & -z & 0 & x \\ 0 & 0 & 1 & 0 & -x & 0 \end{bmatrix} \left\{ \begin{array}{c} u^0(y) \\ v^0(y) \\ w^0(y) \\ \phi_x(y) \\ \phi_y(y) \\ \phi_z(y) \end{array} \right\}
\tag{7.2}
$$

For the case study, the strain components of interest are ϵ_{yy}, ϵ_{xy} and ϵ_{yz}. The geometric relations between strain and displacement components are the following

$$
\epsilon_{yy} = \frac{\partial v}{\partial y}, \qquad \epsilon_{xy} = \frac{\partial u}{\partial y} + \frac{\partial v}{\partial x}, \qquad \epsilon_{yz} = \frac{\partial w}{\partial y} + \frac{\partial v}{\partial z}
\tag{7.3}
$$

Given the displacement components of Eq. (7.1), the following expressions for the strain components are obtained

$$
\begin{aligned}
\epsilon_{yy} &= \frac{\partial v^0(y)}{\partial y} - z\frac{\partial \phi_x(y)}{\partial y} + x\frac{\partial \phi_z(y)}{\partial y} \\
\epsilon_{xy} &= -\frac{\partial u^0(y)}{\partial y} + z\frac{\partial \phi_y(y)}{\partial y} + \phi_z(y) \\
\epsilon_{yz} &= \frac{\partial w^0(y)}{\partial y} - x\frac{\partial \phi_y(y)}{\partial y} - \phi_x(y)
\end{aligned}
\tag{7.4}
$$

Introducing the matrix form

$$
\boldsymbol{\epsilon} = \left\{ \begin{array}{c} \epsilon_{yy} \\ \epsilon_{xy} \\ \epsilon_{yz} \end{array} \right\} = \begin{bmatrix} 0 & \dfrac{\partial}{\partial y} & 0 & -z\dfrac{\partial}{\partial y} & 0 & x\dfrac{\partial}{\partial y} \\ -\dfrac{\partial}{\partial y} & 0 & 0 & 0 & z\dfrac{\partial}{\partial y} & 1 \\ 0 & 0 & \dfrac{\partial}{\partial y} & -1 & -x\dfrac{\partial}{\partial y} & 0 \end{bmatrix} \left\{ \begin{array}{c} u^0(y) \\ v^0(y) \\ w^0(y) \\ \phi_x(y) \\ \phi_y(y) \\ \phi_z(y) \end{array} \right\}
\tag{7.5}
$$

The differential operator matrix $\mathbf{b}$ can be expressed as

$$
\mathbf{b} = \begin{bmatrix} 0 & \dfrac{\partial}{\partial y} & 0 & -z\dfrac{\partial}{\partial y} & 0 & x\dfrac{\partial}{\partial y} \\ -\dfrac{\partial}{\partial y} & 0 & 0 & 0 & z\dfrac{\partial}{\partial y} & 1 \\ 0 & 0 & \dfrac{\partial}{\partial y} & -1 & -x\dfrac{\partial}{\partial y} & 0 \end{bmatrix}
\tag{7.6}
$$

The constitutive relations are those between the strain components ϵ_{yy}, ϵ_{xy} and ϵ_{yz} the stress components σ_{yy}, σ_{xy} and σ_{yz}. These relations are based on the material coefficient and the Hooke's law. The constitutive relations of interest are the following

$$\sigma_{yy} = C_{11}\epsilon_{yy}$$

$$\sigma_{xy} = C_{44}\epsilon_{xy} \tag{7.7}$$

$$\sigma_{yz} = C_{44}\epsilon_{yz}$$

The expression of C_{11} is given in the Eqs. (2.6) and (2.7), while the expression of C_{44} is equal to

$$C_{44} = G \tag{7.8}$$

Equation (5.7) can be written in matrix form as follows

$$\sigma = C\epsilon \tag{7.9}$$

where

$$\sigma = \left\{ \begin{array}{c} \sigma_{yy} \\ \sigma_{xy} \\ \sigma_{yz} \end{array} \right\} \tag{7.10}$$

$$\epsilon = \left\{ \begin{array}{c} \epsilon_{yy} \\ \epsilon_{xy} \\ \epsilon_{yz} \end{array} \right\} \tag{7.11}$$

$$C = \begin{bmatrix} C_{11} & 0 & 0 \\ 0 & C_{44} & 0 \\ 0 & 0 & C_{44} \end{bmatrix} \tag{7.12}$$

7.2 Finite Element Approximation

The finite element approximations of the beam is considered here. This example refers to the two-node beam element of Fig. 7.1, where subscripts 1 and 2 refer to node 1 and node 2 of the beam. The coordinate system of the beam is also indicated in Fig. 7.1: the y axis lays along the elements length, and as a consequence the x and z axes define the cross-section of the beam. w_1^0 and w_2^0 are the displacements,

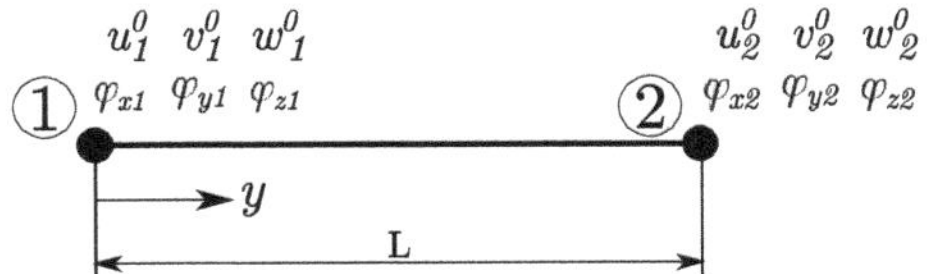

Fig. 7.1 Schematic representation of a complete beam with 2 nodes

along the coordinate z of the two nodes 1 and 2 respectively, whereas ϕ_{x1} and ϕ_{x2} refer to the rotation around the x axis, of the cross-section of the beam at nodes 1 and 2 respectively. Note that the subscripts 1 and 2 refer to the beam nodes, while the exponent 0 refers to the degree of the displacement function over the bar cross-section, and therefore it means that the displacement over the cross section is constant. The geometric characteristics and the dimensions of the beam are the same as in the previous chapters.

7.3 Matrix Notation

In this section the stiffness matrix of the beam under torsion is calculated by the means of PVD. The displacement of interest are the displacements $u^0(y)$ along the x axis, $v^0(y)$ along the y axis, $w^0(y)$, along the z axis and the rotations $\phi_x(y)$ around the x axis, $\phi_y(y)$ around the y axis and $\phi_z(y)$ around the z axis. The variation, along the beam length, of these displacements and rotations can be expressed as a function of the displacements of the nodes 1 and 2 using the shape functions (i.e. u_1^0, v_1^0, w_1^0, $\phi_{x1}, \phi_{y1}, \phi_{z1}$ and $u_2^0, v_2^0, w_2^0, \phi_{x2}, \phi_{y2}, \phi_{z2}$). The shape functions used for this purpose are based on Lagrange polynomials and are reported below

$$N_1(y) = 1 - \frac{y}{L}$$
$$N_2(y) = \frac{y}{L} \tag{7.13}$$

The expressions that describe the variation, along the beam, of the displacements and the rotations, areù the following

$$
\begin{aligned}
u^0(y) &= N_1(y)u_1^0 + N_2(y)u_2^0 \\
v^0(y) &= N_1(y)v_1^0 + N_2(y)v_2^0 \\
w^0(y) &= N_1(y)w_1^0 + N_2(y)w_2^0 \\
\phi_x(y) &= N_1(y)\phi_{x1} + N_2(y)\phi_{x2} \\
\phi_y(y) &= N_1(y)\phi_{y1} + N_2(y)\phi_{y2} \\
\phi_z(y) &= N_1(y)\phi_{z1} + N_2(y)\phi_{z2}
\end{aligned}
\tag{7.14}
$$

The matrix form of Eq. (7.14)

$$
\mathbf{s}^0 = \begin{Bmatrix} u^0(y) \\ v^0(y) \\ w^0(y) \\ \phi_x(y) \\ \phi_y(y) \\ \phi_z(y) \end{Bmatrix} = \begin{bmatrix} 0 & 0 & N_1 & 0 & N_2 & 0 & 0 & 0 & 0 & 0 & 0 & 0 \\ N_1 & N_2 & 0 & 0 & 0 & 0 & 0 & 0 & 0 & 0 & 0 & 0 \\ 0 & 0 & 0 & 0 & 0 & 0 & 0 & N_1 & 0 & N_2 & 0 & 0 \\ 0 & 0 & 0 & 0 & 0 & 0 & N_1 & 0 & N_2 & 0 & 0 & 0 \\ 0 & 0 & 0 & 0 & 0 & 0 & 0 & 0 & 0 & 0 & N_1 & N_2 \\ 0 & 0 & 0 & N_1 & 0 & N_2 & 0 & 0 & 0 & 0 & 0 & 0 \end{bmatrix} \begin{Bmatrix} v_1^0 \\ v_2^0 \\ u_1^0 \\ \phi_{z1} \\ u_2^0 \\ \phi_{z2} \\ \phi_{x1} \\ w_1^0 \\ \phi_{x2} \\ w_2^0 \\ \phi_{y1} \\ \phi_{y2} \end{Bmatrix}
$$

$$(7.15)$$

Introducing the *displacement differentiation matrix* $\mathbf{B}$, it is obtained by deriving the shape functions used to describe the beam displacements. It is expressed in the following

$$
\mathbf{B} = \mathbf{b}\,\mathbf{N} =
$$

$$
= \begin{bmatrix} 0 & \dfrac{\partial}{\partial y} & 0 & -z\dfrac{\partial}{\partial y} & 0 & x\dfrac{\partial}{\partial y} \\ -\dfrac{\partial}{\partial y} & 0 & 0 & 0 & z\dfrac{\partial}{\partial y} & 1 \\ 0 & 0 & \dfrac{\partial}{\partial y} & -1 & -x\dfrac{\partial}{\partial y} & 0 \end{bmatrix} \begin{bmatrix} 0 & 0 & N_1 & 0 & N_2 & 0 & 0 & 0 & 0 & 0 & 0 & 0 \\ N_1 & N_2 & 0 & 0 & 0 & 0 & 0 & 0 & 0 & 0 & 0 & 0 \\ 0 & 0 & 0 & 0 & 0 & 0 & 0 & N_1 & 0 & N_2 & 0 & 0 \\ 0 & 0 & 0 & 0 & 0 & 0 & N_1 & 0 & N_2 & 0 & 0 & 0 \\ 0 & 0 & 0 & 0 & 0 & 0 & 0 & 0 & 0 & 0 & N_1 & N_2 \\ 0 & 0 & 0 & N_1 & 0 & N_2 & 0 & 0 & 0 & 0 & 0 & 0 \end{bmatrix} =
$$

$$
= \begin{bmatrix} N_{1,y} & N_{2,y} & 0 & xN_{1,y} & 0 & xN_{2,y} & -zN_{1,y} & 0 & -zN_{2,y} & 0 & 0 & 0 \\ 0 & 0 & -N_{1,y} & N_1 & -N_{2,y} & N_2 & 0 & 0 & 0 & 0 & zN_{1,y} & zN_{2,y} \\ 0 & 0 & 0 & 0 & 0 & 0 & -N_1 & N_{1,y} & -N_2 & N_{2,y} & -xN_{1,y} & -xN_{2,y} \end{bmatrix}
$$

$$(7.16)$$

where

$$
N_{1,y} = \frac{\partial N_1(y)}{\partial y}
$$
$$
N_{2,y} = \frac{\partial N_2(y)}{\partial y}
$$

$$(7.17)$$

Recalling the stiffness matrix $\mathbf{K}$ evaluated using PVD as shown in previous chapters, it is

$$
\mathbf{K} = \int_V \mathbf{B}^T \mathbf{C} \mathbf{B}\, dV
$$

$$(7.18)$$

Using the definition expressed in Eqs. (7.12) and (7.16), the explicit form of the stiffness matrix $\mathbf{K}$ becomes

$$\mathbf{K} = \int_V \left(\begin{bmatrix} N_{1,y} & 0 & 0 \\ N_{2,y} & 0 & 0 \\ 0 & -N_{1,y} & 0 \\ xN_{1,y} & N_1 & 0 \\ 0 & -N_{2,y} & 0 \\ xN_{2,y} & N_2 & 0 \\ -zN_{1,y} & 0 & -N_1 \\ 0 & 0 & N_{1,y} \\ -zN_{2,y} & 0 & -N_2 \\ 0 & 0 & N_{2,y} \\ 0 & zN_{1,y} & -xN_{1,y} \\ 0 & zN_{2,y} & -xN_{2,y} \end{bmatrix} \begin{bmatrix} C_{11} & 0 & 0 \\ 0 & C_{44} & 0 \\ 0 & 0 & C_{44} \end{bmatrix} * \right.$$

$$\left. * \begin{bmatrix} N_{1,y} & N_{2,y} & 0 & xN_{1,y} & 0 & xN_{2,y} & -zN_{1,y} & 0 & -zN_{2,y} & 0 & 0 & 0 \\ 0 & 0 & -N_{1,y} & N_1 & -N_{2,y} & N_2 & 0 & 0 & 0 & 0 & zN_{1,y} & zN_{2,y} \\ 0 & 0 & 0 & 0 & 0 & 0 & -N_1 & N_{1,y} & -N_2 & N_{2,y} & -xN_{1,y} & -xN_{2,y} \end{bmatrix} \right) dV \tag{7.19}$$

$$\mathbf{K} = \int_V \left(\begin{bmatrix} N_{1,y} & 0 & 0 \\ N_{2,y} & 0 & 0 \\ 0 & -N_{1,y} & 0 \\ xN_{1,y} & N_1 & 0 \\ 0 & -N_{2,y} & 0 \\ xN_{2,y} & N_2 & 0 \\ -zN_{1,y} & 0 & -N_1 \\ 0 & 0 & N_{1,y} \\ -zN_{2,y} & 0 & -N_2 \\ 0 & 0 & N_{2,y} \\ 0 & zN_{1,y} & -xN_{1,y} \\ 0 & zN_{2,y} & -xN_{2,y} \end{bmatrix} * \right.$$

$$\left. * \begin{bmatrix} C_{11}N_{1,y} & C_{11}N_{2,y} & 0 & C_{11}xN_{1,y} & 0 & C_{11}xN_{2,y} & -C_{11}zN_{1,y} & 0 & -C_{11}zN_{2,y} & 0 & 0 & 0 \\ 0 & 0 & -C_{44}N_{1,y} & C_{44}N_1 & -C_{44}N_{2,y} & C_{44}N_2 & 0 & 0 & 0 & 0 & C_{44}zN_{1,y} & C_{44}zN_{2,y} \\ 0 & 0 & 0 & 0 & 0 & 0 & -C_{44}N_1 & C_{44}N_{1,y} & -C_{44}N_2 & C_{44}N_{2,y} & -C_{44}xN_{1,y} & -C_{44}xN_{2,y} \end{bmatrix} \right) dV \tag{7.20}$$

By computing the matrix product of Eq. (7.20), the following relation is obtained

$$\mathbf{K} = \int_V \begin{bmatrix}
C_{11}N_{1,y}N_{1,y} & C_{11}N_{1,y}N_{2,y} & 0 & C_{11}xN_{1,y}N_{1,y} & 0 & C_{11}xN_{1,y}N_{2,y} & -C_{11}zN_{1,y}N_{1,y} & 0 & -C_{11}zN_{1,y}N_{2,y} & 0 & 0 & 0 \\[4pt]
C_{11}N_{1,y}N_{2,y} & C_{11}N_{2,y}N_{2,y} & 0 & C_{11}xN_{1,y}N_{2,y} & 0 & C_{11}xN_{2,y}N_{2,y} & -C_{11}zN_{1,y}N_{2,y} & 0 & -C_{11}zN_{2,y}N_{2,y} & 0 & 0 & 0 \\[4pt]
0 & 0 & C_{44}N_{1,y}N_{1,y} & -C_{44}N_1N_{1,y} & C_{44}N_{1,y}N_{2,y} & -C_{44}N_2N_{1,y} & -C_{44}zN_{1,y}N_{1,y} & 0 & -C_{44}zN_{1,y}N_{2,y} & 0 & 0 & 0 \\[4pt]
C_{11}xN_{1,y}N_{1,y} & C_{11}xN_{1,y}N_{2,y} & -C_{44}N_1N_{1,y} & C_{11}x^2N_{1,y}N_{1,y}+C_{44}N_1N_1 & -C_{44}N_1N_{2,y} & C_{11}x^2N_{1,y}N_{2,y}+C_{44}N_1N_2 & -C_{11}xzN_{1,y}N_{1,y}+C_{44}zN_1N_{1,y} & 0 & -C_{11}xzN_{1,y}N_{2,y}+C_{44}zN_1N_{2,y} & 0 & 0 & 0 \\[4pt]
0 & 0 & C_{44}N_{1,y}N_{2,y} & -C_{44}N_1N_{2,y} & C_{44}N_{2,y}N_{2,y} & -C_{44}N_2N_{2,y} & -C_{44}zN_{1,y}N_{2,y} & 0 & -C_{44}zN_{2,y}N_{2,y} & 0 & 0 & 0 \\[4pt]
C_{11}xN_{1,y}N_{2,y} & C_{11}xN_{2,y}N_{2,y} & -C_{44}N_{1,y}N_2 & C_{11}x^2N_{1,y}N_{2,y}+C_{44}N_1N_2 & -C_{44}N_2N_{2,y} & C_{11}x^2N_{2,y}N_{2,y}+C_{44}N_2N_2 & -C_{11}xzN_{1,y}N_{2,y}+C_{44}zN_{1,y}N_2 & 0 & -C_{11}xzN_{2,y}N_{2,y}+C_{44}zN_{2,y}N_2 & 0 & 0 & 0 \\[4pt]
-C_{11}zN_{1,y}N_{1,y} & -C_{11}zN_{1,y}N_{2,y} & 0 & -C_{11}xzN_{1,y}N_{1,y} & 0 & C_{11}xzN_{1,y}N_{2,y} & C_{11}z^2N_{1,y}N_{1,y}+C_{44}N_1N_1 & -C_{44}N_1N_{1,y} & C_{11}z^2N_{1,y}N_{2,y}+C_{44}N_1N_2 & -C_{44}N_1N_{2,y} & C_{44}xN_1N_{1,y} & C_{44}xN_1N_{2,y} \\[4pt]
0 & 0 & 0 & 0 & 0 & 0 & -C_{44}N_1N_{1,y} & C_{44}N_{1,y}N_{1,y} & -C_{44}N_2N_{1,y} & C_{44}N_{1,y}N_{2,y} & -C_{44}xN_{1,y}N_{1,y} & -C_{44}xN_{1,y}N_{2,y} \\[4pt]
-C_{11}zN_{1,y}N_{2,y} & -C_{11}zN_{2,y}N_{2,y} & 0 & -C_{11}xzN_{1,y}N_{2,y} & 0 & C_{11}xzN_{2,y}N_{2,y} & C_{11}z^2N_{1,y}N_{2,y}+C_{44}N_1N_2 & -C_{44}N_2N_{1,y} & C_{11}z^2N_{2,y}N_{2,y}+C_{44}N_2N_2 & -C_{44}N_2N_{2,y} & C_{44}xN_2N_{1,y} & C_{44}xN_2N_{2,y} \\[4pt]
0 & 0 & 0 & 0 & 0 & 0 & -C_{44}N_1N_{2,y} & C_{44}N_{1,y}N_{2,y} & -C_{44}N_2N_{2,y} & C_{44}N_{2,y}N_{2,y} & -C_{44}xN_{1,y}N_{2,y} & -C_{44}xN_{2,y}N_{2,y} \\[4pt]
0 & 0 & -C_{44}zN_{1,y}N_{1,y} & C_{44}zN_1N_{1,y} & -C_{44}zN_{1,y}N_{2,y} & C_{44}zN_{1,y}N_2 & C_{44}xN_1N_{1,y} & -C_{44}xN_{1,y}N_{1,y} & C_{44}xN_{1,y}N_2 & -C_{44}xN_{1,y}N_{2,y} & C_{44}z^2N_{1,y}N_{1,y}+C_{44}x^2N_{1,y}N_{1,y} & C_{44}z^2N_{1,y}N_{2,y}+C_{44}x^2N_{1,y}N_{2,y} \\[4pt]
0 & 0 & -C_{44}zN_{1,y}N_{2,y} & C_{44}zN_1N_{2,y} & -C_{44}zN_{2,y}N_{2,y} & C_{44}zN_{2,y}N_2 & C_{44}xN_1N_{2,y} & -C_{44}xN_{1,y}N_{2,y} & C_{44}xN_{2,y}N_2 & -C_{44}xN_{2,y}N_{2,y} & C_{44}z^2N_{1,y}N_{2,y}+C_{44}x^2N_{1,y}N_{2,y} & C_{44}z^2N_{2,y}N_{2,y}+C_{44}x^2N_{2,y}N_{2,y}
\end{bmatrix} dV \tag{7.21}$$

It is important to consider that if a symmetric (with respect to the x and z axes) cross-section is considered, the first moment of area and the product of inertia are zero. Then, the following terms are null.

$$
\begin{aligned}
&C_{44}xN_{1,y}N_{2,y}=0 \;\; C_{44}xN_{2,y}N_{2,y}=0 \;\; C_{44}xN_{1,y}N_{1,y}=0 \;\; C_{44}xN_1N_1=0 \;\;\;\; C_{44}xN_1N_2=0 \;\;\;\; C_{44}xN_2N_2=0\\
&C_{44}xN_2N_{1,y}=0 \;\; C_{44}xN_2N_{2,y}=0 \;\; C_{44}xN_1N_{1,y}=0 \;\; C_{44}xN_1N_{2,y}=0 \;\; C_{44}zN_{1,y}N_{2,y}=0 \;\; C_{44}zN_{2,y}N_{2,y}=0\\
&C_{44}zN_{1,y}N_{1,y}=0 \;\; C_{44}zN_1N_1=0 \;\;\;\; C_{44}zN_1N_2=0 \;\;\;\; C_{44}zN_2N_2=0 \;\;\;\; C_{44}zN_2N_{1,y}=0 \;\; C_{44}zN_2N_{2,y}=0\\
&C_{44}N_1N_{1,y}=0 \;\;\;\; C_{44}zN_1N_{2,y}=0 \;\; C_{11}xN_{1,y}N_{2,y}=0 \;\; C_{11}xN_{2,y}N_{2,y}=0 \;\; C_{11}xN_{1,y}N_{1,y}=0 \;\; C_{11}zN_{1,y}N_{2,y}=0\\
&C_{11}zN_{2,y}N_{2,y}=0 \;\; C_{11}zN_{1,y}N_{1,y}=0 \;\; C_{11}xzN_{1,y}N_{1,y}=0 \;\; C_{11}xzN_{1,y}N_{2,y}=0 \;\; C_{11}xzN_{2,y}N_{2,y}=0
\end{aligned}
\tag{7.22}
$$

Considering these relations, the stiffness matrix for a complete beam results as follows

$$
K=\int_V
\begin{bmatrix}
C_{11}N_{1,y}N_{1,y} & C_{11}N_{1,y}N_{2,y} & 0 & 0 & 0 & 0 & 0 & 0 & 0 & 0 & 0 & 0\\[2pt]
C_{11}N_{1,y}N_{2,y} & C_{11}N_{2,y}N_{2,y} & 0 & 0 & 0 & 0 & 0 & 0 & 0 & 0 & 0 & 0\\[2pt]
0 & 0 & C_{44}N_{1,y}N_{1,y} & -C_{44}N_1N_{1,y} & C_{44}N_{1,y}N_{2,y} & -C_{44}N_2N_{1,y} & 0 & 0 & 0 & 0 & 0 & 0\\[2pt]
0 & 0 & -C_{44}N_1N_{1,y} & C_{11}x^2N_{1,y}N_{1,y}+C_{44}N_1N_1 & -C_{44}N_1N_{2,y} & C_{11}x^2N_{1,y}N_{2,y}+C_{44}N_1N_1 & 0 & 0 & 0 & 0 & 0 & 0\\[2pt]
0 & 0 & C_{44}N_{1,y}N_{2,y} & -C_{44}N_1N_{2,y} & C_{44}N_{2,y}N_{2,y} & -C_{44}N_2N_{2,y} & 0 & 0 & 0 & 0 & 0 & 0\\[2pt]
0 & 0 & -C_{44}N_{1,y}N_2 & C_{11}x^2N_{1,y}N_{2,y}+C_{44}N_2N_1 & -C_{44}N_2N_{2,y} & C_{11}x^2N_{2,y}N_{2,y}+C_{44}N_2N_2 & 0 & 0 & 0 & 0 & 0 & 0\\[2pt]
0 & 0 & 0 & 0 & 0 & 0 & C_{11}z^2N_{1,y}N_{1,y}+C_{44}N_1N_1 & -C_{44}N_1N_{1,y} & C_{11}z^2N_{1,y}N_{2,y}+C_{44}N_1N_1 & -C_{44}N_1N_{2,y} & 0 & 0\\[2pt]
0 & 0 & 0 & 0 & 0 & 0 & C_{44}N_1N_{1,y} & -C_{44}N_{1,y}N_{1,y} & C_{44}N_2N_{1,y} & -C_{44}N_{1,y}N_{2,y} & 0 & 0\\[2pt]
0 & 0 & 0 & 0 & 0 & 0 & C_{11}z^2N_{1,y}N_{2,y}+C_{44}N_1N_2 & -C_{44}N_2N_{1,y} & C_{11}z^2N_{2,y}N_{2,y}+C_{44}N_2N_2 & -C_{44}N_2N_{2,y} & 0 & 0\\[2pt]
0 & 0 & 0 & 0 & 0 & 0 & -C_{44}N_1N_{2,y} & C_{44}N_{1,y}N_{2,y} & -C_{44}N_2N_{2,y} & C_{44}N_{2,y}N_{2,y} & 0 & 0\\[2pt]
0 & 0 & 0 & 0 & 0 & 0 & 0 & 0 & 0 & 0 & C_{44}z^2N_{1,y}N_{1,y}+C_{44}x^2N_{1,y}N_{1,y} & C_{44}z^2N_{1,y}N_{2,y}+C_{44}x^2N_{1,y}N_{2,y}\\[2pt]
0 & 0 & 0 & 0 & 0 & 0 & 0 & 0 & 0 & 0 & C_{44}z^2N_{1,y}N_{2,y}+C_{44}x^2N_{1,y}N_{2,y} & C_{44}z^2N_{2,y}N_{2,y}+C_{44}x^2N_{2,y}N_{2,y}
\end{bmatrix}
\, dV
\tag{7.23}
$$

In the matrix of Eq. (7.23), the red terms represent the part of the stiffness matrix related to the axial displacement of the beam (i.e. they relate to the displacements components v_1^0, v_2^0), the blue terms represent the part of the stiffness matrix related to the bending of the beam around the z axis (i.e. they relate to the displacements components u_1^0, ϕ_{z1}, u_2^0, ϕ_{z2}), the green terms represent the part of the stiffness matrix related to the bending of the beam around the x axis (i.e. they relate to the displacements components w_1^0, ϕ_{x1}, w_2^0, ϕ_{x2}), the purple terms represent the part of the stiffness matrix related to the torsion of the beam around the y axis (i.e. they relate to the displacements components ϕ_{y1}, ϕ_{y2}).

It can be pointed out that the order of the rows and the columns depends on the order of the displacement components in the unknown vector. For this stiffness matrix, the order is related to the following unknown vector

$$\left\{ \begin{array}{c} v_1^0 \\ v_2^0 \\ u_1^0 \\ \phi_{z1} \\ u_2^0 \\ \phi_{z2} \\ \phi_{x1} \\ w_1^0 \\ \phi_{x2} \\ w_2^0 \\ \phi_{y1} \\ \phi_{y2} \end{array} \right\} \tag{7.24}$$

If the order of the displacement components in the unknown vector changes, the rows and columns of the stiffness matrix change accordingly. Finally, the stiffness matrix for a complete beam is shown hereafter.

$$\mathbf{K} = \begin{bmatrix}
C_{11}\frac{4ab}{L} & -C_{11}\frac{4ab}{L} & 0 & 0 & 0 & 0 & 0 & 0 & 0 & 0 & 0 & 0 \\[2ex]
-C_{11}\frac{4ab}{L} & C_{11}\frac{4ab}{L} & 0 & 0 & 0 & 0 & 0 & 0 & 0 & 0 & 0 & 0 \\[2ex]
0 & 0 & C_{44}\frac{4ab}{L} & C_{44}2ab & -C_{44}\frac{4ab}{L} & C_{44}2ab & 0 & 0 & 0 & 0 & 0 & 0 \\[2ex]
0 & 0 & C_{44}2ab & C_{11}\frac{4a^3b}{3L}+C_{44}\frac{4abL}{3} & -C_{44}2ab & -C_{11}\frac{4a^3b}{3L}+C_{44}\frac{2abL}{3} & 0 & 0 & 0 & 0 & 0 & 0 \\[2ex]
0 & 0 & -C_{44}\frac{4ab}{L} & -C_{44}2ab & C_{44}\frac{4ab}{L} & -C_{44}2ab & 0 & 0 & 0 & 0 & 0 & 0 \\[2ex]
0 & 0 & C_{44}2ab & -C_{11}\frac{4a^3b}{3L}+C_{44}\frac{2abL}{3} & -C_{44}2ab & C_{11}\frac{4a^3b}{3L}+C_{44}\frac{4abL}{3} & 0 & 0 & 0 & 0 & 0 & 0 \\[2ex]
0 & 0 & 0 & 0 & 0 & 0 & C_{11}\frac{4ab^3}{3L}+C_{44}\frac{4abL}{3} & C_{44}2ab & -C_{11}\frac{4ab^3}{3L}+C_{44}\frac{2abL}{3} & -C_{44}2ab & 0 & 0 \\[2ex]
0 & 0 & 0 & 0 & 0 & 0 & C_{44}2ab & C_{44}\frac{4ab}{L} & C_{44}2ab & -C_{44}\frac{4ab}{L} & 0 & 0 \\[2ex]
0 & 0 & 0 & 0 & 0 & 0 & -C_{11}\frac{4ab^3}{3L}+C_{44}\frac{2abL}{3} & C_{44}2ab & C_{11}\frac{4ab^3}{3L}+C_{44}\frac{4abL}{3} & -C_{44}2ab & 0 & 0 \\[2ex]
0 & 0 & 0 & 0 & 0 & 0 & -C_{44}2ab & -C_{44}\frac{4ab}{L} & -C_{44}2ab & C_{44}\frac{4ab}{L} & 0 & 0 \\[2ex]
0 & 0 & 0 & 0 & 0 & 0 & 0 & 0 & 0 & 0 & C_{44}\frac{4ab^3}{3L}+C_{44}\frac{4a^3b}{3L} & C_{44}\frac{4ab^3}{3L}-C_{44}\frac{4a^3b}{3L} \\[2ex]
0 & 0 & 0 & 0 & 0 & 0 & 0 & 0 & 0 & 0 & -C_{44}\frac{4ab^3}{3L}-C_{44}\frac{4a^3b}{3L} & C_{44}\frac{4ab^3}{3L}+C_{44}\frac{4a^3b}{3L}
\end{bmatrix} \tag{7.25}$$

The comprehensive procedure for deriving the matrix presented in Eq. (7.25) is detailed in Appendix F.1, and the step-by-step implementation is provided as a MATLAB script.

7.4 Recursive Notation Against Shape Functions

In this section the stiffness matrix of the beam is calculated using the recursive notation against shape functions. The displacements field of Eq. (7.14) can be written as follows.

$$
\begin{aligned}
u^0(y) &= N_i(y)u_i^0 \\
v^0(y) &= N_i(y)v_i^0 \\
w^0(y) &= N_i(y)w_i^0 \\
\phi_x(y) &= N_i(y)\phi_{xi} \\
\phi_y(y) &= N_i(y)\phi_{yi} \\
\phi_z(y) &= N_i(y)\phi_{zi}
\end{aligned}
\tag{7.26}
$$

where i is either 1 or 2, according to which node of the beam is considered.

The matrix $\mathbf{B}$ of Eq. (7.16) in this case is equal to

$$
\mathbf{B} = \begin{bmatrix} \mathbf{B}_1 & \mathbf{B}_2 \end{bmatrix}
\tag{7.27}
$$

where

$$
\mathbf{B}_1 = \begin{bmatrix}
N_{1,y} & 0 & xN_{1,y} & 0 & -zN_{1,y} & 0 \\
0 & -N_{1,y} & N_1 & 0 & 0 & zN_{1,y} \\
0 & 0 & 0 & N_{1,y} & -N_1 & -xN_{1,y}
\end{bmatrix}
\tag{7.28}
$$

and

$$
\mathbf{B}_2 = \begin{bmatrix}
N_{2,y} & 0 & xN_{2,y} & 0 & -zN_{2,y} & 0 \\
0 & -N_{2,y} & N_2 & 0 & 0 & zN_{2,y} \\
0 & 0 & 0 & N_{2,y} & -N_2 & -xN_{2,y}
\end{bmatrix}
\tag{7.29}
$$

Using the recursive notation, **B** matrix can be written as follows

$$
\mathbf{B}_i = \begin{bmatrix} N_{i,y} & 0 & x N_{i,y} & 0 & -z N_{i,y} & 0 \\ 0 & -N_{i,y} & N_i & 0 & 0 & z N_{i,y} \\ 0 & 0 & 0 & N_{i,y} & -N_i & -x N_{i,y} \end{bmatrix} \tag{7.30}
$$

In order to distinguish the variables from their virtual variation, the second index j is introduced

$$
\mathbf{B}_j = \begin{bmatrix} N_{j,y} & 0 & x N_{j,y} & 0 & -z N_{j,y} & 0 \\ 0 & -N_{j,y} & N_j & 0 & 0 & z N_{j,y} \\ 0 & 0 & 0 & N_{j,y} & -N_j & -x N_{j,y} \end{bmatrix} \tag{7.31}
$$

The stiffness the stiffness matrix can be written as follows

$$
\mathbf{K}_{ij} = \int_V \mathbf{B}_j^T \mathbf{C} \mathbf{B}_i \, dV = \int_V \begin{bmatrix} N_{j,y} & 0 & 0 \\ 0 & -N_{j,y} & 0 \\ x N_{j,y} & N_j & 0 \\ 0 & 0 & N_{j,y} \\ -z N_{j,y} & 0 & -N_j \\ 0 & z N_{j,y} & -x N_{j,y} \end{bmatrix} \begin{bmatrix} C_{11} N_{i,y} & 0 & C_{11} x N_{i,y} & 0 & -C_{11} z N_{i,y} & 0 \\ 0 & -C_{44} N_{i,y} & C_{44} N_i & 0 & 0 & z C_{44} N_{i,y} \\ 0 & 0 & 0 & C_{44} N_{i,y} & -C_{44} N_i & -x C_{44} N_{i,y} \end{bmatrix} dV \tag{7.32}
$$

By computing the matrix product, the previous relation becomes

$$
\mathbf{K}_{ij} = \int_V \begin{bmatrix} C_{11} N_{i,y} N_{j,y} & 0 & C_{11} x N_{i,y} N_{j,y} & 0 & -C_{11} z N_{i,y} N_{j,y} & 0 \\[1ex] 0 & C_{44} N_{i,y} N_{j,y} & -C_{44} N_i N_{j,y} & 0 & 0 & -C_{44} z N_{i,y} N_{j,y} \\[1ex] C_{11} x N_{i,y} N_{j,y} & -C_{44} N_j N_{i,y} & C_{11} x^2 N_{i,y} N_{j,y} + C_{44} N_i N_j & 0 & -C_{11} xz N_{i,y} N_{j,y} & C_{44} x N_j N_{i,y} \\[1ex] 0 & 0 & 0 & C_{44} N_{i,y} N_{j,y} & -C_{44} N_i N_{j,y} & -C_{44} x N_{i,y} N_{j,y} \\[1ex] -C_{11} z N_{i,y} N_{j,y} & 0 & -C_{11} xz N_{i,y} N_{j,y} & -C_{44} N_j N_{i,y} & C_{11} z^2 N_{i,y} N_{j,y} + C_{44} N_i N_j & C_{44} z N_j N_{i,y} \\[1ex] 0 & -C_{44} z N_{i,y} N_{j,y} & C_{44} z N_i N_{j,y} & -C_{44} x N_{i,y} N_{j,y} & C_{44} x N_i N_{j,y} & C_{44} z^2 N_{i,y} N_{j,y} + C_{44} x^2 N_{i,y} N_{j,y} \end{bmatrix} dV \tag{7.33}
$$

Considering the assumption in Eqs. (7.22), (7.33) can be written as follows

$$
\mathbf{K}_{ij} = \int_V
\begin{bmatrix}
C_{11}N_{i,y}N_{j,y} & 0 & 0 & 0 & 0 & 0 \\
0 & C_{44}N_{i,y}N_{j,y} & -C_{44}N_iN_{j,y} & 0 & 0 & 0 \\
0 & -C_{44}N_jN_{i,y} & \begin{matrix}C_{11}x^2N_{i,y}N_{j,y}+\\C_{44}N_iN_j\end{matrix} & 0 & 0 & 0 \\
0 & 0 & 0 & C_{44}N_{i,y}N_{j,y} & -C_{44}N_iN_{j,y} & 0 \\
0 & 0 & 0 & -C_{44}N_jN_{i,y} & \begin{matrix}C_{11}z^2N_{i,y}N_{j,y}+\\C_{44}N_iN_j\end{matrix} & 0 \\
0 & 0 & 0 & 0 & 0 & \begin{matrix}C_{44}z^2N_{i,y}N_{j,y}+\\C_{44}x^2N_{i,y}N_{j,y}\end{matrix}
\end{bmatrix}
dV
$$

$$(7.34)$$

The following submatrices can be evaluated by expanding the indices i and j from 1 to 2.

$$
\mathbf{K}_{11} = \int_V
\begin{bmatrix}
C_{11}N_{1,y}N_{1,y} & 0 & 0 & 0 & 0 & 0 \\
0 & C_{44}N_{1,y}N_{1,y} & -C_{44}N_1N_{1,y} & 0 & 0 & 0 \\
0 & -C_{44}N_1N_{1,y} & \begin{matrix}C_{11}x^2N_{1,y}N_{j,y}+\\C_{44}N_1N_1\end{matrix} & 0 & 0 & 0 \\
0 & 0 & 0 & C_{44}N_{1,y}N_{1,y} & -C_{44}N_1N_{1,y} & 0 \\
0 & 0 & 0 & -C_{44}N_1N_{1,y} & \begin{matrix}(C_{11}z^2N_{1,y}N_{1,y}+\\C_{44}N_1N_1\end{matrix} & 0 \\
0 & 0 & 0 & 0 & 0 & \begin{matrix}C_{44}z^2N_{1,y}N_{1,y}+\\C_{44}x^2N_{1,y}N_{1,y}\end{matrix}
\end{bmatrix}
dV
$$

$$(7.35)$$

$$
\mathbf{K}_{12} = \int_V
\begin{bmatrix}
C_{11} N_{1,y} N_{2,y} & 0 & 0 & 0 & 0 & 0 \\
0 & C_{44} N_{1,y} N_{2,y} & -C_{44} N_1 N_{2,y} & 0 & 0 & 0 \\
0 & -C_{44} N_2 N_{1,y} & \begin{array}{c} C_{11} x^2 N_{1,y} N_{2,y} + \\ C_{44} N_1 N_2 \\ + \\ C_{44} N_1 N_2) \end{array} & 0 & 0 & 0 \\
0 & 0 & 0 & C_{44} N_{1,y} N_{2,y} & -C_{44} N_1 N_{2,y} & 0 \\
0 & 0 & 0 & -C_{44} N_2 N_{1,y} & \begin{array}{c} C_{11} z^2 N_{1,y} N_{2,y} + \\ C_{44} N_1 N_2 \end{array} & 0 \\
0 & 0 & 0 & 0 & 0 & \begin{array}{c} C_{44} z^2 N_{1,y} N_{2,y} + \\ C_{44} x^2 N_{1,y} N_{2,y} \end{array}
\end{bmatrix}
dV
\tag{7.36}
$$

$$
\mathbf{K}_{21} = \int_V
\begin{bmatrix}
C_{11} N_{2,y} N_{1,y} & 0 & 0 & 0 & 0 & 0 \\
0 & C_{44} N_{2,y} N_{1,y} & -C_{44} N_2 N_{1,y} & 0 & 0 & 0 \\
0 & -C_{44} N_1 N_{2,y} & \begin{array}{c} C_{11} x^2 N_{2,y} N_{1,y} + \\ C_{44} N_2 N_1 \end{array} & 0 & 0 & 0 \\
0 & 0 & 0 & C_{44} N_{2,y} N_{1,y} & -C_{44} N_2 N_{1,y} & 0 \\
0 & 0 & 0 & -C_{44} N_1 N_{2,y} & \begin{array}{c} C_{11} z^2 N_{2,y} N_{1,y} + \\ C_{44} N_2 N_1 \end{array} & 0 \\
0 & 0 & 0 & 0 & 0 & \begin{array}{c} C_{44} z^2 N_{2,y} N_{1,y} + \\ C_{44} x^2 N_{2,y} N_{1,y} \end{array}
\end{bmatrix}
dV
\tag{7.37}
$$

$$\mathbf{K}_{22} = \int_{V} \begin{bmatrix} C_{11}N_{2,y}N_{2,y} & 0 & 0 & 0 & 0 & 0 \\ 0 & C_{44}N_{2,y}N_{2,y} & -C_{44}N_2N_{2,y} & 0 & 0 & 0 \\ 0 & -C_{44}N_2N_{2,y} & C_{11}x^2N_{2,y}N_{2,y}+ \atop C_{44}N_2N_2 & 0 & 0 & 0 \\ 0 & 0 & 0 & C_{44}N_{2,y}N_{2,y} & -C_{44}N_2N_{2,y} & 0 \\ 0 & 0 & 0 & -C_{44}N_2N_{2,y} & C_{11}z^2N_{2,y}N_{2,y}+ \atop C_{44}N_2N_2 & 0 \\ 0 & 0 & 0 & 0 & 0 & C_{44}z^2N_{2,y}N_{2,y}+ \atop C_{44}x^2N_{2,y}N_{2,y} \end{bmatrix} dV$$

$$(7.38)$$

Using these submatrices, the global stiffness matrix can be then written as follows

$$\mathbf{K} = \begin{bmatrix} \mathbf{K}_{11} & \mathbf{K}_{21} \\ \mathbf{K}_{12} & \mathbf{K}_{22} \end{bmatrix} \tag{7.39}$$

The complete stiffness matrix of the beam using the recursive notation against shape functions results then in the following expression

$$
\mathbf{K} =
\begin{bmatrix}
C_{11}N_{1,y}N_{1,y} & 0 & 0 & 0 & 0 & 0 & C_{11}N_{2,y}N_{1,y} & 0 & 0 & 0 & 0 & 0 \\[4pt]
0 & C_{44}N_{1,y}N_{1,y} & -C_{44}N_1 N_{1,y} & 0 & 0 & 0 & 0 & C_{44}N_{2,y}N_{1,y} & -C_{44}N_2 N_{1,y} & 0 & 0 & 0 \\[4pt]
0 & -C_{44}N_1 N_{1,y} & C_{11}x^2 N_{1,y}N_{j,y}+C_{44}N_1 N_1 & 0 & 0 & 0 & 0 & -C_{44}N_1 N_{2,y} & C_{11}x^2 N_{2,y}N_{1,y}+C_{44}N_2 N_1 & 0 & 0 & 0 \\[4pt]
0 & 0 & 0 & C_{44}N_{1,y}N_{1,y} & -C_{44}N_1 N_{1,y} & 0 & 0 & 0 & 0 & C_{44}N_{2,y}N_{1,y} & -C_{44}N_2 N_{1,y} & 0 \\[4pt]
0 & 0 & 0 & -C_{44}N_1 N_{1,y} & C_{11}z^2 N_{1,y}N_{1,y}+C_{44}N_1 N_1 & 0 & 0 & 0 & 0 & -C_{44}N_1 N_{2,y} & C_{11}z^2 N_{2,y}N_{1,y}+C_{44}N_2 N_1 & 0 \\[4pt]
0 & 0 & 0 & 0 & 0 & C_{44}z^2 N_{1,y}N_{1,y}+C_{44}x^2 N_{1,y}N_{1,y} & 0 & 0 & 0 & 0 & 0 & C_{44}z^2 N_{2,y}N_{1,y}+C_{44}x^2 N_{2,y}N_{1,y} \\[4pt]
C_{11}N_{1,y}N_{2,y} & 0 & 0 & 0 & 0 & 0 & C_{11}N_{2,y}N_{2,y} & 0 & 0 & 0 & 0 & 0 \\[4pt]
0 & C_{44}N_{1,y}N_{2,y} & -C_{44}N_1 N_{2,y} & 0 & 0 & 0 & 0 & C_{44}N_{2,y}N_{2,y} & -C_{44}N_2 N_{2,y} & 0 & 0 & 0 \\[4pt]
0 & -C_{44}N_2 N_{1,y} & C_{11}x^2 N_{1,y}N_{2,y}+C_{44}N_1 N_2 & 0 & 0 & 0 & 0 & -C_{44}N_2 N_{2,y} & C_{11}x^2 N_{2,y}N_{2,y}+C_{44}N_2 N_2 & 0 & 0 & 0 \\[4pt]
0 & 0 & 0 & -C_{44}N_{1,y}N_{2,y} & C_{44}N_1 N_{2,y} & 0 & 0 & 0 & 0 & -C_{44}N_{2,y}N_{2,y} & C_{44}N_2 N_{2,y} & 0 \\[4pt]
0 & 0 & 0 & -C_{44}N_2 N_{1,y} & C_{11}z^2 N_{1,y}N_{2,y}+C_{44}N_1 N_2 & 0 & 0 & 0 & 0 & -C_{44}N_2 N_{2,y} & C_{11}z^2 N_{2,y}N_{2,y}+C_{44}N_2 N_2 & 0 \\[4pt]
0 & 0 & 0 & 0 & 0 & C_{44}z^2 N_{1,y}N_{2,y}+C_{44}x^2 N_{1,y}N_{2,y} & 0 & 0 & 0 & 0 & 0 & C_{44}z^2 N_{2,y}N_{2,y}+C_{44}x^2 N_{2,y}N_{2,y}
\end{bmatrix}
\tag{7.40}
$$

Thus, it results in the following

$$\mathbf{K} = \begin{bmatrix}
C_{11}\frac{4ab}{L} & 0 & 0 & 0 & 0 & 0 & -C_{11}\frac{4ab}{L} & 0 & 0 & 0 & 0 & 0 \\[6pt]
0 & C_{44}\frac{4ab}{L} & C_{44}2ab & 0 & 0 & 0 & 0 & -C_{44}\frac{4ab}{L} & C_{44}2ab & 0 & 0 & 0 \\[6pt]
0 & C_{44}2ab & C_{11}\frac{4a^3b}{3L}+C_{44}\frac{4abL}{3} & 0 & 0 & 0 & 0 & -C_{44}2ab & -C_{11}\frac{4a^3b}{3L}+C_{44}\frac{2abL}{3}2 & 0 & 0 & 0 \\[6pt]
0 & 0 & 0 & C_{44}2ab & C_{44}\frac{4ab}{L} & 0 & 0 & 0 & 0 & C_{44}2ab & -C_{44}\frac{4ab}{L} & 0 \\[6pt]
0 & 0 & 0 & C_{44}2ab & C_{11}\frac{4ab^3}{3L}+C_{44}\frac{2abL}{3} & 0 & 0 & 0 & 0 & -C_{44}2ab & -C_{11}\frac{4ab^3}{3L}+C_{44}\frac{4abL}{3} & 0 \\[6pt]
0 & 0 & 0 & 0 & 0 & C_{44}\frac{4ab^3}{3L}+C_{44}\frac{4a^3b}{3L} & 0 & 0 & 0 & 0 & 0 & -C_{44}\frac{4ab^3}{3L}+{-C_{44}\frac{4a^3b}{3L}} \\[6pt]
-C_{11}\frac{4ab}{L} & 0 & 0 & 0 & 0 & 0 & C_{11}\frac{4ab}{L} & 0 & 0 & 0 & 0 & 0 \\[6pt]
0 & -C_{44}\frac{4ab}{L} & -C_{44}2ab & 0 & 0 & 0 & 0 & C_{44}\frac{4ab}{L} & -C_{44}2ab & 0 & 0 & 0 \\[6pt]
0 & C_{44}2ab & -C_{11}\frac{4a^3b}{3L}+C_{44}\frac{2abL}{3} & 0 & 0 & 0 & 0 & -C_{44}2ab & C_{11}\frac{4a^3b}{3L}+C_{44}\frac{4abL}{3} & 0 & 0 & 0 \\[6pt]
0 & 0 & 0 & -C_{44}2ab & -C_{44}\frac{4ab}{L} & 0 & 0 & 0 & 0 & -C_{44}2ab & C_{44}\frac{4ab}{L} & 0 \\[6pt]
0 & 0 & 0 & C_{44}2ab & -C_{11}\frac{4ab^3}{3L}+C_{44}\frac{4abL}{3} & 0 & 0 & 0 & 0 & -C_{44}2ab & C_{11}\frac{4ab^3}{3L}+C_{44}\frac{2abL}{3} & 0 \\[6pt]
0 & 0 & 0 & 0 & 0 & -C_{44}\frac{4ab^3}{3L}+{-C_{44}\frac{4a^3b}{3L}} & 0 & 0 & 0 & 0 & 0 & C_{44}\frac{4ab^3}{3L}+C_{44}\frac{4a^3b}{3L}
\end{bmatrix}$$

$$(7.41)$$

The comprehensive procedure for deriving the matrix presented in Eq. (7.41) is detailed in Appendix F.2, and the step-by-step implementation is provided as a MATLAB script.

7.5 Recursive Notation Against the Theory of Structure

In this section the stiffness matrix of the beam is calculated using the recursive notation against TOS. In this case, the displacements field is written introducing $\mathbf{s}_n = s_x, s_y, s_z$.

$$
\begin{aligned}
s_x &= u(x, y, z) = -u^0(y) + z\phi_y(y) \\
s_y &= v(x, y, z) = v^0(y) - z\phi_x(y) + x\phi_z(y) \\
s_z &= w(x, y, z) = w^0(y) - x\phi_y(y)
\end{aligned}
\tag{7.42}
$$

Then, the expansion functions F, are introduced, so that

$$
\begin{aligned}
s_x &= u(x, y, z) = F_{1x}u^0(y) + F_{2x}\phi_y(y) \\
s_y &= v(x, y, z) = F_{1y}v^0(y) + F_{2y}\phi_x(y) + F_{3y}\phi_z(y) \\
s_z &= w(x, y, z) = F_{1z}w^0(y) + F_{2z}\phi_y(y)
\end{aligned}
\tag{7.43}
$$

In this case

$$
\begin{aligned}
F_{1x} &= -1 & F_{2x} &= z & & \\
F_{1y} &= 1 & F_{2y} &= -z & F_{3y} &= x \\
F_{1z} &= 1 & F_{2z} &= -x & &
\end{aligned}
\tag{7.44}
$$

Equation (7.43) can be expressed in a generic form introducing the index τ which ranges from to the number of the terms in the expansion function. This case can be then expressed as

$$
s_n = F_{\tau n}s_{\tau n}
\tag{7.45}
$$

where

$$
\begin{aligned}
s_{1x} &= u^0(y) & s_{2x} &= \phi_y(y) & & \\
s_{1y} &= v^0(y) & s_{2y} &= \phi_x(y) & s_{3y} &= \phi_y(y) \\
s_{1z} &= w^0(y) & s_{2z} &= \phi_y(y) & &
\end{aligned}
\tag{7.46}
$$

where the generic index n introduced for the directions x, y and z. The same procedure can be used for the virtual variation, by introducing the index s

$$\delta s_n = F_{sn} s_{sn} \tag{7.47}$$

By applying the recursive notation against TOS, $\mathbf{B}_{\tau i}$ and $\mathbf{B}_{sj}$ can be written as follows

$$\mathbf{B}_{\tau i} = \mathbf{b}\,\mathbf{N} = \begin{bmatrix} 0 & F_{1y}\frac{\partial}{\partial y} & 0 & F_{2y}\frac{\partial}{\partial y} & 0 & F_{3y}\frac{\partial}{\partial y} \\[2mm] F_{1x}\frac{\partial}{\partial y} & 0 & 0 & 0 & F_{2x}\frac{\partial}{\partial y} & F_{3y,x} \\[2mm] 0 & 0 & F_{1z}\frac{\partial}{\partial y} & F_{2y,z} & F_{2z}\frac{\partial}{\partial y} & 0 \end{bmatrix} \begin{bmatrix} 0 & N_i & 0 & 0 & 0 & 0 \\ N_i & 0 & 0 & 0 & 0 & 0 \\ 0 & 0 & 0 & N_i & 0 & 0 \\ 0 & 0 & 0 & 0 & N_i & 0 \\ 0 & 0 & 0 & 0 & 0 & N_i \\ 0 & 0 & N_i & 0 & 0 & 0 \end{bmatrix} =$$

$$= \begin{bmatrix} F_{1y}N_{i,y} & 0 & F_{3y}N_{i,y} & 0 & F_{2y}N_{i,y} & 0 \\[3mm] 0 & F_{1x}N_{i,y} & F_{3y,x}N_i & 0 & 0 & F_{2x}N_{i,y} \\[3mm] 0 & 0 & 0 & F_{1z}N_{i,y} & F_{2y,z}N_i & F_{2z}N_{i,y} \end{bmatrix} \tag{7.48}$$

and

$$\mathbf{B}_{sj} = \begin{bmatrix} F_{1y}N_{j,y} & 0 & F_{3y}N_{j,y} & 0 & F_{2y}N_{j,y} & 0 \\[4mm] 0 & F_{1x}N_{j,y} & F_{3y,x}N_j & 0 & 0 & F_{2x}N_{j,y} \\[4mm] 0 & 0 & 0 & F_{1z}N_{j,y} & F_{2y,z}N_j & F_{2z}N_{j,y} \end{bmatrix} \tag{7.49}$$

The stiffness matrix can be expressed as follows

$$
\mathbf{K}_{ij}^{ts} = \int_V \mathbf{B}_j^T \mathbf{C} \mathbf{B}_i \, dV = \int_V \left(\begin{bmatrix} F_{1y}N_{j,y} & 0 & 0 \\ 0 & F_{1x}N_{j,y} & 0 \\ F_{3y}N_{j,y} & F_{3y,x}N_j & 0 \\ 0 & 0 & F_{1z}N_{j,y} \\ F_{2y}N_{j,y} & 0 & F_{2y,z}N_j \\ 0 & F_{2x}N_{j,y} & F_{2z}N_{j,y} \end{bmatrix} \begin{bmatrix} C_{11} & 0 & 0 \\ 0 & C_{44} & 0 \\ 0 & & 0C_{44} \end{bmatrix} \right.
$$

$$
\left. \begin{bmatrix} F_{1y}N_{i,y} & 0 & F_{3y}N_{i,y} & 0 & F_{2y}N_{i,y} & 0 \\ 0 & F_{1x}N_{i,y} & F_{3y,x}N_i & 0 & 0 & F_{2x}N_{i,y} \\ 0 & 0 & 0 & F_{1z}N_{i,y} & F_{2y,z}N_i & F_{2z}N_{i,y} \end{bmatrix} \right) dV \tag{7.50}
$$

By performing the matrix triple product, the previous matrix becomes

$$
\mathbf{K}_{ij} = \int_V \begin{bmatrix} F_{1y}F_{1y}C_{11} \\ N_{i,y}N_{j,y} & 0 & 0 & 0 & 0 & 0 \\[4pt] 0 & \begin{array}{c} F_{1x}F_{1x}C_{44} \\ N_{i,y}N_{j,y} \end{array} & \begin{array}{c} F_{1x}F_{3y,x}C_{44} \\ N_i N_{j,y} \end{array} & 0 & 0 & 0 \\[6pt] 0 & \begin{array}{c} F_{1x}F_{3y,x}C_{44} \\ N_j N_{i,y} \end{array} & \begin{array}{c} F_{3y}F_{3y}C_{11} \\ N_{i,y}N_{j,y} \\ + \\ F_{3y,x}F_{3y,x}C_{44} \\ N_i N_j \end{array} & 0 & 0 & 0 \\[6pt] 0 & 0 & 0 & \begin{array}{c} F_{1z}F_{1z}C_{44} \\ N_{i,y}N_{j,y} \end{array} & \begin{array}{c} F_{1z}F_{2y,z}C_{44} \\ N_i N_{j,y} \end{array} & 0 \\[6pt] 0 & 0 & 0 & \begin{array}{c} F_{1z}F_{2y,z}C_{44} \\ N_j N_{i,y} \end{array} & \begin{array}{c} F_{2y}F_{2y}C_{11} \\ N_{i,y}N_{j,y} \\ + \\ F_{2y,z}F_{2y,z}C_{44} \\ N_i N_j \end{array} & 0 \\[6pt] 0 & 0 & 0 & 0 & 0 & \begin{array}{c} F_{2x}F_{2x}C_{44} \\ N_{i,y}N_{j,y} \\ + \\ F_{2z}F_{2z}C_{44} \\ N_{i,y}N_{j,y} \end{array} \end{bmatrix} dV \tag{7.51}
$$

The complete stiffness matrix for the results then in the following expression

$$\mathbf{K} = \begin{bmatrix}
C_{11}N_{1,y}N_{1,y} & 0 & 0 & 0 & 0 & 0 & C_{11}N_{2,y}N_{1,y} & 0 & 0 & 0 & 0 & 0 \\
0 & C_{44}N_{1,y}N_{1,y} & -C_{44}N_1 N_{1,y} & 0 & 0 & 0 & 0 & C_{44}N_{2,y}N_{1,y} & -C_{44}N_2 N_{1,y} & 0 & 0 & 0 \\
0 & -C_{44}N_1 N_{1,y} & C_{11}x^2 N_{1,y}N_{j,y}+C_{44}N_1 N_1 & 0 & 0 & 0 & 0 & -C_{44}N_1 N_{2,y} & C_{11}x^2 N_{2,y}N_{1,y}+C_{44}N_2 N_1 & 0 & 0 & 0 \\
0 & 0 & 0 & C_{44}N_{1,y}N_{1,y} & -C_{44}N_1 N_{1,y} & 0 & 0 & 0 & 0 & C_{44}N_{2,y}N_{1,y} & -C_{44}N_2 N_{1,y} & 0 \\
0 & 0 & 0 & -C_{44}N_1 N_{1,y} & C_{11}z^2 N_{1,y}N_{1,y}+C_{44}N_1 N_1+C_{44}N_1 N_1 & 0 & 0 & 0 & 0 & -C_{44}N_1 N_{2,y} & C_{11}z^2 N_{2,y}N_{1,y}+C_{44}N_2 N_1+C_{44}N_2 N_1 & 0 \\
0 & 0 & 0 & 0 & 0 & C_{44}z^2 N_{1,y}N_{1,y}+C_{44}x^2 N_{1,y}N_{1,y} & 0 & 0 & 0 & 0 & 0 & C_{44}z^2 N_{2,y}N_{1,y}+C_{44}x^2 N_{2,y}N_{1,y} \\
C_{11}N_{1,y}N_{2,y} & 0 & 0 & 0 & 0 & 0 & C_{11}N_{2,y}N_{2,y} & 0 & 0 & 0 & 0 & 0 \\
0 & C_{44}N_{1,y}N_{2,y} & -C_{44}N_1 N_{2,y} & 0 & 0 & 0 & 0 & C_{44}N_{2,y}N_{2,y} & -C_{44}N_2 N_{2,y} & 0 & 0 & 0 \\
0 & -C_{44}N_2 N_{1,y} & C_{11}x^2 N_{1,y}N_{2,y}+C_{44}N_1 N_2 & 0 & 0 & 0 & 0 & -C_{44}N_2 N_{2,y} & C_{11}x^2 N_{2,y}N_{2,y}+C_{44}N_2 N_2 & 0 & 0 & 0 \\
0 & 0 & 0 & C_{44}N_{1,y}N_{2,y} & -C_{44}N_1 N_{2,y} & 0 & 0 & 0 & 0 & C_{44}N_{2,y}N_{2,y} & -C_{44}N_2 N_{2,y} & 0 \\
0 & 0 & 0 & -C_{44}N_2 N_{1,y} & C_{11}z^2 N_{1,y}N_{2,y}+C_{44}N_1 N_2 & 0 & 0 & 0 & 0 & -C_{44}N_2 N_{2,y} & C_{11}z^2 N_{2,y}N_{2,y}+C_{44}N_2 N_2 & 0 \\
0 & 0 & 0 & 0 & 0 & C_{44}z^2 N_{1,y}N_{2,y}+C_{44}x^2 N_{1,y}N_{2,y} & 0 & 0 & 0 & 0 & 0 & C_{44}z^2 N_{2,y}N_{2,y}+C_{44}x^2 N_{2,y}N_{2,y}
\end{bmatrix} \quad (7.52)$$

Thus, it results in the following

$$\mathbf{K} =
\begin{bmatrix}
C_{11}\frac{4ab}{L} & 0 & 0 & 0 & 0 & 0 & -C_{11}\frac{4ab}{L} & 0 & 0 & 0 & 0 & 0 \\[2mm]
0 & C_{44}\frac{4ab}{L} & C_{44}2ab & 0 & 0 & 0 & 0 & -C_{44}\frac{4ab}{L} & C_{44}2ab & 0 & 0 & 0 \\[2mm]
0 & C_{44}2ab & C_{11}\frac{4a^3b}{3L}+C_{44}\frac{4abL}{3} & 0 & 0 & 0 & 0 & -C_{44}2ab & -C_{11}\frac{4a^3b}{3L}+C_{44}\frac{2abL}{3}2 & 0 & 0 & 0 \\[2mm]
0 & 0 & 0 & C_{44}2ab & C_{44}\frac{4ab}{L} & 0 & 0 & 0 & 0 & C_{44}2ab & -C_{44}\frac{4ab}{L} & 0 \\[2mm]
0 & 0 & 0 & C_{44}2ab & C_{11}\frac{4ab^3}{3L}+C_{44}\frac{2abL}{3} & 0 & 0 & 0 & 0 & -C_{44}2ab & -C_{11}\frac{4ab^3}{3L}+C_{44}\frac{4abL}{3} & 0 \\[2mm]
0 & 0 & 0 & 0 & 0 & C_{44}\frac{4ab^3}{3L}+C_{44}\frac{4a^3b}{3L} & 0 & 0 & 0 & 0 & 0 & -C_{44}\frac{4ab^3}{3L}+C_{44}\frac{4a^3b}{3L} \\[2mm]
-C_{11}\frac{4ab}{L} & 0 & 0 & 0 & 0 & 0 & C_{11}\frac{4ab}{L} & 0 & 0 & 0 & 0 & 0 \\[2mm]
0 & -C_{44}\frac{4ab}{L} & -C_{44}2ab & 0 & 0 & 0 & 0 & C_{44}\frac{4ab}{L} & -C_{44}2ab & 0 & 0 & 0 \\[2mm]
0 & C_{44}2ab & -C_{11}\frac{4a^3b}{3L}+C_{44}\frac{2abL}{3} & 0 & 0 & 0 & 0 & -C_{44}2ab & C_{11}\frac{4a^3b}{3L}+C_{44}\frac{4abL}{3} & 0 & 0 & 0 \\[2mm]
0 & 0 & 0 & -C_{44}2ab & -C_{44}\frac{4ab}{L} & 0 & 0 & 0 & 0 & -C_{44}2ab & C_{44}\frac{4ab}{L} & 0 \\[2mm]
0 & 0 & 0 & C_{44}2ab & -C_{11}\frac{4ab^3}{3L}+C_{44}\frac{4abL}{3} & 0 & 0 & 0 & 0 & -C_{44}2ab & C_{11}\frac{4ab^3}{3L}+C_{44}\frac{2abL}{3} & 0 \\[2mm]
0 & 0 & 0 & 0 & 0 & -C_{44}\frac{4ab^3}{3L}-C_{44}\frac{4a^3b}{3L} & 0 & 0 & 0 & 0 & 0 & C_{44}\frac{4ab^3}{3L}+C_{44}\frac{4a^3b}{3L}
\end{bmatrix}
\tag{7.53}$$

The comprehensive procedure for deriving the matrix presented in Eq. (7.53) is detailed in Appendix F.3, and the step-by-step implementation is provided as a MATLAB script.

Chapter 8
Fundamental Nucleus Using Unified Formulation

Graphical Abstract

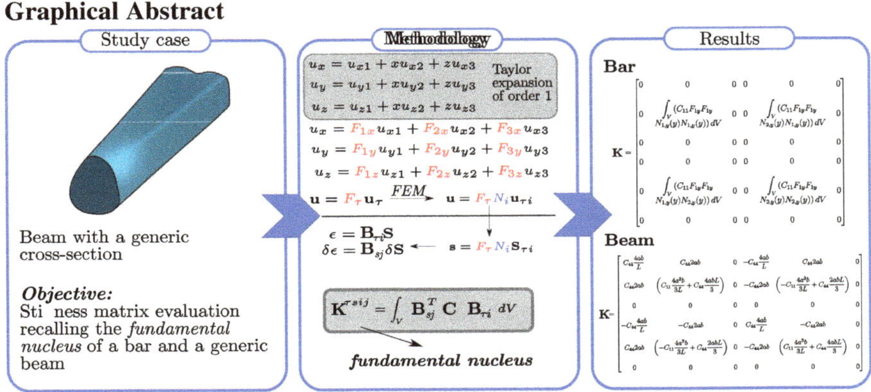

Introduction

In the previous chapters, the stiffness matrices of beams subjected to different deformations were calculated. These calculations are performed by using the displacement fields corresponding to the associated deformation. Figure 8.1 shows the considered deformations and the related displacement fields.

This chapter focuses on the challenge that, as the deformation of a structure becomes more complex, the corresponding displacement field also increases in complexity. This issue becomes particularly evident when the cross-sectional geometry of a beam changes along its axis, a scenario that necessitates the inclusion of additional terms in the displacement field and the application of higher-order expansion functions F. To address these requirements, in this chapter the displacement field of the cross-section is described through a complete linear expansion model based on a first-order Taylor-like polynomial. In this approach, the functions $F\tau x$, $F\tau y$ and $F\tau z$ are introduced to extend the solution from the nodal points to the beam's cross-section. These functions are associated with the Theory of Structures (TOS)

275

E. Carrera et al., *Implementation of Beam-Type Finite Elements Based on Carrera Unified Formulation*, https://doi.org/10.1007/978-3-031-95856-4_8

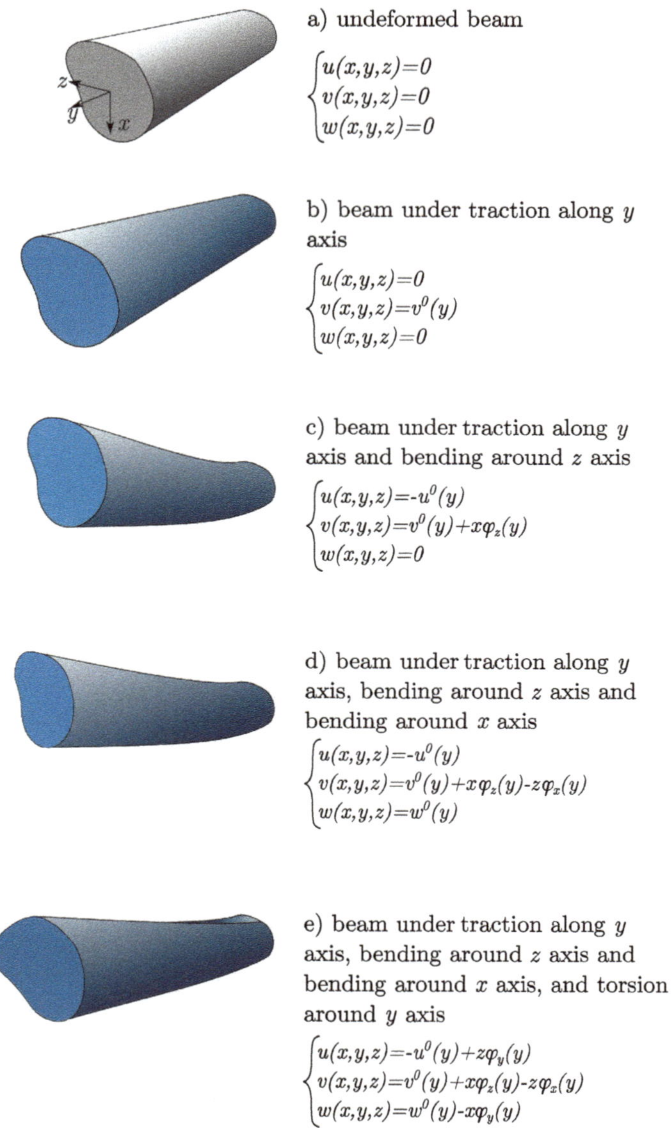

$$\begin{cases} u(x,y,z)=0 \\ v(x,y,z)=0 \\ w(x,y,z)=0 \end{cases}$$

$$\begin{cases} u(x,y,z)=0 \\ v(x,y,z)=v^0(y) \\ w(x,y,z)=0 \end{cases}$$

$$\begin{cases} u(x,y,z)=-u^0(y) \\ v(x,y,z)=v^0(y)+x\varphi_z(y) \\ w(x,y,z)=0 \end{cases}$$

$$\begin{cases} u(x,y,z)=-u^0(y) \\ v(x,y,z)=v^0(y)+x\varphi_z(y)-z\varphi_x(y) \\ w(x,y,z)=w^0(y) \end{cases}$$

$$\begin{cases} u(x,y,z)=-u^0(y)+z\varphi_y(y) \\ v(x,y,z)=v^0(y)+x\varphi_z(y)-z\varphi_x(y) \\ w(x,y,z)=w^0(y)-x\varphi_y(y) \end{cases}$$

Fig. 8.1 Displacement fields of a beam defined using classical beam theories

and reflect the specific kinematic attributes of the case study. By incorporating the relevant kinematic and constitutive relations, it is possible to derive the matrices $\mathbf{B}_{\tau i}$ and $\mathbf{B}_{sj}$ which connect the strain field to the displacements, as outlined in Sect. 8.1. Subsequently, the Principle of Virtual Displacements (PVD) is applied in Sect. 8.2 to obtain the matrix $\mathbf{K}^{\tau sij}$. This 3×3 matrix, referred to as the Fundamental Nucleus

(FN), appears in this chapter in its expanded form and serves as the essential building block of the Carrera Unified Formulation (CUF). The FN itself is composed of a limited number of mathematical expressions whose forms are independent of the chosen TOS, i.e from the employed F. Additional details regarding the computation of the FN of a bar and a beam are provided in Sects. 8.3 and 8.4, respectively. Finally, Sect. 8.5 presents a specific computation of the FN for a bar, along with illustrative numerical examples.

8.1 Structural Model of a Generic Beam

The complete linear expansion model involving a Taylor-like polynomial of first order is shown below

$$u_x = u_{x1} + x u_{x2} + z u_{x3}$$
$$u_y = u_{y1} + x u_{y2} + z u_{y3} \tag{8.1}$$
$$u_z = u_{z1} + x u_{z2} + z u_{z3}$$

In Eq. 8.1 there are nine unknowns. Introducing the s vector,

$$u(x, y, z) = s_x$$
$$v(x, y, z) = s_y \tag{8.2}$$
$$w(x, y, z) = s_z$$

and using TOS approximation thus introducing the expansion functions F, the previous relations can be written as follows

$$s_x = F_{\tau x} s_{\tau x}$$
$$s_y = F_{\tau y} s_{\tau y} \tag{8.3}$$
$$s_z = F_{\tau z} s_{\tau z}$$

Introducing the index n, Eq. (8.3) can be written as follows

$$s_n = F_{\tau n} s_{\tau n} \tag{8.4}$$

Using the same procedure for the virtual variations and introducing the index s, one has

$$\delta s_n = F_{sn} s_{sn} \tag{8.5}$$

Using the finite element approximation, and introducing the shape functions

$$\begin{aligned}
s_{\tau x} &= N_i(y) S_{\tau x} \\
s_{\tau y} &= N_i(y) S_{\tau y} \\
s_{\tau z} &= N_i(y) S_{\tau z}
\end{aligned} \tag{8.6}$$

From Eqs. (8.3) and (8.6) the following relations are derived

$$\begin{aligned}
s_x &= F_{\tau x} s_{\tau x} = F_{\tau x} N_i(y) S_{\tau x} \\
s_y &= F_{\tau y} s_{\tau y} = F_{\tau y} N_i(y) S_{\tau y} \\
s_z &= F_{\tau z} s_{\tau z} = F_{\tau z} N_i(y) S_{\tau z}
\end{aligned} \tag{8.7}$$

The geometrical relations are

$$\epsilon_{xx} = \frac{\partial u}{\partial x} = \frac{\partial}{\partial x} s_x$$

$$\epsilon_{yy} = \frac{\partial v}{\partial y} = \frac{\partial}{\partial y} s_y$$

$$\epsilon_{zz} = \frac{\partial w}{\partial z} = \frac{\partial}{\partial z} s_z$$

$$\epsilon_{xz} = \frac{\partial u}{\partial z} + \frac{\partial w}{\partial x} = \frac{\partial}{\partial z} s_x + \frac{\partial}{\partial x} s_z \tag{8.8}$$

$$\epsilon_{yz} = \frac{\partial v}{\partial z} + \frac{\partial w}{\partial y} = \frac{\partial}{\partial z} s_y + \frac{\partial}{\partial y} s_z$$

$$\epsilon_{xy} = \frac{\partial u}{\partial y} + \frac{\partial v}{\partial x} = \frac{\partial}{\partial y} s_x + \frac{\partial}{\partial x} s_y$$

Introducing Eq. (8.7) into Eq. (8.8), the following expression arises

$$\epsilon_{xx} = \frac{\partial u}{\partial x} = \frac{\partial}{\partial x} s_x = \frac{\partial}{\partial x}(F_{\tau x} N_i(y) S_{\tau x}) = F_{\tau x,x} N_i(y) S_{\tau x}$$

$$\epsilon_{yy} = \frac{\partial v}{\partial y} = \frac{\partial}{\partial y} s_y = \frac{\partial}{\partial y}(F_{\tau y} N_i(y) S_{\tau y}) = F_{\tau y} N_{i,y}(y) S_{\tau y}$$

$$\epsilon_{zz} = \frac{\partial w}{\partial z} = \frac{\partial}{\partial z} s_z = \frac{\partial}{\partial z}(F_{\tau z} N_i(y) S_{\tau z}) = F_{\tau z,z} N_i(y) S_{\tau z}$$

$$\epsilon_{xz} = \frac{\partial u}{\partial z} + \frac{\partial w}{\partial x} = \frac{\partial}{\partial z} s_x + \frac{\partial}{\partial x} s_z = \frac{\partial u}{\partial z} + \frac{\partial w}{\partial x}$$
$$= \frac{\partial}{\partial z}(F_{\tau x} N_i(y) S_{\tau x}) + \frac{\partial}{\partial x}(F_{\tau z} N_i(y) S_{\tau z}) = F_{\tau x,z} N_i(y) S_{\tau x} + F_{\tau z,x} N_i(y) S_{\tau z}$$

$$\epsilon_{yz} = \frac{\partial v}{\partial z} + \frac{\partial w}{\partial y} = \frac{\partial}{\partial z} s_y + \frac{\partial}{\partial y} s_z = \frac{\partial v}{\partial z} + \frac{\partial w}{\partial y}$$
$$= \frac{\partial}{\partial z}(F_{\tau y} N_i(y) S_{\tau y}) + \frac{\partial}{\partial z}(F_{\tau z} N_i(y) S_{\tau z}) = F_{\tau y,z} N_i(y) S_{\tau y} + F_{\tau z} N_{i,y}(y) S_{\tau z}$$

$$\epsilon_{xy} = \frac{\partial u}{\partial y} + \frac{\partial v}{\partial x} = \frac{\partial}{\partial y} s_x + \frac{\partial}{\partial x} s_y = \frac{\partial u}{\partial y} + \frac{\partial v}{\partial x}$$
$$= \frac{\partial}{\partial y}(F_{\tau x} N_i(y) S_{\tau x}) + \frac{\partial}{\partial x}(F_{\tau y} N_i(y) S_{\tau y}) = F_{\tau x} N_{i,y}(y) S_{\tau x} + F_{\tau y,x} N_i(y) S_{\tau y}$$

$$(8.9)$$

Equation (8.9) can be expressed in matrix form as follows

$$\boldsymbol{\epsilon} = \mathbf{B}_{\tau i} \mathbf{S} \tag{8.10}$$

where

$$\mathbf{S} = \left\{ \begin{array}{c} S_{\tau x} \\ S_{\tau y} \\ S_{\tau z} \end{array} \right\} \tag{8.11}$$

and

$$\mathbf{B}_{\tau i} = \begin{bmatrix} F_{\tau x,x} N_i(y) & 0 & 0 \\ 0 & F_{\tau y} N_{i,y}(y) & 0 \\ 0 & 0 & F_{\tau z,z} N_i(y) \\ F_{\tau x,z} N_i(y) & 0 & F_{\tau z,x} N_i(y) \\ 0 & F_{\tau y,z} N_i(y) & F_{\tau z} N_{i,y}(y) \\ F_{\tau x} N_{i,y}(y) & F_{\tau y,x} N_i(y) & 0 \end{bmatrix} \tag{8.12}$$

As far as the virtual variations are concerned, the following relations can be written

$$\delta\boldsymbol{\epsilon} = \mathbf{B}_{sj}\delta\mathbf{S} \tag{8.13}$$

where

$$\delta\mathbf{S} = \left\{ \begin{array}{c} \delta S_{\tau x} \\ \delta S_{\tau y} \\ \delta S_{\tau z} \end{array} \right\} \tag{8.14}$$

and

$$\mathbf{B}_{sj} = \begin{bmatrix} F_{sx,x}N_j(y) & 0 & 0 \\ 0 & F_{sy}N_{j,y}(y) & 0 \\ 0 & 0 & F_{sz,z}N_j(y) \\ F_{sx,z}N_j(y) & 0 & F_{sz,x}N_j(y) \\ 0 & F_{sy,z}N_j(y) & F_{sz}N_{j,y}(y) \\ F_{sx}N_{j,y}(y) & F_{sy,x}N_j(y) & 0 \end{bmatrix} \tag{8.15}$$

The constitutive relations, in the case of a complete strain tensor, are those between the strain components ϵ_{xx}, ϵ_{yy}, ϵ_{zz}, ϵ_{xz}, ϵ_{yz}, ϵ_{xy} and the stress components σ_{xx}, σ_{yy}, σ_{zz}, σ_{xz}, σ_{yz}, σ_{xy}. These relations are based on the material coefficient and the Hooke's law. For an isotropic material, the physical relationship between stress and strain components, in terms of stiffness coefficients, is the following

$$\boldsymbol{\sigma} = \mathbf{C}\boldsymbol{\epsilon} \tag{8.16}$$

where

$$\boldsymbol{\sigma} = \left\{ \begin{array}{c} \sigma_{xx} \\ \sigma_{yy} \\ \sigma_{zz} \\ \sigma_{xz} \\ \sigma_{yz} \\ \sigma_{xy} \end{array} \right\} \qquad \boldsymbol{\epsilon} = \left\{ \begin{array}{c} \epsilon_{xx} \\ \epsilon_{yy} \\ \epsilon_{zz} \\ \epsilon_{xz} \\ \epsilon_{yz} \\ \epsilon_{xy} \end{array} \right\} \tag{8.17}$$

$$\mathbf{C} = \begin{bmatrix} C_{11} & C_{12} & C_{12} & 0 & 0 & 0 \\ C_{21} & C_{11} & C_{12} & 0 & 0 & 0 \\ C_{21} & C_{21} & C_{11} & 0 & 0 & 0 \\ 0 & 0 & 0 & C_{44} & 0 & 0 \\ 0 & 0 & 0 & 0 & C_{44} & 0 \\ 0 & 0 & 0 & 0 & 0 & C_{44} \end{bmatrix} \tag{8.18}$$

For an isotropic material, $C_{12} = C_{21}$, C_{44} is equal to G and

$$C_{12} = \frac{\nu E}{(1+\nu)(1-2\nu)}. \tag{8.19}$$

8.2 Stiffness Matrix

Using PVD, the stiffness matrix is defined as follows

$$\mathbf{K} = \int_V \mathbf{B}^T \mathbf{C} \mathbf{B} dV \tag{8.20}$$

Considering Eqs. (8.12) and (8.15), (8.20) can be written using the recursive notation. $\mathbf{B}_{\tau i}$ and $\mathbf{B}_{sj}$ are expressed by means of four indexes that can vary, accordingly to the problem under study. Then, the stiffness matrix can be written as follows

$$\mathbf{K}^{\tau s i j} = \int_V \mathbf{B}_{sj}^T \mathbf{C} \mathbf{B}_{\tau i} dV \tag{8.21}$$

$\mathbf{K}^{\tau s i j}$ denotes the so-called Fundamental Nucleus (FN), which serves as the essential building block in the formation of the finite element (FE) matrix. Its structure can be adapted to different kinematic assumptions. Specifically, the indices τ and s account for the displacement kinematics across the beam's cross-section, while the indices i and j capture the displacement kinematics along the beam's axis (see Fig. 2.3). Thanks to this indexing scheme, the construction of the FE matrix does not depend on the order of the problem. Once the FN has been calculated for each relevant combination of indices, it is simply a matter of arranging every computed nucleus in the correct position to assemble the complete FE matrix. The process is akin to assembling a mosaic, where each FN represents a separate piece of the matrix. The expanded form of the matrix is shown below.

$$\mathbf{K}^{\tau s i j} = \int_V
\begin{bmatrix}
F_{sx,x} N_j(y) & 0 & 0 \\
0 & F_{sy} N_{j,y}(y) & 0 \\
0 & 0 & F_{sz,z} N_j(y) \\
F_{sx,z} N_j(y) & 0 & F_{sz,x} N_j(y) \\
0 & F_{sy,z} N_j(y) & F_{sz} N_{j,y}(y) \\
F_{sx} N_{j,y}(y) & F_{sy,x} N_j(y) & 0
\end{bmatrix}^T
\begin{bmatrix}
C_{11} & C_{12} & C_{12} & 0 & 0 & 0 \\
C_{21} & C_{11} & C_{12} & 0 & 0 & 0 \\
C_{21} & C_{21} & C_{11} & 0 & 0 & 0 \\
0 & 0 & 0 & C_{44} & 0 & 0 \\
0 & 0 & 0 & 0 & C_{44} & 0 \\
0 & 0 & 0 & 0 & 0 & C_{44}
\end{bmatrix}
\begin{bmatrix}
F_{\tau x,x} N_i(y) & 0 & 0 \\
0 & F_{\tau y} N_{i,y}(y) & 0 \\
0 & 0 & F_{\tau z,z} N_i(y) \\
F_{\tau x,z} N_i(y) & 0 & F_{\tau z,x} N_i(y) \\
0 & F_{\tau y,z} N_i(y) & F_{\tau z} N_{i,y}(y) \\
F_{\tau x} N_{i,y}(y) & F_{\tau y,x} N_i(y) & 0
\end{bmatrix}
dV \tag{8.22}$$

$$\mathbf{K}^{\tau s i j} = \int_V \begin{bmatrix} F_{sx,x} N_j(y) & 0 & 0 & F_{sx,z} N_j(y) & 0 & F_{sx} N_{j,y}(y) \\ 0 & F_{sy} N_{j,y}(y) & 0 & 0 & F_{sy,z} N_j(y) & F_{sy,x} N_j(y) \\ 0 & 0 & F_{sz,z} N_j(y) & F_{sz,x} N_j(y) & F_{sz} N_{j,y}(y) & 0 \end{bmatrix} \begin{bmatrix} C_{11} F_{\tau x,x} N_i(y) & C_{12} F_{\tau y} N_{i,y}(y) & C_{12} F_{\tau z,z} N_i(y) \\ C_{21} F_{\tau x,x} N_i(y) & C_{11} F_{\tau y} N_{i,y}(y) & C_{12} F_{\tau z,z} N_i(y) \\ C_{21} F_{\tau x,x} N_i(y) & C_{21} F_{\tau y} N_{i,y}(y) & C_{11} F_{\tau z,z} N_i(y) \\ C_{44} F_{\tau x,z} N_i(y) & 0 & C_{44} F_{\tau z,x} N_i(y) \\ 0 & C_{44} F_{\tau y,z} N_i(y) & C_{44} F_{\tau z} N_{i,y}(y) \\ C_{44} F_{\tau x} N_{i,y}(y) & C_{44} F_{\tau y,x} N_i(y) & 0 \end{bmatrix} dV \tag{8.23}$$

The FN of the stiffness matrix can be written in a generic form as follows

$$\mathbf{K}^{\tau s i j} = \begin{bmatrix} K_{xx}^{\tau s i j} & K_{xy}^{\tau s i j} & K_{xz}^{\tau s i j} \\ K_{yx}^{\tau s i j} & K_{yy}^{\tau s i j} & K_{yz}^{\tau s i j} \\ K_{zx}^{\tau s i j} & K_{zy}^{\tau s i j} & K_{zz}^{\tau s i j} \end{bmatrix} \tag{8.24}$$

where

$$K_{xx}^{\tau s i j} = \int_V [(C_{11} F_{\tau x,x} F_{sx,x} N_i(y) N_j(y)) + (C_{44} F_{\tau x,z} F_{sx,z} N_i(y) N_j(y)) + \\ + (C_{44} F_{\tau x} F_{sx} N_{i,y}(y) N_{j,y}(y))] dV$$

$$K_{xy}^{\tau s i j} = \int_V \left[(C_{12} F_{\tau y} F_{sx,x} N_{i,y}(y) N_j(y)) + (C_{44} F_{\tau y,x} F_{sx} N_i(y) N_{j,y}(y)) \right] dV$$

$$K_{xz}^{\tau s i j} = \int_V \left[(C_{12} F_{\tau z,z} F_{sx,x} N_i(y) N_j(y)) + (C_{44} F_{\tau z,x} F_{sx,z} N_i(y) N_j(y)) \right] dV$$

$$K_{yx}^{\tau s i j} = \int_V \left[(C_{21} F_{\tau x,x} F_{sy} N_i(y) N_{j,y}(y)) + (C_{44} F_{\tau x} F_{sy,x} N_{i,y}(y) N_j(y)) \right] dV$$

$$K_{yy}^{\tau s i j} = \int_V [(C_{11} F_{\tau y} F_{sy} N_{i,y}(y) N_{j,y}(y)) + (C_{44} F_{\tau y,z} F_{sy,z} N_i(y) N_j(y)) \\ + (C_{44} F_{\tau y,x} F_{sy,x} N_i(y) N_j(y))] dV \tag{8.25}$$

$$K_{yz}^{\tau s i j} = \int_V \left[(C_{12} F_{\tau z,z} F_{sy} N_i(y) N_{j,y}(y)) + (C_{44} F_{\tau z} F_{sy,z} N_{i,y}(y) N_j(y)) \right] dV$$

$$K_{zx}^{\tau s i j} = \int_V \left[(C_{21} F_{\tau x,x} F_{sz,z} N_i(y) N_j(y)) + (C_{44} F_{\tau x,z} F_{sz,x} N_i(y) N_j(y)) \right] dV$$

$$K_{zy}^{\tau s i j} = \int_V \left[(C_{21} F_{\tau y} F_{sz,z} N_{i,y}(y) N_j(y)) + (C_{44} F_{\tau y,z} F_{sz} N_i(y) N_{j,y}(y)) \right] dV$$

$$K_{zz}^{\tau s i j} = \int_V [(C_{11} F_{\tau z,z} F_{sz,z} N_i(y) N_j(y)) \, dV + (C_{44} F_{\tau z,x} F_{sz,x} N_i(y) N_j(y)) \, dV \\ + (C_{44} F_{\tau z} F_{sz} N_{i,y}(y) N_{j,y}(y))] dV.$$

8.3 Fundamental Nucleus for a Bar

The FN of a bar is evaluated here. The bar is a structural element only subjected to axial loads (i.e. a tensile or compressive load acting along the y axis of the bar). The displacement field for the case of a bar, as introduced in Eq. (2.1), is reported below

$$u(x, y, z) = 0$$
$$v(x, y, z) = v^0(y) \qquad\qquad (8.26)$$
$$w(x, y, z) = 0$$

Only the displacement $v^0(y)$ is present in the case of the bar. It is constant on the cross-section and varies along the bar with the y coordinate. The values of the expansion function F are the following

$$F_{\tau x} = 0$$
$$F_{\tau y} = F_{1y} = 1$$
$$F_{\tau z} = 0$$
$$F_{sx} = 0 \qquad\qquad (8.27)$$
$$F_{sy} = F_{1y} = 1$$
$$F_{sz} = 0$$

The expansion function F has only one term. Thus,

$$\tau = 1$$
$$s = 1 \qquad\qquad (8.28)$$

and considering a two-node B2 bar, the indices i and j assume the following values

$$i = \text{from } 1 \text{ to } 2$$
$$j = \text{from } 1 \text{ to } 2 \qquad\qquad (8.29)$$

As a result, the four indexes τ, s, i and j, assume the following values

$$(a) \quad \tau = 1 \;\; ; \;\; s = 1 \;\; ; \;\; i = 1 \;\; ; \;\; j = 1 \qquad\qquad (8.30)$$

$$(b) \quad \tau = 1 \;\; ; \;\; s = 1 \;\; ; \;\; i = 2 \;\; ; \;\; j = 1 \qquad\qquad (8.31)$$

$$(c) \quad \tau = 1 \;\; ; \;\; s = 1 \;\; ; \;\; i = 1 \;\; ; \;\; j = 2 \qquad\qquad (8.32)$$

$$(d) \quad \tau = 1 \;\; ; \;\; s = 1 \;\; ; \;\; i = 2 \;\; ; \;\; j = 2 \qquad\qquad (8.33)$$

The FNs to be calculated for the case of the bar are the following

$$\mathbf{K}^{1111} = \begin{bmatrix} K_{xx}^{1111} & K_{xy}^{1111} & K_{xz}^{1111} \\ K_{yx}^{1111} & K_{yy}^{1111} & K_{yz}^{1111} \\ K_{zx}^{1111} & K_{zy}^{1111} & K_{zz}^{1111} \end{bmatrix} \tag{8.34}$$

$$\mathbf{K}^{1121} = \begin{bmatrix} K_{xx}^{1121} & K_{xy}^{1121} & K_{xz}^{1121} \\ K_{yx}^{1121} & K_{yy}^{1121} & K_{yz}^{1121} \\ K_{zx}^{1121} & K_{zy}^{1121} & K_{zz}^{1121} \end{bmatrix} \tag{8.35}$$

$$\mathbf{K}^{1112} = \begin{bmatrix} K_{xx}^{1112} & K_{xy}^{1112} & K_{xz}^{1112} \\ K_{yx}^{1112} & K_{yy}^{1112} & K_{yz}^{1112} \\ K_{zx}^{1112} & K_{zy}^{1112} & K_{zz}^{1112} \end{bmatrix} \tag{8.36}$$

$$\mathbf{K}^{1122} = \begin{bmatrix} K_{xx}^{1122} & K_{xy}^{1122} & K_{xz}^{1122} \\ K_{yx}^{1122} & K_{yy}^{1122} & K_{yz}^{1122} \\ K_{zx}^{1122} & K_{zy}^{1122} & K_{zz}^{1122} \end{bmatrix} \tag{8.37}$$

Using the expressions of the FN of the Eq. (8.25) and the values of the expansion function F of Eq. (8.27), the following results are obtained

$$K_{xx}^{1111} = \int_V [(C_{11} F_{1x,x} F_{1x,x} N_1(y) N_1(y)) + (C_{44} F_{1x,z} F_{1x,z} N_1(y) N_1(y)) + \\ + (C_{44} F_{1x} F_{1x} N_{1,y}(y) N_{1,y}(y))] dV = 0$$

$$K_{xy}^{1111} = \int_V [(C_{12} F_{1y} F_{1x,x} N_{1,y}(y) N_1(y)) + (C_{44} F_{1y,x} F_{1x} N_1(y) N_{1,y}(y))] dV = 0$$

$$K_{xz}^{1111} = \int_V [(C_{12} F_{1z,z} F_{1x,x} N_1(y) N_1(y)) + (C_{44} F_{1z,x} F_{1x,z} N_1(y) N_1(y))] dV = 0$$

$$K_{yx}^{1111} = \int_V [(C_{21} F_{1x,x} F_{1y} N_1(y) N_{1,y}(y)) + (C_{44} F_{1x} F_{1y,x} N_{1,y}(y) N_1(y))] dV = 0 \tag{8.38}$$

$$K_{yy}^{1111} = \int_V [(C_{11}F_{1y}F_{1y}N_{1,y}(y)N_{1,y}(y))\,dV + (C_{44}F_{1y,z}F_{1y,z}N_1(y)N_1(y)) +$$
$$+ (C_{44}F_{1y,x}F_{1y,x}N_1(y)N_1(y))]dV = \int_V [(C_{11}F_{1y}F_{1y}N_{1,y}(y)N_{1,y}(y))\,dV]dV$$

$$K_{yz}^{1111} = \int_V \left[(C_{12}F_{1z,z}F_{1y}N_1(y)N_{1,y}(y)) + (C_{44}F_{1z}F_{1y,z}N_{1,y}(y)N_1(y))\right]dV = 0$$

$$K_{zx}^{1111} = \int_V \left[(C_{21}F_{1x,x}F_{1z,z}N_1(y)N_1(y)) + (C_{44}F_{1x,z}F_{1z,x}N_1(y)N_1(y))\right]dV = 0$$

$$K_{zy}^{1111} = \int_V \left[(C_{21}F_{1y}F_{1z,z}N_{1,y}(y)N_1(y)) + (C_{44}F_{1y,z}F_{1z}N_1(y)N_{1,y}(y))\right]dV = 0$$

$$K_{zz}^{1111} = \int_V [(C_{11}F_{1z,z}F_{1z,z}N_1(y)N_1(y)) + (C_{44}F_{1z,x}F_{1z,x}N_1(y)N_1(y)) +$$
$$+ (C_{44}F_{1z}F_{1z}N_{1,y}(y)N_{1,y}(y))]dV = 0$$

$$K_{xx}^{1121} = \int_V [(C_{11}F_{1x,x}F_{1x,x}N_2(y)N_1(y)) + (C_{44}F_{1x,z}F_{1x,z}N_2(y)N_1(y)) +$$
$$+ (C_{44}F_{1x}F_{1x}N_{2,y}(y)N_{1,y}(y))]dV = 0$$

$$K_{xy}^{1121} = \int_V \left[(C_{12}F_{1y}F_{1x,x}N_{2,y}(y)N_1(y)) + (C_{44}F_{1y,x}F_{1x}N_2(y)N_{1,y}(y))\right]dV = 0$$

$$K_{xz}^{1111} = \int_V \left[(C_{12}F_{1z,z}F_{1x,x}N_2(y)N_1(y)) + (C_{44}F_{1z,x}F_{1x,z}N_2(y)N_1(y))\right]dV = 0$$

$$K_{yx}^{1121} = \int_V \left[(C_{21}F_{1x,x}F_{1y}N_2(y)N_{1,y}(y)) + (C_{44}F_{1x}F_{1y,x}N_{2,y}(y)N_1(y))\right]dV = 0$$

$$K_{yy}^{1121} = \int_V [(C_{11}F_{1y}F_{1y}N_{2,y}(y)N_{1,y}(y)) + (C_{44}F_{1y,z}F_{1y,z}N_2(y)N_1(y)) +$$
$$+ (C_{44}F_{1y,x}F_{1y,x}N_2(y)N_1(y))]dV = \int_V [(C_{11}F_{1y}F_{1y}N_{2,y}(y)N_{1,y}(y))]dV$$

$$K_{yz}^{1121} = \int_V \left[(C_{12}F_{1z,z}F_{1y}N_2(y)N_{1,y}(y)) + (C_{44}F_{1z}F_{1y,z}N_{2,y}(y)N_1(y))\right]dV = 0$$

$$K_{zx}^{1121} = \int_V \left[(C_{21}F_{1x,x}F_{1z,z}N_2(y)N_1(y)) + (C_{44}F_{1x,z}F_{1z,x}N_2(y)N_1(y))\right]dV = 0$$

$$K_{zy}^{1121} = \int_V \left[(C_{21}F_{1y}F_{1z,z}N_{2,y}(y)N_1(y)) + (C_{44}F_{1y,z}F_{1z}N_2(y)N_{1,y}(y))\right]dV = 0$$

$$K_{zz}^{1121} = \int_V [(C_{11}F_{1z,z}F_{1z,z}N_2(y)N_1(y)) + (C_{44}F_{1z,x}F_{1z,x}N_2(y)N_1(y)) +$$
$$+ (C_{44}F_{1z}F_{1z}N_{2,y}(y)N_{1,y}(y))]dV = 0$$

$$(8.39)$$

$$K_{xx}^{1112} = \int_V [(C_{11} F_{1x,x} F_{1x,x} N_1(y) N_2(y)) + (C_{44} F_{1x,z} F_{1x,z} N_1(y) N_2(y)) + \\ + (C_{44} F_{1x} F_{1x} N_{1,y}(y) N_{2,y}(y))] dV = 0$$

$$K_{xy}^{1112} = \int_V \left[(C_{12} F_{1y} F_{1x,x} N_{1,y}(y) N_2(y)) + (C_{44} F_{1y,x} F_{1x} N_1(y) N_{2,y}(y)) \right] dV = 0$$

$$K_{xz}^{1112} = \int_V \left[(C_{12} F_{1z,z} F_{1x,x} N_2(y) N_1(y)) + (C_{44} F_{1z,x} F_{1x,z} N_1(y) N_2(y)) \right] dV = 0$$

$$K_{yx}^{1112} = \int_V \left[(C_{21} F_{1x,x} F_{1y} N_1(y) N_{2,y}(y)) + (C_{44} F_{1x} F_{1y,x} N_{1,y}(y) N_2(y)) \right] dV = 0$$

$$K_{yy}^{1112} = \int_V [(C_{11} F_{1y} F_{1y} N_{1,y}(y) N_{2,y}(y)) + (C_{44} F_{1y,z} F_{1y,z} N_1(y) N_2(y)) + \\ + (C_{44} F_{1y,x} F_{1y,x} N_1(y) N_2(y))] dV = \int_V [(C_{11} F_{1y} F_{1y} N_{1,y}(y) N_{2,y}(y))] dV$$

$$K_{yz}^{1112} = \int_V \left[(C_{12} F_{1z,z} F_{1y} N_1(y) N_{2,y}(y)) + (C_{44} F_{1z} F_{1y,z} N_{1,y}(y) N_2(y)) \right] dV = 0$$

$$K_{zx}^{1112} = \int_V \left[(C_{21} F_{1x,x} F_{1z,z} N_1(y) N_2(y)) + (C_{44} F_{1x,z} F_{1z,x} N_1(y) N_2(y)) \right] dV = 0$$

$$K_{zy}^{1112} = \int_V \left[(C_{21} F_{1y} F_{1z,z} N_{1,y}(y) N_2(y)) + (C_{44} F_{1y,z} F_{1z} N_1(y) N_{2,y}(y)) \right] dV = 0$$

$$K_{zz}^{1112} = \int_V [(C_{11} F_{1z,z} F_{1z,z} N_1(y) N_2(y)) + (C_{44} F_{1z,x} F_{1z,x} N_1(y) N_2(y)) + \\ + (C_{44} F_{1z} F_{1z} N_{1,y}(y) N_{2,y}(y))] dV = 0$$

$$K_{xx}^{1122} = \int_V [(C_{11} F_{1x,x} F_{1x,x} N_2(y) N_2(y)) + (C_{44} F_{1x,z} F_{1x,z} N_2(y) N_2(y)) + \\ + (C_{44} F_{1x} F_{1x} N_{2,y}(y) N_{2,y}(y))] dV = 0$$

$$K_{xy}^{1122} = \int_V \left[(C_{12} F_{1y} F_{1x,x} N_{2,y}(y) N_2(y)) + (C_{44} F_{1y,x} F_{1x} N_2(y) N_{2,y}(y)) \right] dV = 0$$

$$K_{xz}^{1122} = \int_V \left[(C_{12} F_{1z,z} F_{1x,x} N_2(y) N_2(y)) + (C_{44} F_{1z,x} F_{1x,z} N_2(y) N_2(y)) \right] dV = 0$$

$$K_{yx}^{1122} = \int_V \left[(C_{21} F_{1x,x} F_{1y} N_2(y) N_{2,y}(y)) + (C_{44} F_{1x} F_{1y,x} N_{2,y}(y) N_2(y)) \right] dV = 0$$

$$(8.40)$$

$$K_{yy}^{1122} = \int_V [(C_{11} F_{1y} F_{1y} N_{2,y}(y) N_{2,y}(y))\, dV + (C_{44} F_{1y,z} F_{1y,z} N_2(y) N_2(y)) +$$
$$+ (C_{44} F_{1y,x} F_{1y,x} N_2(y) N_2(y))] dV = \int_V \left[(C_{11} F_{1y} F_{1y} N_{2,y}(y) N_{2,y}(y)) \right] dV$$

$$K_{yz}^{1122} = \int_V \left[(C_{12} F_{1z,z} F_{1y} N_2(y) N_{2,y}(y)) + (C_{44} F_{1z} F_{1y,z} N_{2,y}(y) N_2(y)) \right] dV = 0$$

$$K_{zx}^{1122} = \int_V \left[(C_{21} F_{1x,x} F_{1z,z} N_2(y) N_2(y))\, dV + (C_{44} F_{1x,z} F_{1z,x} N_2(y) N_2(y)) \right] dV = 0$$

$$K_{zy}^{1122} = \int_V \left[(C_{21} F_{1y} F_{1z,z} N_{2,y}(y) N_2(y)) + (C_{44} F_{1y,z} F_{1z} N_2(y) N_{2,y}(y)) \right] dV = 0$$

$$K_{zz}^{1122} = \int_V [(C_{11} F_{1z,z} F_{1z,z} N_2(y) N_2(y)) + (C_{44} F_{1z,x} F_{1z,x} N_2(y) N_2(y)) +$$
$$+ (C_{44} F_{1z} F_{1z} N_{2,y}(y) N_{2,y}(y))] dV = 0$$

$$(8.41)$$

From the previous results, the following FNs are written

$$\mathbf{K}^{1111} = \begin{bmatrix} 0 & 0 & 0 \\ 0 & \left[\int_V (C_{11} F_{1y} F_{1y} N_{1,y}(y) N_{1,y}(y)) \, dV \right. & 0 \\ 0 & 0 & 0 \end{bmatrix} \quad (8.42)$$

$$\mathbf{K}^{1121} = \begin{bmatrix} 0 & 0 & 0 \\ 0 & \left[\int_V (C_{11} F_{1y} F_{1y} N_{2,y}(y) N_{1,y}(y)) \, dV \right. & 0 \\ 0 & 0 & 0 \end{bmatrix} \quad (8.43)$$

$$\mathbf{K}^{1112} = \begin{bmatrix} 0 & 0 & 0 \\ 0 & \left[\int_V (C_{11} F_{1y} F_{1y} N_{1,y}(y) N_{2,y}(y)) \, dV \right. & 0 \\ 0 & 0 & 0 \end{bmatrix} \quad (8.44)$$

$$\mathbf{K}^{1122} = \begin{bmatrix} 0 & 0 & 0 \\ 0 & \left[\int_V (C_{11} F_{1y} F_{1y} N_{2,y}(y) N_{2,y}(y)) \, dV \right. & 0 \\ 0 & 0 & 0 \end{bmatrix} \quad (8.45)$$

The stiffness matrix of a bar can be written, in a general form, as follows

$$\mathbf{K}_{bar} = \begin{bmatrix} K_{1111} & K_{1121} \\ K_{1112} & K_{2222} \end{bmatrix} \tag{8.46}$$

and

$$\mathbf{K}_{bar} = \begin{bmatrix} 0 & 0 & 0 & 0 & 0 & 0 \\ 0 & \int_V (C_{11} F_{1y} F_{1y} N_{1,y}(y) N_{1,y}(y))\, dV & 0 & 0 & \int_V (C_{11} F_{1y} F_{1y} N_{2,y}(y) N_{1,y}(y))\, dV & 0 \\ 0 & 0 & 0 & 0 & 0 & 0 \\ 0 & 0 & 0 & 0 & 0 & 0 \\ 0 & \int_V (C_{11} F_{1y} F_{1y} N_{1,y}(y) N_{2,y}(y))\, dV & 0 & 0 & \int_V (C_{11} F_{1y} F_{1y} N_{2,y}(y) N_{2,y}(y))\, dV & 0 \\ 0 & 0 & 0 & 0 & 0 & 0 \end{bmatrix} \tag{8.47}$$

The previous matrix is related to the following unknown vector

$$\begin{Bmatrix} 0 \\ v_1^0 \\ 0 \\ 0 \\ v_2^0 \\ 0 \end{Bmatrix} \tag{8.48}$$

The resultant matrix of Eq. (8.47) is equal to that of Eq. (2.57) of Chap. 2, considering $v = 0$ and C_{11} equal to E.

8.4 Fundamental Nucleus for a Complete Case Study

In this section, the FN of the most generic case is considered. The beam is considered as with arbitrary cross-section, and could have:

- Different kinematic descriptions on the nodes, i.e. the index i is added to the expansion function F, so that F^i represents the expansion function at the ith node.
- Different shape function for each term of the expansion of the displacement field, i.e. the index τ is added to the shape function N, so that N^τ represents the shape function of the τth term.

. The components of its displacements field can be written as follows

$$
\begin{aligned}
u(x, y, z) &= s_x \\
v(x, y, z) &= s_y \\
w(x, y, z) &= s_z
\end{aligned}
\tag{8.49}
$$

Using the recursive notation against TOS, the previous relation can be written as follows

$$
\begin{aligned}
s_x &= F^i_{\tau x} s^i_{\tau x} \\
s_y &= F^i_{\tau y} s^i_{\tau y} \\
s_z &= F^i_{\tau z} s^i_{\tau z}
\end{aligned}
\tag{8.50}
$$

In Eq. (8.50) the index i accounts for the different kinematics for different FE Nodes. In a more compact form, Eq. (8.50) is written as follows

$$
s_n = F^i_{\tau n} s^i_{\tau n}
\tag{8.51}
$$

Introducing the indexes s and j for the virtual variations,

$$
\delta s_n = F^j_{sn} s^j_{sn}
\tag{8.52}
$$

In order to consider different shape function $N_i(y)$ for different unknowns, the index τ is introduced for the shape functions. Then each different shape function can be defined as varying the value of τ in the expression of $N^\tau_i(y)$.

The vector of the unknowns for each FE node holds

$$
\mathbf{S} = \left\{
\begin{array}{c}
S^i_{\tau x} \\[4pt]
S^i_{\tau y} \\[4pt]
S^i_{\tau z}
\end{array}
\right\}
\tag{8.53}
$$

and using the FE approximation for TOS

$$
\begin{aligned}
s_{\tau x} &= N_i^\tau(y) S_{\tau x}^i \\
s_{\tau y} &= N_i^\tau(y) S_{\tau y}^i \\
s_{\tau z} &= N_i^\tau(y) S_{\tau z}^i
\end{aligned}
\tag{8.54}
$$

From Eqs. 8.50 and 8.54 the following relations are derived

$$
\begin{aligned}
s_x &= F_{\tau x}^i s_{\tau x} = F_{\tau x}^i N_i^\tau(y) S_{\tau x}^i \\
s_y &= F_{\tau y}^i s_{\tau y} = F_{\tau y}^i N_i^\tau(y) S_{\tau y}^i \\
s_z &= F_{\tau z}^i s_{\tau z} = F_{\tau z}^i N_i^\tau(y) S_{\tau z}^i.
\end{aligned}
\tag{8.55}
$$

8.4.1　Geometrical Relations

The geometrical relations between the strains and the displacements are

$$
\epsilon_{xx} = \frac{\partial u}{\partial x} = \frac{\partial}{\partial x} s_x
$$

$$
\epsilon_{yy} = \frac{\partial v}{\partial y} = \frac{\partial}{\partial y} s_y
$$

$$
\epsilon_{zz} = \frac{\partial w}{\partial z} = \frac{\partial}{\partial z} s_z
$$

$$
\epsilon_{xz} = \frac{\partial u}{\partial z} + \frac{\partial w}{\partial x} = \frac{\partial}{\partial z} s_x + \frac{\partial}{\partial x} s_z
\tag{8.56}
$$

$$
\epsilon_{yz} = \frac{\partial v}{\partial z} + \frac{\partial w}{\partial y} = \frac{\partial}{\partial z} s_y + \frac{\partial}{\partial y} s_z
$$

$$
\epsilon_{xy} = \frac{\partial u}{\partial y} + \frac{\partial v}{\partial x} = \frac{\partial}{\partial y} s_x + \frac{\partial}{\partial x} s_y
$$

Introducing Eq. (8.55) into Eq. (8.56), the following expressions are obtained

$$\epsilon_{xx} = \frac{\partial}{\partial x} s_x = \frac{\partial}{\partial x}(F^i_{\tau x} N^\tau_i(y) S^i_{\tau x}) = F^i_{\tau x,x} N^\tau_i(y) S^i_{\tau x}$$

$$\epsilon_{yy} = \frac{\partial}{\partial y} s_y = \frac{\partial}{\partial y}(F^i_{\tau y} N^\tau_i(y) S^i_{\tau y}) = F^i_{\tau y} N^\tau_{i,y}(y) S^i_{\tau y}$$

$$\epsilon_{zz} = \frac{\partial}{\partial z} s_z = \frac{\partial}{\partial z}(F^i_{\tau z} N^\tau_i(y) S^i_{\tau z}) = F^i_{\tau z,z} N^\tau_i y) S^i_{\tau z}$$

$$\epsilon_{xz} = \frac{\partial}{\partial z} s_x + \frac{\partial}{\partial x} s_z = \frac{\partial}{\partial z}(F^i_{\tau x} N^\tau_i(y) S^i_{\tau x}) + \frac{\partial}{\partial x}(F^i_{\tau z} N^\tau_i(y) S^i_{\tau z}) =$$

$$= F^i_{\tau x,z} N^\tau_i(y) S^i_{\tau x} + F^i_{\tau z,x} N^\tau_i(y) S^i_{\tau z}$$

$$\epsilon_{yz} = \frac{\partial}{\partial z} s_y + \frac{\partial}{\partial y} s_z = \frac{\partial}{\partial z}(F^i_{\tau y} N^\tau_i(y) S^i_{\tau y}) + \frac{\partial}{\partial z}(F^i_{\tau z} N^\tau_i(y) S^i_{\tau z}) =$$

$$= F^i_{\tau y,z} N^\tau_i(y) S^i_{\tau y} + F^i_{\tau z} N^\tau_{i,y}(y) S^i_{\tau z}$$

$$\epsilon_{xy} = \frac{\partial}{\partial y} s_x + \frac{\partial}{\partial x} s_y = \frac{\partial}{\partial y}(F^i_{\tau x} N^\tau_i(y) S^i_{\tau x}) + \frac{\partial}{\partial x}(F^i_{\tau y} N^\tau_i(y) S^i_{\tau y}) =$$

$$= F^i_{\tau x} N^\tau_{i,y}(y) S^i_{\tau x} + F^i_{\tau y,x} N^\tau_i(y) S^i_{\tau y}.$$

$$(8.57)$$

8.4.2 Matrices $\mathbf{B}_{\tau i}$ and $\mathbf{B}_{sj}$

Equation (8.57) can be written in matrix form as

$$\epsilon = \mathbf{B}_{\tau i}\mathbf{S} \tag{8.58}$$

where $\mathbf{S}$ is the vector of the unknowns

$$\mathbf{S} = \begin{Bmatrix} S^i_{\tau x} \\ S^i_{\tau y} \\ S^i_{\tau z} \end{Bmatrix} \tag{8.59}$$

and

$$\mathbf{B}_{\tau i} = \begin{bmatrix} F^i_{\tau x,x} N^\tau_i(y) & 0 & 0 \\[2ex] 0 & F^i_{\tau y} N^\tau_{i,y}(y) & 0 \\[2ex] 0 & 0 & F^i_{\tau z,z} N^\tau_i(y) \\[2ex] F^i_{\tau x,z} N^\tau_i(y) & 0 & F^i_{\tau z,x} N^\tau_i(y) \\[2ex] 0 & F^i_{\tau y,z} N^\tau_i(y) & F^i_{\tau z} N^\tau_{i,y}(y) \\[2ex] F^i_{\tau x} N^\tau_{i,y}(y) & F^i_{\tau y,x} N^\tau_i(y) & 0 \end{bmatrix} \tag{8.60}$$

The same procedure can be applied for the virtual variations.

$$\delta\boldsymbol{\epsilon} = \mathbf{B}_{sj}\delta\mathbf{S} \tag{8.61}$$

where

$$\delta\boldsymbol{\epsilon} = \begin{Bmatrix} \delta S^j_{\tau x} \\[1.5ex] \delta S^j_{\tau y} \\[1.5ex] \delta S^j_{\tau z} \end{Bmatrix} \tag{8.62}$$

and

$$\mathbf{B}_{sj} = \begin{bmatrix} F^j_{sx,x} N^s_j(y) & 0 & 0 \\[2ex] 0 & F^j_{sy} N^s_{j,y}(y) & 0 \\[2ex] 0 & 0 & F^j_{sz,z} N^s_j(y) \\[2ex] F^j_{sx,z} N^s_j(y) & 0 & F^j_{sz,x} N^s_j(y) \\[2ex] 0 & F^j_{sy,z} N^s_j(y) & F^j_{sz} N^s_{j,y}(y) \\[2ex] F^j_{sx} N^s_{j,y}(y) & F^j_{sy,x} N^s_j(y) & 0 \end{bmatrix}. \tag{8.63}$$

8.4.3 *Fundamental Nucleus*

Using the PVD approach, it can be written

$$\mathbf{K}^{\tau s i j} = \int_{v} \mathbf{B}_{sj}^{T} \mathbf{C} \mathbf{B}_{\tau i} dV \tag{8.64}$$

Considering the matrix $\mathbf{C}$ of Eq. (8.18) and considering the matrices $\mathbf{B}_{\tau i}$ and $\mathbf{B}_{sj}$, the matrix of Eq. (8.64), after evaluating the volume integrals, can be written as follows

$$\mathbf{K}^{\tau s i j} = \begin{bmatrix} K_{xx}^{\tau s i j} & K_{xy}^{\tau s i j} & K_{xz}^{\tau s i j} \\ K_{yx}^{\tau s i j} & K_{yy}^{\tau s i j} & K_{yz}^{\tau s i j} \\ K_{zx}^{\tau s i j} & K_{zy}^{\tau s i j} & K_{zz}^{\tau s i j} \end{bmatrix} \tag{8.65}$$

where the terms are the following

$$K_{xx}^{\tau s i j} = \int_{V} \left[\left(C_{11} F_{\tau x,x}^{i} F_{sx,x}^{j} N_{i}^{\tau}(y) N_{j}^{s}(y) \right) + \left(C_{44} F_{\tau x,z}^{i} F_{sx,z}^{j} N_{i}^{\tau}(y) N_{j}^{s}(y) \right) + \right.$$
$$\left. + \left(C_{44} F_{\tau x}^{i} F_{sx}^{j} N_{i,y}^{\tau}(y) N_{j,y}^{s}(y) \right) \right] dV$$

$$K_{xy}^{\tau s i j} = \int_{V} \left[\left(C_{12} F_{\tau y}^{i} F_{sx,x}^{j} N_{i,y}^{\tau}(y) N_{j}^{s}(y) \right) + \left(C_{44} F_{\tau y,x}^{i} F_{sx}^{j} N_{i}^{\tau}(y) N_{j,y}^{s}(y) \right) \right] dV$$

$$K_{xz}^{\tau s i j} = \int_{V} \left[\left(C_{12} F_{\tau z,z}^{i} F_{sx,x}^{j} N_{i}^{\tau}(y) N_{j}^{s}(y) \right) + \left(C_{44} F_{\tau z,x}^{i} F_{sx,z}^{j} N_{i}^{\tau}(y) N_{j}^{s}(y) \right) \right] dV$$

$$\tag{8.66}$$

$$K_{yx}^{\tau s i j} = \int_{V} \left[\left(C_{21} F_{\tau x,x}^{i} F_{sy}^{j} N_{i}^{\tau}(y) N_{j,y}^{s}(y) \right) + \left(C_{44} F_{\tau x}^{i} F_{sy,x}^{j} N_{i,y}^{\tau}(y) N_{j}^{s}(y) \right) \right] dV$$

$$K_{yy}^{\tau s i j} = \int_{V} \left[\left(C_{11} F_{\tau y}^{i} F_{sy}^{j} N_{i,y}^{\tau}(y) N_{j,y}^{s}(y) \right) + \left(C_{44} F_{\tau y,z}^{i} F_{sy,z}^{j} N_{i}^{\tau}(y) N_{j}^{s}(y) \right) + \right.$$
$$\left. + \left(C_{44} F_{\tau y,x}^{i} F_{sy,x}^{j} N_{i}^{\tau}(y) N_{j}^{s}(y) \right) \right] dV$$

$$K_{yz}^{\tau s i j} = \int_{V} \left[\left(C_{12} F_{\tau z,z}^{i} F_{sy}^{j} N_{i}^{\tau}(y) N_{j,y}^{s}(y) \right) + \left(C_{44} F_{\tau z}^{i} F_{sy,z}^{j} N_{i,y}^{\tau}(y) N_{j}^{s}(y) \right) \right] dV$$

$$K_{zx}^{\tau s i j} = \int_V \left[\left(C_{21} F_{\tau x,x}^i F_{sz,z}^j N_i^\tau(y) N_j^s(y) \right) + \left(C_{44} F_{\tau x,z}^i F_{sz,x}^j N_i^\tau(y) N_j^s(y) \right) \right] dV$$

$$K_{zy}^{\tau s i j} = \int_V \left[\left(C_{21} F_{\tau y}^i F_{sz,z}^j N_{i,y}^\tau(y) N_j^s(y) \right) + \left(C_{44} F_{\tau y,z}^i F_{sz}^j N_i^\tau(y) N_{j,y}^s(y) \right) \right] dV$$

$$(8.67)$$

$$K_{zz}^{\tau s i j} = \int_V \left[\left(C_{11} F_{\tau z,z}^i F_{sz,z}^j N_i^\tau(y) N_j^s(y) \right) dV + \left(C_{44} F_{\tau z,x}^i F_{sz,x}^j N_i^\tau(y) N_j^s(y) \right) dV + \right.$$

$$\left. + \left(C_{44} F_{\tau z}^i F_{sz}^j N_{i,y}^\tau(y) N_{j,y}^s(y) \right) \right] dV.$$

8.4.4   FN for a Beam Bending Around z Axis

Here, the specific FN for a beam bending around z axis is considered. The related expansion function F is considered constant for the two nodes, and similarly the shape function N is considered constant for each term of the expansion. In this case, the displacements field is as follows

$$
\begin{aligned}
s_x &= u(x, y, z) = -u^0(y) \\
s_y &= v(x, y, z) = x\phi_z(y) \\
s_z &= w(x, y, z) = 0
\end{aligned}
$$

$$(8.68)$$

Introducing the expansion functions F_τ, it can be written

$$
\begin{aligned}
F_{\tau x}^i &= -1 \\
F_{\tau y}^i &= x \\
F_{\tau z}^i &= 0
\end{aligned}
$$

$$(8.69)$$

and

$$
\begin{aligned}
F_{sx}^j &= -1 \\
F_{sy}^j &= x \\
F_{sz}^j &= 0
\end{aligned}
$$

$$(8.70)$$

Introducing Eqs. (8.69) and (8.70) into Eq. (8.67),

$$
K_{xx}^{\tau sij} = \int_V \left[\left(C_{11} F_{\tau x,x}^i F_{sx,x}^j N_i^\tau(y) N_j^s(y) \right) + \left(C_{44} F_{\tau x,z}^i F_{sx,z}^j N_i^\tau(y) N_j^s(y) \right) + \right.
$$
$$
\left. + \left(C_{44} F_{\tau x}^i F_{sx}^j N_{i,y}^\tau(y) N_{j,y}^s(y) \right) \right] dV = 0 + 0 + \int_V \left(C_{44} F_{\tau x}^i F_{sx}^j N_{i,y}^\tau(y) N_{j,y}^s(y) \right) dV =
$$
$$
= C_{44} A \int_y \left(N_{i,y}^\tau(y) N_{j,y}^s(y) \right) dy
$$

$$
K_{xy}^{\tau sij} = \int_V \left[\left(C_{12} F_{\tau y}^i F_{sx,x}^j N_{i,y}^\tau(y) N_j^s(y) \right) + \left(C_{44} F_{\tau y,x}^i F_{sx}^j N_i^\tau(y) N_{j,y}^s(y) \right) \right] dV
$$
$$
= 0 + \int_V \left(C_{44} F_{\tau y,x}^i F_{sx}^j N_i^\tau(y) N_{j,y}^s(y) \right) dV = -C_{44} A \int_y \left(N_i^\tau(y) N_{j,y}^s(y) \right) dy
$$

$$
K_{xz}^{\tau sij} = 0
$$

$$
K_{yx}^{\tau sij} = \int_V \left[\left(C_{21} F_{\tau x,x}^i F_{sy}^j N_i^\tau(y) N_{j,y}^s(y) \right) + \left(C_{44} F_{\tau x}^i F_{sy,x}^j N_{i,y}^\tau(y) N_j^s(y) \right) \right] dV
$$
$$
= 0 + \int_V \left(C_{44} F_{\tau x}^i F_{sy,x}^j N_{i,y}^\tau(y) N_j^s(y) \right) dV = -C_{44} A \int_y \left(N_{i,y}^\tau(y) N_j^s(y) \right) dy
$$

$$
K_{yy}^{\tau sij} = \int_V \left[\left(C_{11} F_{\tau y}^i F_{sy}^j N_{i,y}^\tau(y) N_{j,y}^s(y) \right) + \left(C_{44} F_{\tau y,z}^i F_{sy,z}^j N_i^\tau(y) N_j^s(y) \right) + \right.
$$
$$
\left. + \left(C_{44} F_{\tau y,x}^i F_{sy,x}^j N_i^\tau(y) N_j^s(y) \right) \right] dV = \int_V \left(C_{44} F_{\tau y,x}^i F_{sy,x}^j N_i^\tau(y) N_j^s(y) \right) dV + 0 +
$$
$$
+ \int_V \left(C_{11} F_{\tau y}^i F_{sy}^j N_{i,y}^\tau(y) N_{j,y}^s(y) \right) = C_{44} A \int_y \int_y \left(N_i^\tau(y) N_j^s(y) \right) dy +
$$
$$
+ C_{11} I_z \int_y \left(N_{i,y}^\tau(y) N_{j,y}^s(y) \right) dy
$$

$$
K_{yz}^{\tau sij} = 0
$$

$$
K_{zx}^{\tau sij} = 0
$$

$$
K_{zy}^{\tau sij} = 0
$$

$$
K_{zz}^{\tau sij} = 0
$$

$$
\tag{8.71}
$$

Therefore, the FN results as follows

$$
\mathbf{K}^{\tau sij} =
\begin{bmatrix}
\displaystyle\int_V \begin{pmatrix} C_{44} F^i_{\tau x} F^j_{sx} \\ N^\tau_{i,y}(y) N^s_{j,y}(y) \end{pmatrix} dV & \displaystyle\int_V \begin{pmatrix} C_{44} F^i_{\tau y,x} F_{sx} \\ {}^j N^\tau_i(y) N^s_{j,y}(y) \end{pmatrix} dV & 0 \\[2em]
\displaystyle\int_V \begin{pmatrix} C_{44} F^i_{\tau x} F^j_{sy,x} \\ N^\tau_{i,y}(y) N^s_j(y) \end{pmatrix} dV & \displaystyle\int_V \begin{bmatrix} \begin{pmatrix} C_{44} F^i_{\tau y,x} F^j_{sy,x} N^\tau_i(y) N^s_j(y) \end{pmatrix} + \\ + \begin{pmatrix} C_{11} F^i_{\tau y} F^j_{sy} N^\tau_{i,y}(y) N^s_{j,y}(y) \end{pmatrix} \end{bmatrix} dV & 0 \\[2em]
0 & 0 & 0
\end{bmatrix}
$$

$$(8.72)$$

From Eq. (8.72), the user can evaluate the FN based on the adopted formulation, considering

- different kinematics for different FE Nodes. In this case one must vary the indexes i and j for F^i_τ and F^j_s)
- different shape functions for different unknowns. In this case one must vary the indexes τ and s, for $N^\tau_i(y)$ and for the $N^\tau_j(y)$).

This approach is recalled to as Node-Dependent Kinematic (NDK), which is an approach that will be analyzed in details in Chap. 9. Here, constant expansion and shape functions are considered. Thus, the following relation apply for the shape functions

$$
\begin{aligned}
N^\tau_1(y) &= 1 - \frac{y}{L} & N^s_1(y) &= 1 - \frac{y}{L} \\[1em]
N^\tau_2(y) &= \frac{y}{L} & N^s_2(y) &= \frac{y}{L} \\[1em]
N^\tau_{1,y}(y) &= \frac{\partial}{\partial y}\left(1 - \frac{y}{L}\right) = -\frac{1}{L} & N^s_{1,y}(y) &= \frac{\partial}{\partial y}\left(1 - \frac{y}{L}\right) = -\frac{1}{L} \\[1em]
N^\tau_{2,y}(y) &= \frac{\partial}{\partial y}\left(\frac{y}{L}\right) = \frac{1}{L} & N^s_{2,y}(y) &= \frac{\partial}{\partial y}\left(\frac{y}{L}\right) = \frac{1}{L}
\end{aligned}
$$

$$(8.73)$$

and for the expansion functions

$$
\begin{aligned}
F^1_{\tau x} &= -1 \\
F^1_{\tau y} &= x \\
F^1_{\tau z} &= 0
\end{aligned}
$$

$$(8.74)$$

$$
\begin{aligned}
F^2_{\tau x} &= -1 \\
F^2_{\tau y} &= x \\
F^2_{\tau z} &= 0
\end{aligned}
$$

$$(8.75)$$

and

$$F^1_{sx} = -1$$
$$F^1_{sy} = x$$
$$F^1_{sz} = 0 \tag{8.76}$$

$$F^2_{sx} = -1$$
$$F^2_{sy} = x$$
$$F^2_{sz} = 0 \tag{8.77}$$

Looping the indexes i and j from 1 to 2, the FNs corresponding to the Eq. (8.72) result the following

$$\mathbf{K}^{\tau s 11} = \begin{bmatrix} C_{44}\dfrac{4ab}{L} & C_{44}2ab & 0 \\[2ex] C_{44}2ab & \left(C_{11}\dfrac{4a^3 b}{3L} + C_{44}\dfrac{4abL}{3}\right) & 0 \\[2ex] 0 & 0 & 0 \end{bmatrix} \tag{8.78}$$

$$\mathbf{K}^{\tau s 12} = \begin{bmatrix} -C_{44}\dfrac{4ab}{L} & C_{44}2ab & 0 \\[2ex] -C_{44}2ab & \left(-C_{11}\dfrac{4a^3 b}{3L} + C_{44}\dfrac{2abL}{3}\right) & 0 \\[2ex] 0 & 0 & 0 \end{bmatrix} \tag{8.79}$$

$$\mathbf{K}^{\tau s 21} = \begin{bmatrix} -C_{44}\dfrac{4ab}{L} & -C_{44}2ab & 0 \\[2ex] C_{44}2ab & \left(-C_{11}\dfrac{4a^3 b}{3L} + C_{44}\dfrac{2abL}{3}\right) & 0 \\[2ex] 0 & 0 & 0 \end{bmatrix} \tag{8.80}$$

$$\mathbf{K}^{\tau s 22} = \begin{bmatrix} C_{44}\dfrac{4ab}{L} & -C_{44}2ab & 0 \\[2ex] -C_{44}2ab & \left(C_{11}\dfrac{4a^3 b}{3L} + C_{44}\dfrac{4abL}{3}\right) & 0 \\[2ex] 0 & 0 & 0 \end{bmatrix} \tag{8.81}$$

The complete stiffness matrix, for a beam under bending around the z axis, obtained using the FNs shown in Eqs. (8.78) to (8.81), is the following

$$\mathbf{K} = \begin{bmatrix} k_{\tau s 11} & k_{\tau s 12} \\ k_{\tau s 21} & k_{\tau s 22} \end{bmatrix} \tag{8.82}$$

$$\mathbf{K} = \begin{bmatrix} C_{44}\dfrac{4ab}{L} & C_{44}2ab & 0 & -C_{44}\dfrac{4ab}{L} & C_{44}2ab & 0 \\[2ex] C_{44}2ab & \left(C_{11}\dfrac{4a^3b}{3L} + C_{44}\dfrac{4abL}{3}\right) & 0 & -C_{44}2ab & \left(-C_{11}\dfrac{4a^3b}{3L} + C_{44}\dfrac{2abL}{3}\right) & 0 \\[2ex] 0 & 0 & 0 & 0 & 0 & 0 \\[2ex] -C_{44}\dfrac{4ab}{L} & -C_{44}2ab & 0 & C_{44}\dfrac{4ab}{L} & -C_{44}2ab & 0 \\[2ex] C_{44}2ab & \left(-C_{11}\dfrac{4a^3b}{3L} + C_{44}\dfrac{2abL}{3}\right) & 0 & -C_{44}2ab & \left(C_{11}\dfrac{4a^3b}{3L} + C_{44}\dfrac{4abL}{3}\right) & 0 \\[2ex] 0 & 0 & 0 & 0 & 0 & 0 \end{bmatrix} \tag{8.83}$$

The previous matrix corresponds to that of the Eq. (3.62).

8.5 Numerical Examples

Although a first-order Taylor polynomial was used thus far, there is no theoretical limit on the order of the polynomial that can be employed. In practice, more complex geometries and higher degrees of deformation can be captured only by adopting higher-order Taylor expansions. Compared with the first-order expansion presented in Eq. (8.1), higher-order polynomials simply add additional terms, as described in Table 8.1.

The following section presents the results obtained by applying higher-order polynomials to analyze beams with complex geometries, highlighting in particular the influence of these expansions on shear locking behavior. The geometries considered, shown in Fig. 8.2, are expressed in millimeters. In each geometry, the transverse displacement is measured at point A, which is emphasized in the figure. These examples are taken from [1] and illustrate how higher-order polynomial expansions can offer enhanced accuracy and flexibility when dealing with sophisticated cross-sectional shapes and advanced structural analyses.

The material properties are reported in Table 8.2.

Tables 8.3, 8.4 and 8.5 show the effect of shear locking using B2, B3 and B4 finite elements and different orders of Taylor expansion, i.e. $N = 1, 5$ and 15.

Table 8.1 Taylor-like polynomials. N is the order of the Taylor polynomial and m is the number of terms in the expansions

N	M	F_τ
0	1	$F_1 = 1$
1	3	$F_2 = x \quad F_3 = z$
2	6	$F_4 = x^2 \quad F_5 = xz \quad F_6 = z^2$
3	10	$F_7 = x^3 \quad F_8 = x^2 z \quad F_9 = xz^2 \quad F_{10} = z^3$
...	...	...
N	$\frac{(N+1)(N+2)}{2}$	$F_{\frac{(N^2+N+2)}{2}} = x^N \quad \ldots\ldots \quad F_{\frac{(N+1)(N+2)}{2}} = z^N$

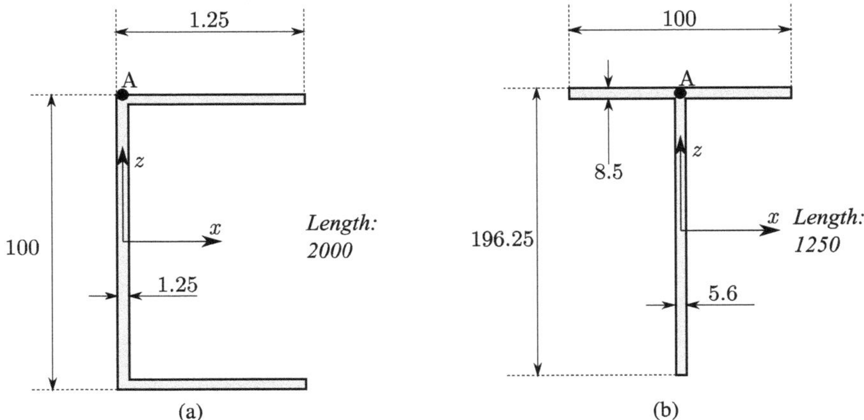

Fig. 8.2 Geometric properties of the C-shaped (**a**) and T-shaped (**b**) cross-section beam

Table 8.2 Material properties of the two considered beams, which cross-sections are described in Fig. 8.2a and b

Material parameters	Fig. 8.2a	Fig. 8.2b
E, MPa	216.4×10^3	210×10^3
ν	0.3	0.3
$\rho, \dfrac{\text{Kg}}{\text{m}^3}$	7850	7800

Table 8.3 Evaluation of the effect of shear locking on the vertical displacements using B2 FEs at the point A of the beam of C-shaped cross-section beam of Fig. 8.2a

DOF	TE	B2	FI	URI	SRI	MITC
18	1	1	0.24	5.45	5.45	5.45
126	5	1	0.26	6.56	5.88	5.82
816	15	1	3.09	12.08	10.56	10.61
99	1	10	5.57	7.23	7.23	7.23
693	5	10	5.73	8.32	8.32	8.63
4488	15	10	6.47	15.44	14.56	14.84
279	1	30	7.01	7.24	7.24	7.24
1953	5	30	7.87	8.34	8.25	8.25
12648	15	30	12.68	15.29	15.22	15.22
459	1	50	7.16	7.24	7.24	7.24
3213	5	50	8.14	8.34	8.29	8.29
20800	15	50	14.21	15.39	15.29	15.29

Table 8.4 Evaluation of the effect of shear locking on the vertical displacements using B3 FEs at the point A of the beam of C-shaped cross-section beam of Fig. 8.2a

DOF	TE	B3	FI	URI	SRI	MITC
27	1	1	5.71	7.24	7.24	7.24
189	5	1	6.23	8.34	7.14	7.14
1224	15	1	11.28	15.52	13.23	13.18
189	1	10	7.24	7.24	7.24	7.24
1323	5	10	8.23	8.34	8.24	8.24
8568	15	10	15.18	15.40	15.20	15.74
369	1	20	7.24	7.24	7.24	7.24
2583	5	20	8.29	8.34	8.29	8.24
16728	15	20	15.30	15.45	15.31	15.30
549	1	30	7.24	7.24	7.24	7.24
3843	5	30	8.31	8.338	8.31	8.31
24888	15	30	15.32	15.39	15.32	15.33

Figures 8.3 and 8.4 display the convergence rate and the distribution of the evaluated transverse displacement over the degrees of freedom for Taylor expansions of order 1 and 15, respectively. Clearly, the higher is the order of the finite elements (B2 to B3 to B4), the faster is the convergence of the displacement. The influence of the Taylor order is also shown, since with $N = 1$ the converged value is equal to 7.24 mm, whereas to accurately evaluate the displacement the order of the polynomial must be increased (with $N = 15$ the value is 15.30). Moreover, for $N = 15$, FULL, URI, SRI and MITC approaches lead to the same results.

Table 8.5 Evaluation of the effect of shear locking on the vertical displacements using B4 FEs at the point A of the C-shaped cross-section beam of Fig. 8.2a

DOF	TE	B4	FI	URI	SRI	MITC
36	1	1	7.24	7.24	7.24	7.24
252	5	1	7.64	8.34	7.69	7.69
1632	15	1	14.11	15.40	14.08	14.21
144	1	5	7.24	7.24	7.24	7.24
1008	5	5	8.22	8.34	8.23	8.23
6528	15	5	15.17	15.38	12.87	14.68
279	1	10	7.24	7.24	7.24	7.24
1953	5	10	8.28	8.34	8.29	8.29
12648	15	10	15.35	15.21	15.31	15.26
549	1	20	7.24	7.24	7.24	7.24
3843	5	20	8.31	8.34	8.32	8.83
24888	15	20	15.34	15.39	15.35	15.30

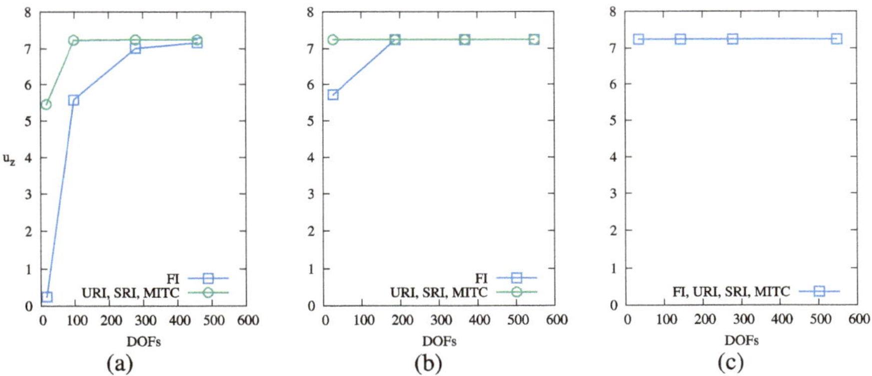

Fig. 8.3 Values of vertical displacement of the point A of the C-shaped cross-section beam of Fig. 8.2a TE1 theory and B2 (**a**), B3 (**b**) and B4 (**c**) elements

The same investigation is conducted for the T-shaped beam of Fig. 8.2b. The numerical values are reported in Tables 8.6, 8.7 and 8.8.

Finally, the evaluated transverse displacements are depicted in Fig. 8.5 for Taylor polynomial of order 1. The same conclusions as for the previous example can be drawn.

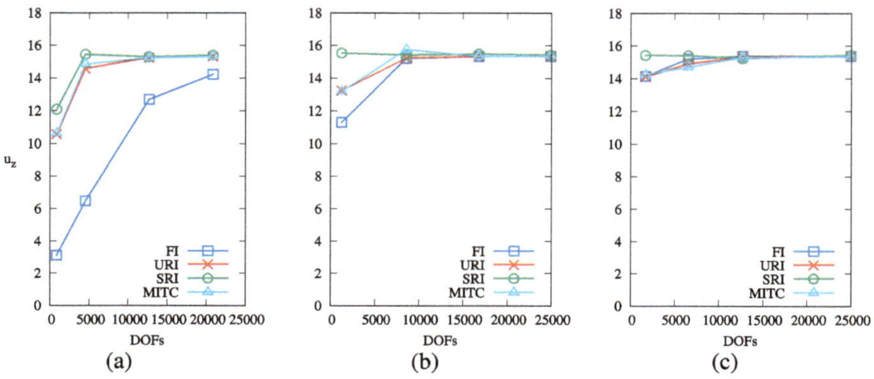

Fig. 8.4 Values of vertical displacement of the point A of the C-shaped cross-section beam of Fig. 8.2a using TE15 theory and B2 (**a**), B3 (**b**) and B4 (**c**) elements

Table 8.6 Evaluation of the effect of shear locking on the vertical displacements using B2 FEs at the point A of the beam of T-shaped cross-section beam of Fig. 8.2b

DOF	TE	B2	FI	URI	SRI	MITC
18	1	1	0.76	18.94	18.94	18.94
126	5	1	0.76	19.17	17.67	17.67
816	15	1	0.76	19.19	17.68	17.68
99	1	10	19.08	25.12	25.13	25.13
693	5	10	18.17	25.23	24.68	24.68
4488	15	10	18.83	25.26	24.70	24.70
279	1	30	24.32	25.18	25.18	25.18
1953	5	30	24.27	25.29	25.13	25.13
12648	15	30	24.28	25.31	20.14	25.14
459	1	50	24.87	25.18	25.18	25.18
3213	5	50	24.89	25.30	25.32	25.21
20800	15	50	24.90	25.32	25.22	25.22

Table 8.7 Evaluation of the effect of shear locking on the vertical displacements using B3 FEs at the point A of the beam of T-shaped cross-section beam of Fig. 8.2b

DOF	TE	B3	FI	URI	SRI	MITC
27	1	1	19.79	25.19	25.19	25.29
189	5	1	19.15	25.33	22.87	22.88
1224	15	1	19.17	25.35	22.88	22.88
189	1	10	25.18	25.29	25.19	25.19
1323	5	10	25.09	25.30	25.10	25.09
8568	15	10	25.17	25.33	25.11	25.11
369	1	20	25.21	25.10	25.19	25.19
2583	5	20	25.21	25.30	25.21	25.21
16728	15	20	25.22	25.31	25.23	25.23
549	1	30	25.29	25.29	25.19	25.19
3843	5	30	25.25	25.30	25.25	25.25
24888	15	30	25.26	25.32	25.26	25.26

Table 8.8 Evaluation of the effect of shear locking on the vertical displacements using B4 FEs at the point A of the beam of T-shaped cross-section beam of Fig. 8.2b

DOF	TE	B4	FI	URI	SRI	MITC
36	1	1	25.29	25.19	25.19	25.19
252	5	1	23.95	25.42	24.02	24.02
1632	15	1	23.97	25.44	24.04	24.04
144	1	5	25.19	25.19	25.19	25.19
1008	5	5	25.07	25.31	25.08	25.08
6528	15	5	25.09	25.35	25.10	25.10
279	1	10	25.19	25.19	25.19	25.19
1953	5	10	25.20	25.30	25.20	25.20
12648	15	10	25.22	25.32	25.20	25.22
549	1	20	25.19	25.19	25.19	25.19
3843	5	20	25.26	25.30	25.26	25.16
24888	15	20	25.28	25.32	25.28	25.28

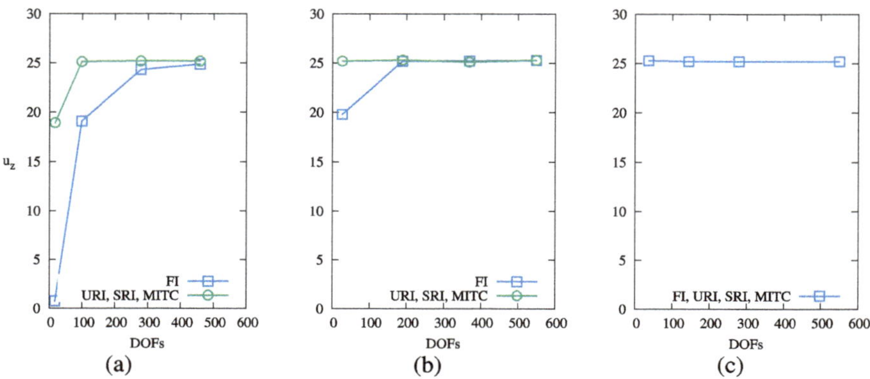

Fig. 8.5 Values of vertical displacement of the point A of the T-shaped cross-section beam of Fig. 8.2b using TE1 theory and B2 (**a**), B3 (**b**) and B4 (**c**) elements

Reference

1. Xu, X., Carrera, E., Augello, R., Daneshkhah, E., Yang, H.: Benchmarks for higher-order modes evaluation in the free vibration response of open thin-walled beams due to the cross-sectional deformations. Thin-Walled Structures **166**, 107965 (2021)

Chapter 9
Stiffness Matrix with Node-Dependent Kinematic

Graphical Abstract

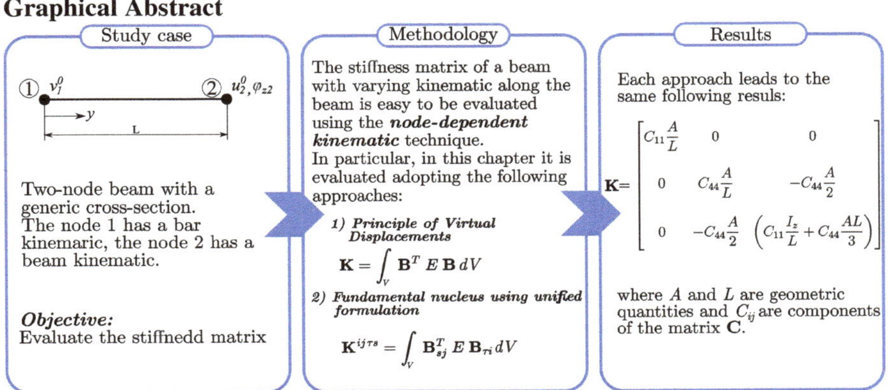

Introduction

This chapter introduces the concept of Node-Dependent Kinematics (NDK) in Finite Element (FE) modeling of beam structures. In a conventional beam model, the same kinematic assumptions are applied uniformly along the entire beam axis. However, there are situations where certain regions experience localized effects, such as stress concentrations or complex deformations, that necessitate a more sophisticated kinematic description than the other domains of the structure. In these cases, it may be advantageous to refine the kinematic assumptions at specific nodes, while retaining a simpler, lower-order formulation (e.g., a standard bar model) elsewhere. NDK approach enables this flexibility by allowing the kinematic description of each node to vary according to local requirements. When one portion of the beam is subjected to complicated local phenomena, a higher-order kinematic approximation can be employed at the corresponding nodes, capturing the intricate deformation more accurately. Meanwhile, other sections of the beam that do not require such refinement can be modeled using a lower-order kinematic description.

This chapter is structured as follows:

Section 9.1 provides an overview of the NDK approach and its theoretical background and advantages in structural modeling.

Section 9.2 presents the derivation of the stiffness matrix for a beam when different kinematic assumptions are applied at each node. The Principle of Virtual Displacements (PVD) is used to formulate the underlying equations.

Section 9.3 introduces the associated Fundamental Nucleus (FN) for a beam incorporating NDK.

Section 9.4 provides a numerical example that illustrates the practical benefits of the NDK concept, demonstrating how varying nodal kinematics can help overcoming the shear locking issue.

Through this approach, it becomes possible to selectively enrich the kinematic description only where necessary, striking a balance between precision and computational cost in beam analyses.

9.1 Structural Model of a Two-Node Beam with Different Kinematics

Consider a two-node beam element to illustrate the application of the Node-Dependent Kinematics (NDK) concept, as reported in Fig. 3.3.

In this figure, the numbers 1 and 2 refer to the node 1 and node 2 of the beam. The coordinate system of the beam is also indicated in the figure. v_1^0 is related to the displacements of node 1 along the beam y axis, U_2^0 is related to the displacements of node 2 along the beam transverse x axis and ϕ_{z2} refers to the rotation, of the cross-section of the beam at node 2, around the transverse z axis. The subscripts 1 and 2 refer to the beam nodes, while the exponent 0 refers to the degree of the displacement function over the bar cross-section (0 means that the displacement on the cross-section is constant and then do not change over this domain). The geometric characteristics and the dimensions of the beam are the same as those of the bar reported in the previous chapters. Node 1 adopts the kinematic assumptions of a bar, while node 2 employs the assumptions of a beam that can bend about the z axis. Consequently, node 1 can only stretch or contract along the element's longitudinal axis, whereas node 2 can also rotate around the z axis, thus capturing bending behavior in addition to axial deformation. By design, the displacement fields at the two nodes differ because they stem from different kinematic assumptions. Specifically, the displacement at node 1 is limited to an axial component (bar kinematics), whereas the displacement at node 2 incorporates both axial deformation and a rotational degree of freedom (beam kinematics). This selective enrichment at one node reflects the central idea behind NDK: different nodes of a structure can be modeled with distinct kinematics, optimizing the balance between computational cost and accuracy (Fig. 9.1).

The displacement field at node 1, governed solely by bar-like behavior, can be expressed as follows:

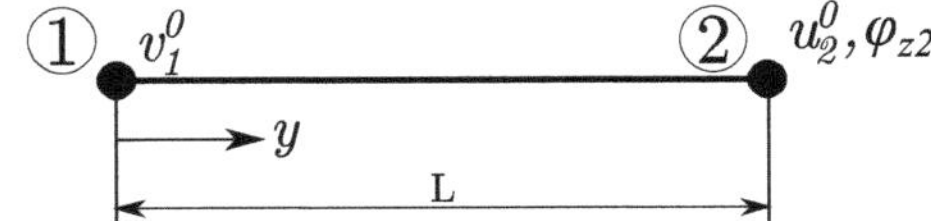

Fig. 9.1 Schematic representation of a two-node beam with NDK application

$$u(x, y, z) = 0$$
$$v(x, y, z) = v^0(y) \qquad (9.1)$$
$$w(x, y, z) = 0$$

while the displacement field for the node 2, governed solely by beam-like behavior, is the following

$$u(x, y, z) = -u^0(y)$$
$$v(x, y, z) = x\phi_z(y) \qquad (9.2)$$
$$w(x, y, z) = 0$$

The matrix form can be recalled to write the previous displacement fields, so that for node 1 it has

$$\mathbf{s}_1 = \{v\} = [1]\left\{v^0(y)\right\} \qquad (9.3)$$

and the node 2

$$\mathbf{s}_2 = \begin{Bmatrix} u \\ v \end{Bmatrix} = \begin{bmatrix} -1 & 0 \\ 0 & x \end{bmatrix} \begin{Bmatrix} u^0(y) \\ \phi_z(y) \end{Bmatrix} \qquad (9.4)$$

Introducing the vector s^0,

$$\mathbf{s}_1^0 = \left\{v^0(y)\right\} \qquad (9.5)$$

$$\mathbf{s}_2^0 = \begin{Bmatrix} u^0(y) \\ \phi_z(y) \end{Bmatrix} \qquad (9.6)$$

and, consequently

$$\mathbf{s}^0 = \begin{Bmatrix} \mathbf{s}_1^0 \\ \mathbf{s}_2^0 \end{Bmatrix} = \begin{Bmatrix} v^0(y) \\ u^0(y) \\ \phi_z(y) \end{Bmatrix} \qquad (9.7)$$

For the case study, the strain components of interest are ϵ_{yy} and ϵ_{xy}. The geometric relations are the following

$$\epsilon_{yy} = \frac{\partial v}{\partial y}$$

$$\epsilon_{xy} = \frac{\partial u}{\partial y} + \frac{\partial v}{\partial x}$$

(9.8)

By introducing the displacement components of Eqs. (9.1) and (9.2), the following expressions for the strain components are obtained

$$\epsilon_{yy} = \frac{\partial}{\partial y}\left(v^0(y) + x\phi_z(y)\right) = \frac{\partial v^0(y)}{\partial y} + x\frac{\partial \phi_z(y)}{\partial y}$$

$$\epsilon_{xy} = \frac{\partial}{\partial y}\left(-u^0(y)\right) + \frac{\partial}{\partial y}\left(v^0(y) + x\phi_z(y)\right) = -\frac{\partial u^0(y)}{\partial y} + 0 + \phi_z(y)$$

(9.9)

and the matricial form is

$$\epsilon = \left\{ \begin{array}{c} \epsilon_{yy} \\ \epsilon_{xy} \end{array} \right\} = \left[\begin{array}{ccc} \dfrac{\partial}{\partial y} & 0 & x\dfrac{\partial}{\partial y} \\ 0 & -\dfrac{\partial}{\partial y} & 1 \end{array} \right] \left\{ \begin{array}{c} v^0(y) \\ u^0(y) \\ \phi_z(y) \end{array} \right\}$$

(9.10)

From the previous relation, the following differential operator matrix $\mathbf{b}$ can be introduced

$$\mathbf{b} = \left[\begin{array}{ccc} \dfrac{\partial}{\partial y} & 0 & x\dfrac{\partial}{\partial y} \\ 0 & -\dfrac{\partial}{\partial y} & 1 \end{array} \right]$$

(9.11)

Equation (9.10) can then be written as follows

$$\epsilon = \mathbf{b}\,\mathbf{s}^0$$

(9.12)

For the variations of the strains, it can be written

$$\epsilon = \mathbf{b}\delta\mathbf{s}^0$$

(9.13)

The constitutive relations for the beam with different kinematic assumptions at the two nodes are those between the strain components ϵ_{yy} and ϵ_{xy} and the stress components σ_{yy} and σ_{xy}. These relations are based on the material coefficient and the Hooke's law. They are the following

$$\sigma_{yy} = C_{11}\epsilon_{yy}$$

$$\sigma_{xy} = C_{44}\epsilon_{xy} \tag{9.14}$$

Writing Eq. (3.7) in a matricial form

$$\sigma = \mathbf{C}\epsilon \tag{9.15}$$

where

$$\sigma = \left\{ \begin{array}{c} \sigma_{yy} \\ \sigma_{xy} \end{array} \right\} \tag{9.16}$$

$$\epsilon = \left\{ \begin{array}{c} \epsilon_{yy} \\ \epsilon_{xy} \end{array} \right\} \tag{9.17}$$

$$\mathbf{C} = \left[\begin{array}{cc} C_{11} & 0 \\ 0 & C_{44} \end{array} \right] \tag{9.18}$$

9.2 Stiffness Matrix via PVD

In this section the stiffness matrix of the previously described beam is calculated by the means of PVD. The displacement model of the Eqs. (9.1) and (9.2) are considered. The variation, along the beam length, of the displacement $v^0(y)$ can be expressed as a function of the displacements of the node 1 of the beam, i.e. v_1^0, and the shape functions of the node 1 of the beam. In the same manner the displacement $u^0(y)$ and the rotation ϕ_z can be expressed as a function of the displacements of the node 2 of the beam, i.e. u_2^0 and $\phi_z 2$, and the shape functions of the node 2 of the beam. The shape functions used for this purpose are based on Lagrange polynomials and are the following

$$N_1(y) = 1 - \frac{y}{L}$$

$$N_2(y) = \frac{y}{L} \tag{9.19}$$

The expressions that describe the variation along the beam axis of the displacements $u^0(y)$, $v^0(y)$ and the rotation $\phi_z(y)$, are the following

$$v^0(y) = N_1(y)V_1^0$$

$$u^0(y) = N_2(y)U_1^0 \tag{9.20}$$

$$\phi_z(y) = N_2(y)\phi_{z2}$$

Its matrix form is the following

$$\mathbf{s}^0 = \begin{bmatrix} N_1 & 0 & 0 \\ 0 & N_2 & 0 \\ 0 & 0 & N_2 \end{bmatrix} \begin{bmatrix} v_1^0 \\ u_2^0 \\ \phi_{z2} \end{bmatrix} \tag{9.21}$$

Introducing $\mathbf{S}$, equal to

$$\mathbf{S} = \begin{Bmatrix} V_1^0 \\ U_2^0 \\ \phi_{z2} \end{Bmatrix} \tag{9.22}$$

and the shape functions matrix $\mathbf{N}$ equal to

$$\mathbf{N} = \begin{bmatrix} N_1 & 0 & 0 \\ 0 & N_2 & 0 \\ 0 & 0 & N_2 \end{bmatrix} \tag{9.23}$$

the expression of Eq. (9.21) can then be written as follows

$$\mathbf{s}^0 = \mathbf{NS} \tag{9.24}$$

As far as the variation is concerned,

$$\delta\mathbf{s}^0 = \mathbf{N}\delta\mathbf{S} \tag{9.25}$$

Using Eqs. (9.24) and (9.25), Eqs. (9.12) and (9.13) can be written as follows

$$\epsilon = \mathbf{b}\,\mathbf{N}\delta\mathbf{S} \tag{9.26}$$

$$\delta\epsilon = \mathbf{b}\,\mathbf{N}\delta\mathbf{S} \tag{9.27}$$

Matrix $\mathbf{B}$ is equal to product of the differential operator matrix $\mathbf{b}$ by the shape functions matrix $\mathbf{N}$. The complete expression of the matrix $\mathbf{B}$ is the following

$$\mathbf{B} = \mathbf{b}\,\mathbf{N} = \begin{bmatrix} \dfrac{\partial}{\partial y} & 0 & x\dfrac{\partial}{\partial y} \\[2mm] 0 & -\dfrac{\partial}{\partial y} & 1 \end{bmatrix} \begin{bmatrix} N_1 & 0 & 0 \\ 0 & N_2 & 0 \\ 0 & 0 & N_2 \end{bmatrix} = \begin{bmatrix} N_{1,y} & 0 & x N_{2,y} \\ 0 & -N_{2,y} & N_2 \end{bmatrix} \tag{9.28}$$

where

$$N_{1,y} = \frac{\partial N_1(y)}{\partial y}$$

$$\tag{9.29}$$

$$N_{2,y} = \frac{\partial N_2(y)}{\partial y}$$

As demonstrated in the previous chapters, using PVD the stiffness matrix is defined as follows

$$\mathbf{K} = \int_V \mathbf{B}^T \mathbf{C} \mathbf{B}\, dV \tag{9.30}$$

Using the previous definitions of $\mathbf{B}$ and $\mathbf{C}$, the explicit form of the stiffness matrix $\mathbf{K}$ becomes

$$\mathbf{K} = \int_V \begin{bmatrix} N_{1,y} & 0 & x N_{2,y} \\ 0 & -N_{2,y} & N_2 \end{bmatrix}^T \begin{bmatrix} C_{11} & 0 \\ 0 & C_{44} \end{bmatrix} \begin{bmatrix} N_{1,y} & 0 & x N_{2,y} \\ 0 & -N_{2,y} & N_2 \end{bmatrix} dV \tag{9.31}$$

$$\mathbf{K} = \int_V \begin{bmatrix} N_{1,y} & 0 \\ 0 & -N_{2,y} \\ x N_{2,y} & N_2 \end{bmatrix} \begin{bmatrix} C_{11} N_{1,y} & 0 & C_{11} x N_{2,y} \\ 0 & -C_{44} N_{2,y} & C_{44} N_2 \end{bmatrix} dV \tag{9.32}$$

$$\mathbf{K} = \int_V \begin{bmatrix} C_{11} N_{1,y} N_{1,y} & 0 & C_{11} x N_{1,y} N_{2,y} \\ 0 & C_{44} N_{2,y} N_{2,y} & -C_{44} N_2 N_{2,y} \\ C_{11} x N_{1,y} N_{2,y} & -C_{44} N_2 N_{2,y} & C_{11} x^2 N_{2,y} N_{2,y} + C_{44} N_2 N_2 \end{bmatrix} dV \tag{9.33}$$

$$\mathbf{K} = \begin{bmatrix} \int_V \left(C_{11} N_{1,y} N_{1,y} \right) dV & 0 & \int_V \left(C_{11} x N_{1,y} N_{2,y} \right) dV \\[2ex] 0 & \int_V \left(C_{44} N_{2,y} N_{2,y} \right) dV & -\int_V \left(C_{44} N_2 N_{2,y} \right) dV \\[2ex] \int_V \left(C_{11} x N_{1,y} N_{2,y} \right) dV & -\int_V \left(C_{44} N_2 N_{2,y} \right) dV & \int_V \left(C_{11} x^2 N_{2,y} N_{2,y} + C_{44} N_2 N_2 \right) dV \end{bmatrix}$$

$$(9.34)$$

The previous matrix can be written, in a generic form, as follows

$$\mathbf{K} = \begin{bmatrix} k_{11} & k_{12} & k_{13} \\ k_{21} & k_{22} & k_{23} \\ k_{31} & k_{32} & k_{33} \end{bmatrix} \tag{9.35}$$

Introducing the following relations

$$dA = dx\,dz$$

$$A = \int_A dA = \int_{-a}^{a} dx \int_{-b}^{b} dz = 4ab$$

$$dV = dA\,dy = dx\,dz\,dy$$

$$V = \int_{-a}^{a} dx \int_{-b}^{b} dz \int_{0}^{L} dy = 4abL = AL \tag{9.36}$$

$$I_z = \int_A x^2 dA$$

$$I_z = \int_{-a}^{a} x^2 dx \int_{-b}^{b} dz = \frac{4a^3 b}{3}$$

and using the shape functions of Eq. 9.19 and their derivatives shown below

$$N_{1,y} = \frac{\partial}{\partial y}\left(1 - \frac{y}{L}\right) = -\frac{1}{L}$$

$$\text{(9.37)}$$

$$N_{2,y} = \frac{\partial}{\partial y}\left(\frac{y}{L}\right) = \frac{1}{L}$$

the volume integral of each term of matrix $\mathbf{K}$ of Eq. (9.35) can be written as follows

$$
\begin{aligned}
k_{11} &= \int_V (C_{11} N_{1,y} N_{1,y}) dV = C_{11} \int_{-a}^{a} dx \int_{-b}^{b} dz \int_0^L N_{1,y} N_{1,y} dy \\
&= C_{11} \int_{-a}^{a} dx \int_{-b}^{b} dz \int_0^L \left[\left(-\frac{1}{L}\right)\left(-\frac{1}{L}\right)\right] dy = C_{11} \int_{-a}^{a} dx \int_{-b}^{b} dz \int_0^L \frac{1}{L^2} dy \\
&= C_{11} \frac{A}{L} = C_{11} \frac{4ab}{L} = C_{11} \frac{A}{L}
\end{aligned}
$$

$$k_{12} = 0$$

$$
\begin{aligned}
k_{13} &= \int_V (C_{11} x N_{1,y} N_{2,y}) dV = C_{11} \int_{-a}^{a} x\, dx \int_{-b}^{b} dz \int_0^L N_{1,y} N_{2,y} dy \\
&= C_{11} \int_{-a}^{a} x\, dx \int_{-b}^{b} dz \int_0^L \left[\left(-\frac{1}{L}\right)\left(\frac{1}{L}\right)\right] dy = C_{11} \int_{-a}^{a} x\, dx \int_{-b}^{b} dz \int_0^L -\frac{1}{L^2} dy \\
&= 0
\end{aligned}
$$

$$k_{21} = 0$$

$$
\begin{aligned}
k_{22} &= \int_V (C_{44} N_{2,y} N_{2,y}) dV = C_{44} \int_{-a}^{a} dx \int_{-b}^{b} dz \int_0^L N_{2,y} N_{2,y} dy \\
&= C_{44} \int_{-a}^{a} dx \int_{-b}^{b} dz \int_0^L \left[\left(\frac{1}{L}\right)\left(\frac{1}{L}\right)\right] dy = C_{44} \int_{-a}^{a} dx \int_{-b}^{b} dz \int_0^L \frac{1}{L^2} dy \\
&= C_{44} \frac{A}{L} = C_{44} \frac{4ab}{L} = C_{44} \frac{A}{L}
\end{aligned}
$$

$$
\begin{aligned}
k_{23} &= \int_V (-C_{44} N_2 N_{2,y}) dV = -C_{44} \int_{-a}^{a} dx \int_{-b}^{b} dz \int_0^L N_2 N_{1,y} dy \\
&= -C_{44} \int_{-a}^{a} dx \int_{-b}^{b} dz \int_0^L \left[\left(\frac{y}{L}\right)\left(-\frac{1}{L}\right)\right] dy = -C_{44} \int_{-a}^{a} dx \int_{-b}^{b} dz \int_0^L \left(-\frac{y}{L} + \frac{y}{L^2}\right) dy \\
&= -C_{44} \frac{A}{2} = -C_{44} \frac{4ab}{2} = -C_{44} 2ab
\end{aligned}
$$

$$\text{(9.38)}$$

$$k_{31} = \int_V (C_{11} x N_{1,y} N_{2,y}) dV = C_{11} \int_{-a}^{a} x\,dx \int_{-b}^{b} dz \int_{0}^{L} N_{1,y} N_{2,y}\,dy$$

$$= C_{11} \int_{-a}^{a} x\,dx \int_{-b}^{b} dz \int_{0}^{L} \left[\left(-\frac{1}{L}\right) \left(\frac{1}{L}\right) \right] dy = C_{11} \int_{-a}^{a} x\,dx \int_{-b}^{b} dz \int_{0}^{L} -\frac{1}{L^2}\,dy$$

$$= 0$$

$$k_{32} = k_{23}$$

$$k_{33} = \int_V \left(C_{11} x^2 N_{2,y} N_{2,y} + C_{44} N_2 N_2 \right) dV$$

$$= C_{11} \int_{-a}^{a} x^2\,dx \int_{-b}^{b} dz \int_{0}^{L} N_{2,y} N_{2,y}\,dy + C_{44} \int_{-a}^{a} dx \int_{-b}^{b} dz \int_{0}^{L} N_2 N_2\,dy$$

$$= C_{11} \int_{-a}^{a} x^2\,dx \int_{-b}^{b} dz \int_{0}^{L} \left[\left(\frac{1}{L}\right)\left(\frac{1}{L}\right) \right] dy + C_{44} \int_{-a}^{a} dx \int_{-b}^{b} dz \int_{0}^{L} \left(\frac{y}{L}\right)\left(\frac{y}{L}\right) dy$$

$$= C_{11} \int_{-a}^{a} x^2\,dx \int_{-b}^{b} dz \int_{0}^{L} \left(\frac{1}{L^2}\right) dy + C_{44} \int_{-a}^{a} dx \int_{-b}^{b} dz \int_{0}^{L} \left(\frac{y^2}{L^2}\right) dy$$

$$= C_{11} \frac{I_z}{L} + C_{44} \frac{AL}{3} = C_{11} \frac{4a^3 b}{3L} + C_{44} \frac{4abL}{3}$$

Finally, the stiffness matrix of the beam with different kinematic assumptions at the nodes is the following

$$\mathbf{K} = \begin{bmatrix} C_{11} \dfrac{4ab}{L} & 0 & 0 \\[2em] 0 & C_{44} \dfrac{4ab}{L} & -C_{44} 2ab \\[2em] 0 & -C_{44} 2ab & \left(C_{11} \dfrac{4a^3 b}{3L} + C_{44} \dfrac{4abL}{3} \right) \end{bmatrix} \tag{9.39}$$

Considering Eq. (9.36) the previous matrix can be written as follows

$$\mathbf{K} = \begin{bmatrix} C_{11} \dfrac{A}{L} & 0 & -C_{11} \dfrac{S_z}{L} \\[2em] 0 & C_{44} \dfrac{A}{L} & -C_{44} \dfrac{A}{2} \\[2em] -C_{11} \dfrac{S_z}{L} & -C_{44} \dfrac{A}{2} & \left(C_{11} \dfrac{I_z}{L} + C_{44} \dfrac{AL}{3} \right) \end{bmatrix} \tag{9.40}$$

In the previous matrix, the value S_z is zero due to the symmetry of the cross-section of the beam around the x axis. Furthermore, the final form reads

$$
\mathbf{K} = \begin{bmatrix} C_{11}\dfrac{A}{L} & 0 & 0 \\[2em] 0 & C_{44}\dfrac{A}{L} & -C_{44}\dfrac{A}{2} \\[2em] 0 & -C_{44}\dfrac{A}{2} & \left(C_{11}\dfrac{I_z}{L} + C_{44}\dfrac{AL}{3} \right) \end{bmatrix}
\tag{9.41}
$$

9.3 Fundamental Nucleus for NDK

In Chap. 8, the Fundamental Nucleus (FN) was derived using the unified formulation, taking into account how the expansion functions F_τ^i and F_s^j depend on the indices i and j, as well as how the shape functions N_i^τ and N_j^s depend on the indices τ and s. In that context, the matrix was evaluated under the assumption that both the expansion and shape functions are constant.

In this chapter, by contrast, the dependence of the expansion functions on the specific nodes of the finite element is explicitly addressed. As a starting point for the calculations, Eq. 8.66 is employed. Here, the two indices i and j each range from 1 to the total number of structural nodes in the finite element. In the particular example discussed, a two-node element, these indices thus vary from 1 to 2.

9.3.1 Computation of F_τ^i and F_s^i and F_τ^j and F_s^j

To compute the expansion functions F_τ^i and F_s^j, it is necessary to apply the considerations presented in Sect. 9.4.2. In particular, the index i pertains to the expansion functions F_τ^i, which are derived with respect to the kinematic assumptions made for node i, whereas the expansion functions F_s^j are associated with the variations in the displacement field assumed for node j. In the specific example covered in this chapter, the kinematic descriptions for Nodes 1 and 2 differ, leading to distinct displacement fields at each node. As a result, the forms of the expansion functions F_τ^i and F_s^j will likewise differ, reflecting the Node-Dependent Kinematic (NDK) approach. Specifically, the displacement fields adopted for each node are as follows:

- The displacement field for node 1 (i.e. $i = 1$) is the following

$$
\begin{aligned}
u(x, y, z) &= 0 \\
v(x, y, z) &= v^0(y) \\
w(x, y, z) &= 0
\end{aligned}
\tag{9.42}
$$

For this displacement field, it can be written

$$
\begin{aligned}
F^i_{\tau x} &= F^1_{\tau x} = 0 \\
F^i_{\tau y} &= F^1_{\tau y} = 1 \\
F^i_{\tau z} &= F^1_{\tau z} = 0
\end{aligned}
\tag{9.43}
$$

The variations of the displacement field for node 1 (i.e. $j = 1$) are the following

$$
\begin{aligned}
\delta\left(u(x, y, z)\right) &= 0 \\
\delta\left(v(x, y, z)\right) &= \delta v^0(y) \\
\delta\left(w(x, y, z)\right) &= 0
\end{aligned}
\tag{9.44}
$$

For the previous variations of the displacement field, it can be written

$$
\begin{aligned}
F^j_{sx} &= F^1_{sx} = 0 \\
F^j_{sy} &= F^1_{sy} = 1 \\
F^j_{sz} &= F^1_{sz} = 0
\end{aligned}
\tag{9.45}
$$

- The displacement field for node 2 (i.e. $i = 2$) is the following

$$
\begin{aligned}
u(x, y, z) &= -u^0(y) \\
v(x, y, z) &= x\phi_z(y) \\
w(x, y, z) &= 0
\end{aligned}
\tag{9.46}
$$

For the previous displacement field, it can be written

$$
\begin{aligned}
F^i_{\tau x} &= F^2_{\tau x} = -1 \\
F^i_{\tau y} &= F^2_{\tau y} = x \\
F^i_{\tau z} &= F^2_{\tau z} = 0
\end{aligned}
\tag{9.47}
$$

The variations of the displacement field for node 1 (i.e. $j = 1$) are the following

$$
\begin{aligned}
\delta\left(u(x, y, z)\right) &= -\delta\left(u^0(y)\right) \\
\delta\left(v(x, y, z)\right) &= \delta\left(x\phi_z(y)\right) = x\delta\left(\phi_z(y)\right) \\
\delta\left(w(x, y, z)\right) &= 0
\end{aligned}
\tag{9.48}
$$

For the previous variations of the displacement field, it can be written

$$\begin{aligned}
F_{sx}^{j} &= F_{sx}^{2} = -1 \\
F_{sy}^{j} &= F_{sy}^{2} = x \\
F_{sz}^{j} &= F_{sz}^{2} = 0
\end{aligned} \tag{9.49}$$

9.3.2   FN for $i = 1$ and $j = 1$

The expansion functions hold the following value.

For $i = 1$ (see Eq. (9.43))

$$\begin{aligned}
F_{\tau x}^{i} &= F_{\tau x}^{1} = 0 \\
F_{\tau y}^{i} &= F_{\tau y}^{1} = 1 \\
F_{\tau z}^{i} &= F_{\tau z}^{1} = 0
\end{aligned} \tag{9.50}$$

For $j = 1$ (see Eq. (9.45))

$$\begin{aligned}
F_{sx}^{j} &= F_{sx}^{1} = 0 \\
F_{sy}^{j} &= F_{sy}^{1} = 1 \\
F_{sz}^{j} &= F_{sz}^{1} = 0
\end{aligned} \tag{9.51}$$

The terms of the 3×3, for $i = 1$ and $j = 1$ are expressed in the following

$$K_{xx}^{\tau s 11} = 0$$

$$K_{xy}^{\tau s 11} = 0$$

$$\tag{9.52}$$

$$K_{xz}^{\tau s 11} = 0$$

$$K_{yx}^{\tau s 11} = 0$$

$$
\begin{aligned}
K_{yy}^{\tau s11} &= \int_V \left[\left(C_{11} F_{\tau y}^1 F_{sy}^1 N_{1,y}(y) N_{1,y}(y) \right) + \left(C_{44} F_{\tau y,z}^1 F_{sy,z}^1 N_1(y) N_1(y) \right) + \right. \\
&= \left. \left(C_{44} F_{\tau y,x}^1 F_{sy,x}^1 N_1(y) N_1(y) \right) \right] dV = \int_V \left(C_{11} F_{\tau y}^1 F_{sy}^1 N_{1,y}(y) N_{1,y}(y) \right) dV + 0 + 0 = \\
&= C_{11} \int_{-a}^{a} (1 \cdot 1)\, dx \int_{-b}^{b} dz \int_0^L N_{1,y} N_{1,y}\, dy = C_{11} \int_{-a}^{a} dx \int_{-b}^{b} dz \int_0^L \left(-\frac{1}{L} \right) \left(-\frac{1}{L} \right) dy \\
&= C_{11} \int_{-a}^{a} dx \int_{-b}^{b} dz \int_0^L \frac{1}{L^2}\, dy = C_{11} \frac{A}{L} = C_{11} \frac{4ab}{L} = C_{11} \frac{A}{L}
\end{aligned}
$$

$$
K_{yz}^{\tau s11} = 0
$$

$$
K_{zx}^{\tau s11} = 0
$$

$$
K_{zy}^{\tau s11} = 0
$$

$$
K_{zz}^{\tau s11} = 0
$$

(9.53)

Finally, the FN for the previous state can be written as follows

$$
\mathbf{K}^{\tau s11} = C_{11} \frac{A}{L} \tag{9.54}
$$

9.3.3   FN for $i = 1$ and $j = 2$

The expansion functions have the following value. For $i = 1$ (see Eq. (9.43))

$$
\begin{aligned}
F_{\tau x}^i &= F_{\tau x}^1 = 0 \\
F_{\tau y}^i &= F_{\tau y}^1 = 1 \\
F_{\tau z}^i &= F_{\tau z}^1 = 0
\end{aligned} \tag{9.55}
$$

For $j = 2$ (see Eq. (9.49))

$$
\begin{aligned}
F_{sx}^j &= F_{sx}^2 = -1 \\
F_{sy}^j &= F_{sy}^2 = x \\
F_{sz}^j &= F_{sz}^2 = 0
\end{aligned} \tag{9.56}
$$

The terms of the 3×3, for $i = 1$ and $j = 2$ are the following

$$K_{xx}^{\tau s12} = 0$$

$$K_{xy}^{\tau s12} = 0$$

$$K_{xz}^{\tau s12} = 0$$

$$K_{yx}^{\tau s12} = 0$$

$$K_{yy}^{\tau s12} = \int_V \left[\left(C_{11} F_{\tau y}^1 F_{sy}^2 N_{1,y}(y) N_{2,y}(y) \right) + \left(C_{44} F_{\tau y,z}^1 F_{sy,z}^2 N_1(y) N_2(y) \right) + \right.$$

$$\left. + \left(C_{44} F_{\tau y,x}^1 F_{sy,x}^2 N_1(y) N_2(y) \right) \right] dV = \int_V \left(C_{11} F_{\tau y}^1 F_{sy}^2 N_{1,y}(y) N_{2,y}(y) \right) + 0 + 0 =$$

$$= C_{11} \int_{-a}^a (1 \cdot x)\, dx \int_{-b}^b dz \int_0^L N_{1,y} N_{2,y} dy = C_{11} \int_{-a}^a x\, dx \int_{-b}^b dz \int_0^L \left(-\frac{1}{L} \right) \left(-\frac{1}{L} \right) dy =$$

$$= C_{11} \int_{-a}^a x\, dx \int_{-b}^b dz \int_0^L \frac{1}{L^2} dy = 0$$

$$K_{yz}^{\tau s12} = 0$$

$$K_{zx}^{\tau s12} = 0$$

$$K_{zy}^{\tau s12} = 0$$

$$K_{zz}^{\tau s12} = 0$$

$$\tag{9.57}$$

Then, the FN can be written as follows

$$\mathbf{K}^{\tau s12} = \begin{bmatrix} K_{xx}^{\tau s12} \\ K_{yy}^{\tau s12} \end{bmatrix} = \begin{bmatrix} 0 \\ 0 \end{bmatrix} \tag{9.58}$$

9.3.4   *FN for i = 2 and j = 1*

For $i = 2$ (see Eq. (9.47)), the expansion function has the following value

$$
\begin{aligned}
F_{\tau x}^i &= F_{\tau x}^2 = -1 \\
F_{\tau y}^i &= F_{\tau y}^2 = x \\
F_{\tau z}^i &= F_{\tau z}^2 = 0
\end{aligned}
\tag{9.59}
$$

For $j = 1$ (see Eq. (9.45))

$$
\begin{aligned}
F_{sx}^j &= F_{sx}^2 = 0 \\
F_{sy}^j &= F_{sy}^2 = 1 \\
F_{sz}^j &= F_{sz}^2 = 0
\end{aligned}
\tag{9.60}
$$

The terms of the 3×3, for $i = 2$ and $j = 1$ are the following

$$ K_{xx}^{\tau s21} = 0 $$

$$ K_{xy}^{\tau s21} = 0 $$

$$ K_{xz}^{\tau s21} = 0 $$

$$ K_{yx}^{\tau s21} = 0 $$

$$
\begin{aligned}
K_{yy}^{\tau s21} &= \int_V \left[\left(C_{11} F_{\tau y}^2 F_{sy}^1 N_{2,y}(y) N_{1,y}(y) \right) + \left(C_{44} F_{\tau y,z}^2 F_{sy,z}^1 N_2(y) N_1(y) \right) + \right. \\
&\quad \left. + \left(C_{44} F_{\tau y,x}^2 F_{sy,x}^1 N_2(y) N_1(y) \right) \right] dV = \int_V \left(C_{11} F_{\tau y}^2 F_{sy}^1 N_{2,y}(y) N_{1,y}(y) \right) + 0 + 0 = \\
&= C_{11} \int_{-a}^{a} (1 \cdot x)\, dx \int_{-b}^{b} dz \int_0^L N_{2,y} N_{1,y} dy = C_{11} \int_{-a}^{a} x\, dx \int_{-b}^{b} dz \int_0^L \left(-\frac{1}{L} \right) \left(-\frac{1}{L} \right) dy = \\
&= C_{11} \int_{-a}^{a} x\, dx \int_{-b}^{b} dz \int_0^L \frac{1}{L^2} dy = 0
\end{aligned}
\tag{9.61}
$$

$$K_{yz}^{\tau s21} = 0$$

$$K_{zx}^{\tau s21} = 0$$

$$K_{zy}^{\tau s21} = 0 \tag{9.62}$$

$$K_{zz}^{\tau s21} = 0$$

Then, the FN can be written as follows

$$\mathbf{K}^{\tau s21} = \begin{bmatrix} K_{xx}^{\tau s21} & K_{yy}^{\tau s21} \end{bmatrix} = \begin{bmatrix} 0 & 0 \end{bmatrix} \tag{9.63}$$

9.3.5 FN for i = 2 and j = 2

For $i = 2$ (see Eq. (9.47)), the expansion functions have the following value.

$$\begin{aligned}
F_{\tau x}^{i} &= F_{\tau x}^{2} = -1 \\
F_{\tau y}^{i} &= F_{\tau y}^{2} = x \\
F_{\tau z}^{i} &= F_{\tau z}^{2} = 0
\end{aligned} \tag{9.64}$$

For $j = 1$ (see Eq. (9.49))

$$\begin{aligned}
F_{sx}^{j} &= F_{sx}^{2} = -1 \\
F_{sy}^{j} &= F_{sy}^{2} = x \\
F_{sz}^{j} &= F_{sz}^{2} = 0
\end{aligned} \tag{9.65}$$

The terms of the 3×3, for $i = 2$ and $j = 2$ are the following

$$
\begin{aligned}
K_{xx}^{\tau s22} &= \int_{V} \left[\left(C_{11} F_{\tau x,x}^{2} F_{sx,x}^{2} N_2(y) N_2(y) \right) + \left(C_{44} F_{\tau x,z}^{2} F_{sx,z}^{2} N_2(y) N_2(y) \right) + \right. \\
&\quad \left. + \left(C_{44} F_{\tau x}^{2} F_{sx}^{2} N_{2,y}(y) N_{2,y}(y) \right) \right] dV = 0 + 0 + \int_{V} \left(C_{44} F_{\tau x}^{2} F_{sx}^{2} N_{2,y}(y) N_{2,y}(y) \right) dV \\
&= C_{44} \int_{-a}^{a} (1 \cdot 1)\, dx \int_{-b}^{b} dz \int_{0}^{L} \left[\left(\frac{1}{L} \right) \left(\frac{1}{L} \right) \right] dy = C_{44} \int_{-a}^{a} dx \int_{-b}^{b} dz \int_{0}^{L} \frac{1}{L^2} dy \\
&= C_{44} \frac{A}{L} = C_{44} \frac{4ab}{L} = C_{44} \frac{A}{L}
\end{aligned} \tag{9.66}
$$

$$K_{xy}^{\tau s 22} = \int_V \left[\left(C_{12} F_{\tau y}^2 F_{sx,x}^2 N_{2,y}(y) N_2(y) \right) + \left(C_{44} F_{\tau y,x}^2 F_{sx}^j N_2(y) N_{2,y}(y) \right) \right] dV$$

$$= 0 + 0 + \int_V \left(C_{44} F_{\tau y,x}^2 F_{sx}^j N_2(y) N_{2,y}(y) \right) dV = C_{44} \int_{-a}^{a} (1 \cdot -1) \, dx \int_{-b}^{b} dz \int_0^L N_2 N_{2,y} dy$$

$$= -C_{44} \int_{-a}^{a} dx \int_{-b}^{b} dz \int_0^L \left[\left(\frac{y}{L} \right) \left(\frac{1}{L} \right) \right] dy = -C_{44} \int_{-a}^{a} dx \int_{-b}^{b} dz \int_0^L \left(\frac{y}{L^2} \right) dy$$

$$= -C_{44} \frac{4ab}{2} = -C_{44} 2ab = -C_{44} \frac{A}{2}$$

$$K_{xz}^{\tau s 22} = 0$$

$$K_{yx}^{\tau s 22} = \int_V \left[\left(C_{21} F_{\tau x,x}^2 F_{sy}^2 N_2(y) N_{2,y}(y) \right) + \left(C_{44} F_{\tau x}^2 F_{sy,x}^2 N_{2,y}(y) N_2(y) \right) \right] dV =$$

$$= 0 + \int_V \left(C_{44} F_{\tau x}^2 F_{sy,x}^2 N_{2,y}(y) N_2(y) \right) dV = C_{44} \int_{-a}^{a} (-1 \cdot 1) \, dx \int_{-b}^{b} dz \int_0^L N_2 N_{2,y} dy$$

$$= -C_{44} \int_{-a}^{a} dx \int_{-b}^{b} dz \int_0^L \left[\left(\frac{y}{L} \right) \left(\frac{1}{L} \right) \right] dy = -C_{44} \int_{-a}^{a} dx \int_{-b}^{b} dz \int_0^L \left(\frac{y}{L^2} \right) dy$$

$$= -C_{44} \frac{4ab}{2} = -C_{44} 2ab = -C_{44} \frac{A}{2}$$

$$K_{yy}^{\tau s 22} = \int_V \left[\left(C_{11} F_{\tau y}^2 F_{sy}^2 N_{2,y}(y) N_{2,y}(y) \right) + \left(C_{44} F_{\tau y,z}^2 F_{sy,z}^2 N_2(y) N_2(y) \right) + \right.$$

$$= \left(C_{44} F_{\tau y,x}^2 F_{sy,x}^2 N_2(y) N_2(y) \right) \Big] dV = \int_V \left(C_{11} F_{\tau y}^2 F_{sy}^2 N_{2,y}(y) N_{2,y}(y) \right) dV + 0 +$$

$$+ \int_V \left(C_{44} F_{\tau y,x}^2 F_{sy,x}^2 N_2(y) N_2(y) \right) dV = C_{11} \int_{-a}^{a} (x \cdot x) \, dx \int_{-b}^{b} dz \int_0^L N_{2,y} N_{2,y} dy$$

$$+ C_{44} \int_{-a}^{a} (1 \cdot 1) \, dx \int_{-b}^{b} dz \int_0^L N_2 N_{2,y} dy = C_{11} \int_{-a}^{a} (x^2) \, dx \int_{-b}^{b} dz \int_0^L \left(\frac{1}{L} \right) \left(\frac{1}{L} \right) dy +$$

$$+ C_{44} \int_{-a}^{a} dx \int_{-b}^{b} dz \int_0^L \left(\frac{y}{L} \right) \left(\frac{y}{L} \right) dy = C_{11} \int_{-a}^{a} (x^2) \, dx \int_{-b}^{b} dz \int_0^L \left(\frac{1}{L^2} \right) dy +$$

$$+ C_{44} \int_{-a}^{a} dx \int_{-b}^{b} dz \int_0^L \left(\frac{y}{L^2} \right) dy = C_{11} \frac{4a^3 b}{3L} + C_{44} \frac{4ab}{3L} = C_{11} \frac{I_z}{L} + C_{44} \frac{A}{3L}$$

$$(9.67)$$

$$K_{yz}^{\tau s 22} = 0$$

$$K_{zx}^{\tau s 22} = 0$$

$$K_{zy}^{\tau s 22} = 0 \tag{9.68}$$

$$K_{zz}^{\tau s 22} = 0$$

Finally, the FN can be written as follows

$$
\mathbf{K}^{\tau s 22} =
\begin{bmatrix}
k_{xx}^{\tau s 22} & k_{xy}^{\tau s 22} \\
k_{yx}^{\tau s 22} & k_{yy}^{\tau s 22}
\end{bmatrix}
=
\begin{bmatrix}
C_{44}\dfrac{A}{L} & -C_{44}\dfrac{A}{2} \\
-C_{44}\dfrac{A}{2} & \left(C_{11}\dfrac{I_z}{L} + C_{44}\dfrac{A}{3L}\right)
\end{bmatrix}
=
$$
$$
=
\begin{bmatrix}
C_{44}\dfrac{4ab}{L} & -C_{44}2ab \\
-C_{44}2ab & \left(C_{11}\dfrac{4a^3 b}{3L} + C_{44}\dfrac{4ab}{3L}\right)
\end{bmatrix}
\tag{9.69}
$$

9.3.6 Complete Stiffness Matrix

Based on the previously derived Fundamental Nucleus (FN), it is possible to construct the stiffness matrix for a two-node beam element whose nodes are governed by different kinematic assumptions. For node 1 (bar-like behavior), the only possible deformation is along the beam's longitudinal axis. By contrast, node 2 (beam-like behavior) can not only stretch but also rotate about the z axis, capturing bending deformations in that plane. Consequently, the global vector of unknowns of the element must reflect these distinct kinematic assumptions at each node.

$$
\{S\} =
\begin{Bmatrix}
v_1^0 \\
u_2^0 \\
\phi_{z2}
\end{Bmatrix}
\tag{9.70}
$$

The stiffness matrix is equal to the following expression

$$\mathbf{K} = \begin{bmatrix} k_{\tau s 11} & k_{\tau s 21} \\ k_{\tau s 12} & k_{\tau s 22} \end{bmatrix} \tag{9.71}$$

Using the expressions of Eqs. (9.54) to (9.69), the complete expression is obtained

$$\mathbf{K} = \begin{bmatrix} C_{11}\dfrac{A}{L} & 0 & 0 \\ 0 & C_{44}\dfrac{A}{L} & -C_{44}\dfrac{A}{2} \\ 0 & -C_{44}\dfrac{A}{2} & \left(C_{11}\dfrac{I_z}{L} + C_{44}\dfrac{AL}{3} \right) \end{bmatrix} \tag{9.72}$$

The previous stiffness Matrix is equal to that obtained in Eq. (9.2).

9.4 Different Shape Functions Against Shear Locking

In Chap. 8, the FN was originally evaluated by assigning constant expansion functions, F_τ and F_s and constant shape functions N_i and N_j. In the first part of this chapter, the FN is evaluated considering different expansion functions applied at the two nodes of each finite element while the shape functions remained constant and this approach is referred to as NDK. The present chapter explores a different approach: using the same expansion function but different shape functions for various unknowns in the formulation. In particular, the bending about the z axis is considered, and it permits the assumption of a constant rotation angle, ϕ_z. This assumption allows for the use of constant shape functions solely for the rotation angle, while linear shape functions based on Lagrange polynomials are applied to the bending terms of the problem. This mixed formulation, i.e. constant shape functions for shear-related terms and linear shape functions for bending-related terms, corresponds naturally to a Gauss integration of order 1 for the shear components. Consequently, the solution proceeds by applying two distinct shape functions for the different unknown fields. As will be shown, this methodology offers a practical means of circumventing shear-locking problem, thereby improving the accuracy and robustness of finite element simulations. Below, the specific shape functions adopted for the bending term $u^0(y)$ are detailed.

$$N_1(y) = 1 - \frac{y}{L}$$

$$\tag{9.73}$$

$$N_2(y) = \frac{y}{L}$$

For the description of the constant distribution of the term $\phi_z(y)$, the following constant shape functions are used

$$N_1^{\phi_z}(y) = \frac{1}{2}$$

$$N_2^{\phi_z}(y) = \frac{1}{2}$$

(9.74)

The expression that describe the variation, along the beam length, of the displacement $u^0(y)$ and of the rotation $\phi_z(y)$ is then written as follows

$$u^0(y) = N_1(y)u_1^0 + N_2(y)u_2^0$$

$$\phi_z(y) = N_1^{\phi}(y)\phi_{z1} + N_2^{\phi}(y)\phi_{z2} = \frac{\phi_{z1} + \phi_{z2}}{2} = \text{constant}$$

(9.75)

9.4.1 Geometric Relation

The geometric relations are written in the following expressions.

$$\begin{aligned}
\epsilon_{xx} &= \frac{\partial u}{\partial x} = \frac{\partial}{\partial x}\left(-1u^0(y)\right) = \frac{\partial}{\partial x}\left(F_{\tau x}u^0(y)\right) = F_{\tau x,x}u^0(y) = \\
&= F_{\tau x,x}N_1^{\tau}u_1^0 + F_{\tau x,x}N_2^{\tau}u_2^0
\end{aligned}$$

$$\begin{aligned}
\epsilon_{yy} &= \frac{\partial v}{\partial y} = \frac{\partial}{\partial y}\left(x\phi_z(y)\right) = \frac{\partial}{\partial y}\left(F_{\tau y}\phi_z(y)\right) = F_{\tau y}\frac{\partial}{\partial y}\left(\phi_z(y)\right) = \\
&= F_{\tau y}\frac{\partial}{\partial y}\left(N_1^{\tau}\phi_{z1}(y) + N_2^{\tau}\phi_{z2}(y)\right) = F_{\tau y}N_{1,y}^{\tau}\phi_{z1} + F_{\tau y}N_{2,y}^{\tau}\phi_{z2}
\end{aligned}$$

(9.76)

$$\begin{aligned}
\epsilon_{zz} &= \frac{\partial w}{\partial z} = \frac{\partial}{\partial z}\left(w^0(y)\right) = \frac{\partial}{\partial z}\left(F_{\tau z}w^0(y)\right) = F_{\tau z,z}w^0(y) = \\
&= F_{\tau z,z}N_1^{\tau}w_1^0 + F_{\tau z,z}N_2^{\tau}w_2^0
\end{aligned}$$

In Eq. (9.76), the ϵ_{yy} relates to the bending part of the matrix and, then, the linear shape functions of Eq. (9.73) are used.

$$
\begin{aligned}
\epsilon_{xz} &= \frac{\partial u}{\partial z} + \frac{\partial w}{\partial x} = \frac{\partial}{\partial z}\left(F_{\tau x} u^0(y)\right) + \frac{\partial}{\partial x}\left(F_{\tau z} w^0(y)\right) = \\
&= F_{\tau x,z} u^0(y) + F_{\tau z,x} w^0(y) = \\
&= F_{\tau x,z} N_1^\tau u_1^0 + F_{\tau x,z} N_2^\tau u_2^0 + F_{\tau z,x} N_1^\tau w_1^0 + F_{\tau z,x} N_2^\tau w_2^0
\end{aligned}
$$

$$
\begin{aligned}
\epsilon_{yz} &= \frac{\partial v}{\partial z} + \frac{\partial w}{\partial y} = \frac{\partial}{\partial z}\left(F_{\tau y} \phi_z(y)\right) + \frac{\partial}{\partial x}\left(F_{\tau z} w^0(y)\right) = \\
&= F_{\tau x,z} \phi_z(y) + F_{\tau z} \frac{\partial}{\partial y}\left(w^0(y)\right) = \\
&= F_{\tau y,z} N_1^\phi \phi_{z1} + F_{\tau y,z} N_2^\phi \phi_{z2} + F_{\tau z} N_{1,y}^\tau w_1^0 + F_{\tau z} N_{2,y}^\tau w_2^0
\end{aligned}
\tag{9.77}
$$

$$
\begin{aligned}
\epsilon_{xy} &= \frac{\partial u}{\partial y} + \frac{\partial v}{\partial x} = \frac{\partial}{\partial y}\left(F_{\tau x} u^0(y)\right) + \frac{\partial}{\partial x}\left(F_{\tau y} \phi_z(y)\right) = \\
&= F_{\tau x} \frac{\partial}{\partial y}\left(u^0(y)\right) + F_{\tau y,x} \phi_z(y) = \\
&= F_{\tau x} N_{1,y}^\tau u_1^0 + F_{\tau x} N_{2,y}^\tau u_2^0 + F_{\tau y,x} N_1^\phi \phi_{z1} + F_{\tau y,x} N_2^\phi \phi_{z2}
\end{aligned}
$$

Introducing the following relations

$$
\mathbf{S} = \left\{ \begin{matrix} u_0^i \\ \phi_{zi} \\ w_0^i \end{matrix} \right\} = \left\{ \begin{matrix} S_{\tau x}^i \\ S_{\tau y}^i \\ S_{\tau z}^i \end{matrix} \right\}
\tag{9.78}
$$

and using the recursive notation, Eqs. (9.76) and (9.77) can be written as follows

$$\epsilon_{xx} = \frac{\partial}{\partial x} s_x = \frac{\partial}{\partial x}(F_{\tau x} N_i^\tau(y) S_{\tau x}^i) = F_{\tau x,x} N_i^\tau(y) S_{\tau x}^i$$

$$\epsilon_{yy} = \frac{\partial}{\partial y} s_y = \frac{\partial}{\partial y}(F_{\tau y} N_i^\tau(y) S_{\tau y}^i) = F_{\tau y} N_{i,y}^\tau(y) S_{\tau y}^i$$

$$\epsilon_{zz} = \frac{\partial}{\partial z} s_z = \frac{\partial}{\partial z}(F_{\tau z} N_i^\tau(y) S_{\tau z}^i) = F_{\tau z,z} N_i^\tau y) S_{\tau z}^i$$

$$\epsilon_{xz} = \frac{\partial}{\partial z} s_x + \frac{\partial}{\partial x} s_z = \frac{\partial}{\partial z}(F_{\tau x} N_i^\tau(y) S_{\tau x}^i) + \frac{\partial}{\partial x}(F_{\tau z} N_i^\tau(y) S_{\tau z}^i) =$$

$$= F_{\tau x,z} N_i^\tau(y) S_{\tau x}^i + F_{\tau z,x} N_i^\tau(y) S_{\tau z}^i$$

$$\epsilon_{yz} = \frac{\partial}{\partial z} s_y + \frac{\partial}{\partial y} s_z = \frac{\partial}{\partial z}(F_{\tau y} N_i^\tau(y) S_{\tau y}^i) + \frac{\partial}{\partial z}(F_{\tau z} N_i^\tau(y) S_{\tau z}^i) =$$

$$= F_{\tau y,z} N_i^\phi(y) S_{\tau y}^i + F_{\tau z} N_{i,y}^\tau(y) S_{\tau z}^i$$

$$(9.79)$$

$$\epsilon_{xy} = \frac{\partial}{\partial y} s_x + \frac{\partial}{\partial x} s_y = \frac{\partial}{\partial y}(F_{\tau x} N_i^\tau(y) S_{\tau x}^i) + \frac{\partial}{\partial x}(F_{\tau y} N_i^\tau(y) S_{\tau y}^i) =$$

$$= F_{\tau x} N_{i,y}^\tau(y) S_{\tau x}^i + F_{\tau y,x} N_i^\phi(y) S_{\tau y}^i$$

where the generic term $N_i^\phi(y)$ refers to the constant shape function used for the unknown $\phi_{zi}(y)$.

9.4.2   Calculation of Matrices B_i and B_j

Equation (9.4.1) can be written in matrix form as follows

$$\epsilon = \mathbf{B}_{\tau i} \mathbf{S}$$

$$(9.80)$$

where

$$
\mathbf{B}_{\tau i} =
\begin{bmatrix}
F_{\tau x,x} N_i^{\tau}(y) & 0 & 0 \\[2mm]
0 & F_{\tau y} N_{i,y}^{\tau}(y) & 0 \\[2mm]
0 & 0 & F_{\tau z,z} N_i^{\tau}(y) \\[2mm]
F_{\tau x,z} N_i^{\tau}(y) & 0 & F_{\tau z,x} N_i^{\tau}(y) \\[2mm]
0 & F_{\tau y,z} N_i^{\phi}(y) & F_{\tau z} N_{i,y}^{\tau}(y) \\[2mm]
F_{\tau x} N_{i,y}^{\tau}(y) & F_{\tau y,x} N_i^{\phi}(y) & 0
\end{bmatrix}
\tag{9.81}
$$

For the virtual variations, introducing the indices j and s, the following relations
holds

$$
\delta\epsilon = \mathbf{B}_{sj}\delta\mathbf{S}
\tag{9.82}
$$

where

$$
\delta\mathbf{S} =
\begin{Bmatrix}
\delta S_{\tau x}^{j} \\[2mm]
\delta S_{\tau y}^{j} \\[2mm]
\delta S_{\tau z}^{j}
\end{Bmatrix}
\tag{9.83}
$$

and

$$
\mathbf{B}_{sj} =
\begin{bmatrix}
F_{sx,x} N_j^{s}(y) & 0 & 0 \\[2mm]
0 & F_{sy} N_{j,y}^{s}(y) & 0 \\[2mm]
0 & 0 & F_{sz,z} N_j^{s}(y) \\[2mm]
F_{sx,z} N_j^{s}(y) & 0 & F_{sz,x} N_j^{s}(y) \\[2mm]
0 & F_{sy,z} N_j^{\phi}(y) & F_{sz} N_{j,y}^{s}(y) \\[2mm]
F_{sx} N_{j,y}^{s}(y) & F_{sy,x} N_j^{\phi}(y) & 0
\end{bmatrix}
\tag{9.84}
$$

9.4.3   Calculation of FN for Different Shape Functions to Mitigate Shear Locking Effects

Introducing the constitutive matrix for the complete strains stress relations reported below

$$\mathbf{C} = \begin{bmatrix} C_{11} & C_{12} & C_{12} & 0 & 0 & 0 \\ C_{21} & C_{11} & C_{12} & 0 & 0 & 0 \\ C_{21} & C_{21} & C_{11} & 0 & 0 & 0 \\ 0 & 0 & 0 & C_{44} & 0 & 0 \\ 0 & 0 & 0 & 0 & C_{44} & 0 \\ 0 & 0 & 0 & 0 & 0 & C_{44} \end{bmatrix} \tag{9.85}$$

and considering Eq. (9.84), it can be written

$$\mathbf{B}_{sj}^{T} = \begin{bmatrix} F_{sx,x}N_{j}^{s}(y) & 0 & 0 & F_{sx,z}N_{j}^{s}(y) & 0 & F_{sx}N_{j,y}^{s}(y) \\[2mm] 0 & F_{sy}N_{j,y}^{s}(y) & 0 & 0 & F_{sy,z}N_{j}^{\phi}(y) & F_{sy,x}N_{j}^{\phi}(y) \\[2mm] 0 & 0 & F_{sz,z}N_{j}^{s}(y) & F_{sz,x}N_{j}^{s}(y) & F_{sz}N_{j,y}^{s}(y) & 0 \end{bmatrix} \tag{9.86}$$

Furthermore,

$$\mathbf{C} \times \mathbf{B}_{\tau i} = \begin{bmatrix} C_{11} & C_{12} & C_{12} & 0 & 0 & 0 \\ C_{21} & C_{11} & C_{12} & 0 & 0 & 0 \\ C_{21} & C_{21} & C_{11} & 0 & 0 & 0 \\ 0 & 0 & 0 & C_{44} & 0 & 0 \\ 0 & 0 & 0 & 0 & C_{44} & 0 \\ 0 & 0 & 0 & 0 & 0 & C_{44} \end{bmatrix} * \begin{bmatrix} F_{\tau x,x}N_{i}^{\tau}(y) & 0 & 0 \\[1mm] 0 & F_{\tau y}N_{i,y}^{\tau}(y) & 0 \\[1mm] 0 & 0 & F_{\tau z,z}N_{i}^{\tau}(y) \\[1mm] F_{\tau x,z}N_{i}^{\tau}(y) & 0 & F_{\tau z,x}N_{i}^{\tau}(y) \\[1mm] 0 & F_{\tau y,z}N_{i}^{\phi}(y) & F_{\tau z}N_{i,y}^{\tau}(y) \\[1mm] F_{\tau x}N_{i,y}^{\tau}(y) & F_{\tau y,x}N_{i}^{\phi}(y) & 0 \end{bmatrix} = \tag{9.87}$$

$$= \begin{bmatrix} C_{11}F_{\tau x,x}N_{i}^{\tau}(y) & C_{12}F_{\tau y}N_{i,y}^{\tau}(y) & C_{12}F_{\tau z,z}N_{i}^{\tau}(y) \\[2mm] C_{21}F_{\tau x,x}N_{i}^{\tau}(y) & C_{11}F_{\tau y}N_{i,y}^{\tau}(y) & C_{12}F_{\tau z,z}N_{i}^{\tau}(y) \\[2mm] C_{21}F_{\tau x,x}N_{i}^{\tau}(y) & C_{21}F_{\tau y}N_{i,y}^{\tau}(y) & C_{11}F_{\tau z,z}N_{i}^{\tau}(y) \\[2mm] C_{44}F_{\tau x,z}N_{i}^{\tau}(y) & 0 & C_{44}F_{\tau z,x}N_{i}^{\tau}(y) \\[2mm] 0 & C_{44}F_{\tau y,z}N_{i}^{\phi}(y) & C_{44}F_{\tau z}N_{i,y}^{\tau}(y) \\[2mm] C_{44}F_{\tau x}N_{i,y}^{\tau}(y) & C_{44}F_{\tau y,x}N_{i}^{\phi}(y) & 0 \end{bmatrix}$$

Then, the following expressions arise

$$K_{xx}^{\tau sij} = \int_V \left[\left(C_{11} F_{\tau x,x} F_{sx,x} N_i^\tau(y) N_j^s(y) \right) + \left(C_{44} F_{\tau x,z} F_{sx,z} N_i^\tau(y) N_j^s(y) \right) + \right.$$
$$\left. + \left(C_{44} F_{\tau x} F_{sx} N_{i,y}^\tau(y) N_{j,y}^s(y) \right) \right] dV$$

$$K_{xy}^{\tau sij} = \int_V \left[\left(C_{12} F_{\tau y} F_{sx,x} N_{i,y}^\tau(y) N_j^s(y) \right) + \left(C_{44} F_{\tau y,x} F_{sx} \underline{N_i^\phi(y)} N_{j,y}^s(y) \right) \right] dV$$

$$K_{xz}^{\tau sij} = \int_V \left[\left(C_{12} F_{\tau z,z} F_{sx,x} N_i^\tau(y) N_j^s(y) \right) + \left(C_{44} F_{\tau z,x} F_{sx,z} N_i^\tau(y) N_j^s(y) \right) \right] dV$$

$$(9.88)$$

$$K_{xx}^{\tau sij} = \int_V \left[\left(C_{11} F_{\tau x,x} F_{sx,x} N_i^\tau(y) N_j^s(y) \right) + \left(C_{44} F_{\tau x,z} F_{sx,z} N_i^\tau(y) N_j^s(y) \right) + \right.$$
$$\left. + \left(C_{44} F_{\tau x} F_{sx} N_{i,y}^\tau(y) N_{j,y}^s(y) \right) \right] dV$$

$$K_{xy}^{\tau sij} = \int_V \left[\left(C_{12} F_{\tau y} F_{sx,x} N_{i,y}^\tau(y) N_j^s(y) \right) + \left(C_{44} F_{\tau y,x} F_{sx} \underline{N_i^\phi(y)} N_{j,y}^s(y) \right) \right] dV$$

$$K_{xz}^{\tau sij} = \int_V \left[\left(C_{12} F_{\tau z,z} F_{sx,x} N_i^\tau(y) N_j^s(y) \right) + \left(C_{44} F_{\tau z,x} F_{sx,z} N_i^\tau(y) N_j^s(y) \right) \right] dV$$

$$K_{yx}^{\tau sij} = \int_V \left[\left(C_{21} F_{\tau x,x} F_{sy} N_i^\tau(y) N_{j,y}^s(y) \right) + \left(C_{44} F_{\tau x} F_{sy,x} N_{i,y}^\tau(y) \underline{N_j^\phi(y)} \right) \right] dV$$

$$K_{yy}^{\tau sij} = \int_V \left[\left(C_{11} F_{\tau y} F_{sy} N_{i,y}^\tau(y) N_{j,y}^s(y) \right) + \left(C_{44} F_{\tau y,z} F_{sy,z} \underline{N_i^\phi(y) N_j^\phi(y)} \right) + \right.$$
$$\left. + \left(C_{44} F_{\tau y,x} F_{sy,x} \underline{N_i^\phi(y) N_j^\phi(y)} \right) \right] dV$$

$$(9.89)$$

$$K_{yz}^{\tau sij} = \int_V \left[\left(C_{12} F_{\tau z,z} F_{sy} N_i^\tau(y) N_{j,y}^s(y) \right) + \left(C_{44} F_{\tau z} F_{sy,z} N_{i,y}^\tau(y) \underline{N_j^\phi(y)} \right) \right] dV$$

$$K_{zx}^{\tau sij} = \int_V \left[\left(C_{21} F_{\tau x,x} F_{sz,z} N_i^\tau(y) N_j^s(y) \right) + \left(C_{44} F_{\tau x,z} F_{sz,x} N_i^\tau(y) N_j^s(y) \right) \right] dV$$

$$K_{zy}^{\tau sij} = \int_V \left[\left(C_{21} F_{\tau y} F_{sz,z} N_{i,y}^\tau(y) N_j^s(y) \right) + \left(C_{44} F_{\tau y,z} F_{sz} \underline{N_i^\phi(y)} N_{j,y}^s(y) \right) \right] dV$$

$$K_{zz}^{\tau sij} = \int_V \left[\left(C_{11} F_{\tau z,z} F_{sz,z} N_i^\tau(y) N_j^s(y) \right) dV + \left(C_{44} F_{\tau z,x} F_{sz,x} N_i^\tau(y) N_j^s(y) \right) dV + \right.$$
$$\left. + \left(C_{44} F_{\tau z} F_{sz} N_{i,y}^\tau(y) N_{j,y}^s(y) \right) \right] dV$$

$$
K_{xx}^{\tau sij} = \int_V \left[\left(C_{11} F_{\tau x,x} F_{sx,x} N_i^\tau(y) N_j^s(y) \right) + \left(C_{44} F_{\tau x,z} F_{sx,z} N_i^\tau(y) N_j^s(y) \right) + \right.
$$
$$
\left. + \left(C_{44} F_{\tau x} F_{sx} N_{i,y}^\tau(y) N_{j,y}^s(y) \right) \right] dV = 0 + 0 + \int_V \left(C_{44} F_{\tau x} F_{sx} N_{i,y}^\tau(y) N_{j,y}^s(y) \right) dV =
$$
$$
= C_{44} A \int_y \left(N_{i,y}^\tau(y) N_{j,y}^s(y) \right) dy
$$

$$
K_{xy}^{\tau sij} = \int_V \left[\left(C_{12} F_{\tau y} F_{sx,x} N_{i,y}^\tau(y) N_j^s(y) \right) + \left(C_{44} F_{\tau y,x} F_{sx} N_i^\phi(y) N_{j,y}^s(y) \right) \right] dV
$$
$$
= 0 + \int_V \left(C_{44} F_{\tau y,x} F_{sx} N_i^\phi(y) N_{j,y}^s(y) \right) dV = -C_{44} A \int_y \left(N_i^\phi(y) N_{j,y}^s(y) \right) dy
$$

$$
K_{xz}^{\tau sij} = 0
$$

$$
K_{yx}^{\tau sij} = \int_V \left[\left(C_{21} F_{\tau x,x} F_{sy} N_i^\tau(y) N_{j,y}^s(y) \right) + \left(C_{44} F_{\tau x} F_{sy,x} N_{i,y}^\tau(y) N_j^\phi(y) \right) \right] dV
$$
$$
= 0 + \int_V \left(C_{44} F_{\tau x} F_{sy,x} N_{i,y}^\tau(y) N_j^\phi(y) \right) dV = -C_{44} A \int_y \left(N_{i,y}^\tau(y) N_j^\phi(y) \right) dy
$$

$$
K_{yy}^{\tau sij} = \int_V \left[\left(C_{11} F_{\tau y} F_{sy} N_{i,y}^\tau(y) N_{j,y}^s(y) \right) + \left(C_{44} F_{\tau y,z} F_{sy,z} N_i^\phi(y) N_j^\phi(y) \right) + \right.
$$
$$
\left. + \left(C_{44} F_{\tau y,x} F_{sy,x} N_i^\phi(y) N_j^\phi(y) \right) \right] dV = \int_V \left(C_{11} F_{\tau y} F_{sy} N_{i,y}^\tau(y) N_{j,y}^s(y) \right) + 0 +
$$
$$
+ \int_V \left(C_{44} F_{\tau y,x} F_{sy,x} N_i^\phi(y) N_j^\phi(y) \right) dV = C_{11} I_z \int_y \left(N_{i,y}^\tau(y) N_{j,y}^s(y) \right) dy +
$$
$$
+ C_{44} A \int_y \left(N_i^\phi(y) N_j^\phi(y) \right) dy
$$

$$
K_{yz}^{\tau sij} = 0
$$

$$
K_{zx}^{\tau sij} = 0
$$

$$
K_{zy}^{\tau sij} = 0
$$

$$
K_{zz}^{\tau sij} = 0
$$

(9.90)

Therefore, the FN results as follows

$$\mathbf{K}^{\tau s i j} = \begin{bmatrix} C_{44}A \int_y \left(N^\tau_{i,y}(y) N^s_{j,y}(y) \right) dy & -C_{44}A \int_y \left(\underline{N^\phi_i(y)} N^s_{j,y}(y) \right) dy & 0 \\[2ex] -C_{44}A \int_y \left(N^\tau_{i,y}(y) \underline{N^\phi_j(y)} \right) dy & \begin{aligned} & C_{11}I_z \int_y \left(N^\tau_{i,y}(y) N^s_{j,y}(y) \right) dy + \\ & C_{44}A \int_y \left(\underline{N^\phi_i(y)} N^\phi_j(y) \right) dy \end{aligned} & 0 \\[2ex] 0 & 0 & 0 \end{bmatrix} \tag{9.91}$$

Introducing the following shape functions and their derivatives

$$N^\tau_1(y) = 1 - \frac{y}{L}$$

$$N^s_1(y) = 1 - \frac{y}{L}$$

$$N^\tau_2(y) = \frac{y}{L}$$

$$N^s_2(y) = \frac{y}{L}$$

$$N^\tau_{1,y}(y) = \frac{\partial}{\partial y}\left(1 - \frac{y}{L}\right) = -\frac{1}{L} \tag{9.92}$$

$$N^s_{1,y}(y) = \frac{\partial}{\partial y}\left(1 - \frac{y}{L}\right) = -\frac{1}{L}$$

$$N^\tau_{2,y}(y) = \frac{\partial}{\partial y}\left(\frac{y}{L}\right) = \frac{1}{L}$$

$$N^s_{2,y}(y) = \frac{\partial}{\partial y}\left(\frac{y}{L}\right) = \frac{1}{L}$$

$$N^\phi_1(y) = \frac{1}{2}$$

$$N^\phi_2(y) = \frac{1}{2}$$

$$N^\phi_{1,y}(y) = \frac{\partial}{\partial y}\left(\frac{1}{2}\right) = 0 \tag{9.93}$$

$$N^\phi_{2,y}(y) = \frac{\partial}{\partial y}\left(\frac{1}{2}\right) = 0$$

Considering that the indices i and j vary from 1 to 2, Eq. (9.91) becomes the following

$$
\mathbf{K}^{\tau s 11} = \begin{bmatrix} C_{44}A\int_y\left(-\frac{1}{L}\right)\left(-\frac{1}{L}\right)dy & -C_{44}A\int_y\left(\frac{1}{2}\right)\left(-\frac{1}{L}\right)dy & 0 \\ -C_{44}A\int_y\left(\frac{1}{2}\right)\left(-\frac{1}{L}\right)dy & C_{11}I_z\int_y\left(-\frac{1}{L}\right)\left(-\frac{1}{L}\right)dy + C_{44}A\int_y\left(\frac{1}{2}\frac{1}{2}\right)dy & 0 \\ 0 & 0 & 0 \end{bmatrix} =
$$

$$
= \begin{bmatrix} C_{44}\dfrac{4ab}{L} & C_{44}2ab & 0 \\ C_{44}2ab & \left(C_{11}\dfrac{4a^3b}{3L} + C_{44}\dfrac{4abL}{4}\right) & 0 \\ 0 & 0 & 0 \end{bmatrix}
$$

$$(9.94)$$

$$
\mathbf{K}^{\tau s 12} = \begin{bmatrix} -C_{44}\dfrac{4ab}{L} & C_{44}2ab & 0 \\ -C_{44}2ab & \left(-C_{11}\dfrac{4a^3b}{3L} + C_{44}\dfrac{4abL}{4}\right) & 0 \\ 0 & 0 & 0 \end{bmatrix} \tag{9.95}
$$

$$
\mathbf{K}^{\tau s 21} = \begin{bmatrix} -C_{44}\dfrac{4ab}{L} & -C_{44}2ab & 0 \\ C_{44}2ab & \left(-C_{11}\dfrac{4a^3b}{3L} + C_{44}\dfrac{4abL}{4}\right) & 0 \\ 0 & 0 & 0 \end{bmatrix} \tag{9.96}
$$

$$
\mathbf{K}^{\tau s 22} = \begin{bmatrix} C_{44}\dfrac{4ab}{L} & -C_{44}2ab & 0 \\ -C_{44}2ab & \left(C_{11}\dfrac{4a^3b}{3L} + C_{44}\dfrac{4abL}{4}\right) & 0 \\ 0 & 0 & 0 \end{bmatrix} \tag{9.97}
$$

The complete stiffness matrix, obtained using the FNs $\mathbf{K}^{\tau s 11}$, $\mathbf{K}^{\tau s 12}$, $\mathbf{K}^{\tau s 21}$ and $\mathbf{K}^{\tau s 22}$, can be written as follows

$$
\mathbf{K} = \begin{bmatrix} \mathbf{K}^{\tau s 11} & \mathbf{K}^{\tau s 12} \\ \mathbf{K}^{\tau s 21} & \mathbf{K}^{\tau s 22} \end{bmatrix} \tag{9.98}
$$

$$\mathbf{K} = \begin{bmatrix} C_{44}\dfrac{4ab}{L} & C_{44}2ab & 0 & -C_{44}\dfrac{4ab}{L} & C_{44}2ab & 0 \\[2ex] C_{44}2ab & \left(C_{11}\dfrac{4a^3b}{3L} + C_{44}\dfrac{4abL}{4}\right) & 0 & -C_{44}2ab & \left(-C_{11}\dfrac{4a^3b}{3L} + C_{44}\dfrac{4abL}{4}\right) & 0 \\[2ex] 0 & 0 & 0 & 0 & 0 & 0 \\[2ex] -C_{44}\dfrac{4ab}{L} & -C_{44}2ab & 0 & C_{44}\dfrac{4ab}{L} & -C_{44}2ab & 0 \\[2ex] C_{44}2ab & \left(-C_{11}\dfrac{4a^3b}{3L} + C_{44}\dfrac{4abL}{4}\right) & 0 & -C_{44}2ab & \left(C_{11}\dfrac{4a^3b}{3L} + C_{44}\dfrac{4abL}{4}\right) & 0 \\[2ex] 0 & 0 & 0 & 0 & 0 & 0 \end{bmatrix} \tag{9.99}$$

The previous matrix corresponds to that obtained using the Uniform Reduced Integration of Eq. (4.25). Thus, the same conclusions can be drawn, i.e. that the present matrix does not suffer by the shear locking effect.

Chapter 10
Application to the Analysis of Real Structures

In the previous chapters, it was introduced how the three-dimensional displacement field $\mathbf{u}$ of a generic structure can be written succinctly within the framework of the Carrera Unified Formulation (CUF). The core idea is summarized by the following relation:

$$\mathbf{u}(x, y, z) = F_\tau(x, z)\mathbf{u}_\tau(y), \qquad \tau = 1, 2, ..., M \tag{10.1}$$

where F_τ vary above the cross-section, $\mathbf{u}_\tau$ is the displacement vector and M stands for the number of terms of the expansion. The choice of the expansion functions F_τ determines the class of CUF model being employed. In earlier chapters, we focused primarily on Taylor Polynomial expansions, i.e., terms of the form $x^i z^j$, to represent the displacement field over the cross-section of the structure (i and j being non-negative integers). For instance, a second-order model might include polynomial terms up to x^2, xz, z^2, thereby enriching the cross-sectional kinematics.

$$u_x(x, y, z) = u_{x_1}(y) + xu_{x_2}(y) + zu_{x_3}(y) + x^2u_{x_4}(y) + xzu_{x_5}(y) + z^2u_{x_6}(y)$$
$$u_y(x, y, z) = u_{y_1}(y) + xu_{y_2}(y) + zu_{y_3}(y) + x^2u_{y_4}(y) + xzu_{y_5}(y) + z^2u_{y_6}(y)$$
$$u_z(x, y, z) = u_{z_1}(y) + xu_{z_2}(y) + zu_{z_3}(y) + x^2u_{z_4}(y) + xzu_{z_5}(y) + z^2u_{z_6}(y)$$
$$\tag{10.2}$$

Crucially, the polynomial choice is not restricted to Taylor polynomials. A broad range of function families can be utilized for F_τ, including Lagrange polynomials. These alternative expansions allow for refined structural modeling by employing interpolation points (often called Lagrange Points, LPs) distributed within a reference domain. Figure 10.1 provides an example of two such cross-sectional elements, namely the six-point L6 element and nine-point L9 element

Both configurations rely on an isoparametric formulation, meaning that the same Lagrange polynomials can describe the geometry of the cross-section and the displacement field over that geometry. This approach simplifies the treatment of arbitrary shaped geometries by mapping a standard $(-1, +1) \times (-1, +1)$ domain (in local

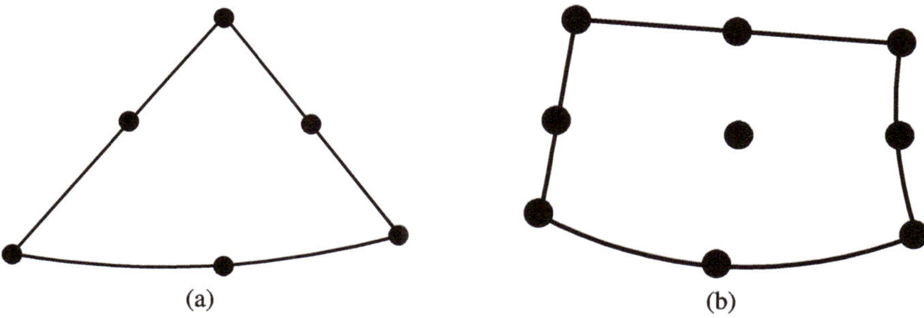

Fig. 10.1   Nine-point elements, L9 and six-point elements, L6 based on Lagrange polynomials

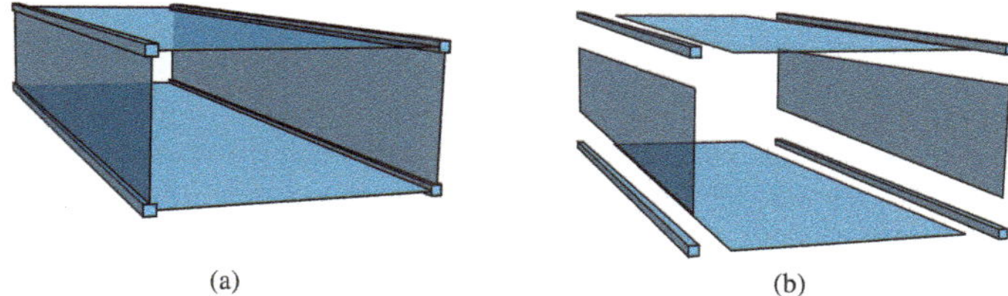

Fig. 10.2   Four stringer wing-box (**a**) split into its components (**b**)

coordinates r and s) to the actual cross-section. As an example, for an L9 element, the interpolation polynomials are:

$$u_x(x, y, z) = F_1(x, z)u_{x_1}(y) + F_2(x, z)u_{x_2}(y) + \cdots + F_9(x, z)u_{x_9}(y)$$
$$u_y(x, y, z) = F_1(x, z)u_{y_1}(y) + F_2(x, z)u_{y_2}(y) + \cdots + F_9(x, z)u_{y_9}(y) \quad (10.3)$$
$$u_z(x, y, z) = F_1(x, z)u_{z_1}(y) + F_2(x, z)u_{z_2}(y) + \cdots + F_9(x, z)u_{z_9}(y)$$

In which $u_{x_1}, \ldots, u_{z_9}$ represent the displacement components of each of the nine LPs and

$$F_\tau = \frac{1}{4}(r^2 + rr_\tau)(s^2 + ss_\tau) \qquad\qquad \tau = 1, 3, 5, 7$$
$$F_\tau = \frac{1}{2}s_\tau^2(s^2 - ss_\tau)(1 - r^2) + \frac{1}{2}r_\tau^2(r^2 - rr_\tau)(1 - s^2) \quad \tau = 2, 4, 6, 8 \qquad (10.4)$$
$$F_\tau = (1 - r^2)(1 - s^2) \qquad\qquad \tau = 9$$

where r and s vary from -1 to $+1$, whereas r_τ and s_τ are the predefined coordinates of each LPs in the natural coordinate system.

One of the key advantages of Lagrange expansions is their flexibility. Multiple polynomial sets can be combined within the same structural model to capture varying geometric and material characteristics. This strategy is especially beneficial for complex cross-sections or built-up structures. A representative example is the four-stringer wing box shown in Fig. 10.2.

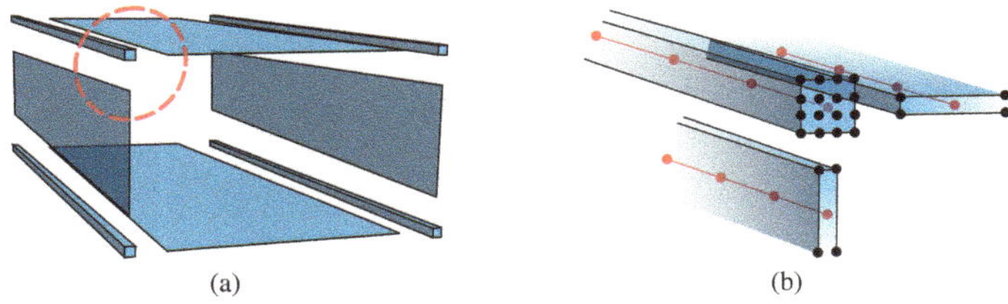

Fig. 10.3 Component-wise approach to simultaneously model panels, stringers and ribs

As illustrated in Fig. 10.2b, its cross-section can be divided into multiple subdomains—stringers, panels, corner reinforcements, etc. The ability to treat each subdomain with its own polynomial expansion is referred to as the Component-Wise (CW) approach [1]. Using CW modeling with Lagrange expansions (hereafter CW-Lagrange) offers several benefits:

- **Geometrical Fidelity**: Each structural component (e.g., stiffener, panel) can be represented accurately with polynomials that best fit its shape and thickness.
- **Material Tailoring**: Different components may have different materials. The CW approach allows each subdomain to retain its own material properties without forcing uniform assumptions across the cross-section.
- **Adaptive Refinement**: By selectively increasing (or decreasing) the polynomial order in certain subdomains, one can balance accuracy and computational cost. Highly stressed or geometrically complex regions might employ higher-order expansions, while simpler regions use lower-order ones.

Figure 10.3 illustrates how each subdomain in the CW approach can be modeled separately, thereby capturing the local deformation patterns of stringers, panels, and other essential components with high fidelity.

This chapter demonstrates the CW Lagrange methodology in action. We present three detailed examples that highlight the versatility and computational efficiency of mixing Lagrange expansions with a CW splitting of the cross-section. Together, these examples show how refined CUF-based models (using Lagrange polynomials in a CW sense) can deliver deeper insight into the structural response, paving the way for optimized designs and reliable simulations in engineering practice.

10.1 Industrial Building

A prototypical industrial building is depicted in Fig. 10.4, where the principal dimensions and boundary conditions are clearly illustrated. The model used in this study is based on [2]. The structure consists of square-section columns and roof frames, all fabricated from a single metallic material. At the base of each column, four clamped boundary conditions are imposed to replicate the restraining effect of a fixed foundation. Despite the complexity of the structural arrangement, the model is discretized

Fig. 10.4 Industrial building

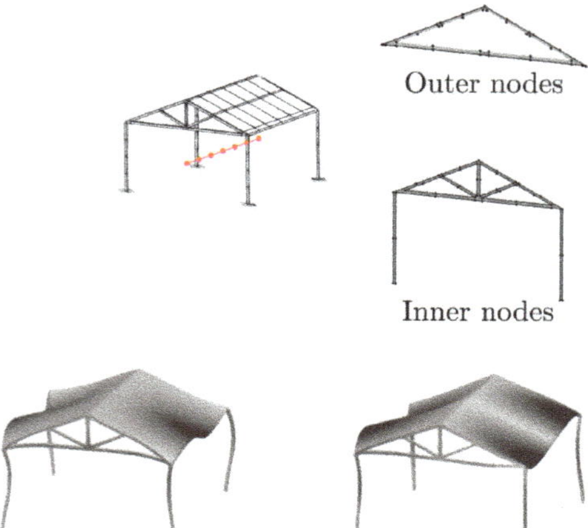

Fig. 10.5 Representative mode shapes of the undamaged and damaged industrial building: **a** f = 11.11 Hz; **b** f = 11.60 Hz

using a single finite element beam (shown in red in Fig. 10.4). The key insight here is that each node of this beam element can employ a different cross-sectional expansions or, in the terminology of advanced structural modeling, each node can be assigned a specific cross-section model. As a result:

- **Outer Nodes**: Represent both the columns (pillars) and the roof.
- **Inner** Nodes: Represent only the roof profile.

This approach allows for local variations in cross-sectional properties within a single element, thereby capturing the essential features of columns in the lower portion of the structure and the roof elements in the upper portion. By doing so, the finite element remains topologically simple (one element with multiple nodes), while still accommodating the geometric and mechanical distinctions between the columns and roof sections. To assess the dynamic behavior, a modal analysis is performed. Specifically, the first two natural frequencies are evaluated, providing insight into the fundamental vibration characteristics of the building. The corresponding mode shapes are shown in Fig. 10.5.

By examining these initial modes, engineers can identify critical vibration patterns that might lead to serviceability or even structural safety concerns. If necessary, design modifications, such as stiffening certain sections, adding braces, or adjusting connections, can be introduced to shift these modes or reduce the amplitude of vibrations. Overall, this example demonstrates how a single finite element beam, augmented by different cross-sectional expansions at various nodes, can effectively model a realistic industrial building.

10.2 Composite Empty Launcher

A second illustrative example, drawn from [3], concerns a composite, empty launcher. The overall geometry, depicted in Fig. 10.6, is loosely based on the European Ariane 5 launcher and features three cylindrical bodies. Each cylindrical stage consists of a skin reinforced by four longitudinal stringers. Despite the launcher's notable geometric complexity, the structure is modeled by a single beam element (indicated by the red arrow in Fig. 10.6). As in previous cases, each node of the beam employs a different Lagrange Expansion to capture the varying cross-sectional properties, thereby accurately representing the three cylindrical bodies and their respective stringers. The main cryogenic storage device has an empty mass of about 15 tons and a gross-mass of about 185 tons. The upper cryogenic stage has an empty mass of about 5 tons and a gross mass of about 20 tons.

To validate the accuracy and efficiency of the one-dimensional (1D) Lagrange-based approach, the computed natural frequencies and mode shapes are compared against a high-fidelity 3D finite element model containing 197,436 degrees of freedom (DOFs). Table 10.1 summarizes the first five natural frequencies predicted by both models.

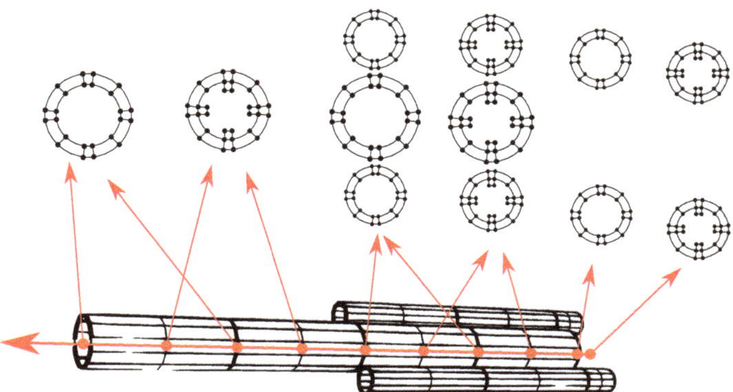

Fig. 10.6 Reinforce launcher model

Table 10.1 First 10 modes in Hz for empty launcher

DOF	Present (29628)	FE-3D (197436)
Mode 1	0.74	0.73
Mode 2	0.92	0.90
Mode 3	4.70	4.55
Mode 4	6.84	6.60
Mode 5	7.94	7.76

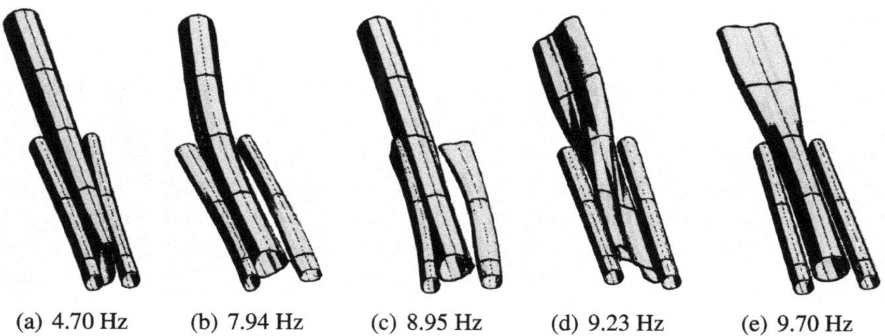

Fig. 10.7 Different modal shapes of the composite launcher

Notably, the mismatch in the first five mode frequencies is below 4%, a strong indication that the 1D model, despite its relative simplicity, captures the essential dynamics with high accuracy. The present 1D model utilizes only about 15% of the DOFs required by the refined 3D model, offering substantial computational savings. Figure 10.7 illustrates representative mode shapes obtained via the present 1D framework. Both global bending/torsional modes and local modes (often associated with the stringers and cylindrical skins) appear correctly in the predicted response, validating the model's ability to capture complex structural behaviors with significantly reduced computational effort.

10.3 Boat Structure

As a final example, boat structures featuring tapered or slanted flat faces at the fore and aft sections are analyzed, based on the configurations described in [4]. These geometries introduce further complexity, particularly when the tapers are inclined in multiple planes. Figure 10.8 shows geometries being considered for structural analysis of boats with varying tapered configurations. The boat in Fig. 10.8a has taper inclined only to vertical plane, whereas the boat in Fig. 10.8b has taper inclined to both the vertical and horizontal planes. The configuration shown in Fig. 10.8c has a cabin added to the former configurations. Each slanted wall is modeled as a separate component within a Component-Wise (CW) formulation, with Lagrange polynomials defining the cross-sectional behavior. Consequently, each component (e.g., slanted walls, decks, cabins) can be assigned its own thickness, material properties, and polynomial interpolation scheme. These structures cannot be modeled with typical beam theories and warrant the use of expensive 2D/3D solid elements.

Modal analyses were conducted for each configuration in Fig. 10.8, yielding natural frequencies and associated mode shapes (Fig. 10.9 and Table 10.2). Models with thin shells typically require a very fine mesh when employing 3D elements. By

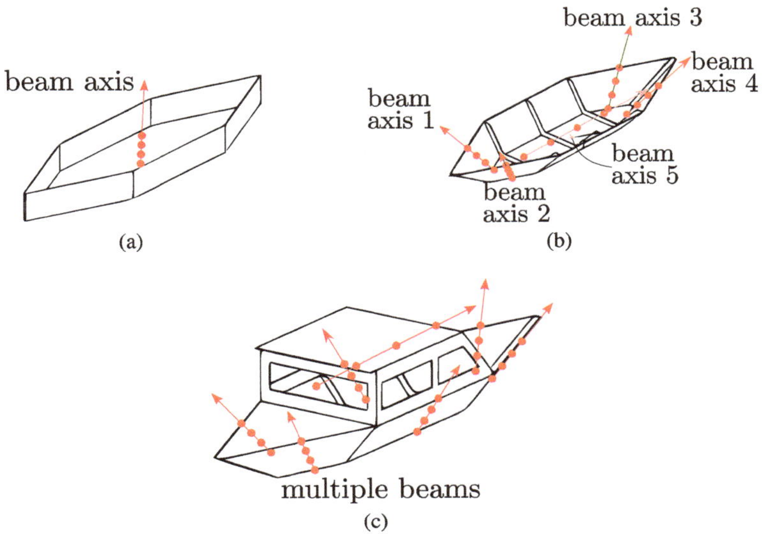

Fig. 10.8 Boat geometries with tapered walls incorporated in front and rear

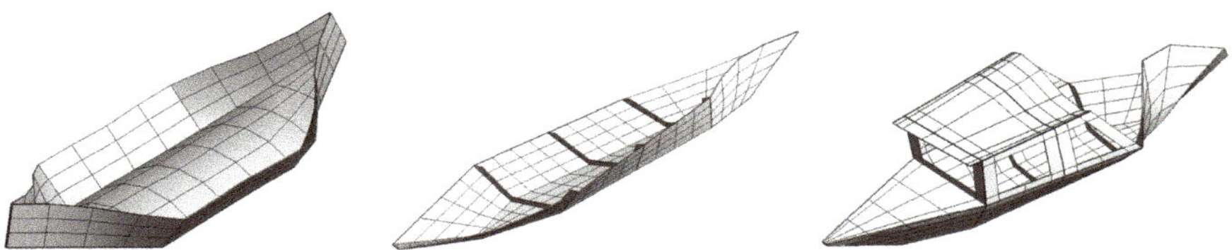

Fig. 10.9 First deformed mode for each of the three configurations of Fig. 10.8

Table 10.2 First three natural frequencies (in Hz) of the three boat configurations of Fig. 10.8

DOF	Model 1		Model 2		Model 3	
	Present (67, 77)	FE-3D (193, 425)	Present (12, 825)	FE-3D (217, 539)	Present (19, 125)	FE-3D (434, 130)
Mode 1	7.522	7.433	15.83	14.94	22.84	19.40
Mode 2	10.50	10.84	20.13	18.43	26.49	25.02
Mode 3	13.46	13.49	25.48	25.13	50.08	48.72

exploiting the CW-Lagrange formulation, the total number of elements and DOFs remains markedly lower than in an equivalent 3D solid analysis. Despite using fewer DOFs, the CW approach accurately captures both global hull deformations and more localized vibrations at the inclined panels or cabin joints.

These diverse examples, from a composite rocket launcher to tapered boat hulls, showcase the versatility and efficiency of one-dimensional high-fidelity modeling via Lagrange polynomials in a Component-Wise framework. By tailoring the cross-sectional expansions at each node (or for each component),

10.4   Concluding Remarks

These diverse examples—from a composite rocket launcher to tapered boat hulls—showcase the versatility and efficiency of one-dimensional high-fidelity modeling via Lagrange polynomials in a Component-Wise framework. By tailoring the cross-sectional expansions at each node (or for each component), a single beam element or a small set of components can represent complex geometries. The models match benchmark 3D FE solutions for frequencies and mode shapes with only minor errors. DOFs are drastically curtailed relative to traditional 2D/3D analyses, making this an attractive option for preliminary design or large-scale parametric studies. Overall, these results highlight the potential of refined 1D CUF-based models to tackle a wide range of industrial and academic problems without incurring the full computational burden of 3D modeling.

References

1. Carrera, E., Alfonso, P., Petrolo, M.: Component-wise method applied to vibration of wing structures. J. Appl. Mech. **80**(4), 1–15 (2013)
2. Cavallo, T., Pagani, A., Zappino, E., Carrera, E.: Effect of localized damages on the free vibration analysis of civil structures by component-wise approach. J. Struct. Eng. **144**(8), 04018113 (2018)
3. Carrera, E., Cavallo, T., Zappino, E.: Effect of solid mass consumption on the free-vibration analysis of launchers. J. Spacecr. Rocket. **54**(3), 773–780 (2017)
4. Pagani, A., Carrera, E., Jamshed, R.: Application of aerospace structural models to marine engineering advances in aircraft and spacecraft. Science **4**(3), 219–235 (2017)

Appendix A
Bar Element Under Axial Deformation

A.1 Evaluation of the Stiffness Matrix via the Matrix Notation

The following MATLAB program enables both symbolic and numerical computation of a bar's stiffness matrix using the matrix notation.

To run the program, the user must provide two input files ((a) and (b)) for the main script (c), as listed below.

(a) INPUT_SYMBOLIC.m
In this script the user provides the shape functions.

```
1   % ================================================================================
2   %                        THIS IS THE INPUT FILE
3   % ================================================================================
4   syms y L;
5   syms a b;
6   syms E;
7   %   --------------------------------------------------------------------------------
8   %                        SYMBOLIC VARIABLES
9   %   --------------------------------------------------------------------------------
10  % 1) Enter shape function in symbolic form:
11  %
12  % Note : shape functions must have the following characterisctics :
13  %   - must be a function of y
14  %   - the length of the bar must be named L
15  %
16  % Shape Function for Node 1 :
17  N1 = 1-y/L;
18  %
19  % Shape Function for Node 2 :
20  N2 = y/L;
21  %
22  %   --------------------------------------------------------------------------------
```

(b) INPUT_NUMERIC.m
In this script the user provides the bar cross-section dimensions (a, b), the bar length (L) and the values of the Young Modulus (E) and the Poisson's ratio (v).

© The Editor(s) (if applicable) and The Author(s), under exclusive license to Springer Nature Switzerland AG 2026
E. Carrera et al., *Implementation of Beam-Type Finite Elements Based on Carrera Unified Formulation*, https://doi.org/10.1007/978-3-031-95856-4

```
1   % =========================================================
2   %                  THIS IS THE NUMERIC VALUES INPUT FILE
3   % =========================================================
4   % -------------------------------------------------------
5   % 1) Cross Section Dimensions
6   %
7   % The dimensions must be expressed in terms a = and b =, without
8   % dimensions
9   %
10  a = 10;
11  b = 10;
12  % 2) Enter Cross Section Dimensions
13  %
14  % The Bar length must be expressed in terms L = a number,without
15  % dimensions
16  %
17  L = 500;
18  %
19  % 3) Enter Young Modulus Value
20  %
21  % The dimensions must be expressed in terms of E = a number, without
22  % dimensions
23  %
24  E = 210000;
25  %
```

(c) Matrix_APPROACH.m
The following is the main script to run.

```
1   %
2   syms N1 N2;
3   syms D1 D2;
4   syms B BT;
5   syms x z;
6   syms y L;
7   syms E;
8   syms A;
9   syms K;
10  syms a b;
11  %
12  % =========================================================
13  % =========================================================
14  %
15  % =========================================================
16  %                       READ INPUT
17  % =========================================================
18  %
19  cd ..\
20  run INPUT_SYMBOLIC\INPUT_SYMBOLIC.m;
21  %
22  % Computation of the derivative of N1 and N2 with respect to y
23  %
24  D1 = diff(N1,y,1);
25  D2 = diff(N2,y,1);
26  %
27  % =========================================================
28  % B Matrix
29  %
30  B = [D1 D2];
31  BT = transpose(B);
32  %
33  % Computation of the Stiffness Matrix
34  %
35  K = E * int(1,z,-b,b)*int(1,x,-a,a) * int(BT*B,y,0,L);
```

```
36   %
37   % NOTE : the following integral computes the cross section Area
38   %
39   A = int(1,z,-b,b)*int(1,x,-a,a);  % Computation of Bar Cros Section Area
40   %
41   fprintf(1,'\n\nSymbolic Stiffness Matrix of a Bar\n\n\n');
42   %
43   pretty(K)
44   %
45   run INPUT_NUMERIC\INPUT_NUMERIC.m;
46   %
47   K_NUM = subs(K);
48   %
49   %
50   fprintf(1,'\n\n\nNumeric Stiffness Matrix of a Bar\n\n\n');
51   %
52   pretty(K_NUM)
53   %
```

In this script the command `diff` is used to compute the symbolic derivatives of the shape functions and the command **int** is used to compute the symbolic integrals, for the calculation of the stiffness matrix **K**.

Once the main code is executed, the following symbolic and numeric results are obtained:

```
                Symbolic Stiffness Matrix of a Bar


                   /  4 E a b       4 E a b \
                   |  -------,   - ------- |
                   |     L             L     |
                   |                         |
                   |    4 E a b   4 E a b   |
                   | - -------,   -------   |
                   \      L           L    /




                Numeric Stiffness Matrix of a Bar


                   /  168000, -168000 \
                   |                   |
                   \ -168000,  168000 /

                   >>
```

The user can change the input data and verify the variation of the results.

A.2 Evaluation of the Stiffness Matrix via the Recursive Notation Against Shape Functions

The following MATLAB program enables both symbolic and numerical computation of a bar's stiffness matrix using the recursive notation against shape functions.

To run the program, the user must provide the same input files ((a) and (b)) as for the matrix notation, and the following main gscript (c).

(c) FE_APPROACH.m
The following is the main script to run.

```matlab
1   %
2   syms N1 N2;
3   syms D1 D2;
4   syms B BT;
5   syms x z;
6   syms y L;
7   syms E;
8   syms A;
9   syms K;
10  syms a b;
11  %
12  %=================================================================
13  %=================================================================
14  %
15  %  ==============================================================
16  %                         READ INPUT
17  %  ==============================================================
18  %
19  cd ..\
20  run INPUT_SYMBOLIC\INPUT_SYMBOLIC.m;
21  %
22  % Computation of the derivative of N(1) and N(2) with respect to y
23  %
24  for i = 1:2;
25  B(i) = diff(N(i),y,1);
26  end
27  %
28  % ================================================================
29  % B Matrix
30  %
31  B = [B(1) B(2)];
32  %
33  % Calculation of the Stiffness Matrix using the recursive notation K(i
        ,j)
34  %
35  for i = 1:2;
36      for j = 1:2;
37      K(i,j) = E * int(1,z,-b,b)*int(1,x,-a,a) * int((diff(N(i),y,1))...
38          *(diff(N(j),y,1)),y,0,L);
39      end
40  end
41  %
42  % Calculation Stiffness Matrix
43  %
44  K = [K(1,1) K(1,2); K(2,1) K(2,2)];
45  %
46  %
47  %
48  % NOTE : the following integral computes the cross section Area
49  %
50  A = int(1,z,-b,b)*int(1,x,-a,a); % Computation of Bar Cros Section
        Area
51  %
52  fprintf(1,'\n\nSymbolic Stiffness Matrix of a Bar\n\n\n');
```

```
53   %
54   pretty(K)
55   %
56   run  INPUT_NUMERIC\INPUT_NUMERIC.m;
57   %
58   K_NUM = subs(K);
59   %
60   %
61   fprintf(1,'\n\n\nNumeric Stiffness Matrix of a Bar\n\n\n');
62   %
63   pretty(K_NUM)
64   %
```

In this script the command diff is used to compute the symbolic derivatives of
the shape functions and the command **int** is used to compute the symbolic integrals,
for the calculation of the stiffness matrix **K**.

Once the main code is executed, the following symbolic and numeric results are
obtained:

```
                   Symbolic Stiffness Matrix of a Bar


          /   4 E a b        4 E a b \
          |   -------,    -  ------- |
          |      L              L    |
          |                          |
          |     4 E a b     4 E a b  |
          |  -  -------,     -------  |
          \        L           L     /




                   Numeric Stiffness Matrix of a Bar


          /   168000,  -168000 \
          |                     |
          \  -168000,   168000 /

          >>
```

The user can change the input data and verify the variation of the results.

A.3 Evaluation of the Stiffness Matrix via the Recursive Notation Against Theory of Structures

The MATLAB program enables both symbolic and numerical computation of a bar's
stiffness matrix using the recursive notation against theory of structures.

To run the main script (c), the user must provide the same input (b) as for the
matrix notation end the recursive notation against shape functions.

Regarding the input (a), the user must provide the shape functions ($N(1)$ and
$N(2)$) and the expansion functions ($F_{\tau y}$ and F_{sy}). The file INPUT_SYMBOLIC.m
for is shown below.

(a) INPUT_SYMBOLIC.m

```
1    %                              SYMBOLIC VARIABLES
2    %
3    % 1) Enter shape function in symbolic form and recursive notation:
4    %
5    % Note : shape functions must have the following characterisctics :
6    % - must be a function of y
7    % - the length of the bar must be named L
8    %
9    % The shape functions are entered using Recursive Notation N(i)
10   % and considering Node 1 and Node 2.
11   % N(1) is the shape notation for Node 1 and N(2) is the shape function
         for
12   % Node 2.
13   %
14   % Shape Function for Node 1 :
15   %
16   N(1) = 1-y/L;
17   %
18   % Shape Function for Node 2 :
19   %
20   N(2) = y/L;
21   %
22   % - - - - - - - - - - - - - - - - - - - - - - - - - - - - - - - - - - - -
23   % 2) Enter the Fty(tau) (Ftauy) and Fsy(s) (Fsy) values using
         recursive
24   %    notation:
25   %    For the problem of a Bar tau = 1 and s = 1 then :
26   %    Fty(tau) = Fty(1) and Fsy(s) = Fsy(1).
27   %
28   Fty(1) = 1;
29   %
30   Fsy(1) = 1;
31   %
```

(c) TS_APPROACH.m
The following is the main script to run.

```
1    %
2    syms N1 N2;
3    syms D1 D2;
4    syms B BT;
5    syms x z;
6    syms y L;
7    syms E;
8    syms A;
9    syms K;
10   syms a b;
11   syms Fty Fsy;
12   %
13   %==================================================================
14   %==================================================================
15   %
16   %  ================================================================
17   %                              READ INPUT
18   %  ================================================================
19   %
20   cd ..\
21   run INPUT_SYMBOLIC\INPUT_SYMBOLIC.m;
22   %
```

```matlab
23    %
24        for i = 1:1;
25        B(i) = Fty(i) * diff(N(i),y,1);
26        end
27    %
28    for i = 1:1;
29        for j = 2:2;
30        B(j) = Fsy(i) * diff(N(j),y,1);
31        end
32    end
33    %
34    %
35    %
36    for t = 1:1;
37        for s = 1:1;
38    A = int(int(Fty(t)*Fsy(s),z,-b,b),x,-a,a); % Bar Cross Section Area
39        end
40    end
41    %
42    % Calculation of Kijts
43    %
44    for j = 1:2;
45       for s = 1:1;
46          for i = 1:2;
47             for t = 1:1;
48    K(i,j)   = E*int(int(Fty(t)*Fsy(s),z,-b,b),x,-a,a)...
49                *int(diff(N(i),y,1)*diff(N(j),y,1),y,0,L);
50             end
51          end
52       end
53    end
54    %
55    % Construction of overall Stiffness matrix
56    %
57    K = [K(1,1) K(1,2); K(2,1) K(2,2)];
58    %
59    %
60    fprintf(1,'\n\nSymbolic Stiffness Matrix of a Bar\n\n\n');
61    %
62    pretty(K)
63    %
64    run INPUT_NUMERIC\INPUT_NUMERIC.m;
65    %
66    K_NUM = subs(K);
67    %
68    %
69    fprintf(1,'\n\n\nNumeric Stiffness Matrix of a Bar\n\n\n');
70    %
71    pretty(K_NUM)
72    %
```

In this script the command diff is used to compute the symbolic derivatives of the shape functions to obtain B_i and B_j and the command **int** is used to compute the symbolic integrals, for the calculation of the stiffness matrix **K**.

Once the main code is executed, the following symbolic and numeric results are obtained:

```
Symbolic Stiffness Matrix of a Bar


/  4 E a b        4 E a b \
|  -------,    -  ------- |
|     L              L    |
|                         |
|   4 E a b    4 E a b    |
| - -------,   -------    |
\      L          L       /




Numeric Stiffness Matrix of a Bar


/  168000, -168000 \
|                  |
\ -168000,  168000 /

>>
```

The user can change the input data and verify the variation of the results.

Appendix B
Beam Element Bending Around the z Axis

B.1 Evaluation of the Stiffness Matrix via the Matrix Notation

The following MATLAB program enables both symbolic and numerical computation of a bar's stiffness matrix using the matrix notation.

To run the program, the user must provide two input files ((a) and (b)) for the main script (c), as listed below.

(1) INPUT_SYMBOLIC.m In this script the user provides the shape functions. In this script the user provides the shape functions.

```
1   %===========================================================================
2   %                          THIS IS THE INPUT FILE
3   %===========================================================================
4   syms y L;
5   syms a b;
6   syms C C11 C44;
7   %---------------------------------------------------------------------------
8   %                          SYMBOLIC VARIABLES
9   %===========================================================================
10  %===========================================================================
11  %
12  % 1) Enter shape function in symbolic form:
13  %
14  % Note : shape functions must have the following characterisctics :
15  %  - must be a function of y
16  %  - the length of the beam must be named L
17  %
18  % Shape Function for Node 1 :
19  N1 = 1-y/L;
20  %
21  % Shape Function for Node 2 :
22  N2 = y/L;
23  %
24  %===========================================================================
25  % NOTE : for the present example case the Matrix C = [C11 0; 0 C44]
26  %
27  C = [C11 0; 0 C44];
28  %===========================================================================
```

© The Editor(s) (if applicable) and The Author(s), under exclusive license to Springer 351
Nature Switzerland AG 2026
E. Carrera et al., *Implementation of Beam-Type Finite Elements Based on Carrera Unified Formulation*, https://doi.org/10.1007/978-3-031-95856-4

(2) INPUT_NUMERIC.m

In this script the user provides the bar cross-section dimensions (a, b), the bar length (L) and the values of the Young Modulus (E) and the Poisson's ratio (ν).

```matlab
% ===============================================================
%                    THIS IS THE NUMERIC VALUES INPUT FILE
% ===============================================================
% ===============================================================
% ===============================================================
% -----------------------------------------------------------
% 1) Enter Cross Section Dimensions
%
% The dimensions must be expressed in terms a = and b =, without
% dimensions
%
a = 10;
b = 10;
%
% 2) Enter Beam length
%
% The Beam length must be expressed in terms L = a number, without
% dimensions
%
L = 500;
%
% 3) Enter C Matrix values
%
% The dimensions must be expressed in terms of C11 = a number and
% C44 = a number, without dimensions
%
% For isotropic material the elastic constants have the following values :
%
% For Steel material :
%
E = 210000;
%
%nu = 0.29;
%
% The Poisson ratio is set equal to 0 for this example case
%
nu = 0.0;
%
% ===============================================================
% See pae 30 of REF_2 for the following data:
% ===============================================================
%
G = E/(2*(1+nu));
Lambda = (nu*E)/((1+nu)*(1-2*nu));
% ===============================================================
% NOTE :
%            C11 = C22 = C33
%            C44 = C55 = C66
%            C12 = C13 = C21 = C23 = C31 = C32
%
% ===============================================================
C11 = 2*G + Lambda;
C22 = 2*G + Lambda;
C33 = 2*G + Lambda;
%
C44 = G;
C55 = G;
C66 = G;
%
C12 = Lambda;
C13 = Lambda;
C21 = Lambda;
```

```
63   C23  =  Lambda;
64   C31  =  Lambda;
65   C32  =  Lambda;
66   %
67   % NOTE : in the present example case the Matrix C = [C11 0; 0 C44]
68   %
69   C  =  [C11  0;  0  C44];
70   % =============================================================
```

(3) Matrix_APPROACH.m
The following is the main script to run.

```
1    %
2    syms  N1  N2;
3    syms  D1  D2;
4    syms  B  BT;
5    syms  x  z;
6    syms  y  L;
7    syms  E;
8    syms  A  Iz;
9    syms  KK  K;
10   syms  a  b;
11   syms  C11  C44;
12   %
13   % =============================================================
14   %                        READ INPUT
15   %=============================================================
16   %=============================================================
17   %
18   cd  ..\
19   run  INPUT_SYMBOLIC\INPUT_SYMBOLIC.m;
20   %
21   % Computation of the derivative of N1 and N2 with respect to y
22   %
23   D1  =  diff(N1,y,1);
24   D2  =  diff(N2,y,1);
25   %
26   %=============================================================
27   %
28   B  =  [0  x*D1  0  x*D2;  -D1   N1  -D2  N2];
29   %
30   BT  =  transpose(B);
31   %
32   %
33   BTCB  =  BT*C*B
34   %
35   %=============================================================
36   %
37   %
38   KK(1,1)  =  C44*int(1,z,-b,b)*int(1,x,-a,a)*int(D1*D1,y,0,L);
39   KK(1,2)  =  -C44*int(1,z,-b,b)*int(1,x,-a,a)*int(N1*D1,y,0,L);
40   KK(1,3)  =  C44*int(1,z,-b,b)*int(1,x,-a,a)*int(D1*D2,y,0,L);
41   KK(1,4)  =  -C44*int(1,z,-b,b)*int(1,x,-a,a)*int(D1*N2,y,0,L);
42   %
43   %
44   KK(2,1)  =  -C44*int(1,z,-b,b)*int(1,x,-a,a)*int(N1*D1,y,0,L);
45   KK(2,2)  =  C11*int(1,z,-b,b)*int(x^2,x,-a,a)*int(D1*D1,y,0,L)...
46              +C44*int(1,z,-b,b)*int(1,x,-a,a)*int(N1*N1,y,0,L);
47   KK(2,3)  =  -C44*int(1,z,-b,b)*int(1,x,-a,a)*int(N1*D2,y,0,L);
48   KK(2,4)  =   C11*int(1,z,-b,b)*int(x^2,x,-a,a)*int(D1*D2,y,0,L)...
49              +C44*int(1,z,-b,b)*int(1,x,-a,a)*int(N1*N2,y,0,L);
50   %
51   %
52   KK(3,1)  =  C44*int(1,z,-b,b)*int(1,x,-a,a)*int(D1*D2,y,0,L);
53   KK(3,2)  =  -C44*int(1,z,-b,b)*int(1,x,-a,a)*int(N1*D2,y,0,L);
```

```matlab
54   KK(3,3) = C44*int(1,z,-b,b)*int(1,x,-a,a)*int(D2*D2,y,0,L);
55   KK(3,4) = -C44*int(1,z,-b,b)*int(1,x,-a,a)*int(D2*N2,y,0,L);
56   %
57   %
58   KK(4,1) = -C44*int(1,z,-b,b)*int(1,x,-a,a)*int(D1*N2,y,0,L);
59   KK(4,2) = C11*int(1,z,-b,b)*int(x^2,x,-a,a)*int(D1*D2,y,0,L)...
60           +C44*int(1,z,-b,b)*int(1,x,-a,a)*int(N2*N1,y,0,L);
61   KK(4,3) = -C44*int(1,z,-b,b)*int(1,x,-a,a)*int(N2*D2,y,0,L);
62   KK(4,4) = C11*int(1,z,-b,b)*int(x^2,x,-a,a)*int(D2*D2,y,0,L)...
63           +C44*int(1,z,-b,b)*int(1,x,-a,a)*int(N2*N2,y,0,L);
64   %
65   %
66   A = int(1,z,-b,b)*int(1,x,-a,a); % Cross section area
67   Iz = int(1,z,-b,b)*int(x^2,x,-a,a); % Cross section z Moment of Inertia
68   %
69   %
70   K(1,1) = C44*A*int(D1*D1,y,0,L);
71   K(1,2) = -C44*A*int(N1*D1,y,0,L);
72   K(1,3) = C44*A*int(D1*D2,y,0,L);
73   K(1,4) = -C44*A*int(D1*N2,y,0,L);
74   %
75   %
76   K(2,1) = -C44*A*int(N1*D1,y,0,L);
77   K(2,2) = C11*Iz*int(D1*D1,y,0,L)...
78           +C44*A*int(N1*N1,y,0,L);
79   K(2,3) = -C44*A*int(N1*D2,y,0,L);
80   K(2,4) =  C11*Iz*int(D1*D2,y,0,L)...
81           +C44*A*int(N1*N2,y,0,L);
82   %
83   %
84   K(3,1) = C44*A*int(D1*D2,y,0,L);
85   K(3,2) = -C44*A*int(N1*D2,y,0,L);
86   K(3,3) = C44*A*int(D2*D2,y,0,L);
87   K(3,4) = -C44*A*int(D2*N2,y,0,L);
88   %
89   %
90   K(4,1) = -C44*A*int(D1*N2,y,0,L);
91   K(4,2) = C11*Iz*int(D1*D2,y,0,L)...
92           +C44*A*int(N2*N1,y,0,L);
93   K(4,3) = -C44*A*int(N2*D2,y,0,L);
94   K(4,4) = C11*Iz*int(D2*D2,y,0,L)...
95           +C44*A*int(N2*N2,y,0,L);
96   %
97   %
98   fprintf(1,'\n\nSymbolic Stiffness Matrix of a Beam under z axis bending\n\n\n
         ');
99   %
100  %
101  K = [K(1,1) K(1,2) K(1,3) K(1,4);K(2,1) K(2,2) K(2,3) K(2,4);...
102       K(3,1) K(3,2) K(3,3) K(3,4);K(4,1) K(4,2) K(4,3) K(4,4)]
103  %
104  pretty(K)
105  %
106  run INPUT_NUMERIC\INPUT_NUMERIC.m;
107  %
108  K_NUM = subs(K);
109  %
110  %
111  fprintf(1,'\n\n\nNumeric Stiffness Matrix of a Beam under z axis bending\n\n\
         n');
112  %
113  pretty(K_NUM)
114  %
```

In this script the command **diff** is used to compute the symbolic derivatives of the shape functions and the command **int** is used to compute the symbolic integrals, for the calculation of the stiffness matrix **K**.

Once the main code is executed, the following symbolic and numeric results are obtained:

```
Symbolic Stiffness Matrix of a Beam under z axis bending


/        #4,      2 C44 a b,      -#4,       #3 \
|                                                |
| 2 C44 a b,        #1,       2 C44 a b, #2 |
|                                                |
|       -#4,      2 C44 a b,        #4,       #3 |
|                                                |
\        #3,           #2,            #3,      #1 /


where


                     3
             C11 b a   4    C44 L b a 4
     #1 ==  ----------  +  -----------
                 3 L             3


                                       3
              2 C44 L a b    4 C11 a  b
     #2 ==  -----------  -  ----------
                  3               3 L


     #3 == -2 C44 a b


              4 C44 a b
     #4 ==  ---------
                  L
```

```
Numeric Stiffness Matrix of a Beam under z axis bending


/   84000,      21000000,      -84000,      -21000000 \
|                                                       |
|  21000000, 7005600000,   21000000, 3494400000 |
|                                                       |
|   -84000,      21000000,       84000,      -21000000 |
|                                                       |
\ -21000000, 3494400000, -21000000, 7005600000 /


     >>
```

The user can change the input data and verify the variation of the results.

B.2 Evaluation of the Stiffness Matrix via the Recursive Notation Against Shape Functions

The following MATLAB program enables both symbolic and numerical computation of a bar's stiffness matrix using the recursive notation against shape functions.

To run the program, the user must provide the same input files ((a) and (b)) as for the matrix notation, and the following main gscript (c).

(c) FE_APPROACH.m
The following is the main script to run.

```matlab
%
syms N1 N2;
syms D1 D2;
syms B BT;
syms x z;
syms y L;
syms E;
syms A Iz;
syms KK K;
syms K11 K12 K21 K22;
syms a b;
syms C11 C44;
syms M M1 M2 M3 M4;
%
%
% =========================================================================
% =========================================================================
%
%                              READ INPUT
% =========================================================================
%
cd ..\
run INPUT_SYMBOLIC\INPUT_SYMBOLIC.m;
%
% Computation of the derivative of N1 and N2 with respect to y
%
% The derivatives are entered using Recursive Notation D(i)
% and considering Node 1 and Node 2.
% D(1) is the derivative of the shape notation for Node 1 and D(2)
     is the
% derivative of the shape function for Node 2.
%
D(1) = diff(N(1),y,1);
D(2) = diff(N(2),y,1);
%
% =========================================================================
%
%
B1 = [0 x*D(1); -D(1) N(1)];
%
%
B2 = [0 x*D(2); -D(2) N(2)];
%
%
```

```
44   B = [0  x*D(1)  0  x*D(2); -D(1)   N(1)  -D(2)  N(2)];
45   %
46   % =====================================================================
47   %
48   % Computation of the Stiffness Matrix
49   %
50   % ------------------------------------------------------------------
51   %
52   K(1,1)  = C44*int(1,z,-b,b)*int(1,x,-a,a)*int(D(1)*D(1),y,0,L);
53   K(1,2)  = -C44*int(1,z,-b,b)*int(1,x,-a,a)*int(N(1)*D(1),y,0,L);
54   K(2,1)  = -C44*int(1,z,-b,b)*int(1,x,-a,a)*int(N(1)*D(1),y,0,L);
55   K(2,2)  = C11*int(1,z,-b,b)*int(x^2,x,-a,a)*int(D(1)*D(1),y,0,L)...
56             +C44*int(1,z,-b,b)*int(1,x,-a,a)*int(N(1)*N(1),y,0,L);
57   %
58   %
59   K11 = [K(1,1)  K(1,2);  K(2,1)  K(2,2)];
60   %
61   %
62   K(1,3)  = C44*int(1,z,-b,b)*int(1,x,-a,a)*int(D(1)*D(2),y,0,L);
63   K(1,4)  = -C44*int(1,z,-b,b)*int(1,x,-a,a)*int(D(1)*N(2),y,0,L);
64   K(2,3)  = -C44*int(1,z,-b,b)*int(1,x,-a,a)*int(N(1)*D(2),y,0,L);
65   K(2,4)  =  C11*int(1,z,-b,b)*int(x^2,x,-a,a)*int(D(1)*D(2),y,0,L)...
66             +C44*int(1,z,-b,b)*int(1,x,-a,a)*int(N(1)*N(2),y,0,L);
67   %
68   %
69   K12 = [K(1,3)  K(1,4);  K(2,3)  K(2,4)];
70   %
71   %
72   K(3,1)  = C44*int(1,z,-b,b)*int(1,x,-a,a)*int(D(1)*D(2),y,0,L);
73   K(3,2)  = -C44*int(1,z,-b,b)*int(1,x,-a,a)*int(N(1)*D(2),y,0,L);
74   K(4,1)  = -C44*int(1,z,-b,b)*int(1,x,-a,a)*int(D(1)*N(2),y,0,L);
75   K(4,2)  = C11*int(1,z,-b,b)*int(x^2,x,-a,a)*int(D(1)*D(2),y,0,L)...
76             +C44*int(1,z,-b,b)*int(1,x,-a,a)*int(N(2)*N(1),y,0,L);
77   %
78   %
79   %
80   K21 = [K(3,1)  K(3,2);  K(4,1)  K(4,2)];
81   %
82   %
83   K(3,3)  = C44*int(1,z,-b,b)*int(1,x,-a,a)*int(D(2)*D(2),y,0,L);
84   K(3,4)  = -C44*int(1,z,-b,b)*int(1,x,-a,a)*int(D(2)*N(2),y,0,L);
85   K(4,3)  = -C44*int(1,z,-b,b)*int(1,x,-a,a)*int(N(2)*D(2),y,0,L);
86   K(4,4)  = C11*int(1,z,-b,b)*int(x^2,x,-a,a)*int(D(2)*D(2),y,0,L)...
87             +C44*int(1,z,-b,b)*int(1,x,-a,a)*int(N(2)*N(2),y,0,L);
88   %
89   %
90   K22 = [K(3,3)  K(3,4);  K(4,3)  K(4,4)];
91   %
92   % ------------------------------------------------------------------
93   %
94   %
95   %
96   K = [K11 K12; K21 K22];
97   %
98   % ------------------------------------------------------------------
99   %
100  %
101  A = int(1,z,-b,b)*int(1,x,-a,a); % Cross section area
102  Iz = int(1,z,-b,b)*int(x^2,x,-a,a); % Cross section z Moment of
         Inertia
```

```matlab
103    %
104    %
105    count = 1;
106    %
107    for i=1:2
108        for j=1:2
109    K(1:2,1:2,count) = [C44*A*int(D(i)*D(j),y,0,L), -C44*A*int(N(i)*D(j
           ),y,0,L);...
110                -C44*A*int(N(j)*D(i),y,0,L), (C11*Iz*int(D(i)*D(j),y,0,L)
                  ...
111                +C44*A*int(N(i)*N(j),y,0,L))];
112        count = count + 1;
113        end
114    end
115    %
116    %
117    K = [K(1:2,1:2,1) K(1:2,1:2,2) ;K(1:2,1:2,3) K(1:2,1:2,4)];
118    %
119    %
120    fprintf(1,'\n\nSymbolic Stiffness Matrix of a Beam under z axis
           bending\n\n\n');
121    %
122    pretty(K)
123    %
124    run INPUT_NUMERIC\INPUT_NUMERIC.m;
125    %
126    K_NUM = subs(K);
127    %
128    %
129    fprintf(1,'\n\n\nNumeric Stiffness Matrix of a Beam under z axis
           bending\n\n\n');
130    %
131    pretty(K_NUM)
132    %
```

In this script the command **diff** is used to compute the symbolic derivatives of the shape functions and the command **int** is used to compute the symbolic integrals, for the calculation of the stiffness matrix **K**.

Once the main code is executed, the following symbolic and numeric results are obtained:

```
Symbolic Stiffness Matrix of a Beam under z axis bending


/        #4,       2 C44 a b,       -#4,        #3 \
|                                                  |
| 2 C44 a b,        #1,       2 C44 a b,  #2 |
|                                                  |
|       -#4,       2 C44 a b,        #4,        #3 |
|                                                  |
\        #3,             #2,             #3,    #1 /


where


                     3
            C11 b a   4    C44 L b a 4
    #1 ==  ---------- + -----------
               3 L            3


                                       3
            2 C44 L a b    4 C11 a  b
    #2 ==  ----------- - ----------
                 3            3 L


    #3 == -2 C44 a b


            4 C44 a b
    #4 ==  ---------
                L




Numeric Stiffness Matrix of a Beam under z axis bending


/   84000,       21000000,      -84000,      -21000000 \
|                                                        |
|  21000000, 7005600000,  21000000, 3494400000 |
|                                                        |
|   -84000,      21000000,       84000,      -21000000 |
|                                                        |
\ -21000000, 3494400000, -21000000, 7005600000 /

    >>
```

The user can change the input data and verify the variation of the results.

B.3 Evaluation of the Stiffness Matrix via the Recursive Notation Against Theory of Structures

The MATLAB program enables both symbolic and numerical computation of a bar's stiffness matrix using the recursive notation against theory of structures.

To run the main script (c), the user must provide the same input (b) as for the matrix notation and the recursive notation against shape functions.

Regarding the input (a), the user must provide the shape functions ($N(1)$ and $N(2)$) and the expansion functions ($F_{\tau y}$ and F_{sy}). The file INPUT_SYMBOLIC.m for is shown below.

(a) INPUT_SYMBOLIC.m

```
1   %=============================================================================
2   %                          THIS IS THE INPUT FILE
3   %=============================================================================
4   syms y L;
5   syms a b;
6   syms E;
7   syms Ftx Fsx;
8   syms Fty Fsy;
9   %  -------------------------------------------------------------------------
10  %                          SYMBOLIC VARIABLES
11  %=============================================================================
12  %=============================================================================
13  %  -------------------------------------------------------------------------
14  % 1) Enter shape function in symbolic form and recursive notation:
15  %
16  % Note : shape functions must have the following characterisctics :
17  % - must be a function of y
18  % - the length of the bar must be named L
19  %
20  % The shape functions are entered using Recursive Notation N(i)
21  % and considering Node 1 and Node 2.
22  % N(1) is the shape notation for Node 1 and N(2) is the shape function
            for
23  % Node 2.
24  %
25  % Shape Function for Node 1 :
26  %
27  N(1) = 1-y/L;
28  %
29  % Shape Function for Node 2 :
30  %
31  N(2) = y/L;
32  %
33  %  -------------------------------------------------------------------------
34  % 2) Enter the Fty(tau) (Ftauy) and Fsy(s) (Fsy) values using
            recursive notation:
35  %     For the problem of a Beam with bending around z axis
36  %     tau = 1 and s = 1 then :
37  %
38  %     Ftx(tau)  =  Ftx(1)
39  %     Fty(tau)  =  Fty(1)
40  %
41  %     and
```

```matlab
42   %
43   %      Fsx(s)  =  Fsx(1)
44   %      Fsy(s)  =  Fsy(1)
45   %
46   %
47   Ftx(1)  =  -1;
48   Fty(1)  =  x;
49   %
50   %
51   Fsx(1)  =  -1;
52   Fsy(1)  =  x;
53   %
54   %
55   % NOTE : the Matrix C = [C11 0;  0 C44]
56   %
57   C = [C11 0;  0 C44];
58   %
59   % ==========================================================================
```

(c) TS_APPROACH.m
The following is the main script to run.

```matlab
1    %
2    syms N;;
3    syms D;
4    syms Bi Bj;
5    syms x z;
6    syms y L;
7    syms E;
8    syms A;
9    syms K;
10   syms a b;
11   syms Ftx Fty;
12   syms Fsx Fsy;
13   syms C11 C44;
14   syms Iz;
15   syms Res1 Res2 Res3 Res4 Res5;
16   %
17   % ==========================================================================
18   % ==========================================================================
19   %                          READ INPUT
20   % ==========================================================================
21   %
22   cd ..\
23   run INPUT_SYMBOLIC\INPUT_SYMBOLIC.m;
24   %
25   % ==========================================================================
26   %
27   % Computation of the derivative of N1 and N2 with respect to y
28   %
29   % The derivatives are entered using Recursive Notation D(i)
30   % and considering Node 1 and Node 2.
31   % D(1) is the derivative of the shape notation for Node 1 and D(2) is
          the
32   % derivative of the shape function for Node 2.
33   %
34   D(1)  =  diff(N(1),y,1);
35   D(2)  =  diff(N(2),y,1);
36   % ==========================================================================
37   %
38   count1 = 1;
```

```matlab
39   %
40   for i=1:2;
41       for j=1:2;
42   BTCB(1:2,1:2,count1) = transpose([ 0 Fsy(1) * diff(N(j),y,1); Fsx(1) *
         diff(N(j),y,1) diff(Fsy(1),x,1) * N(j)])...
43               *C*[ 0 Fty(1)* diff(N(i),y,1); Ftx(1) * diff(N(i),y,1) diff(
                   Fty(1),x,1) * N(i)];
44       count1 = count1 + 1;
45       end
46   end
47   %
48   %//////////////////////////////////////////////////////////////////////
49   % The previously computed matrices,for the i,j pair,
50   % are the following 4 matrices:
51   %
52   % i =1 , j =1
53   %
54   BTCB(1:2,1:2,1);
55   %
56   % i =1 , j =2
57   %
58   BTCB(1:2,1:2,2);
59   %
60   % i =2 , j =1
61   %
62   BTCB(1:2,1:2,3);
63   %
64   % i =2 , j =2
65   %
66   BTCB(1:2,1:2,4);
67   %
68   %=================================================================================
69   %
70   %
71   % The Res(ult)1 is equal to the Cross section AREA
72   %
73   Res1 = int(int(Fsx(1)*Ftx(1),z,-b,b),x,-a,a);
74   %
75   % The Res(ult)2 is equal to negative value of the Cross section AREA
76   %
77   Res2 = int(int(Fsx(1)*diff(Fty(1),x,1),z,-b,b),x,-a,a);
78   %
79   % The Res(ult)3 is equal to negative value of the Cross section AREA
80   %
81   Res3 = int(int(diff(Fsy(1),x,1)*Ftx(1),z,-b,b),x,-a,a);
82   %
83   % The Res(ult)4 is equal to the Cross section Moment of Inerzia Iz
84   % related to the z axis
85   %
86   Res4 = int(int(Fsy(1)*Fty(1),z,-b,b),x,-a,a);
87   %
88   % The Res(ult)5 is equal to negative value of the Cross section AREA
89   %
90   Res5 = int(int(diff(Fsy(1),x,1)*diff(Fty(1),x,1),z,-b,b),x,-a,a);
91   %
92   %
93   % Equivalence of Res1 to Cross Section Area A
94   %
95   A = Res1;
96   %
97   %
98   % Equivalence of Res4 to Cross Section Moment of Inertia Iz
```

```matlab
99    %
100   Iz = Res4;
101   %
102   %
103   count2 = 1;
104   %
105   for i=1:2
106       for j=1:2
107   K(1:2,1:2,count2) = [C44*A*int(D(i)*D(j),y,0,L), -C44*A*int(N(i)*D(j),
          y,0,L);...
108               -C44*A*int(N(j)*D(i),y,0,L), (C11*Iz*int(D(i)*D(j),y,0,L)...
109               +C44*A*int(N(i)*N(j),y,0,L))];
110       count2 = count2 + 1;
111       end
112   end
113   %
114   K = [K(1:2,1:2,1) K(1:2,1:2,2) ;K(1:2,1:2,3) K(1:2,1:2,4)];
115   %
116   %
117   fprintf(1,'\n\nSymbolic Stiffness Matrix of a Beam under z axis
          bending\n\n\n');
118   %
119   pretty(K)
120   %
121   run INPUT_NUMERIC\INPUT_NUMERIC.m;
122   %
123   K_NUM = subs(K);
124   %
125   %
126   fprintf(1,'\n\n\nNumeric Stiffness Matrix of a Beam under z axis
          bending\n\n\n');
127   %
128   pretty(K_NUM)
129   %
```

In this script the command diff is used to compute the symbolic derivatives of the shape functions to obtain B_i and B_j and the command **int** is used to compute the symbolic integrals, for the calculation of the stiffness matrix **K**.

Once the main code is executed, the following symbolic and numeric results are obtained:

```
Symbolic Stiffness Matrix of a Beam under z axis bending


 /      #4,      2 C44 a b,      -#4,     #3 \
 |                                           |
 | 2 C44 a b,      #1,      2 C44 a b, #2 |
 |                                           |
 |     -#4,      2 C44 a b,      #4,     #3 |
 |                                           |
 \      #3,          #2,          #3,     #1 /

where

                      3
           C11 b a  4    C44 L b a 4
    #1 ==  ---------- + -----------
              3 L             3

                                     3
           2 C44 L a b    4 C11 a  b
    #2 ==  ----------- - ----------
               3              3 L

    #3 == -2 C44 a b

           4 C44 a b
    #4 ==  ---------
               L




Numeric Stiffness Matrix of a Beam under z axis bending


 /   84000,      21000000,      -84000,     -21000000 \
 |                                                     |
 |  21000000, 7005600000,   21000000, 3494400000 |
 |                                                     |
 |   -84000,      21000000,      84000,      -21000000 |
 |                                                     |
 \ -21000000, 3494400000, -21000000, 7005600000 /

    >>
```

The user can change the input data and verify the variation of the results.

Appendix C
Shear Locking Correction

C.1 Full Integration

The following the the script MATLAB for the evaluation of the stiffness matrix of a two-node beam element subjected to bending using the Full Integration technique.

```matlab
%%%%%%%%%%%%%%%%%%%%%%%%%%%%%%%%%%%%%%%%%%%%%%%%%%%%%%%%%%%%%%%%%%%%%%%%%
%
clc
%
%=====================================================================
%=====================================================================
%=====================================================================
%
%                    SYMBOLIC VARIABLES DECLARATION
%
syms a b L y x z C44 C11
syms r Jinv detJ
%
%=====================================================================
%                SAMPLING POINTS POSITION AND WEIGHT DEFINITION
%=====================================================================
%                       1 POINTS GAUSS QUADRATURE
%
r1_SP1 = 0;
alfa1_SP1 = 2;
%=====================================================================
%                       2 POINTS GAUSS QUADRATURE
%
r1_SP2 = -1/sqrt(3);
r2_SP2 = +1/sqrt(3);
%
alfa1_SP2 = 1;
alfa2_SP2 = 1;
%=====================================================================
%                       3 POINTS GAUSS QUADRATURE
%
r1_SP3 = -0.774596669241483;
r2_SP3 = 0.;
r3_SP3 = +0.774596669241483;
%
alfa1_SP3 = 0.555555555555556;
```

© The Editor(s) (if applicable) and The Author(s), under exclusive license to Springer Nature Switzerland AG 2026

E. Carrera et al., *Implementation of Beam-Type Finite Elements Based on Carrera Unified Formulation*, https://doi.org/10.1007/978-3-031-95856-4

```
37   alfa2_SP3 = 0.888888888888889;
38   alfa3_SP3 = 0.555555555555556;
39   %
40   %===============================================================
41   %                         Polynomial expansion
42   %
43   tau = 1;
44   sigma = 1;
45   %
46   % DEFINITION OF tmax AND smax FOR CUF NUCLEI REVOVERY PURPOSE
47   %
48   smax= sigma;
49   %
50   tmax = tau;
51   %
52   % ////////////////////////////////////////////////////////////
53   % ////////////////////////////////////////////////////////////
54   %
55   % Number of Nodes in the element ( Element type : B2)
56   %
57   %
58   Num_nodes = 2;
59   %
60   %////////////////////////////////////////////////////////////
61   %////////////////////////////////////////////////////////////
62   %
63   %                        EXPANSION FUNCTIONS
64   %
65   %
66   % NOTE = Ftx(t,i) where t = 1 and i = 1 to 2
67   %
68   Ftx(1,1) = -1;
69   Ftx(1,2) = -1;
70   %
71   %
72   Fty(1,1) = x;
73   Fty(1,2) = x;
74   %
75   % NOT USED IN THIS CASE
76   %
77   Ftz(1,1) = 0;
78   Ftz(1,2) = 0;
79   %
80   %
81   Fsx(1,1) = -1;
82   Fsx(1,2) = -1;
83   %
84   %
85   Fsy(1,1) = x;
86   Fsy(1,2) = x;
87   %
88   % NOT USED IN THIS CASE
89   %
90   Fsz(1,1) = 0;
91   Fsz(1,2) = 0;
92   %
93   %===============================================================
94   %                         Jacobian Matrix
95   %
96   %                         J = |L/2|
97   %
98   %                         Jacobian Matrix Detrminant
99   %
100  detJ = L/2;
```

```
101  %
102  %                               Inverse of the Jacobian Matrix
103  %
104  Jinv = 2/L;
105  %
106  %=========================================================
107  %=========================================================
108  %
109  %   SHAPE FUNCTION INTERPOLATION OF THE DISPLACEMENTS FOR THE ELEMENT B3
110  %   IN NATURAL COORDINATE
111  %
112  Ndit(1,1) = (1/2)*(1-r);
113  Ndit(1,2) = (1/2)*(1+r);
114  %
115  %
116  %
117  Ndis(1,1) = (1/2)*(1-r);
118  Ndis(1,2) = (1/2)*(1+r);
119  %
120  %
121  %=========================================================
122  %   DERIVATIVES OF THE SHAPE FUNCTION INTERPOLATION OF THE
123  %   DISPLACEMENTS FOR THE ELEMENT B2.
124  %   NATURAL COORDINATES
125  %
126  %
127  % NOTE = DNdt(t,i)
128  %         FOR B3 ELEMENT [ t = 1 , i = 1 TO 2 ]
129  %
130  %
131  %
132  DNdit(1,1) = diff(Ndit(1,1),r,1);
133  DNdit(1,2) = diff(Ndit(1,2),r,1);
134  %
135  % NOTE = DNds(s,j)
136  %         FOR B3 ELEMENT [ s = 1 , j = 1 TO 2 ]
137
138  DNdis(1,1) = diff(Ndis(1,1),r,1);
139  DNdis(1,2) = diff(Ndis(1,2),r,1);
140  %=========================================================
141  % ///////////////////////////////////////////////////////
142  % ///////////////////////////////////////////////////////
143  % ///////////////////////////////////////////////////////
144  %
145  %INITIALIZATION OF COUNTER
146  %
147  count1 = 1;
148  count2 = 1;
149  %
150  % ///////////////////////////////////////////////////////
151  % ///////////////////////////////////////////////////////
152  % ///////////////////////////////////////////////////////
153  %
154  %
155  K_xx = 0;
156  K_xy = 0;
157  K_xz = 0;
158  %
159  K_yx = 0;
160  K_yy = 0;
161  K_yz = 0;
162  %
163  K_zx = 0;
164  K_zy = 0;
```

```
165   K_zz = 0;
166   %
167   RIS_TOT = 0;
168   %
169   %^^^^^^^^^^^^^^^^^^^^^^^^^^^^^^^^^^^^^^^^^^^^^^^^^^^^^^^^^^^^^^^^^^
170   %^^^^^^^^^^^^^^^^^^^^^^^^^^^^^^^^^^^^^^^^^^^^^^^^^^^^^^^^^^^^^^^^^^
171   %^^^^^^^^^^^^^^^^^^^^^^^^^^^^^^^^^^^^^^^^^^^^^^^^^^^^^^^^^^^^^^^^^^
172   %
173   %
174   %                     LOOP FOR MATRIX TERMS CALCULATION
175   %
176   for j = 1:Num_nodes;
177       for s = 1:sigma;
178           for i = 1:Num_nodes;
179               for t = 1:tau;
180   %
181   %================================================================
182   %(((((((((((((((((((((((((((((((((((((((((((((((((((((((((((((((((
183   %///////////////////////////////////////////////////////////////
184   %)))))))))))))))))))))))))))))))))))))))))))))))))))))))))))))))))
185   %================================================================
186   %
187   % COMPUTE K_xx
188   %
189   % COMPLETE ANALYTICAL CALCULATION
190   %
191   % Kxx  = C44*int(int(Ftx(t,i)*Fsx(s,j),z,-b,b),x,-a,a)...
192   %          *int(Jinv*Jinv*detJ*DNdit(t,i)*DNdis(s,j),r,-1,1);
193   %
194   %^^^^^^^^^^^^^^^^^^^^^^^^^^^^^^^^^^^^^^^^^^^^^^^^^^^^^^^^^^^^^^^^^^
195   %^^^^^^^^^^^^^^^^^^^^^^^^^^^^^^^^^^^^^^^^^^^^^^^^^^^^^^^^^^^^^^^^^^
196   %^^^^^^^^^^^^^^^^^^^^^^^^^^^^^^^^^^^^^^^^^^^^^^^^^^^^^^^^^^^^^^^^^^
197   %
198   % GAUSS QUADRATURE CALCULATION
199   % FOR B2 ELEMENT THE GREATEST POLYNOMIAL DEGREE IS 2 (Kyy)
200   % THE EXACT CALCULATION IS DONE BY TWO POINTS GAUSS QUADRATURE
201   %
202   % THE FIRST PART OF THE Kxx TERM IS CALCULATED ANALITICALLY
203   %
204   Kxx_1 = C44*int(int(Ftx(t,i)*Fsx(s,j),z,-b,b),x,-a,a);
205   %
206   % THE SECOND PART OF THE Kxx TERM IS INTEGRATED USING THE TWO POINTS
207   % GAUSS QUADRATURE
208   % THE APPROACH IS THE FOLLOWING :
209   % 1) MAKING OF 2 COPIES OF THE SECOND TERM:
210   %
211   Kxx_21 = Jinv*Jinv*detJ*DNdit(t,i)*DNdis(s,j);
212   Kxx_22 = Jinv*Jinv*detJ*DNdit(t,i)*DNdis(s,j);
213   %
214   % DEFINITION OF r = r1_SP2 AND SUBSTITUTION IN Kxx_1
215   %
216   r = r1_SP2;
217   Kxx_21 = subs(Kxx_21);
218   %
219   r = r2_SP2;
220   Kxx_22 = subs(Kxx_22);
221   %
222   % CALCULATION OF THE TWO POINTS GAUSS QUADRATUE
223   %
224   Kxx_2 = alfa1_SP2*Kxx_21 + alfa2_SP2*Kxx_22;
225   %
226   % CALCULATION OF THE Kxx TERM
227   %
228   Kxx = Kxx_1*Kxx_2;
```

```matlab
229  %
230  % RESET OF THE VARIABLE r
231  %
232  clear r
233  syms r
234  %
235  %=================================================================
236  %((((((((((((((((((((((((((((((((((((((((((((((((((((((((((((((((((
237  %////////////////////////////////////////////////////////////////
238  %))))))))))))))))))))))))))))))))))))))))))))))))))))))))))))))))))
239  %=================================================================
240  % %
241  % % COMPUTE K_xy
242  % %
243  % Kxy   = C44*int(int(diff(Fty(t,i),x,1)*Fsx(s,j),z,-b,b),x,-a,a)...
244  %            *int(Jinv*detJ*Ndit(t,i)*DNdis(s,j),r,-1,1);
245  % %
246  % % Kxy = vpa((Kxy),4);
247  %
248  %^^^^^^^^^^^^^^^^^^^^^^^^^^^^^^^^^^^^^^^^^^^^^^^^^^^^^^^^^^^^^^^^^^^
249  %^^^^^^^^^^^^^^^^^^^^^^^^^^^^^^^^^^^^^^^^^^^^^^^^^^^^^^^^^^^^^^^^^^^
250  %^^^^^^^^^^^^^^^^^^^^^^^^^^^^^^^^^^^^^^^^^^^^^^^^^^^^^^^^^^^^^^^^^^^
251  %
252  %
253  % GAUSS QUADRATURE CALCULATION
254  % FOR B3 ELEMENT THE GREATEST POLYNOMIAL DEGREE IS 2 (Kyy)
255  % THE EXACT CALCULATION IS DONE BY TWO POINTS GAUSS QUADRATURE
256  %
257  % THE FIRST PART OF THE Kxx TERM IS CALCULATED ANALITICALLY
258  %
259  Kxy_1 = C44*int(int(diff(Fty(t,i),x,1)*Fsx(s,j),z,-b,b),x,-a,a);
260  %
261  % THE SECOND PART OF THE Kxx TERM IS INTEGRATED USING THE TWO POINTS
262  % GAUSS QUADRATURE
263  % THE APPROACH IS THE FOLLOWING :
264  % 1) MAKING OF 2 COPIES OF THE SECOND TERM:
265  %
266  Kxy_21 = Jinv*detJ*Ndit(t,i)*DNdis(s,j);
267  Kxy_22 = Jinv*detJ*Ndit(t,i)*DNdis(s,j);
268  %
269  % DEFINITION OF r = r1_SP2 AND SUBSTITUTION IN Kxy_1
270  %
271  r = r1_SP2;
272  Kxy_21 = subs(Kxy_21);
273  %
274  r = r2_SP2;
275  Kxy_22 = subs(Kxy_22);
276  %
277  % ESECUTION OF THE THREE POINTS GAUSS QUADRATUE
278  %
279  Kxy_2 = alfa1_SP2*Kxy_21 + alfa2_SP2*Kxy_22;
280  %
281  % CALCULATION OF THE Kxx TERM
282  %
283  Kxy = Kxy_1*Kxy_2;
284  %
285  % RESET OF THE VARIABLE r
286  %
287  clear r
288  syms r
289  %
290  %
291  %=================================================================
292  %=================================================================
```

```
293   % COMPUTE K_xz
294   %
295   Kxz   = 0;
296   %
297   %
298   %================================================================
299   %
300   %
301   % GAUSS QUADRATURE CALCULATION
302   % FOR B3 ELEMENT THE GREATEST POLYNOMIAL DEGREE IS 2 (Kyy)
303   % THE EXACT CALCULATION IS DONE BY TWO POINTS GAUSS QUADRATURE
304   %
305   % THE FIRST PART OF THE Kxx TERM IS CALCULATED ANALITICALLY
306   %
307   Kyx_1 = C44*int(int(Ftx(t,i)*diff(Fsy(s,i),x,1),z,-b,b),x,-a,a);
308   %
309   % THE SECOND PART OF THE Kxx TERM IS INTEGRATED USING THE TWO POINTS
310   % GAUSS QUADRATURE
311   % THE APPROACH IS THE FOLLOWING :
312   % 1) MAKING OF 2 COPIES OF THE SECOND TERM:
313   %
314   Kyx_21 = Jinv*detJ*DNdit(t,i)*Ndis(s,j);
315   Kyx_22 = Jinv*detJ*DNdit(t,i)*Ndis(s,j);
316   %
317   % DEFINITION OF r = r1_SP2 AND SUBSTITUTION IN Kyx_1
318   %
319   r = r1_SP2;
320   Kyx_21 = subs(Kyx_21);
321   %
322   r = r2_SP2;
323   Kyx_22 = subs(Kyx_22);
324   %
325   % ESECUTION OF THE THREE POINTS GAUSS QUADRATUE
326   %
327   Kyx_2 = alfa1_SP2*Kyx_21 + alfa2_SP2*Kyx_22;
328   %
329   % CALCULATION OF THE Kxx TERM
330   %
331   Kyx = Kyx_1*Kyx_2;
332   %
333   % RESET OF THE VARIABLE r
334   %
335   clear r
336   syms r
337   %
338   %
339   % %
340   % %^^^^^^^^^^^^^^^^^^^^^^^^^^^^^^^^^^^^^^^^^^^^^^^^^^^^^^^^^^^^^^^^^^^^^
341   % %^^^^^^^^^^^^^^^^^^^^^^^^^^^^^^^^^^^^^^^^^^^^^^^^^^^^^^^^^^^^^^^^^^^^^
342   % %^^^^^^^^^^^^^^^^^^^^^^^^^^^^^^^^^^^^^^^^^^^^^^^^^^^^^^^^^^^^^^^^^^^^^
343   %
344   %================================================================
345   %
346   %
347   % GAUSS QUADRATURE CALCULATION
348   % FOR B3 ELEMENT THE GREATEST POLYNOMIAL DEGREE IS 2 (Kyy)
349   % THE EXACT CALCULATION IS DONE BY TWO POINTS GAUSS QUADRATURE
350   %
351   % THE FIRST PART OF THE Kxx TERM IS CALCULATED ANALITICALLY
352   %
353   Kyy_1 = C44*int(int(diff(Fty(t,i),x,1)*diff(Fsy(s,i),x,1),z,-b,b),x,-a,a)
          ;
354   %
355   % THE SECOND PART OF THE Kyy TERM IS INTEGRATED USING THE TWO POINTS
```

```
356   % GAUSS QUADRATURE
357   % THE APPROACH IS THE FOLLOWING :
358   % 1) MAKING OF 2 COPIES OF THE SECOND TERM:
359   %
360   Kyy_21 = detJ*Ndit(t,i)*Ndis(s,j);
361   Kyy_22 = detJ*Ndit(t,i)*Ndis(s,j);
362   %
363   % DEFINITION OF r = r1_SP2 AND SUBSTITUTION IN Kyy_1
364   %
365   r = r1_SP2;
366   Kyy_21 = subs(Kyy_21);
367   %
368   r = r2_SP2;
369   Kyy_22 = subs(Kyy_22);
370   %
371   % ESECUTION OF THE TWO POINTS GAUSS QUADRATUE
372   %
373   Kyy_2 = alfa1_SP2*Kyy_21 + alfa2_SP2*Kyy_22;
374   %
375   % CALCULATION OF THE Kxx TERM
376   %
377   Kyy = Kyy_1*Kyy_2;
378   %
379   % RESET OF THE VARIABLE r
380   %
381   clear r
382   syms r
383   %
384   %
385   % ================================================================
386   % COMPUTE K_yz
387   %
388   Kyz  = 0;
389   %
390   % Kyz = vpa((Kyz),4);
391   % ================================================================
392   % ================================================================
393   % COMPUTE K_zx
394   %
395   Kzx  = 0;
396   %
397   % Kzx = vpa((Kzx),4);
398   % ================================================================
399   % ================================================================
400   % COMPUTE K_zy
401   %
402   Kzy  = 0;
403   %
404   % Kzy = vpa((Kzy),4);
405   % ================================================================
406   % ================================================================
407   % COMPUTE K_zz
408   %
409   Kzz  = 0;
410   %
411   % Kzz = vpa((Kzz),4);
412   % ================================================================
413   % ================================================================
414   % ================================================================
415   % ================================================================
416   % ================================================================
417   % ================================================================
418   % ================================================================
419   % ================================================================
```

```matlab
420   %
421   %
422   %=================================================================
423   %$$$$$$$$$$$$$$$$$$$$$$$$$$$$$$$$$$$$$$$$$$$$$$$$$$$$$$$$$$$$$$$$$$$$$
424   %$$$$$$$$$$$$$$$$$$$$$$$$$$$$$$$$$$$$$$$$$$$$$$$$$$$$$$$$$$$$$$$$$$$$$
425   %
426   % BEFORE WRITING MATRIX CUF NUCLEUS
427   % COMPUTATION OF THE FLEXIONAL TERM USING VALUES OF t, s, i, j
428   %  %
429   %  %^^^^^^^^^^^^^^^^^^^^^^^^^^^^^^^^^^^^^^^^^^^^^^^^^^^^^^^^^^^^^^^^
430   %  %^^^^^^^^^^^^^^^^^^^^^^^^^^^^^^^^^^^^^^^^^^^^^^^^^^^^^^^^^^^^^^^^
431   %  %^^^^^^^^^^^^^^^^^^^^^^^^^^^^^^^^^^^^^^^^^^^^^^^^^^^^^^^^^^^^^^^^
432   %
433   %=================================================================
434   %
435   %
436   % GAUSS QUADRATURE CALCULATION
437   % FOR B2 ELEMENT THE GREATEST POLYNOMIAL DEGREE IS 2 (Kyy)
438   % THE EXACT CALCULATION IS DONE BY TWO POINTS GAUSS QUADRATURE
439   %
440   % THE FIRST PART OF THE Kyyy_FLEX_1 TERM IS CALCULATED ANALITICALLY
441   %
442   Kyy_FLEX_1 = C11*int(int(Fty(t,i)*Fsy(s,j),z,-b,b),x,-a,a);
443   %
444   % THE SECOND PART OF THE Kyy TERM IS INTEGRATED USING THE TWO POINTS
445   % GAUSS QUADRATURE
446   % THE APPROACH IS THE FOLLOWING :
447   % 1) MAKING OF 2 COPIES OF THE SECOND TERM:
448   %
449   Kyy_FLEX_21 = Jinv*Jinv*detJ*DNdit(t,i)*DNdis(s,j);
450   Kyy_FLEX_22 = Jinv*Jinv*detJ*DNdit(t,i)*DNdis(s,j);
451   %
452   % DEFINITION OF r = r1_SP2 AND SUBSTITUTION IN Kyy_FLEX_21 AND
            Kyy_FLEX_22
453   %
454   r = r1_SP2;
455   Kyy_FLEX_21 = subs(Kyy_FLEX_21);
456   %
457   r = r2_SP2;
458   Kyy_FLEX_22 = subs(Kyy_FLEX_22);
459   %
460   % ESECUTION OF THE TWO POINTS GAUSS QUADRATUE
461   %
462   Kyy_FLEX_2 = alfa1_SP2*Kyy_FLEX_21 + alfa2_SP2*Kyy_FLEX_22;
463   %
464   % CALCULATION OF THE Kyy_FLEX TERM
465   %
466   Kyy_FLEX = Kyy_FLEX_1*Kyy_FLEX_2;
467   %
468   % RESET OF THE VARIABLE r
469   %
470   clear r
471   syms r
472   %
473   %
474   %=================================================================
475   %$$$$$$$$$$$$$$$$$$$$$$$$$$$$$$$$$$$$$$$$$$$$$$$$$$$$$$$$$$$$$$$$$$$$$
476   %$$$$$$$$$$$$$$$$$$$$$$$$$$$$$$$$$$$$$$$$$$$$$$$$$$$$$$$$$$$$$$$$$$$$$
477   %
478   % WRITING OF CUF NUCLEUS
479   %
480   K(:,:,count1,count2) = [Kxx Kxy Kxz; Kyx (Kyy+Kyy_FLEX) Kyz; Kzx Kzy Kzz
            ];
481   %
```

```matlab
482    %????????????????????????????????????????????????????????????????????
483    %
484    % WRITING OF CUF NUCLEUS [ ONLY 2 X 2 ]
485    %
486    K_TOT_2(:,:,count1,count2) = [Kxx Kxy ; Kyx (Kyy+Kyy_FLEX)];
487    %
488    %????????????????????????????????????????????????????????????????????
489    % COMPUTE BENDING SYIFFNESS MATRIX
490    %
491    K_BENDING(:,:,count1,count2) = [0 0 0; 0 (0+Kyy_FLEX) 0; 0 0 0];
492    %
493    %????????????????????????????????????????????????????????????????????
494    % COMPUTE BENDING SYIFFNESS MATRIX [ 2 x 2 ]
495    %
496    K_BENDING_2(:,:,count1,count2) = [0 0 ; 0 (0+Kyy_FLEX)];
497    %
498    %????????????????????????????????????????????????????????????????????
499    %
500    % COMPUTE SHEAR SYIFFNESS MATRIX
501    %
502    K_SHEAR(:,:,count1,count2) = [Kxx Kxy Kxz; Kyx (Kyy+0) Kyz; Kzx Kzy Kzz];
503    %
504    %
505    %????????????????????????????????????????????????????????????????????
506    % COMPUTE SHEAR SYIFFNESS MATRIX [ 2 x 2 ]
507    %
508    K_SHEAR_2(:,:,count1,count2) = [Kxx Kxy ; Kyx (Kyy+0)];
509    %
510    %????????????????????????????????????????????????????????????????????
511    % K(:,:,count1,count2) = vpa((K(:,:,count1,count2)),4);
512    %
513    %
514    %$$$$$$$$$$$$$$$$$$$$$$$$$$$$$$$$$$$$$$$$$$$$$$$$$$$$$$$$$$$$$$$$$$$$$$$$
515    %$$$$$$$$$$$$$$$$$$$$$$$$$$$$$$$$$$$$$$$$$$$$$$$$$$$$$$$$$$$$$$$$$$$$$$$$
516    %^^^^^^^^^^^^^^^^^^^^^^^^^^^^^^^^^^^^^^^^^^^^^^^^^^^^^^^^^^^^^^^^^^^^^^^^
517    %^^^^^^^^^^^^^^^^^^^^^^^^^^^^^^^^^^^^^^^^^^^^^^^^^^^^^^^^^^^^^^^^^^^^^^^^
518    %^^^^^^^^^^^^^^^^^^^^^^^^^^^^^^^^^^^^^^^^^^^^^^^^^^^^^^^^^^^^^^^^^^^^^^^^
519    %^^^^^^^^^^^^^^^^^^^^^^^^^^^^^^^^^^^^^^^^^^^^^^^^^^^^^^^^^^^^^^^^^^^^^^^^
520    %
521    % RESET TO 0 OF THE FOLLOWING MATRIX TERMS AFTER THE m n LOOP
522    %
523    K_xx = 0;
524    K_xy = 0;
525    K_xz = 0;
526    %
527    K_yx = 0;
528    K_yy = 0;
529    K_yz = 0;
530    %
531    K_zx = 0;
532    K_zy = 0;
533    K_zz = 0;
534    %
535    RIS_TOT = 0;
536    %
537    %????????????????????????????????????????????????????????????????????
538    %????????????????????????????????????????????????????????????????????
539    %
540    % MANAGE THE COUNTER countmat AND count1
541    %
542    count1 = count1 + 1;
543    %
544    %????????????????????????????????????????????????????????????????????
545    %????????????????????????????????????????????????????????????????????
```

```
546   %???????????????????????????????????????????????????????????
547   %
548             end   % END OF t LOOP
549   %
550   %??????????????????????????????????????????????????????????????
551   %??????????????????????????????????????????????????????????????
552         end       % END OF i LOOP
553   %
554   %???????????????????????????????????????????????????????????????
555   %???????????????????????????????????????????????????????????????
556   %
557   % MANAGE THE COUNTER count11 AND count2
558   %
559   count1 = 1;
560   count2 = count2 +1;
561   %
562   %??????????????????????????????????????????????????????????????????
563   %???????????????????????????????????????????????????????????????
564   %
565     end     % END OF s LOOP
566       %
567   end   % END OF j LOOP
568   %
569   %================================================================
570   % MATR(t,s,i,j)
571   %
572   % COMPLETE MATRIX (BENDING + SHEAR) : CUF BASED MATRIX
573   %
574   MATR = [K(:,:,1,1) K(:,:,2,1); K(:,:,1,2) K(:,:,2,2)]
575   %
576   MATR = vpa(MATR,3)
577   %
578   %
579   %   COMPUTE [ 3 X 3 ] MATRICES : CUF BASED MATRICES
580   %
581   %
582   % BENDING MATRIX
583   %
584   MATR_BENDING = [K_BENDING(:,:,1,1) K_BENDING(:,:,2,1); K_BENDING(:,:,1,2)
          K_BENDING(:,:,2,2)]
585   %
586   %
587   % SHEAR MATRIX
588   %
589   MATR_SHEAR = [K_SHEAR(:,:,1,1) K_SHEAR(:,:,2,1); K_SHEAR(:,:,1,2) K_SHEAR
          (:,:,2,2)]
590   %
591   %
592   %
593   %   COMPUTE [ 2 X 2 ] MATRICES FOR BATHE ALPHA -ALPHA DEMONSTRATION
594   %
595   MATR_TOT_2 = [K_TOT_2(:,:,1,1) K_TOT_2(:,:,2,1); K_TOT_2(:,:,1,2) K_TOT_2
          (:,:,2,2)]
596   %
597   MATR_BENDING_2 = [K_BENDING_2(:,:,1,1) K_BENDING_2(:,:,2,1); K_BENDING_2
          (:,:,1,2) K_BENDING_2(:,:,2,2)]
598   %
599   MATR_SHEAR_2 = [K_SHEAR_2(:,:,1,1) K_SHEAR_2(:,:,2,1); K_SHEAR_2(:,:,1,2)
          K_SHEAR_2(:,:,2,2)]
600   %
601   %^^^^^^^^^^^^^^^^^^^^^^^^^^^^^^^^^^^^^^^^^^^^^^^^^^^^^^^^^^^^^^^^^^^
602   %
603   % Open file : Bar_Symbolic_Stiffness_Matrix.dat
604   %
```

```matlab
605    Fout = fopen('
           ELEMENTAL_Beam_Z_axis_Bending_Stiffness_Matrix_B2_3ELE_EXACT_GAUSS.
           dat','w');
606    %
607    %//////////////////////////////////////////////////////////////////////
608    %//////////////////////////////////////////////////////////////////////
609    %
610    % MAKE A COPY OF MATR FOR PRINTING PURPOSE ONLY
611    %
612    MATPR = MATR;
613    MATPRR = vpa(MATPR,4);
614    %
615        fprintf(Fout,'\n\n\n\nSymbolic Stiffness Matrix of a Beam under Z
               axis bending\n\n\n');
616        fprintf(Fout,'\n\n\n\nNORMAL FORMULATION\n\n\n');
617        Columns = Num_nodes*3;
618        Rows = Num_nodes*3;
619    %
620    for m = 1:Rows;
621    %
622            if m==1
623            formatSpec = '%s                  %s                  %s                  %s
                            %s                  %s                  %s
                            %s                  %s                  %s
                            %s                  %s|\n';
624            fprintf(Fout,formatSpec,MATPRR(m:m,1:Columns));
625            fprintf(Fout,'\n\n');
626            elseif m==3
627            formatSpec = '%s                  %s                  %s                  %s
                            %s                  %s                  %s
                            %s                  %s                  %s
                            %s                  %s|\n';
628            fprintf(Fout,formatSpec,MATPRR(m:m,1:Columns));
629            fprintf(Fout,'\n\n');
630            elseif m==5
631            formatSpec = '%s                  %s                  %s                  %s
                            %s                  %s                  %s
                            %s                  %s                  %s
                            %s                  %s|\n';
632            fprintf(Fout,formatSpec,MATPRR(m:m,1:Columns));
633            fprintf(Fout,'\n\n');
634            elseif m==7
635            formatSpec = '%s                  %s                  %s                  %s
                            %s                  %s                  %s
                            %s                  %s                  %s
                            %s                  %s|\n';
636            fprintf(Fout,formatSpec,MATPRR(m:m,1:Columns));
637            fprintf(Fout,'\n\n');
638            elseif m==9
639            formatSpec = '%s                  %s                  %s                  %s
                            %s                  %s                  %s
                            %s                  %s                  %s
                            %s                  %s|\n';
640            fprintf(Fout,formatSpec,MATPRR(m:m,1:Columns));
641            fprintf(Fout,'\n\n');
642            elseif m==11
643            formatSpec = '%s                  %s                  %s                  %s
                            %s                  %s                  %s
                            %s                  %s                  %s
                            %s                  %s|\n';
644            fprintf(Fout,formatSpec,MATPRR(m:m,1:Columns));
645            fprintf(Fout,'\n\n');
646            end
647            %
```

```matlab
648            if m==2
649            formatSpec = '%s                    %s                  %s              %s
                        %s                %s              %s
                        %s                %s              %s
                        %s                %s|\n';
650            fprintf(Fout,formatSpec,MATPRR(m:m,1:Columns));
651            fprintf(Fout,'\n\n');
652            elseif m==4
653            formatSpec = '%s                    %s                  %s              %s
                        %s                %s              %s
                        %s                %s              %s
                        %s                %s|\n';
654            fprintf(Fout,formatSpec,MATPRR(m:m,1:Columns));
655            fprintf(Fout,'\n\n');
656            elseif m==6
657            formatSpec = '%s                    %s                  %s              %s
                        %s                %s              %s
                        %s                %s              %s
                        %s                %s|\n';
658            fprintf(Fout,formatSpec,MATPRR(m:m,1:Columns));
659            fprintf(Fout,'\n\n');
660            elseif m==8
661            formatSpec = '%s                    %s                  %s              %s
                        %s                %s              %s
                        %s                %s              %s
                        %s                %s|\n';
662            fprintf(Fout,formatSpec,MATPRR(m:m,1:Columns));
663            fprintf(Fout,'\n\n');
664            elseif m==10
665            formatSpec = '%s                    %s                  %s              %s
                        %s                %s              %s
                        %s                %s              %s
                        %s                %s|\n';
666            fprintf(Fout,formatSpec,MATPRR(m:m,1:Columns));
667            fprintf(Fout,'\n\n');
668            elseif m==12
669            formatSpec = '%s                    %s                  %s              %s
                        %s                %s              %s
                        %s                %s              %s
                        %s                %s|\n';
670            fprintf(Fout,formatSpec,MATPRR(m:m,1:Columns));
671            fprintf(Fout,'\n\n');
672            end
673            %
674    end
```

C.2   Uniform Reduced Integration

The following the the script MATLAB for the evaluation of the stiffness matrix of a
two-node beam element subjected to bending using the Uniform Reduced Integration
technique.

```matlab
1    %%%%%%%%%%%%%%%%%%%%%%%%%%%%%%%%%%%%%%%%%%%%%%%%%%%%%%%%%%%%%%%%%%%%%%%%%
2    %
3    clc
4    %
5    %===================================================================
6    %===================================================================
7    %===================================================================
```

```
8    %
9    %                      SYMBOLIC VARIABLES DECLARATION
10   %
11   syms a b L y x z C44 C11
12   syms r Jinv detJ
13   %
14   %==============================================================
15   %           SAMPLING POINTS POSITION AND WEIGHT DEFINITION
16   %==============================================================
17   %                       1 POINTS GAUSS QUADRATURE
18   %
19   r1_SP1 = 0;
20   alfa1_SP1 = 2;
21   %==============================================================
22   %                       2 POINTS GAUSS QUADRATURE
23   %
24   r1_SP2 = -1/sqrt(3);
25   r2_SP2 = +1/sqrt(3);
26   %
27   alfa1_SP2 = 1;
28   alfa2_SP2 = 1;
29   %==============================================================
30   %                       3 POINTS GAUSS QUADRATURE
31   %
32   r1_SP3 = -0.774596669241483;
33   r2_SP3 = 0.;
34   r3_SP3 = +0.774596669241483;
35   %
36   alfa1_SP3 = 0.555555555555556;
37   alfa2_SP3 = 0.888888888888889;
38   alfa3_SP3 = 0.555555555555556;
39   %
40   %==============================================================
41   %                       Polynomial expansion
42   %
43   tau = 1;
44   sigma = 1;
45   %
46   % DEFINITION OF tmax AND smax FOR CUF NUCLEI REVOVERY PURPOSE
47   %
48   smax= sigma;
49   %
50   tmax = tau;
51   %
52   % ////////////////////////////////////////////////////////////
53   % ////////////////////////////////////////////////////////////
54   %
55   % Number of Nodes in the element ( Element type : B2)
56   %
57   %
58   Num_nodes = 2;
59   %
60   %////////////////////////////////////////////////////////////
61   %////////////////////////////////////////////////////////////
62   %
63   %                       EXPANSION FUNCTIONS
64   %
65   %
66   % NOTE = Ftx(t,i) where t = 1 and i = 1 to 2
67   %
68   Ftx(1,1) = -1;
69   Ftx(1,2) = -1;
70   %
71   %
72   Fty(1,1) = x;
73   Fty(1,2) = x;
74   %
```

```matlab
75   % NOT USED IN THIS CASE
76   %
77   Ftz(1,1) = 0;
78   Ftz(1,2) = 0;
79   %
80   %
81   Fsx(1,1) = -1;
82   Fsx(1,2) = -1;
83   %
84   %
85   Fsy(1,1) = x;
86   Fsy(1,2) = x;
87   %
88   % NOT USED IN THIS CASE
89   %
90   Fsz(1,1) = 0;
91   Fsz(1,2) = 0;
92   %
93   %===========================================================
94   %                          Jacobian Matrix
95   %
96   %                          J = |L/2|
97   %
98   %                          Jacobian Matrix Detrminant
99   %
100  detJ = L/2;
101  %
102  %                          Inverse of the Jacobian Matrix
103  %
104  Jinv = 2/L;
105  %
106  %===========================================================
107  %===========================================================
108  %
109  %   SHAPE FUNCTION INTERPOLATION OF THE DISPLACEMENTS FOR THE ELEMENT B3
110  %   IN NATURAL COORDINATE
111  %
112  Ndit(1,1) = (1/2)*(1-r);
113  Ndit(1,2) = (1/2)*(1+r);
114  %
115  %
116  %
117  Ndis(1,1) = (1/2)*(1-r);
118  Ndis(1,2) = (1/2)*(1+r);
119  %
120  %
121  %===========================================================
122  %   DERIVATIVES OF THE SHAPE FUNCTION INTERPOLATION OF THE
123  %   DISPLACEMENTS FOR THE ELEMENT B2.
124  %   NATURAL COORDINATES
125  %
126  %
127  % NOTE = DNdt(t,i)
128  %        FOR B3 ELEMENT [ t = 1 , i = 1 TO 2 ]
129  %
130  %
131  %
132  DNdit(1,1) = diff(Ndit(1,1),r,1);
133  DNdit(1,2) = diff(Ndit(1,2),r,1);
134  %
135  % NOTE = DNds(s,j)
136  %        FOR B3 ELEMENT [ s = 1 , j = 1 TO 2 ]
137
138  DNdis(1,1) = diff(Ndis(1,1),r,1);
139  DNdis(1,2) = diff(Ndis(1,2),r,1);
140  %===========================================================
141  % /////////////////////////////////////////////////////////////////////
```

```
142    % ///////////////////////////////////////////////////////////////
143    % ///////////////////////////////////////////////////////////////
144    %
145    %INITIALIZATION OF COUNTER
146    %
147    count1 = 1;
148    count2 = 1;
149    %
150    % ///////////////////////////////////////////////////////////////
151    % ///////////////////////////////////////////////////////////////
152    % ///////////////////////////////////////////////////////////////
153    %
154    % PUT 0 THE FOLLOWING MATRICES
155    %
156    K_xx = 0;
157    K_xy = 0;
158    K_xz = 0;
159    %
160    K_yx = 0;
161    K_yy = 0;
162    K_yz = 0;
163    %
164    K_zx = 0;
165    K_zy = 0;
166    K_zz = 0;
167    %
168    RIS_TOT = 0;
169    %
170    %^^^^^^^^^^^^^^^^^^^^^^^^^^^^^^^^^^^^^^^^^^^^^^^^^^^^^^^^^^^^^^^^^^^
171    %^^^^^^^^^^^^^^^^^^^^^^^^^^^^^^^^^^^^^^^^^^^^^^^^^^^^^^^^^^^^^^^^^^^
172    %^^^^^^^^^^^^^^^^^^^^^^^^^^^^^^^^^^^^^^^^^^^^^^^^^^^^^^^^^^^^^^^^^^^
173    %
174    %
175    %                    LOOP FOR MATRIX TERMS CALCULATION
176    %
177    for j = 1:Num_nodes;
178       for s = 1:sigma;
179          for i = 1:Num_nodes;
180             for t = 1:tau;
181    %
182    %===============================================================
183    %(((((((((((((((((((((((((((((((((((((((((((((((((((((((((((((((
184    %///////////////////////////////////////////////////////////////
185    %)))))))))))))))))))))))))))))))))))))))))))))))))))))))))))))))
186    %===============================================================
187    %
188    % COMPUTE K_xx
189    %
190    % COMPLETE ANALYTICAL CALCULATION
191    %
192    % Kxx  = C44*int(int(Ftx(t,i)*Fsx(s,j),z,-b,b),x,-a,a)...
193    %          *int(Jinv*Jinv*detJ*DNdit(t,i)*DNdis(s,j),r,-1,1);
194    %
195    %^^^^^^^^^^^^^^^^^^^^^^^^^^^^^^^^^^^^^^^^^^^^^^^^^^^^^^^^^^^^^^^^^^^
196    %^^^^^^^^^^^^^^^^^^^^^^^^^^^^^^^^^^^^^^^^^^^^^^^^^^^^^^^^^^^^^^^^^^^
197    %^^^^^^^^^^^^^^^^^^^^^^^^^^^^^^^^^^^^^^^^^^^^^^^^^^^^^^^^^^^^^^^^^^^
198    %                    GENERAL PRELIMINARY INFORMATION
199    %
200    % GAUSS QUADRATURE CALCULATION FOR B2 ELEMENT:
201    %
202    % 1) THE SHEAR TERMS OF THE OVERALL MATRIX HAVE A
203    %    MAXIMUM POLYNOMIAL DEGREGE OF 2
204    % 2) THE BENDING TERM OF THE MATRIX IS A CONSTANT TERM THEN THE POLYNOMIAL
205    %    DEGREE IS 0.
206    %
207    % THE GREATEST POLYNOMIAL DEGREE IS 2 FOR Kyy MATRIX TERM.
208    %
```

```matlab
209   % THE EXACT CALCULATION IS DONE BY TWO POINTS GAUSS QUADRATURE
210   %
211   % THE UNIFORM REDUCED INTEGRATION IS DONE BY USING A
212   % 1 POINT GAUSS QUADRATURE
213   % THE SELECTIVE REDUCED INTEGRATION IS EXPLAINED BELOW.
214   %
215   %
216   %                   SELECTIVE REDUCED INTEGRATION
217   %
218   % A SELECTIVE REDUCED INTEGRATION RULE IS USED WHEN THE SHEAR TERMS OF
219   % THE  MATRIX IS INTEGRATED USING A REDUCED INTEGRATION (i.e. A LOWER
220   % ORDER INTEGRATION RULE IS USED) WHEREAS THE BENDIN TERM OF THW MATRIX IS
221   % INTEGRATED USING THE NORMAL INTEGRATION RULE.
222   % DUE TO THE FACT THAT THE NORMAL INTEGRATION RULE IS BASED ON TWO GAUSS
223   % POINTS QUADRATURE, THEN THE SELECTIVE REDUCED INTEGRATION CONSISTS
224   % IN THE FOLLOWING RULE:
225   % 1) THE SHEAR TERMS OF THE  MATRIX IS INTEGRATED USING A 1 POINT
226   %    GAUSS QUADRATURE INSTEAD OF THE NORMAL INTEGRATION RULE BASED ON
227   %    A 2 POINTS GAUSS QUADRATURE
228   % 2) THE BENDING TERMS OF THE  MATRIX IS INTEGRATED USING A 2 POINT
229   %    GAUSS QUADRATURE THAT IS THE NORMAL INTEGRATION RULE
230   %
231   % NOTE :
232   %
233   % THE FIRST PART OF THE Kxx TERM IS CALCULATED ANALITICALLY
234   %
235   Kxx_1 = C44*int(int(Ftx(t,i)*Fsx(s,j),z,-b,b),x,-a,a);
236   %
237   % THE SECOND PART OF THE Kxx TERM (THE SHEAR TERM)IS INTEGRATED USING
238   % THE 1 POINT GAUSS QUADRATURE (REDUCED INTEGRATION RULE).
239   %
240   % THE APPROACH IS THE FOLLOWING :
241   % 1) MAKING OF 1 COPIES OF THE SECOND TERM:
242   %
243   % NOTE : COPIES OF MATRIX PARTS ARE DONE IN ORDER TO AVOID CONFLICT
244   % PROBLEMS
245   %
246   Kxx_21 = Jinv*Jinv*detJ*DNdit(t,i)*DNdis(s,j);
247   %
248   % DEFINITION OF r = r1_SP1 AND SUBSTITUTION IN Kxx_1
249   %
250   r = r1_SP1;
251   Kxx_21 = subs(Kxx_21);
252   %
253   %
254   % ESECUTION OF THE TWO POINTS GAUSS QUADRATUE
255   %
256   Kxx_2 = alfa1_SP1*Kxx_21;
257   %
258   % CALCULATION OF THE Kxx TERM
259   %
260   Kxx = Kxx_1*Kxx_2;
261   %
262   % RESET OF THE VARIABLE r
263   %
264   clear r
265   syms r
266   %
267   %================================================================
268   %{{{{{{{{{{{{{{{{{{{{{{{{{{{{{{{{{{{{{{{{{{{{{{{{{{{{{{{{{{{{{{{{
269   %////////////////////////////////////////////////////////////////
270   %}}}}}}}}}}}}}}}}}}}}}}}}}}}}}}}}}}}}}}}}}}}}}}}}}}}}}}}}}}}}}}}}
271   %================================================================
272   % %
273   % % COMPUTE K_xy
274   % %
275   % Kxy  = C44*int(int(diff(Fty(t,i),x,1)*Fsx(s,j),z,-b,b),x,-a,a)...
```

```
276  %          *int(Jinv*detJ*Ndit(t,i)*DNdis(s,j),r,-1,1);
277  % %
278  % %  Kxy  =  vpa((Kxy),4);
279  %
280  %^^^^^^^^^^^^^^^^^^^^^^^^^^^^^^^^^^^^^^^^^^^^^^^^^^^^^^^^^^^^^^^^^^^^
281  %^^^^^^^^^^^^^^^^^^^^^^^^^^^^^^^^^^^^^^^^^^^^^^^^^^^^^^^^^^^^^^^^^^^^
282  %^^^^^^^^^^^^^^^^^^^^^^^^^^^^^^^^^^^^^^^^^^^^^^^^^^^^^^^^^^^^^^^^^^^^
283  %
284  % THE FIRST PART OF THE KxY TERM IS CALCULATED ANALITICALLY
285  %
286  Kxy_1 = C44*int(int(diff(Fty(t,i),x,1)*Fsx(s,j),z,-b,b),x,-a,a);
287  %
288  %
289  % THE SECOND PART OF THE Kxy TERM IS INTEGRATED USING THE 1 POINT
290  % GAUSS QUADRATURE (REDUCED INTEGRATION RULE)
291  % THE APPROACH IS THE FOLLOWING :
292  % 1) MAKING OF 1 COPIES OF THE SECOND TERM:
293  % NOTE : COPIES OF MATRIX PARTS ARE DONE IN ORDER TO AVOID CONFLICT
294  % PROBLEMS
295  %
296  %
297  Kxy_21 = Jinv*detJ*Ndit(t,i)*DNdis(s,j);
298  %
299  % DEFINITION OF r = r1_SP1 AND SUBSTITUTION IN Kxy_1
300  %
301  r = r1_SP1;
302  Kxy_21 = subs(Kxy_21);
303  %
304  %
305  % ESECUTION OF THE ONE POINTS GAUSS QUADRATUE
306  %
307  Kxy_2 = alfa1_SP1*Kxy_21;
308  %
309  % CALCULATION OF THE Kxy TERM
310  %
311  Kxy = Kxy_1*Kxy_2;
312  %
313  % RESET OF THE VARIABLE r
314  %
315  clear r
316  syms r
317  %
318  %
319  %================================================================
320  %================================================================
321  % COMPUTE K_xz
322  %
323  Kxz  = 0;
324  %
325  % Kxz = vpa((Kxz),4);
326  %
327  %^^^^^^^^^^^^^^^^^^^^^^^^^^^^^^^^^^^^^^^^^^^^^^^^^^^^^^^^^^^^^^^^^^^^
328  %^^^^^^^^^^^^^^^^^^^^^^^^^^^^^^^^^^^^^^^^^^^^^^^^^^^^^^^^^^^^^^^^^^^^
329  %^^^^^^^^^^^^^^^^^^^^^^^^^^^^^^^^^^^^^^^^^^^^^^^^^^^^^^^^^^^^^^^^^^^^
330  %
331  %
332  %================================================================
333  %================================================================
334  % %  COMPUTE  K_yx
335  % %
336  % Kyx   = C44*int(int(Ftx(t,i)*diff(Fsy(s,i),x,1),z,-b,b),x,-a,a)...
337  %          *int(Jinv*detJ*DNdit(t,i)*Ndis(s,j),r,-1,1);
338  % %
339  % %  Kyx =  vpa((Kyx),4);
340  % %
341  % %^^^^^^^^^^^^^^^^^^^^^^^^^^^^^^^^^^^^^^^^^^^^^^^^^^^^^^^^^^^^^^^^^^^
342  % %^^^^^^^^^^^^^^^^^^^^^^^^^^^^^^^^^^^^^^^^^^^^^^^^^^^^^^^^^^^^^^^^^^^
```

```
343   % %^^^^^^^^^^^^^^^^^^^^^^^^^^^^^^^^^^^^^^^^^^^^^^^^^^^^^^^^^^^^^^^^^^^^^^^
344   %
345   % =================================================================
346   %
347   %
348   % THE FIRST PART OF THE Kyx TERM IS CALCULATED ANALITICALLY
349   %
350   Kyx_1 = C44*int(int(Ftx(t,i)*diff(Fsy(s,i),x,1),z,-b,b),x,-a,a);
351   %
352   %
353   % THE SECOND PART OF THE Kyx TERM IS INTEGRATED USING THE ONE POINT
354   % GAUSS QUADRATURE (REDUCED INTEGRATION RULE)
355   % THE APPROACH IS THE FOLLOWING :
356   % 1) MAKING OF 1 COPIES OF THE SECOND TERM:
357   % NOTE : COPIES OF MATRIX PARTS ARE DONE IN ORDER TO AVOID CONFLICT
358   % PROBLEMS
359   %
360   Kyx_21 = Jinv*detJ*DNdit(t,i)*Ndis(s,j);
361   %
362   %
363   % DEFINITION OF r = r1_SP1 AND SUBSTITUTION IN Kyx_1
364   %
365   r = r1_SP1;
366   Kyx_21 = subs(Kyx_21);
367   %
368   %
369   % ESECUTION OF THE ONE POINT GAUSS QUADRATUE
370   %
371   Kyx_2 = alfa1_SP1*Kyx_21;
372   %
373   % CALCULATION OF THE Kyx TERM
374   %
375   Kyx = Kyx_1*Kyx_2;
376   %
377   % RESET OF THE VARIABLE r
378   %
379   clear r
380   syms r
381   %
382   %
383   % =================================================================
384   % COMPUTE K_yy
385   % %
386   % Kyy  = C44*int(int(diff(Fty(t,i),x,1)*diff(Fsy(s,i),x,1),z,-b,b),x,-a,a)...
387   %        *int(detJ*Ndit(t,i)*Ndis(s,j),r,-1,1);
388   % %
389   % % Kyy = vpa((Kyy),4);
390   % =================================================================
391   % %
392   % %^^^^^^^^^^^^^^^^^^^^^^^^^^^^^^^^^^^^^^^^^^^^^^^^^^^^^^^^^^^^^^^^^^^^
393   % %^^^^^^^^^^^^^^^^^^^^^^^^^^^^^^^^^^^^^^^^^^^^^^^^^^^^^^^^^^^^^^^^^^^^
394   % %^^^^^^^^^^^^^^^^^^^^^^^^^^^^^^^^^^^^^^^^^^^^^^^^^^^^^^^^^^^^^^^^^^^^
395   %
396   % =================================================================
397   %
398   %
399   % THE FIRST PART OF THE Kyy TERM IS CALCULATED ANALITICALLY
400   %
401   Kyy_1 = C44*int(int(diff(Fty(t,i),x,1)*diff(Fsy(s,i),x,1),z,-b,b),x,-a,a);
402   %
403   %
404   % THE SECOND PART OF THE Kyy TERM IS INTEGRATED USING THE ONE POINT
405   % GAUSS QUADRATURE (REDUCED INTEGRATION RULE)
406   % THE APPROACH IS THE FOLLOWING :
407   % 1) MAKING OF 1 COPIES OF THE SECOND TERM:
408   % NOTE : COPIES OF MATRIX PARTS ARE DONE IN ORDER TO AVOID CONFLICT
409   % PROBLEMS
```

```
410   %
411   %
412   Kyy_21 = detJ*Ndit(t,i)*Ndis(s,j);
413   %
414   % DEFINITION OF r = r1_SP1 AND SUBSTITUTION IN Kyy_1
415   %
416   r = r1_SP1;
417   Kyy_21 = subs(Kyy_21);
418   %
419   %
420   % ESECUTION OF THE ONE POINTS GAUSS QUADRATUE
421   %
422   Kyy_2 = alfa1_SP1*Kyy_21;
423   %
424   % CALCULATION OF THE Kyy TERM
425   %
426   Kyy = Kyy_1*Kyy_2;
427   %
428   % RESET OF THE VARIABLE r
429   %
430   clear r
431   syms r
432   %
433   %
434   %=================================================================
435   % COMPUTE K_yz
436   %
437   Kyz  = 0;
438   %
439   %=================================================================
440   %=================================================================
441   % COMPUTE K_zx
442   %
443   Kzx  = 0;
444   %
445   %=================================================================
446   %=================================================================
447   % COMPUTE K_zy
448   %
449   Kzy  = 0;
450   %
451   %=================================================================
452   %=================================================================
453   % COMPUTE K_zz
454   %
455   Kzz  = 0;
456   %
457   %=================================================================
458   %=================================================================
459   %=================================================================
460   %=================================================================
461   %=================================================================
462   %=================================================================
463   %=================================================================
464   %=================================================================
465   %
466   %
467   %=================================================================
468   %$$$$$$$$$$$$$$$$$$$$$$$$$$$$$$$$$$$$$$$$$$$$$$$$$$$$$$$$$$$$$$$$$$$
469   %$$$$$$$$$$$$$$$$$$$$$$$$$$$$$$$$$$$$$$$$$$$$$$$$$$$$$$$$$$$$$$$$$$$
470   %
471   % BEFORE WRITING MATRIX CUF NUCLEUS
472   % COMPUTATION OF THE FLEXIONAL TERM USING VALUES OF t, s, i, j
473   % %
474   % %
475   % Kyy_FLEX  = C11*int(int(Fty(t,i)*Fsy(s,j),z,-b,b),x,-a,a)...
476   %           *int(Jinv*Jinv*detJ*DNdit(t,i)*DNdis(s,j),r,-1,1);
```

```matlab
477  % %
478  %==========================================================================
479  % %
480  %  %^^^^^^^^^^^^^^^^^^^^^^^^^^^^^^^^^^^^^^^^^^^^^^^^^^^^^^^^^^^^^^^^^^^^^^^^
481  %  %^^^^^^^^^^^^^^^^^^^^^^^^^^^^^^^^^^^^^^^^^^^^^^^^^^^^^^^^^^^^^^^^^^^^^^^^
482  %  %^^^^^^^^^^^^^^^^^^^^^^^^^^^^^^^^^^^^^^^^^^^^^^^^^^^^^^^^^^^^^^^^^^^^^^^^
483  %
484  %==========================================================================
485  %
486  %
487  % THE FIRST PART OF THE Kyyy_FLEX_1 TERM IS CALCULATED ANALITICALLY
488  %
489  Kyy_FLEX_1 = C11*int(int(Fty(t,i)*Fsy(s,j),z,-b,b),x,-a,a);
490  %
491  %
492  % THE SECOND PART OF THE Kyy_FLEX TERM IS INTEGRATED USING THE 2 POINTS
493  % GAUSS QUADRATURE (NORMAL INTEGRATION RULE)
494  % THE APPROACH IS THE FOLLOWING :
495  % 1) MAKING OF 1 COPIES OF THE SECOND TERM:
496  % NOTE : COPIES OF MATRIX PARTS ARE DONE IN ORDER TO AVOID CONFLICT
497  % PROBLEMS
498  %
499  %
500  Kyy_FLEX_21 = Jinv*Jinv*detJ*DNdit(t,i)*DNdis(s,j);
501  %
502  % DEFINITION OF r = r1_SP1 AND SUBSTITUTION IN Kyy_FLEX_21
503  %
504  r = r1_SP1;
505  Kyy_FLEX_21 = subs(Kyy_FLEX_21);
506  %
507  %
508  % ESECUTION OF THE ONE POINTS GAUSS QUADRATUE
509  %
510  Kyy_FLEX_2 = alfa1_SP1*Kyy_FLEX_21;
511  %
512  % CALCULATION OF THE Kyy_FLEX TERM
513  %
514  Kyy_FLEX = Kyy_FLEX_1*Kyy_FLEX_2;
515  %
516  % RESET OF THE VARIABLE r
517  %
518  clear r
519  syms r
520  %
521  %
522  %==========================================================================
523  %$$$$$$$$$$$$$$$$$$$$$$$$$$$$$$$$$$$$$$$$$$$$$$$$$$$$$$$$$$$$$$$$$$$$$$$$$$$
524  %$$$$$$$$$$$$$$$$$$$$$$$$$$$$$$$$$$$$$$$$$$$$$$$$$$$$$$$$$$$$$$$$$$$$$$$$$$$
525  %
526  % WRITING OF CUF NUCLEUS
527  %
528  K(:,:,count1,count2) = [Kxx Kxy Kxz; Kyx (Kyy+Kyy_FLEX) Kyz; Kzx Kzy Kzz];
529  %
530  %
531  % K(:,:,count1,count2) = vpa((K(:,:,count1,count2)),4);
532  %
533  %????????????????????????????????????????????????????????????????????????
534  %
535  % WRITING OF CUF NUCLEUS [ ONLY 2 X 2 ]
536  %
537  K_TOT_2(:,:,count1,count2) = [Kxx Kxy ; Kyx (Kyy+Kyy_FLEX)];
538  %
539  %????????????????????????????????????????????????????????????????????????
540  % COMPUTE BENDING SYIFFNESS MATRIX
541  %
542  K_BENDING(:,:,count1,count2) = [0 0 0; 0 (0+Kyy_FLEX) 0; 0 0 0];
543  %
```

```matlab
544     %??????????????????????????????????????????????????????????????????
545     % COMPUTE BENDING SYIFFNESS MATRIX [ 2 x 2 ]
546     %
547     K_BENDING_2(:,:,count1,count2) = [0 0 ; 0 (0+Kyy_FLEX)];
548     %
549     %???????????????????????????????????????????????????????????????????
550     %
551     % COMPUTE SHEAR SYIFFNESS MATRIX
552     %
553     K_SHEAR(:,:,count1,count2) = [Kxx Kxy Kxz; Kyx (Kyy+0) Kyz; Kzx Kzy Kzz];
554     %
555     %
556     %?????????????????????????????????????????????????????????????????????
557     % COMPUTE SHEAR SYIFFNESS MATRIX [ 2 x 2 ]
558     %
559     K_SHEAR_2(:,:,count1,count2) = [Kxx Kxy ; Kyx (Kyy+0)];
560     %
561     %?????????????????????????????????????????????????????????????????????
562     %
563     %$$$$$$$$$$$$$$$$$$$$$$$$$$$$$$$$$$$$$$$$$$$$$$$$$$$$$$$$$$$$$$$$$$$$$$$$
564     %$$$$$$$$$$$$$$$$$$$$$$$$$$$$$$$$$$$$$$$$$$$$$$$$$$$$$$$$$$$$$$$$$$$$$$$$
565     %^^^^^^^^^^^^^^^^^^^^^^^^^^^^^^^^^^^^^^^^^^^^^^^^^^^^^^^^^^^^^^^^^^^^
566     %^^^^^^^^^^^^^^^^^^^^^^^^^^^^^^^^^^^^^^^^^^^^^^^^^^^^^^^^^^^^^^^^^^^^
567     %^^^^^^^^^^^^^^^^^^^^^^^^^^^^^^^^^^^^^^^^^^^^^^^^^^^^^^^^^^^^^^^^^^^^
568     %^^^^^^^^^^^^^^^^^^^^^^^^^^^^^^^^^^^^^^^^^^^^^^^^^^^^^^^^^^^^^^^^^^^^
569     %
570     % RESET TO 0 OF THE FOLLOWING MATRIX TERMS AFTER THE m n LOOP
571     %
572     K_xx = 0;
573     K_xy = 0;
574     K_xz = 0;
575     %
576     K_yx = 0;
577     K_yy = 0;
578     K_yz = 0;
579     %
580     K_zx = 0;
581     K_zy = 0;
582     K_zz = 0;
583     %
584     RIS_TOT = 0;
585     %
586     %????????????????????????????????????????????????????????????????????
587     %????????????????????????????????????????????????????????????????????
588     %
589     % MANAGE THE COUNTER countmat AND count1
590     %
591     count1 = count1 + 1;
592     %
593     %??????????????????????????????????????????????????????????????????????
594     %??????????????????????????????????????????????????????????????????????
595     %??????????????????????????????????????????????????????????????????????
596     %
597             end  % END OF t LOOP
598     %
599     %?????????????????????????????????????????????????????????????????????
600     %?????????????????????????????????????????????????????????????????????
601         end       % END OF i LOOP
602     %
603     %??????????????????????????????????????????????????????????????????????
604     %??????????????????????????????????????????????????????????????????????
605     %
606     % MANAGE THE COUNTER count11 AND count2
607     %
608     count1 = 1;
609     count2 = count2 +1;
610     %
```

```matlab
611  %????????????????????????????????????????????????????????????????
612  %????????????????????????????????????????????????????????????????
613  %
614      end    % END OF s LOOP
615      %
616  end   % END OF j LOOP
617  %
618  %================================================================
619  % MATR(t,s,i,j)
620  %
621  MATR = [K(:,:,1,1) K(:,:,2,1); K(:,:,1,2) K(:,:,2,2)]
622  %
623  MATR = vpa(MATR,3)
624  %
625  %
626  %   COMPUTE [ 3 X 3 ] MATRICES : CUF BASED MATRICES
627  %
628  %
629  % BENDING MATRIX
630  %
631  MATR_BENDING = [K_BENDING(:,:,1,1) K_BENDING(:,:,2,1); K_BENDING(:,:,1,2)
         K_BENDING(:,:,2,2)]
632  %
633  %
634  % SHEAR MATRIX
635  %
636  MATR_SHEAR = [K_SHEAR(:,:,1,1) K_SHEAR(:,:,2,1); K_SHEAR(:,:,1,2) K_SHEAR
         (:,:,2,2)]
637  %
638  %
639  %
640  %   COMPUTE [ 2 X 2 ] MATRICES FOR BATHE ALPHA -ALPHA DEMONSTRATION
641  %
642  MATR_TOT_2 = [K_TOT_2(:,:,1,1) K_TOT_2(:,:,2,1); K_TOT_2(:,:,1,2) K_TOT_2
         (:,:,2,2)]
643  %
644  MATR_BENDING_2 = [K_BENDING_2(:,:,1,1) K_BENDING_2(:,:,2,1); K_BENDING_2
         (:,:,1,2) K_BENDING_2(:,:,2,2)]
645  %
646  MATR_SHEAR_2 = [K_SHEAR_2(:,:,1,1) K_SHEAR_2(:,:,2,1); K_SHEAR_2(:,:,1,2)
         K_SHEAR_2(:,:,2,2)]
647  %
648  %^^^^^^^^^^^^^^^^^^^^^^^^^^^^^^^^^^^^^^^^^^^^^^^^^^^^^^^^^^^^^^^^^^^
649  %^^^^^^^^^^^^^^^^^^^^^^^^^^^^^^^^^^^^^^^^^^^^^^^^^^^^^^^^^^^^^^^^^^^
650  %
651  %
652  Fout = fopen('ELEMENTAL_Beam_Z_axis_Bending_Stiffness_Matrix_B2_3ELE_REDUCED.
         dat','w');
653  %
654  %////////////////////////////////////////////////////////////////
655  %////////////////////////////////////////////////////////////////
656  %
657  % MAKE A COPY OF MATR FOR PRINTING PURPOSE ONLY
658  %
659  MATPR = MATR;
660  MATPRR = vpa(MATPR,4);
661  %
662      fprintf(Fout,'\n\n\n\nSymbolic Stiffness Matrix of a Beam under Z axis
             bending\n\n\n');
663      fprintf(Fout,'\n\n\n\nREDUCED FORMULATION\n\n\n');
664      Columns = Num_nodes*3;
665      Rows = Num_nodes*3;
666  %
667  for m = 1:Rows;
668  %
669              if m==1
```

```matlab
670        formatSpec = '%s                    %s                  %s              %s
                       %s                %s              %s              %s
                       %s                %s              %s              %s
           |\n';
671        fprintf(Fout,formatSpec,MATPRR(m:m,1:Columns));
672        fprintf(Fout,'\n\n');
673        elseif m==3
674        formatSpec = '%s                    %s                  %s              %s
                       %s                %s              %s              %s
                       %s                %s              %s              %s
           |\n';
675        fprintf(Fout,formatSpec,MATPRR(m:m,1:Columns));
676        fprintf(Fout,'\n\n');
677        elseif m==5
678        formatSpec = '%s                    %s                  %s              %s
                       %s                %s              %s              %s
                       %s                %s              %s              %s
           |\n';
679        fprintf(Fout,formatSpec,MATPRR(m:m,1:Columns));
680        fprintf(Fout,'\n\n');
681        elseif m==7
682        formatSpec = '%s                    %s                  %s              %s
                       %s                %s              %s              %s
                       %s                %s              %s              %s
           |\n';
683        fprintf(Fout,formatSpec,MATPRR(m:m,1:Columns));
684        fprintf(Fout,'\n\n');
685        elseif m==9
686        formatSpec = '%s                    %s                  %s              %s
                       %s                %s              %s              %s
                       %s                %s              %s              %s
           |\n';
687        fprintf(Fout,formatSpec,MATPRR(m:m,1:Columns));
688        fprintf(Fout,'\n\n');
689        elseif m==11
690        formatSpec = '%s                    %s                  %s              %s
                       %s                %s              %s              %s
                       %s                %s              %s              %s
           |\n';
691        fprintf(Fout,formatSpec,MATPRR(m:m,1:Columns));
692        fprintf(Fout,'\n\n');
693        end
694        %
695        if m==2
696        formatSpec = '%s                    %s                  %s              %s
                       %s                %s              %s              %s
                       %s                %s              %s              %s
           |\n';
697        fprintf(Fout,formatSpec,MATPRR(m:m,1:Columns));
698        fprintf(Fout,'\n\n');
699        elseif m==4
700        formatSpec = '%s                    %s                  %s              %s
                       %s                %s              %s              %s
                       %s                %s              %s              %s
           |\n';
701        fprintf(Fout,formatSpec,MATPRR(m:m,1:Columns));
702        fprintf(Fout,'\n\n');
703        elseif m==6
704        formatSpec = '%s                    %s                  %s              %s
                       %s                %s              %s              %s
                       %s                %s              %s              %s
           |\n';
705        fprintf(Fout,formatSpec,MATPRR(m:m,1:Columns));
706        fprintf(Fout,'\n\n');
707        elseif m==8
708        formatSpec = '%s                    %s                  %s              %s
                       %s                %s              %s              %s
```

```matlab
                                %s              %s            %s           %s
                 |\n';
709             fprintf(Fout,formatSpec,MATPRR(m:m,1:Columns));
710             fprintf(Fout,'\n\n');
711             elseif m==10
712             formatSpec = '%s                %s            %s           %s
                                %s            %s          %s         %s
                                %s            %s          %s         %s
                 |\n';
713             fprintf(Fout,formatSpec,MATPRR(m:m,1:Columns));
714             fprintf(Fout,'\n\n');
715             elseif m==12
716             formatSpec = '%s                %s            %s           %s
                                %s          %s        %s       %s
                                %s          %s        %s       %s
                 |\n';
717             fprintf(Fout,formatSpec,MATPRR(m:m,1:Columns));
718             fprintf(Fout,'\n\n');
719             end
720             %
721     end
```

C.3   Selective Reduced Integration

The following the the script MATLAB for the evaluation of the stiffness matrix of a two-node beam element subjected to bending using the Selective Reduced Integration technique.

```matlab
1    %%%%%%%%%%%%%%%%%%%%%%%%%%%%%%%%%%%%%%%%%%%%%%%%%%%%%%%%%%%%%%%%
2    %
3    clc
4    %
5    %===========================================================
6    %===========================================================
7    %===========================================================
8    %
9    %              SYMBOLIC VARIABLES DECLARATION
10   %
11   syms a b L y x z C44 C11
12   syms r Jinv detJ
13   %
14   %===========================================================
15   %          SAMPLING POINTS POSITION AND WEIGHT DEFINITION
16   %===========================================================
17   %                 1 POINTS GAUSS QUADRATURE
18   %
19   r1_SP1 = 0;
20   alfa1_SP1 = 2;
21   %===========================================================
22   %                 2 POINTS GAUSS QUADRATURE
23   %
24   r1_SP2 = -1/sqrt(3);
25   r2_SP2 = +1/sqrt(3);
26   %
27   alfa1_SP2 = 1;
28   alfa2_SP2 = 1;
29   %===========================================================
30   %                 3 POINTS GAUSS QUADRATURE
31   %
```

```
32   r1_SP3  =  -0.774596669241483;
33   r2_SP3  =  0.;
34   r3_SP3  =  +0.774596669241483;
35   %
36   alfa1_SP3  =  0.555555555555556;
37   alfa2_SP3  =  0.888888888888889;
38   alfa3_SP3  =  0.555555555555556;
39   %
40   %=================================================================
41   %                        Polynomial expansion
42   %
43   tau  =  1;
44   sigma  =  1;
45   %
46   % DEFINITION OF tmax AND smax FOR CUF NUCLEI REVOVERY PURPOSE
47   %
48   smax=  sigma;
49   %
50   tmax  =  tau;
51   %
52   % ///////////////////////////////////////////////////////////////
53   % ///////////////////////////////////////////////////////////////
54   %
55   % Number of Nodes in the element ( Element type : B2)
56   %
57   %
58   Num_nodes  =  2;
59   %
60   %///////////////////////////////////////////////////////////////
61   %///////////////////////////////////////////////////////////////
62   %
63   %                        EXPANSION FUNCTIONS
64   %
65   %
66   % NOTE = Ftx(t,i) where t = 1 and i = 1 to 2
67   %
68   Ftx(1,1)  =  -1;
69   Ftx(1,2)  =  -1;
70   %
71   %
72   Fty(1,1)  =  x;
73   Fty(1,2)  =  x;
74   %
75   % NOT USED IN THIS CASE
76   %
77   Ftz(1,1)  =  0;
78   Ftz(1,2)  =  0;
79   %
80   %
81   Fsx(1,1)  =  -1;
82   Fsx(1,2)  =  -1;
83   %
84   %
85   Fsy(1,1)  =  x;
86   Fsy(1,2)  =  x;
87   %
88   % NOT USED IN THIS CASE
89   %
90   Fsz(1,1)  =  0;
91   Fsz(1,2)  =  0;
92   %
93   %=================================================================
94   %                        Jacobian Matrix
95   %
```

```matlab
96    %                                    J = |L/2|
97    %
98    %                         Jacobian Matrix Detrminant
99    %
100   detJ = L/2;
101   %
102   %                       Inverse of the Jacobian Matrix
103   %
104   Jinv = 2/L;
105   %
106   %===============================================================
107   %===============================================================
108   %
109   %   SHAPE FUNCTION INTERPOLATION OF THE DISPLACEMENTS FOR THE ELEMENT B2
110   %   IN NATURAL COORDINATE
111   %
112   Ndit(1,1) = (1/2)*(1-r);
113   Ndit(1,2) = (1/2)*(1+r);
114   %
115   %
116   %
117   Ndis(1,1) = (1/2)*(1-r);
118   Ndis(1,2) = (1/2)*(1+r);
119   %
120   %
121   %===============================================================
122   %   DERIVATIVES OF THE SHAPE FUNCTION INTERPOLATION OF THE
123   %   DISPLACEMENTS FOR THE ELEMENT B2.
124   %   NATURAL COORDINATES
125   %
126   %
127   %  NOTE = DNdt(t,i)
128   %         FOR B3 ELEMENT [ t = 1 , i = 1 TO 2 ]
129   %
130   %
131   %
132   DNdit(1,1) = diff(Ndit(1,1),r,1);
133   DNdit(1,2) = diff(Ndit(1,2),r,1);
134   %
135   %  NOTE = DNds(s,j)
136   %         FOR B3 ELEMENT [ s = 1 , j = 1 TO 2 ]
137
138   DNdis(1,1) = diff(Ndis(1,1),r,1);
139   DNdis(1,2) = diff(Ndis(1,2),r,1);
140   %===============================================================
141   % /////////////////////////////////////////////////////////////
142   % /////////////////////////////////////////////////////////////
143   % /////////////////////////////////////////////////////////////
144   %
145   %INITIALIZATION OF COUNTER
146   %
147   count1 = 1;
148   count2 = 1;
149   %
150   % /////////////////////////////////////////////////////////////
151   % /////////////////////////////////////////////////////////////
152   % /////////////////////////////////////////////////////////////
153   %
154   % PUT 0 THE FOLLOWING MATRICES
155   %
156   K_xx = 0;
157   K_xy = 0;
158   K_xz = 0;
159   %
```

```
160   K_yx = 0;
161   K_yy = 0;
162   K_yz = 0;
163   %
164   K_zx = 0;
165   K_zy = 0;
166   K_zz = 0;
167   %
168   RIS_TOT = 0;
169   %
170   %^^^^^^^^^^^^^^^^^^^^^^^^^^^^^^^^^^^^^^^^^^^^^^^^^^^^^^^^^^^^^^^^^^^^^^^^^^^
171   %^^^^^^^^^^^^^^^^^^^^^^^^^^^^^^^^^^^^^^^^^^^^^^^^^^^^^^^^^^^^^^^^^^^^^^^^^^
172   %^^^^^^^^^^^^^^^^^^^^^^^^^^^^^^^^^^^^^^^^^^^^^^^^^^^^^^^^^^^^^^^^^^^^^^^
173   %
174   %
175   %                       LOOP FOR MATRIX TERMS CALCULATION
176   %
177   for j = 1:Num_nodes;
178      for s = 1:sigma;
179         for i = 1:Num_nodes;
180            for t = 1:tau;
181   %
182   %=======================================================================
183   %((((((((((((((((((((((((((((((((((((((((((((((((((((((((((((((((((((((((
184   %//////////////////////////////////////////////////////////////////////
185   %))))))))))))))))))))))))))))))))))))))))))))))))))))))))))))))))))))))))
186   %=======================================================================
187   %
188   %^^^^^^^^^^^^^^^^^^^^^^^^^^^^^^^^^^^^^^^^^^^^^^^^^^^^^^^^^^^^^^^^^^^^^
189   %^^^^^^^^^^^^^^^^^^^^^^^^^^^^^^^^^^^^^^^^^^^^^^^^^^^^^^^^^^^^^^^^^^^^
190   %^^^^^^^^^^^^^^^^^^^^^^^^^^^^^^^^^^^^^^^^^^^^^^^^^^^^^^^^^^^^^^^^
191   %                       GENERAL PRELIMINARY INFORMATION
192   %
193   % GAUSS QUADRATURE CALCULATION FOR B2 ELEMENT:
194   %
195   % 1)  THE SHEAR TERMS OF THE OVERALL MATRIX HAVE A
196   %     MAXIMUM POLYNOMIAL DEGREE OF 2
197   % 2)  THE BENDING TERM OF THE MATRIX IS A CONSTANT TERM THEN THE
        POLYNOMIAL
198   %     DEGREE IS 0.
199   %
200   % THE GREATEST POLYNOMIAL DEGREE IS 2 FOR Kyy MATRIX TERM.
201   %
202   % THE EXACT CALCULATION IS DONE BY TWO POINTS GAUSS QUADRATURE
203   %
204   % THE UNIFORM REDUCED INTEGRATION IS DONE BY USING A
205   % 1 POINT GAUSS QUADRATURE
206   % THE SELECTIVE REDUCED INTEGRATION IS EXPLAINED BELOW.
207   %
208   %
209   %                       SELECTIVE REDUCED INTEGRATION
210   %
211   % A SELECTIVE REDUCED INTEGRATION RULE IS USED WHEN THE SHEAR TERMS OF
212   % THE  MATRIX IS INTEGRATED USING A REDUCED INTEGRATION (i.e. A LOWER
213   % ORDER INTEGRATION RULE IS USED) WHEREAS THE BENDIN TERM OF THW MATRIX
        IS
214   % INTEGRATED USING THE NORMAL INTEGRATION RULE.
215   % DUE TO THE FACT THAT THE NORMAL INTEGRATION RULE IS BASED ON TWO GAUSS
216   % POINTS QUADRATURE, THEN THE SELECTIVE REDUCED INTEGRATION CONSISTS
217   % IN THE FOLLOWING RULE:
218   % 1)  THE SHEAR TERMS OF THE  MATRIX IS INTEGRATED USING A 1 POINT
219   %     GAUSS QUADRATURE INSTEAD OF THE NORMAL INTEGRATION RULE BASED ON
220   %     A 2 POINTS GAUSS QUADRATURE
221   % 2)  THE BENDING TERMS OF THE  MATRIX IS INTEGRATED USING A 2 POINT
```

```
222   %      GAUSS QUADRATURE THAT IS THE NORMAL INTEGRATION RULE
223   %
224   % NOTE :
225   %
226   % THE FIRST PART OF THE Kxx TERM IS CALCULATED ANALITICALLY
227   %
228   Kxx_1 = C44*int(int(Ftx(t,i)*Fsx(s,j),z,-b,b),x,-a,a);
229   %
230   % THE SECOND PART OF THE Kxx TERM (THE SHEAR TERM)IS INTEGRATED USING
231   % THE 1 POINT GAUSS QUADRATURE (REDUCED INTEGRATION RULE).
232   %
233   % THE APPROACH IS THE FOLLOWING :
234   % 1) MAKING OF 1 COPIES OF THE SECOND TERM:
235   %
236   % NOTE : COPIES OF MATRIX PARTS ARE DONE IN ORDER TO AVOID CONFLICT
237   % PROBLEMS
238   %
239   Kxx_21 = Jinv*Jinv*detJ*DNdit(t,i)*DNdis(s,j);
240   %
241   % DEFINITION OF r = r1_SP1 AND SUBSTITUTION IN Kxx_1
242   %
243   r = r1_SP1;
244   Kxx_21 = subs(Kxx_21);
245   %
246   %
247   % ESECUTION OF THE TWO POINTS GAUSS QUADRATUE
248   %
249   Kxx_2 = alfa1_SP1*Kxx_21;
250   %
251   % CALCULATION OF THE Kxx TERM
252   %
253   Kxx = Kxx_1*Kxx_2;
254   %
255   % RESET OF THE VARIABLE r
256   %
257   clear r
258   syms r
259   %
260   %===============================================================
261   %((((((((((((((((((((((((((((((((((((((((((((((((((((((((((((((((
262   %//////////////////////////////////////////////////////////////
263   %))))))))))))))))))))))))))))))))))))))))))))))))))))))))))))))))
264   %===============================================================
265   % %
266   % % COMPUTE K_xy
267   % %
268   % Kxy  = C44*int(int(diff(Fty(t,i),x,1)*Fsx(s,j),z,-b,b),x,-a,a)...
269   %        *int(Jinv*detJ*Ndit(t,i)*DNdis(s,j),r,-1,1);
270   % %
271   % % Kxy = vpa((Kxy),4);
272   %
273   %^^^^^^^^^^^^^^^^^^^^^^^^^^^^^^^^^^^^^^^^^^^^^^^^^^^^^^^^^^^^^^^^^^
274   %^^^^^^^^^^^^^^^^^^^^^^^^^^^^^^^^^^^^^^^^^^^^^^^^^^^^^^^^^^^^^^^^^^
275   %^^^^^^^^^^^^^^^^^^^^^^^^^^^^^^^^^^^^^^^^^^^^^^^^^^^^^^^^^^^^^^^^^^
276   %
277   % THE FIRST PART OF THE KxY TERM IS CALCULATED ANALITICALLY
278   %
279   Kxy_1 = C44*int(int(diff(Fty(t,i),x,1)*Fsx(s,j),z,-b,b),x,-a,a);
280   %
281   %
282   % THE SECOND PART OF THE Kxy TERM IS INTEGRATED USING THE 1 POINT
283   % GAUSS QUADRATURE (REDUCED INTEGRATION RULE)
284   % THE APPROACH IS THE FOLLOWING :
285   % 1) MAKING OF 1 COPIES OF THE SECOND TERM:
```

```
286   % NOTE : COPIES OF MATRIX PARTS ARE DONE IN ORDER TO AVOID CONFLICT
287   % PROBLEMS
288   %
289   %
290   Kxy_21 = Jinv*detJ*Ndit(t,i)*DNdis(s,j);
291   %
292   % DEFINITION OF r = r1_SP1 AND SUBSTITUTION IN Kxy_1
293   %
294   r = r1_SP1;
295   Kxy_21 = subs(Kxy_21);
296   %
297   %
298   % ESECUTION OF THE ONE POINTS GAUSS QUADRATUE
299   %
300   Kxy_2 = alfa1_SP1*Kxy_21;
301   %
302   % CALCULATION OF THE Kxy TERM
303   %
304   Kxy = Kxy_1*Kxy_2;
305   %
306   % RESET OF THE VARIABLE r
307   %
308   clear r
309   syms r
310   %
311   %
312   %=========================================================================
313   %=========================================================================
314   % COMPUTE K_xz
315   %
316   Kxz   = 0;
317   %
318   % Kxz = vpa((Kxz),4);
319   %
320   %^^^^^^^^^^^^^^^^^^^^^^^^^^^^^^^^^^^^^^^^^^^^^^^^^^^^^^^^^^^^^^^^^^^^^^^^^
321   %^^^^^^^^^^^^^^^^^^^^^^^^^^^^^^^^^^^^^^^^^^^^^^^^^^^^^^^^^^^^^^^^^^^^^^^^^
322   %^^^^^^^^^^^^^^^^^^^^^^^^^^^^^^^^^^^^^^^^^^^^^^^^^^^^^^^^^^^^^^^^^^^^^^^^^
323   %
324   %
325   %=========================================================================
326   %=========================================================================
327   % % COMPUTE K_yx
328   % %
329   % Kyx   = C44*int(int(Ftx(t,i)*diff(Fsy(s,i),x,1),z,-b,b),x,-a,a)...
330   %         *int(Jinv*detJ*DNdit(t,i)*Ndis(s,j),r,-1,1);
331   % %
332   % % Kyx = vpa((Kyx),4);
333   % %
334   % %^^^^^^^^^^^^^^^^^^^^^^^^^^^^^^^^^^^^^^^^^^^^^^^^^^^^^^^^^^^^^^^^^^^^^^^
335   % %^^^^^^^^^^^^^^^^^^^^^^^^^^^^^^^^^^^^^^^^^^^^^^^^^^^^^^^^^^^^^^^^^^^^^^^
336   % %^^^^^^^^^^^^^^^^^^^^^^^^^^^^^^^^^^^^^^^^^^^^^^^^^^^^^^^^^^^^^^^^^^^^^^^
337   %
338   %=========================================================================
339   %
340   %
341   % THE FIRST PART OF THE Kyx TERM IS CALCULATED ANALITICALLY
342   %
343   Kyx_1 = C44*int(int(Ftx(t,i)*diff(Fsy(s,i),x,1),z,-b,b),x,-a,a);
344   %
345   %
346   % THE SECOND PART OF THE Kyx TERM IS INTEGRATED USING THE ONE POINT
347   % GAUSS QUADRATURE (REDUCED INTEGRATION RULE)
348   % THE APPROACH IS THE FOLLOWING :
349   % 1) MAKING OF 1 COPIES OF THE SECOND TERM:
```

```matlab
350   % NOTE : COPIES OF MATRIX PARTS ARE DONE IN ORDER TO AVOID CONFLICT
351   % PROBLEMS
352   %
353   %
354   Kyx_21 = Jinv*detJ*DNdit(t,i)*Ndis(s,j);
355   %
356   % DEFINITION OF r = r1_SP1 AND SUBSTITUTION IN Kyx_1
357   %
358   r = r1_SP1;
359   Kyx_21 = subs(Kyx_21);
360   %
361   %
362   % ESECUTION OF THE ONE POINT GAUSS QUADRATUE
363   %
364   Kyx_2 = alfa1_SP1*Kyx_21;
365   %
366   % CALCULATION OF THE Kyx TERM
367   %
368   Kyx = Kyx_1*Kyx_2;
369   %
370   % RESET OF THE VARIABLE r
371   %
372   clear r
373   syms r
374   %
375   %
376   %=============================================================
377   % COMPUTE K_yy
378   % %
379   % Kyy   = C44*int(int(diff(Fty(t,i),x,1)*diff(Fsy(s,i),x,1),z,-b,b),x,-a,a
          )...
380   %            *int(detJ*Ndit(t,i)*Ndis(s,j),r,-1,1);
381   % %
382   % % Kyy = vpa((Kyy),4);
383   %=============================================================
384   % %
385   % %^^^^^^^^^^^^^^^^^^^^^^^^^^^^^^^^^^^^^^^^^^^^^^^^^^^^^^^^^^^^
386   % %^^^^^^^^^^^^^^^^^^^^^^^^^^^^^^^^^^^^^^^^^^^^^^^^^^^^^^^^^^^^
387   % %^^^^^^^^^^^^^^^^^^^^^^^^^^^^^^^^^^^^^^^^^^^^^^^^^^^^^^^^^^^^
388   %
389   %=============================================================
390   %
391   %
392   % THE FIRST PART OF THE Kyy TERM IS CALCULATED ANALITICALLY
393   %
394   Kyy_1 = C44*int(int(diff(Fty(t,i),x,1)*diff(Fsy(s,i),x,1),z,-b,b),x,-a,a)
          ;
395   %
396   %
397   % THE SECOND PART OF THE Kyy TERM IS INTEGRATED USING THE ONE POINT
398   % GAUSS QUADRATURE (REDUCED INTEGRATION RULE)
399   % THE APPROACH IS THE FOLLOWING :
400   % 1) MAKING OF 1 COPIES OF THE SECOND TERM:
401   % NOTE : COPIES OF MATRIX PARTS ARE DONE IN ORDER TO AVOID CONFLICT
402   % PROBLEMS
403   %
404   %
405   Kyy_21 = detJ*Ndit(t,i)*Ndis(s,j);
406   %
407   % DEFINITION OF r = r1_SP1 AND SUBSTITUTION IN Kyy_1
408   %
409   r = r1_SP1;
410   Kyy_21 = subs(Kyy_21);
411   %
```

```
412   %
413   % ESECUTION OF THE ONE POINTS GAUSS QUADRATUE
414   %
415   Kyy_2 = alfa1_SP1*Kyy_21;
416   %
417   % CALCULATION OF THE Kyy TERM
418   %
419   Kyy = Kyy_1*Kyy_2;
420   %
421   % RESET OF THE VARIABLE r
422   %
423   clear r
424   syms r
425   %
426   %
427   %================================================================
428   % COMPUTE K_yz
429   %
430   Kyz   = 0;
431   %
432   % Kyz = vpa((Kyz),4);
433   %================================================================
434   %================================================================
435   % COMPUTE K_zx
436   %
437   Kzx   = 0;
438   %
439   % Kzx = vpa((Kzx),4);
440   %================================================================
441   %================================================================
442   % COMPUTE K_zy
443   %
444   Kzy   = 0;
445   %
446   % Kzy = vpa((Kzy),4);
447   %================================================================
448   %================================================================
449   % COMPUTE K_zz
450   %
451   Kzz   = 0;
452   %
453   % Kzz = vpa((Kzz),4);
454   %================================================================
455   %================================================================
456   %================================================================
457   %================================================================
458   %================================================================
459   %================================================================
460   %================================================================
461   %================================================================
462   %
463   %
464   %================================================================
465   %$$$$$$$$$$$$$$$$$$$$$$$$$$$$$$$$$$$$$$$$$$$$$$$$$$$$$$$$$$$$$$$$$$
466   %$$$$$$$$$$$$$$$$$$$$$$$$$$$$$$$$$$$$$$$$$$$$$$$$$$$$$$$$$$$$$$$$$$
467   %
468   % BEFORE WRITING MATRIX CUF NUMCLEUS
469   % COMPUTATION OF THE FLEXIONAL TERM USING VALUES OF t, s, i, j
470   % %
471   % %
472   % Kyy_FLEX   = C11*int(int(Fty(t,i)*Fsy(s,j),z,-b,b),x,-a,a)...
473   %            *int(Jinv*Jinv*detJ*DNdit(t,i)*DNdis(s,j),r,-1,1);
474   % %
475   %================================================================
```

```
476   %  %
477   %  %^^^^^^^^^^^^^^^^^^^^^^^^^^^^^^^^^^^^^^^^^^^^^^^^^^^^^^^^^^^^^^^^^
478   %  %^^^^^^^^^^^^^^^^^^^^^^^^^^^^^^^^^^^^^^^^^^^^^^^^^^^^^^^^^^^^^^^^^
479   %  %^^^^^^^^^^^^^^^^^^^^^^^^^^^^^^^^^^^^^^^^^^^^^^^^^^^^^^^^^^^^^^^^^
480   %
481   %=================================================================
482   %
483   %
484   % THE FIRST PART OF THE Kyyy_FLEX_1 TERM IS CALCULATED ANALITICALLY
485   %
486   Kyy_FLEX_1 = C11*int(int(Fty(t,i)*Fsy(s,j),z,-b,b),x,-a,a);
487   %
488   %
489   % THE SECOND PART OF THE Kyy_FLEX TERM IS INTEGRATED USING THE 2 POINTS
490   % GAUSS QUADRATURE (NORMAL INTEGRATION RULE)
491   % THE APPROACH IS THE FOLLOWING :
492   % 1) MAKING OF 2 COPIES OF THE SECOND TERM:
493   %
494   Kyy_FLEX_21 = Jinv*Jinv*detJ*DNdit(t,i)*DNdis(s,j);
495   Kyy_FLEX_22 = Jinv*Jinv*detJ*DNdit(t,i)*DNdis(s,j);
496   %
497   % DEFINITION OF r = r1_SP2 AND SUBSTITUTION IN Kyy_FLEX_21 AND
        Kyy_FLEX_22
498   %
499   r = r1_SP2;
500   Kyy_FLEX_21 = subs(Kyy_FLEX_21);
501   %
502   r = r2_SP2;
503   Kyy_FLEX_22 = subs(Kyy_FLEX_22);
504   %
505   % ESECUTION OF THE TWO POINTS GAUSS QUADRATUE
506   %
507   Kyy_FLEX_2 = alfa1_SP2*Kyy_FLEX_21 + alfa2_SP2*Kyy_FLEX_22;
508   %
509   % CALCULATION OF THE Kyy_FLEX TERM
510   %
511   Kyy_FLEX = Kyy_FLEX_1*Kyy_FLEX_2;
512   %
513   % RESET OF THE VARIABLE r
514   %
515   clear r
516   syms r
517   %
518   %
519   %=================================================================
520   %$$$$$$$$$$$$$$$$$$$$$$$$$$$$$$$$$$$$$$$$$$$$$$$$$$$$$$$$$$$$$$$$$$$$
521   %$$$$$$$$$$$$$$$$$$$$$$$$$$$$$$$$$$$$$$$$$$$$$$$$$$$$$$$$$$$$$$$$$$$$
522   %
523   % WRITING OF CUF NUCLEUS
524   %
525   K(:,:,count1,count2) = [Kxx Kxy Kxz; Kyx (Kyy+Kyy_FLEX) Kyz; Kzx Kzy Kzz
        ];
526   %
527   %
528   % K(:,:,count1,count2) = vpa((K(:,:,count1,count2)),4);
529   %
530   %????????????????????????????????????????????????????????????????
531   %
532   % WRITING OF CUF NUCLEUS [ ONLY 2 X 2 ]
533   %
534   K_TOT_2(:,:,count1,count2) = [Kxx Kxy ; Kyx (Kyy+Kyy_FLEX)];
535   %
536   %????????????????????????????????????????????????????????????????
537   % COMPUTE BENDING SYIFFNESS MATRIX
```

```matlab
538    %
539    K_BENDING(:,:,count1,count2) = [0 0 0; 0 (0+Kyy_FLEX) 0; 0 0 0];
540    %
541    %?????????????????????????????????????????????????????????????????
542    % COMPUTE BENDING SYIFFNESS MATRIX [ 2 x 2 ]
543    %
544    K_BENDING_2(:,:,count1,count2) = [0 0 ; 0 (0+Kyy_FLEX)];
545    %
546    %?????????????????????????????????????????????????????????????????
547    %
548    % COMPUTE SHEAR SYIFFNESS MATRIX
549    %
550    K_SHEAR(:,:,count1,count2) = [Kxx Kxy Kxz; Kyx (Kyy+0) Kyz; Kzx Kzy Kzz];
551    %
552    %
553    %?????????????????????????????????????????????????????????????????
554    % COMPUTE SHEAR SYIFFNESS MATRIX [ 2 x 2 ]
555    %
556    K_SHEAR_2(:,:,count1,count2) = [Kxx Kxy ; Kyx (Kyy+0)];
557    %
558    %?????????????????????????????????????????????????????????????????
559    %
560    %$$$$$$$$$$$$$$$$$$$$$$$$$$$$$$$$$$$$$$$$$$$$$$$$$$$$$$$$$$$$$$$$$$$$
561    %$$$$$$$$$$$$$$$$$$$$$$$$$$$$$$$$$$$$$$$$$$$$$$$$$$$$$$$$$$$$$$$$$$$$
562    %^^^^^^^^^^^^^^^^^^^^^^^^^^^^^^^^^^^^^^^^^^^^^^^^^^^^^^^^^^^^^^^^^^^^
563    %^^^^^^^^^^^^^^^^^^^^^^^^^^^^^^^^^^^^^^^^^^^^^^^^^^^^^^^^^^^^^^^^^^^^
564    %^^^^^^^^^^^^^^^^^^^^^^^^^^^^^^^^^^^^^^^^^^^^^^^^^^^^^^^^^^^^^^^^^^^^
565    %^^^^^^^^^^^^^^^^^^^^^^^^^^^^^^^^^^^^^^^^^^^^^^^^^^^^^^^^^^^^^^^^^^^^
566    %
567    % RESET TO 0 OF THE FOLLOWING MATRIX TERMS AFTER THE m n LOOP
568    %
569    K_xx = 0;
570    K_xy = 0;
571    K_xz = 0;
572    %
573    K_yx = 0;
574    K_yy = 0;
575    K_yz = 0;
576    %
577    K_zx = 0;
578    K_zy = 0;
579    K_zz = 0;
580    %
581    RIS_TOT = 0;
582    %
583    %???????????????????????????????????????????????????????????????????
584    %???????????????????????????????????????????????????????????????????
585    %
586    % MANAGE THE COUNTER countmat AND count1
587    %
588    count1 = count1 + 1;
589    %
590    %???????????????????????????????????????????????????????????????????
591    %???????????????????????????????????????????????????????????????????
592    %???????????????????????????????????????????????????????????????????
593    %
594              end  % END OF t LOOP
595    %
596    %???????????????????????????????????????????????????????????????????
597    %?????????????Shear Locking Correction??????????????????????????????
598         end       % END OF i LOOP
599    %
600    %???????????????????????????????????????????????????????????????????
601    %???????????????????????????????????????????????????????????????????
```

```matlab
602  %
603  % MANAGE THE COUNTER count11 AND count2
604  %
605  count1 = 1;
606  count2 = count2 +1;
607  %
608  %???????????????????????????????????????????????????????????????????
609  %???????????????????????????????????????????????????????????????????
610  %
611      end    % END OF s LOOP
612      %
613  end % END OF j LOOP
614  %
615  %======================================================================
616  % MATR(t,s,i,j)
617  %
618  MATR = [K(:,:,1,1) K(:,:,2,1); K(:,:,1,2) K(:,:,2,2)]
619  %
620  MATR = vpa(MATR,3)
621  %
622  %
623  %  COMPUTE [ 3 X 3 ] MATRICES : CUF BASED MATRICES
624  %
625  %
626  % BENDING MATRIX
627  %
628  MATR_BENDING = [K_BENDING(:,:,1,1) K_BENDING(:,:,2,1); K_BENDING(:,:,1,2)
         K_BENDING(:,:,2,2)]
629  %
630  %
631  % SHEAR MATRIX
632  %
633  MATR_SHEAR = [K_SHEAR(:,:,1,1) K_SHEAR(:,:,2,1); K_SHEAR(:,:,1,2) K_SHEAR
         (:,:,2,2)]
634  %
635  %
636  %
637  %  COMPUTE [ 2 X 2 ] MATRICES FOR BATHE ALPHA -ALPHA DEMONSTRATION
638  %
639  MATR_TOT_2 = [K_TOT_2(:,:,1,1) K_TOT_2(:,:,2,1); K_TOT_2(:,:,1,2) K_TOT_2
         (:,:,2,2)]
640  %
641  MATR_BENDING_2 = [K_BENDING_2(:,:,1,1) K_BENDING_2(:,:,2,1); K_BENDING_2
         (:,:,1,2) K_BENDING_2(:,:,2,2)]
642  %
643  MATR_SHEAR_2 = [K_SHEAR_2(:,:,1,1) K_SHEAR_2(:,:,2,1); K_SHEAR_2(:,:,1,2)
         K_SHEAR_2(:,:,2,2)]
644  %
645  %^^^^^^^^^^^^^^^^^^^^^^^^^^^^^^^^^^^^^^^^^^^^^^^^^^^^^^^^^^^^^^^^^^^^^^^^
646  %^^^^^^^^^^^^^^^^^^^^^^^^^^^^^^^^^^^^^^^^^^^^^^^^^^^^^^^^^^^^^^^^^^^^^^^^
647  %^^^^^^^^^^^^^^^^^^^^^^^^^^^^^^^^^^^^^^^^^^^^^^^^^^^^^^^^^^^^^^^^^^^^^^^^
648  %
649  % Open file : Bar_Symbolic_Stiffness_Matrix.dat
650  %
651  Fout = fopen('
         ELEMENTAL_Beam_Z_axis_Bending_Stiffness_Matrix_B2_3ELE_SELECTIVE.dat
         ','w');
652  %
653  %////////////////////////////////////////////////////////////////////////
654  %////////////////////////////////////////////////////////////////////////
655  %
656  % MAKE A COPY OF MATR FOR PRINTING PURPOSE ONLY
657  %
658  MATPR = MATR;
```

```matlab
659    MATPRR = vpa(MATPR,4);
660    %
661        fprintf(Fout,'\n\n\n\nSymbolic Stiffness Matrix of a Beam under Z
                axis bending\n\n\n');
662        fprintf(Fout,'\n\n\n\nSELECTIVE REDUCED FORMULATION\n\n\n');
663        Columns = Num_nodes*3;
664        Rows = Num_nodes*3;
665    %
666    for m = 1:Rows;
667    %
668            if m==1
669            formatSpec = '%s                    %s                %s                %s
                          %s                %s                %s
                          %s                %s                %s
                          %s                %s|\n';
670            fprintf(Fout,formatSpec,MATPRR(m:m,1:Columns));
671            fprintf(Fout,'\n\n');
672            elseif m==3
673            formatSpec = '%s                    %s                %s                %s
                          %s                %s                %s
                          %s                %s                %s
                          %s                %s|\n';
674            fprintf(Fout,formatSpec,MATPRR(m:m,1:Columns));
675            fprintf(Fout,'\n\n');
676            elseif m==5
677            formatSpec = '%s                    %s                %s                %s
                          %s                %s                %s
                          %s                %s                %s
                          %s                %s|\n';
678            fprintf(Fout,formatSpec,MATPRR(m:m,1:Columns));
679            fprintf(Fout,'\n\n');
680            elseif m==7
681            formatSpec = '%s                    %s                %s                %s
                          %s                %s                %s
                          %s                %s                %s
                          %s                %s|\n';
682            fprintf(Fout,formatSpec,MATPRR(m:m,1:Columns));
683            fprintf(Fout,'\n\n');
684            elseif m==9
685            formatSpec = '%s                    %s                %s                %s
                          %s                %s                %s
                          %s                %s                %s
                          %s                %s|\n';
686            fprintf(Fout,formatSpec,MATPRR(m:m,1:Columns));
687            fprintf(Fout,'\n\n');
688            elseif m==11
689            formatSpec = '%s                    %s                %s                %s
                          %s                %s                %s
                          %s                %s                %s
                          %s                %s|\n';
690            fprintf(Fout,formatSpec,MATPRR(m:m,1:Columns));
691            fprintf(Fout,'\n\n');
692            end
693            %
694            if m==2
695            formatSpec = '%s                    %s                %s                %s
                          %s                %s                %s
                          %s                %s                %s
                          %s                %s|\n';
696            fprintf(Fout,formatSpec,MATPRR(m:m,1:Columns));
697            fprintf(Fout,'\n\n');
698            elseif m==4
699            formatSpec = '%s                    %s                %s                %s
                          %s                %s                %s
```

```
700          fprintf(Fout,formatSpec,MATPRR(m:m,1:Columns));
701          fprintf(Fout,'\n\n');
702          elseif m==6
703          formatSpec = '%s                    %s                    %s                    %s
                           %s                    %s                    %s
                           %s                    %s                    %s
                           %s                    %s|\n';
704          fprintf(Fout,formatSpec,MATPRR(m:m,1:Columns));
705          fprintf(Fout,'\n\n');
706          elseif m==8
707          formatSpec = '%s                    %s                    %s                    %s
                           %s                    %s                    %s
                           %s                    %s                    %s
                           %s                    %s|\n';
708          fprintf(Fout,formatSpec,MATPRR(m:m,1:Columns));
709          fprintf(Fout,'\n\n');
710          elseif m==10
711          formatSpec = '%s                    %s                    %s                    %s
                           %s                    %s                    %s
                           %s                    %s                    %s
                           %s                    %s|\n';
712          fprintf(Fout,formatSpec,MATPRR(m:m,1:Columns));
713          fprintf(Fout,'\n\n');
714          elseif m==12
715          formatSpec = '%s                    %s                    %s                    %s
                           %s                    %s                    %s
                           %s                    %s                    %s
                           %s                    %s|\n';
716          fprintf(Fout,formatSpec,MATPRR(m:m,1:Columns));
717          fprintf(Fout,'\n\n');
718          end
719          %
720 end
```

C.4   Mixed Interpolation of Tensorial Components

The following the the script MATLAB for the evaluation of the stiffness matrix of
a two-node beam element subjected to bending using the Mixed Interpolation of
Tensorial Components technique.

```
1  %%%%%%%%%%%%%%%%%%%%%%%%%%%%%%%%%%%%%%%%%%%%%%%%%%%%%%%%%%%%%%%%%%%%%%%%%%%%%%
2  %
3  clc
4  %
5  %==========================================================================
6  %==========================================================================
7  %==========================================================================
8  %
9  %                 SYMBOLIC  VARIABLES  DECLARATION
10 %
11 syms a b L y x z C44 C11
12 %
13 %==========================================================================
14 %==========================================================================
15 %==========================================================================
16 %==========================================================================
```

```
17   % Polynomial expansion
18   %
19   tau = 1;
20   sigma = 1;
21   %
22   % DEFINITION OF tmax AND smax FOR CUF NUCLEI REVOVERY PURPOSE
23   %
24   smax= sigma;
25   %
26   tmax = tau;
27   %
28   % ////////////////////////////////////////////////////////////////////
29   % ////////////////////////////////////////////////////////////////////
30   %
31   % Number of Nodes in the element ( Element type : B2)
32   %
33   %
34   Num_nodes = 2;
35   %
36   % ////////////////////////////////////////////////////////////////////
37   % ////////////////////////////////////////////////////////////////////
38   %
39   % PUT m == mm AND n == nn
40   % TYING POINTS NUMBER FOR B2 ELEMENT
41   %
42   mm = 1;
43   nn = 1;
44   %
45   % ////////////////////////////////////////////////////////////////////
46   % ////////////////////////////////////////////////////////////////////
47   %
48   %                     EXPANSION FUNCTIONS
49   %
50   % F(t,i) CON t = 1 E i = 1 TO 4
51   %
52   Ftx(1,1)  =  -1;
53   Ftx(1,2)  =  -1;
54   %
55   %
56   Fty(1,1)  =  x;
57   Fty(1,2)  =  x;
58   %
59   %
60   Ftz(1,1)  =  0;
61   Ftz(1,2)  =  0;
62   %
63   %
64   Fsx(1,1)  =  -1;
65   Fsx(1,2)  =  -1;
66   %
67   %
68   Fsy(1,1)  =  x;
69   Fsy(1,2)  =  x;
70   %
71   %
72   Fsz(1,1)  =  0;
73   Fsz(1,2)  =  0;
74   %
75   % ====================================================================
76   % ====================================================================
77   %    VALUE OF THE SHAPE FUNCTION FOR STRAINS INTERPOLATION
78   %    AT TYING POINT yT1, yT2 AND yT3
```

```
79   %
80   % yT1 = L / 2
81   %
82   %      SHAPE FUNCTION FOR STRAINS INTERPOLATION
83   %
84   %
85   % Ns1 = 1
86   %
87   %
88   % NOTE :   THE STRAINS SHAPE FUNZION INTERPOLATION Ns1 IS EQUAL TO :
89   %            1 AT yT1
90   %
91   % BASED ON THE ABOVE ASSUMPTIONS ONE WRITES, PUTTING THE STRAINS SHAPE
92   % FUNCTION INTERPOLATION AS Ns, :
93   %
94   % Ns(k,Tyings) > k = SHAPE NUMBER (1 FOR SHAPE 1)
95   %                 Tying = 1( Tying Points)
96   %
97   Ns(1,1) = 1;
98   %
99   %===============================================================
100  %===============================================================
101  %
102  %   SHAPE FUNCTIONS FOR DISPLACEMENTS FOR ELEMENT B2
103  %
104  % Ndit(t,i) WITH t = 1 AND i = 1 TO 2
105  %
106  % Shape Function for Node 1 :
107  Ndit(1,1) = 1-y/L;
108  %
109  % Shape Function for Node 2 :
110  Ndit(1,2) = y/L;
111  %
112  %
113  % Shape Function for Node 1 :
114  Ndis(1,1) = 1-y/L;
115  %
116  % Shape Function for Node 2 :
117  Ndis(1,2) = y/L;
118  %
119  %
120  %
121  %
122  %===============================================================
123  %===============================================================
124  %
125  %   VALUE OF THE SHAPE FUNCTIONS FOR DISPLACEMENTS FOR ELEMENT B2
126  %   AT TYING POINT yT1, yT2 AND yT3
127  %
128  % yT1 = (L/2)
129  %
130  %   Ndt(t,i,m)  > t = 1 ; i = 1 TO 2 ( i.e. N1 TO N2); m = 1 .
131  %
132  Ndt(1,1,1) =   1/2;
133  %
134  Ndt(1,2,1) =   1/2;
135  %
136  %   Ndt(s,j,n)  > s = 1 ; j = 1 TO 2 ; n = 1 .
137  %
138  %
139  %
140  Nds(1,1,1) =   1/2;
```

```
141   %
142   Nds(1,2,1)  =   1/2;
143   %
144   %=================================================================
145   % COMPUTATION OF THE DERIVATIVES, WITH RESPECT TO y, OF THE SHAPE
146           FUNCTIONS
      % FOR DISPLACEMENTS FOR ELEMENT B2
147   % THE SHAPE FUNCTIONS ARE THOSE REPORTED PREVIOUSLY AND ARE
148   %
149   % Ndit(1,1), Ndit(1,2)
150   %
151   % AND
152   %
153   % Ndis(1,1), Ndis(1,2)
154   %
155   %
156   DNdit(1,1)  =  diff(Ndit(1,1),y,1);
157   DNdit(1,2)  =  diff(Ndit(1,2),y,1);
158   %
159   DNdis(1,1)  =  diff(Ndis(1,1),y,1);
160   DNdis(1,2)  =  diff(Ndis(1,2),y,1);
161   %=================================================================
162   %
163   %
164   %   VALUE OF THE DERIVATIVES OF THE SHAPE FUNCTIONS FOR DISPLACEMENTS
165   %   FOR ELEMENT B2 AT TYING POINT yT1
166   %
167   % yT1 = (L/2)
168   %
169   %   DNdt(t,i,m)  > t = 1 ; i = 1 TO 2 (i.e. N1 TO N2); m = 1.
170   %
171   %
172   DNdt(1,1,1)  =  -1/L;
173   %
174   DNdt(1,2,1)  =   1/L;
175   %
176   %
177   %   Ndt(s,j,n)  > s = 1 ; j = 1 TO 2 (i.e. N1 TO N2); n = 1.
178   %
179   %
180   %
181   DNds(1,1,1)  =  -1/L;
182   %
183   DNds(1,2,1)  =   1/L;
184   %
185   %=================================================================
186   % ///////////////////////////////////////////////////////////////
187   % ///////////////////////////////////////////////////////////////
188   % ///////////////////////////////////////////////////////////////
189   %
190   %INITIALIZATION OF COUNTER
191   %
192   count1 = 1;
193   count2 = 1;
194   %
195   % ///////////////////////////////////////////////////////////////
196   % ///////////////////////////////////////////////////////////////
197   % ///////////////////////////////////////////////////////////////
198   %
199   % PUT 0 THE FOLLOWING MATRICES
200   %
201   K_xx = 0;
```

```matlab
202   K_xy  = 0;
203   K_xz  = 0;
204   %
205   K_yx  = 0;
206   K_yy  = 0;
207   K_yz  = 0;
208   %
209   K_zx  = 0;
210   K_zy  = 0;
211   K_zz  = 0;
212   %
213   RIS_TOT = 0;
214   %
215   %^^^^^^^^^^^^^^^^^^^^^^^^^^^^^^^^^^^^^^^^^^^^^^^^^^^^^^^^^^^^^^^^^^^^^^^^^^
216   %^^^^^^^^^^^^^^^^^^^^^^^^^^^^^^^^^^^^^^^^^^^^^^^^^^^^^^^^^^^^^^^^^^^^^^^^^^
217   %^^^^^^^^^^^^^^^^^^^^^^^^^^^^^^^^^^^^^^^^^^^^^^^^^^^^^^^^^^^^^^^^^^^^^^^^^^
218   %
219   %                     LOOP FOR MATRIX TERMS CALCULATION
220   %
221   for  j = 1:Num_nodes;
222      for s = 1:sigma;
223         for i = 1:Num_nodes;
224            for t = 1:tau;
225               for m = 1:mm;
226                  for n = 1:nn;
227   %
228   %=================================================================
229   % COMPUTE K_xx
230   %
231   Kxx   = C44*int(int(Ftx(t,i)*Fsx(s,j),z,-b,b),x,-a,a)...
232         *Ns(m,n)*Ns(n,m)*int(DNdt(t,i,m)*DNds(s,j,n),y,0,L);
233   %
234   Kxx = vpa((Kxx),4);
235   K_xx = vpa((Kxx+K_xx),4);
236   %=================================================================
237   %=================================================================
238   % COMPUTE K_xy
239   %
240   Kxy   = C44*int(int(diff(Fty(t,i),x,1)*Fsx(s,j),z,-b,b),x,-a,a)...
241         *Ns(m,n)*Ns(n,m)*int(Ndt(t,i,m)*DNds(s,j,n),y,0,L);
242   %
243   Kxy = vpa((Kxy),4);
244   K_xy = vpa((Kxy+K_xy),4);
245   %=================================================================
246   %=================================================================
247   % COMPUTE K_xz
248   %
249   Kxz   = 0;
250   %
251   Kxz = vpa((Kxz),4);
252   K_xz = vpa((Kxz+K_xz),4);
253   %=================================================================
254   %=================================================================
255   % COMPUTE K_yx
256   %
257   Kyx   = C44*int(int(Ftx(t,i)*diff(Fsy(s,i),x,1),z,-b,b),x,-a,a)...
258         *Ns(m,n)*Ns(n,m)*int(DNdt(t,i,m)*Nds(s,j,n),y,0,L);
259   %
260   Kyx = vpa((Kyx),4);
261   K_yx = vpa((Kyx+K_yx),4);
262   %=================================================================
263   %=================================================================
```

```
264   % COMPUTE K_yy
265   %
266   Kyy   = C44*int(int(diff(Fty(t,i),x,1)*diff(Fsy(s,i),x,1),z,-b,b),x,-a,
          a)...
267          *Ns(m,n)*Ns(n,m)*int(Ndt(t,i,m)*Nds(s,j,n),y,0,L);
268   %
269   Kyy = vpa((Kyy),4);
270   K_yy = vpa((Kyy+K_yy),4);
271   %==================================================================
272   %==================================================================
273   % COMPUTE K_yz
274   %
275   Kyz   = 0;
276   %
277   Kyz = vpa((Kyz),4);
278   K_yz = vpa((Kyz+K_yz),4);
279   %==================================================================
280   %==================================================================
281   % COMPUTE K_zx
282   %
283   Kzx   = 0;
284   %
285   Kzx = vpa((Kzx),4);
286   K_zx = vpa((Kzx+K_zx),4);
287   %==================================================================
288   %==================================================================
289   % COMPUTE K_zy
290   %
291   Kzy   = 0;
292   %
293   Kzy = vpa((Kzy),4);
294   K_zy = vpa((Kzy+K_zy),4);
295   %==================================================================
296   %==================================================================
297   % COMPUTE K_zz
298   %
299   Kzz   = 0;
300   %
301   Kzz = vpa((Kzz),4);
302   K_zz = vpa((Kzz+K_zz),4);
303   %==================================================================
304   %==================================================================
305   %==================================================================
306   %==================================================================
307   %==================================================================
308   %==================================================================
309   %==================================================================
310   %==================================================================
311   %
312                    end % END OF LOOP ON n
313                  end   % END OF LOOP ON m
314   %
315   %==================================================================
316   %$$$$$$$$$$$$$$$$$$$$$$$$$$$$$$$$$$$$$$$$$$$$$$$$$$$$$$$$$$$$$$$$$$$$
317   %$$$$$$$$$$$$$$$$$$$$$$$$$$$$$$$$$$$$$$$$$$$$$$$$$$$$$$$$$$$$$$$$$$$$
318   %
319   % BEFORE WRITING MATRIX CUL NUMCLEUS
320   % COMPUTATION OF THE FLEXIONAL TERM USING VALUES OF t, s, i, j
321   %
322   %
323   K_yy_FLEX   = C11*int(int(Fty(t,i)*Fsy(s,j),z,-b,b),x,-a,a)...
324          *int(DNdit(t,i)*DNdis(s,j),y,0,L);
```

```
325  %
326  %$$$$$$$$$$$$$$$$$$$$$$$$$$$$$$$$$$$$$$$$$$$$$$$$$$$$$$$$$$$$$$$$$$$$$$$$$$$$$$
327  %$$$$$$$$$$$$$$$$$$$$$$$$$$$$$$$$$$$$$$$$$$$$$$$$$$$$$$$$$$$$$$$$$$$$$$$$$$$$$$
328  %
329  % WRITING OF CUF NUCLEUS
330  %
331  K(:,:,count1,count2) = [K_xx K_xy K_xz; K_yx (K_yy + K_yy_FLEX) K_yz;
        K_zx K_zy K_zz];
332  %
333  %K = vpa(K(:,:,count1,count2),4)
334  %
335  K(:,:,count1,count2) = vpa((K(:,:,count1,count2)),4);
336  %
337  %??????????????????????????????????????????????????????????????????????????
338  %
339  % WRITING OF CUF NUCLEUS [ ONLY 2 X 2 ]
340  %
341  K_TOT_2(:,:,count1,count2) = [K_xx K_xy ; K_yx (K_yy+K_yy_FLEX)];
342  %
343  %??????????????????????????????????????????????????????????????????????????
344  % COMPUTE BENDING SYIFFNESS MATRIX
345  %
346  K_BENDING(:,:,count1,count2) = [0 0 0; 0 (0+K_yy_FLEX) 0; 0 0 0];
347  %
348  %??????????????????????????????????????????????????????????????????????????
349  % COMPUTE BENDING SYIFFNESS MATRIX [ 2 x 2 ]
350  %
351  K_BENDING_2(:,:,count1,count2) = [0 0 ; 0 (0+K_yy_FLEX)];
352  %
353  %??????????????????????????????????????????????????????????????????????????
354  %
355  % COMPUTE SHEAR SYIFFNESS MATRIX
356  %
357  K_SHEAR(:,:,count1,count2) = [K_xx K_xy K_xz; K_yx (K_yy+0) K_yz; K_zx
        K_zy K_zz];
358  %
359  %
360  %??????????????????????????????????????????????????????????????????????????
361  % COMPUTE SHEAR SYIFFNESS MATRIX [ 2 x 2 ]
362  %
363  K_SHEAR_2(:,:,count1,count2) = [K_xx K_xy ; K_yx (K_yy+0)];
364  %
365  %??????????????????????????????????????????????????????????????????????????
366  %
367  %$$$$$$$$$$$$$$$$$$$$$$$$$$$$$$$$$$$$$$$$$$$$$$$$$$$$$$$$$$$$$$$$$$$$$$$$$$$$$$
368  %$$$$$$$$$$$$$$$$$$$$$$$$$$$$$$$$$$$$$$$$$$$$$$$$$$$$$$$$$$$$$$$$$$$$$$$$$$$$$$
369  %^^^^^^^^^^^^^^^^^^^^^^^^^^^^^^^^^^^^^^^^^^^^^^^^^^^^^^^^^^^^^^^^^^^^^^^^^^^^^^
370  %^^^^^^^^^^^^^^^^^^^^^^^^^^^^^^^^^^^^^^^^^^^^^^^^^^^^^^^^^^^^^^^^^^^^^^^^^^^^^^
371  %^^^^^^^^^^^^^^^^^^^^^^^^^^^^^^^^^^^^^^^^^^^^^^^^^^^^^^^^^^^^^^^^^^^^^^^^^^^^^^
372  %^^^^^^^^^^^^^^^^^^^^^^^^^^^^^^^^^^^^^^^^^^^^^^^^^^^^^^^^^^^^^^^^^^^^^^^^^^^^^^
373  %
374  % RESET TO 0 OF THE FOLLOWING MATRIX TERMS AFTER THE m n LOOP
375  %
376  K_xx = 0;
377  K_xy = 0;
378  K_xz = 0;
379  %
380  K_yx = 0;
381  K_yy = 0;
382  K_yz = 0;
383  %
384  K_zx = 0;
```

```matlab
385   K_zy = 0;
386   K_zz = 0;
387   %
388   RIS_TOT = 0;
389   %
390   %??????????????????????????????????????????????????????????????????????????
391   %??????????????????????????????????????????????????????????????????????????
392   %
393   % MANAGE THE COUNTER countmat AND count1
394   %
395   count1 = count1 + 1;
396   %
397   %??????????????????????????????????????????????????????????????????????????
398   %??????????????????????????????????????????????????????????????????????????
399   %??????????????????????????????????????????????????????????????????????????
400   %
401              end  % END OF t LLOP
402   %
403   %??????????????????????????????????????????????????????????????????????????
404   %??????????????????????????????????????????????????????????????????????????
405          end        % END OF i LLOP
406   %
407   %??????????????????????????????????????????????????????????????????????????
408   %??????????????????????????????????????????????????????????????????????????
409   %
410   % MANAGE THE COUNTER count11 AND count2
411   %
412   count1 = 1;
413   count2 = count2 +1;
414   %
415   %??????????????????????????????????????????????????????????????????????????
416   %??????????????????????????????????????????????????????????????????????????
417   %
418     end     % END OF s LLOP
419        %
420   end  % END OF j LLOP
421   %
422   %================================================================================
423   %
424   %
425   MATR = [K(:,:,1,1) K(:,:,2,1);...
426           K(:,:,1,2) K(:,:,2,2)];
427   %
428   %
429   %
430   %  COMPUTE [ 3 X 3 ] MATRICES : CUF BASED MATRICES
431   %
432   %
433   % BENDING MATRIX
434   %
435   MATR_BENDING = [K_BENDING(:,:,1,1) K_BENDING(:,:,2,1); K_BENDING
         (:,:,1,2) K_BENDING(:,:,2,2)]
436   %
437   %
438   % SHEAR MATRIX
439   %
440   MATR_SHEAR = [K_SHEAR(:,:,1,1) K_SHEAR(:,:,2,1); K_SHEAR(:,:,1,2)
         K_SHEAR(:,:,2,2)]
441   %
442   %
443   %
444   %  COMPUTE [ 2 X 2 ] MATRICES FOR BATHE ALPHA -ALPHA DEMONSTRATION
```

```
445   %
446   MATR_TOT_2 = [K_TOT_2(:,:,1,1) K_TOT_2(:,:,2,1); K_TOT_2(:,:,1,2)
          K_TOT_2(:,:,2,2)]
447   %
448   MATR_BENDING_2 = [K_BENDING_2(:,:,1,1) K_BENDING_2(:,:,2,1);
          K_BENDING_2(:,:,1,2) K_BENDING_2(:,:,2,2)]
449   %
450   MATR_SHEAR_2 = [K_SHEAR_2(:,:,1,1) K_SHEAR_2(:,:,2,1); K_SHEAR_2
          (:,:,1,2) K_SHEAR_2(:,:,2,2)]
451   %
452   %^^^^^^^^^^^^^^^^^^^^^^^^^^^^^^^^^^^^^^^^^^^^^^^^^^^^^^^^^^^^^^^^^^
453   %^^^^^^^^^^^^^^^^^^^^^^^^^^^^^^^^^^^^^^^^^^^^^^^^^^^^^^^^^^^^^^^^^^
454   %================================================================
455   %^^^^^^^^^^^^^^^^^^^^^^^^^^^^^^^^^^^^^^^^^^^^^^^^^^^^^^^^^^^^^^^^^^
456   % STAMPA DELLE MATRICI KF, KT E K_TOT = KF + KT IN FORMA SIMBOLICA
457   %^^^^^^^^^^^^^^^^^^^^^^^^^^^^^^^^^^^^^^^^^^^^^^^^^^^^^^^^^^^^^^^^^^
458   %^^^^^^^^^^^^^^^^^^^^^^^^^^^^^^^^^^^^^^^^^^^^^^^^^^^^^^^^^^^^^^^^^^
459   %
460   % Open file : Bar_Symbolic_Stiffness_Matrix.dat
461   %
462   Fout = fopen('ELEMENTAL_Beam_Z_axis_Bending_Stiffness_Matrix_MITC_B2.
          dat','w');
463   %
464   %//////////////////////////////////////////////////////////////////
465   %//////////////////////////////////////////////////////////////////
466   %
467   % MAKE A COPY OF MATR FOR PRINTING PURPOSE ONLY
468   %
469   MATPR = MATR;
470   MATPRR = vpa(MATPR,4);
471   %
472       fprintf(Fout,'\n\n\n\nSymbolic Stiffness Matrix of a Beam under Z
              axis bending\n\n\n');
473       fprintf(Fout,'\n\n\n\nMITC FORMULATION\n\n\n');
474       Columns = Num_nodes*3;
475       Rows = Num_nodes*3;
476   %
477   for m = 1:Rows;
478   %
479           if m==1
480           formatSpec = '%s                %s                %s
                  %s                %s                %s                %s
                          %s                %s                %s
                          %s                %s|\n';
481           fprintf(Fout,formatSpec,MATPRR(m:m,1:Columns));
482           fprintf(Fout,'\n\n');
483           elseif m==3
484           formatSpec = '%s                %s                %s
                  %s                %s                %s                %s
                          %s                %s                %s
                          %s                %s|\n';
485           fprintf(Fout,formatSpec,MATPRR(m:m,1:Columns));
486           fprintf(Fout,'\n\n');
487           elseif m==5
488           formatSpec = '%s                %s                %s
                  %s                %s                %s                %s
                          %s                %s                %s
                          %s                %s|\n';
489           fprintf(Fout,formatSpec,MATPRR(m:m,1:Columns));
490           fprintf(Fout,'\n\n');
491           elseif m==7
```

```matlab
492             formatSpec = '%s                  %s                    %s
                    %s               %s                 %s
                            %s                %s                   %s
                            %s                 %s|\n';
493             fprintf(Fout,formatSpec,MATPRR(m:m,1:Columns));
494             fprintf(Fout,'\n\n');
495             elseif m==9
496             formatSpec = '%s                  %s                    %s
                    %s               %s                 %s
                            %s                %s                   %s
                            %s                 %s|\n';
497             fprintf(Fout,formatSpec,MATPRR(m:m,1:Columns));
498             fprintf(Fout,'\n\n');
499             elseif m==11
500             formatSpec = '%s                  %s                    %s
                    %s               %s                 %s
                            %s                %s                   %s
                            %s                 %s|\n';
501             fprintf(Fout,formatSpec,MATPRR(m:m,1:Columns));
502             fprintf(Fout,'\n\n');
503             end
504             %
505             if m==2
506             formatSpec = '%s                  %s                    %s
                    %s               %s                 %s
                            %s                %s                   %s
                            %s                 %s|\n';
507             fprintf(Fout,formatSpec,MATPRR(m:m,1:Columns));
508             fprintf(Fout,'\n\n');
509             elseif m==4
510             formatSpec = '%s                  %s                    %s
                    %s               %s                 %s
                            %s                %s                   %s
                            %s                 %s|\n';
511             fprintf(Fout,formatSpec,MATPRR(m:m,1:Columns));
512             fprintf(Fout,'\n\n');
513             elseif m==6
514             formatSpec = '%s                  %s                    %s
                    %s               %s                 %s
                            %s                %s                   %s
                            %s                 %s|\n';
515             fprintf(Fout,formatSpec,MATPRR(m:m,1:Columns));
516             fprintf(Fout,'\n\n');
517             elseif m==8
518             formatSpec = '%s                  %s                    %s
                    %s               %s                 %s
                            %s                %s                   %s
                            %s                 %s|\n';
519             fprintf(Fout,formatSpec,MATPRR(m:m,1:Columns));
520             fprintf(Fout,'\n\n');
521             elseif m==10
522             formatSpec = '%s                  %s                    %s
                    %s               %s                 %s
                            %s                %s                   %s
                            %s                 %s|\n';
523             fprintf(Fout,formatSpec,MATPRR(m:m,1:Columns));
524             fprintf(Fout,'\n\n');
525             elseif m==12
526             formatSpec = '%s                  %s                    %s
                    %s               %s                 %s
                            %s                %s                   %s
                            %s                 %s|\n';
```

```matlab
527            fprintf(Fout,formatSpec,MATPRR(m:m,1:Columns));
528            fprintf(Fout,'\n\n');
529            end
530            %
531     end
```

Appendix D
Beam Element Bending Around the x Axis

D.1 Evaluation of the Stiffness Matrix via the Matrix Notation

The MATLAB program enables both symbolic and numerical computation of a beam stiffness matrix using the matrix notation. To run the program, the user must provide two input files for the main script, as listed below.

(1) INPUT_SYMBOLIC.m
In this script the user provides the shape functions.

```
% ================================================================
%                      THIS IS THE INPUT FILE
% ================================================================
syms y L;
syms a b;
syms C C11 C44;
% ----------------------------------------------------------------
%                      SYMBOLIC VARIABLES
% ================================================================
% ================================================================
%
% 1) Enter shape function in symbolic form:
%
% Note : shape functions must have the following characterisctics :
% - must be a function of y
% - the length of the beam must be named L
%
% Shape Function for Node 1 :
N1 = 1-y/L;
%
% Shape Function for Node 2 :
N2 = y/L;
%
% ================================================================
% NOTE : for the present example case the Matrix C = [C11 0; 0 C44]
%
C = [C11 0; 0 C44];
% ================================================================
```

(33332) INPUT_NUMERIC.m

In this script the user provides the bar cross-section dimensions (a, b), the bar length (L) and the values of the Young Modulus (E) and the Poisson's ratio (ν).

```matlab
% ================================================================================
%                        THIS IS THE NUMERIC VALUES INPUT FILE
% ================================================================================
% ================================================================================
% ================================================================================
% --------------------------------------------------------------------------------
% 1) Enter Cross Section Dimensions
%
% The dimensions must be expressed in terms a = and b =, without
% dimensions
%
a = 10;
b = 10;
%
% 2) Enter Beam length
%
% The Beam length must be expressed in terms L = a number, without
% dimensions
%
L = 500;
%
% 3) Enter C Matrix values
%
% The dimensions must be expressed in terms of C11 = a number and
% C44 = a number, without dimensions
%
% For isotropic material the elastic constants have the following values :
%
% For Steel material :
%
E = 210000;
%
%nu = 0.29;
%
% The Poisson ratio is set equal to 0 for this example case
%
nu = 0.0;
%
% ================================================================================
% See pae 30 of REF_2 for the following data:
% ================================================================================
%
G = E/(2*(1+nu));
Lambda = (nu*E)/((1+nu)*(1-2*nu));
% ================================================================================
% NOTE :
%           C11 = C22 = C33
%           C44 = C55 = C66
%           C12 = C13 = C21 = C23 = C31 = C32
%
% ================================================================================
C11 = 2*G + Lambda;
C22 = 2*G + Lambda;
C33 = 2*G + Lambda;
%
C44 = G;
C55 = G;
C66 = G;
%
C12 = Lambda;
C13 = Lambda;
C21 = Lambda;
C23 = Lambda;
C31 = Lambda;
```

```
65   C32  =  Lambda;
66   %
67   % NOTE : in the present example case the Matrix C = [C11 0; 0 C44]
68   %
69   C  =  [C11  0;  0  C44];
70   % =============================================================
```

(3) PVD_APPROACH.m
The following is the main script to run.

```
1    %
2    syms  N1  N2;
3    syms  D1  D2;
4    syms  B  BT;
5    syms  x  z;
6    syms  y  L;
7    syms  E;
8    syms  A  Iz;
9    syms  KK  K;
10   syms  a  b;
11   syms  C11  C44;
12   %
13   % =============================================================
14   %                        READ INPUT
15   % =============================================================
16   % =============================================================
17   %
18   cd  ..\
19   run  INPUT_SYMBOLIC\INPUT_SYMBOLIC.m;
20   %
21   % Computation of the derivative of N1 and N2 with respect to y
22   %
23   D1  =  diff(N1,y,1);
24   D2  =  diff(N2,y,1);
25   %
26   % =============================================================
27   %
28   B  =  [0  x*D1  0  x*D2;  -D1   N1  -D2  N2];
29   %
30   BT  =  transpose(B);
31   %
32   %
33   BTCB  =  BT*C*B;
34   %
35   % =============================================================
36   %
37   %
38   KK(1,1)  =  C44*int(1,z,-b,b)*int(1,x,-a,a)*int(D1*D1,y,0,L);
39   KK(1,2)  =  -C44*int(1,z,-b,b)*int(1,x,-a,a)*int(N1*D1,y,0,L);
40   KK(1,3)  =  C44*int(1,z,-b,b)*int(1,x,-a,a)*int(D1*D2,y,0,L);
41   KK(1,4)  =  -C44*int(1,z,-b,b)*int(1,x,-a,a)*int(D1*N2,y,0,L);
42   %
43   %
44   KK(2,1)  =  -C44*int(1,z,-b,b)*int(1,x,-a,a)*int(N1*D1,y,0,L);
45   KK(2,2)  =  C11*int(1,z,-b,b)*int(x^2,x,-a,a)*int(D1*D1,y,0,L)...
46               +C44*int(1,z,-b,b)*int(1,x,-a,a)*int(N1*N1,y,0,L);
47   KK(2,3)  =  -C44*int(1,z,-b,b)*int(1,x,-a,a)*int(N1*D2,y,0,L);
48   KK(2,4)  =   C11*int(1,z,-b,b)*int(x^2,x,-a,a)*int(D1*D2,y,0,L)...
49               +C44*int(1,z,-b,b)*int(1,x,-a,a)*int(N1*N2,y,0,L);
50   %
51   %
52   KK(3,1)  =  C44*int(1,z,-b,b)*int(1,x,-a,a)*int(D1*D2,y,0,L);
53   KK(3,2)  =  -C44*int(1,z,-b,b)*int(1,x,-a,a)*int(N1*D2,y,0,L);
54   KK(3,3)  =  C44*int(1,z,-b,b)*int(1,x,-a,a)*int(D2*D2,y,0,L);
55   KK(3,4)  =  -C44*int(1,z,-b,b)*int(1,x,-a,a)*int(D2*N2,y,0,L);
56   %
```

```matlab
57   %
58   KK(4,1)  =  -C44*int(1,z,-b,b)*int(1,x,-a,a)*int(D1*N2,y,0,L);
59   KK(4,2)  =  C11*int(1,z,-b,b)*int(x^2,x,-a,a)*int(D1*D2,y,0,L)...
60              +C44*int(1,z,-b,b)*int(1,x,-a,a)*int(N2*N1,y,0,L);
61   KK(4,3)  =  -C44*int(1,z,-b,b)*int(1,x,-a,a)*int(N2*D2,y,0,L);
62   KK(4,4)  =  C11*int(1,z,-b,b)*int(x^2,x,-a,a)*int(D2*D2,y,0,L)...
63              +C44*int(1,z,-b,b)*int(1,x,-a,a)*int(N2*N2,y,0,L);
64   %
65   %
66   A = int(1,z,-b,b)*int(1,x,-a,a); % Cross section area
67   Iz = int(1,z,-b,b)*int(x^2,x,-a,a); % Cross section z Moment of Inertia
68   %
69   %
70   K(1,1)  =  C44*A*int(D1*D1,y,0,L);
71   K(1,2)  =  -C44*A*int(N1*D1,y,0,L);
72   K(1,3)  =  C44*A*int(D1*D2,y,0,L);
73   K(1,4)  =  -C44*A*int(D1*N2,y,0,L);
74   %
75   %
76   K(2,1)  =  -C44*A*int(N1*D1,y,0,L);
77   K(2,2)  =  C11*Iz*int(D1*D1,y,0,L)...
78              +C44*A*int(N1*N1,y,0,L);
79   K(2,3)  =  -C44*A*int(N1*D2,y,0,L);
80   K(2,4)  =   C11*Iz*int(D1*D2,y,0,L)...
81              +C44*A*int(N1*N2,y,0,L);
82   %
83   %
84   K(3,1)  =  C44*A*int(D1*D2,y,0,L);
85   K(3,2)  =  -C44*A*int(N1*D2,y,0,L);
86   K(3,3)  =  C44*A*int(D2*D2,y,0,L);
87   K(3,4)  =  -C44*A*int(D2*N2,y,0,L);
88   %
89   %
90   K(4,1)  =  -C44*A*int(D1*N2,y,0,L);
91   K(4,2)  =  C11*Iz*int(D1*D2,y,0,L)...
92              +C44*A*int(N2*N1,y,0,L);
93   K(4,3)  =  -C44*A*int(N2*D2,y,0,L);
94   K(4,4)  =  C11*Iz*int(D2*D2,y,0,L)...
95              +C44*A*int(N2*N2,y,0,L);
96   %
97   %
98   fprintf(1,'\n\nSymbolic Stiffness Matrix of a Beam under z axis bending\n\n\n');
99   %
100  %
101  K = [K(1,1) K(1,2) K(1,3) K(1,4);K(2,1) K(2,2) K(2,3) K(2,4);...
102       K(3,1) K(3,2) K(3,3) K(3,4);K(4,1) K(4,2) K(4,3) K(4,4)];
103  %
104  pretty(K)
105  %
106  run INPUT_NUMERIC\INPUT_NUMERIC.m;
107  %
108  K_NUM = subs(K);
109  %
110  %
111  fprintf(1,'\n\n\nNumeric Stiffness Matrix of a Beam under z axis bending\n\n\n');
112  %
113  pretty(K_NUM)
114  %
```

In this script the command diff is used to compute the symbolic derivatives of
the shape functions and the command **int** is used to compute the symbolic integrals,
for the calculation of the stiffness matrix **K**.

Once the main code is executed, the following results, symbolic and numeric, are obtained:

```
Symbolic Stiffness Matrix of a Beam under z axis bending


/       #4,      2 C44 a b,      -#4,       #3 \
|                                               |
| 2 C44 a b,       #1,      2 C44 a b,  #2 |
|                                               |
|     -#4,      2 C44 a b,       #4,       #3 |
|                                               |
\       #3,          #2,          #3,      #1 /


where

                      3
           C11 b a   4    C44 L b a 4
     #1 == ---------- + -----------
              3 L             3


                                     3
           2 C44 L a b   4 C11 a  b
     #2 == ----------- - ----------
                3            3 L


     #3 == -2 C44 a b


           4 C44 a b
     #4 == ---------
               L
```

```
Numeric Stiffness Matrix of a Beam under z axis bending


/    84000,     21000000,     -84000,     -21000000 \
|                                                    |
|  21000000, 7005600000,   21000000, 3494400000 |
|                                                    |
|   -84000,     21000000,      84000,     -21000000 |
|                                                    |
\ -21000000, 3494400000, -21000000, 7005600000 /


   >>
```

The user can change the input data and verify the variation of the results.

D.2   Evaluation of the Stiffness Matrix via the Recursive Notation Against Shape Functions

The MATLAB program enables both symbolic and numerical computation of a beam's stiffness matrix using the recursive notation against shape functions. To run the program, the user must provide the same input files as for the matrix notation.

(3) FE_APPROACH.m
The following is the main script to run.

```matlab
%
syms N1 N2;
syms D1 D2;
syms B BT;
syms x z;
syms y L;
syms E;
syms A Iz;
syms KK K;
syms K11 K12 K21 K22;
syms a b;
syms C11 C44;
syms M M1 M2 M3 M4;
%
%
% ================================================================================
% ================================================================================
%
%                              READ INPUT
% ================================================================================
%
cd ..\
run INPUT_SYMBOLIC\INPUT_SYMBOLIC.m;
%
% Computation of the derivative of N1 and N2 with respect to y
%
% The derivatives are entered using Recursive Notation D(i)
% and considering Node 1 and Node 2.
% D(1) is the derivative of the shape notation for Node 1 and D(2) is the
%      the
% derivative of the shape function for Node 2.
%
D(1) = diff(N(1),y,1);
D(2) = diff(N(2),y,1);
%
% ================================================================================
%
%
B1 = [0 x*D(1); -D(1) N(1)];
%
%
B2 = [0 x*D(2); -D(2) N(2)];
%
```

```
43   %
44   B = [0 x*D(1) 0 x*D(2); -D(1)  N(1) -D(2) N(2)];
45   %
46   %==============================================================
47   %
48   % Computation of the Stiffness Matrix
49   %
50   %-------------------------------------------------------------
51   %
52   K(1,1) = C44*int(1,z,-b,b)*int(1,x,-a,a)*int(D(1)*D(1),y,0,L);
53   K(1,2) = -C44*int(1,z,-b,b)*int(1,x,-a,a)*int(N(1)*D(1),y,0,L);
54   K(2,1) = -C44*int(1,z,-b,b)*int(1,x,-a,a)*int(N(1)*D(1),y,0,L);
55   K(2,2) = C11*int(1,z,-b,b)*int(x^2,x,-a,a)*int(D(1)*D(1),y,0,L)...
56            +C44*int(1,z,-b,b)*int(1,x,-a,a)*int(N(1)*N(1),y,0,L);
57   %
58   %
59   K11 = [K(1,1) K(1,2); K(2,1) K(2,2)];
60   %-------------------------------------------------------------
61   %
62   K(1,3) = C44*int(1,z,-b,b)*int(1,x,-a,a)*int(D(1)*D(2),y,0,L);
63   K(1,4) = -C44*int(1,z,-b,b)*int(1,x,-a,a)*int(D(1)*N(2),y,0,L);
64   K(2,3) = -C44*int(1,z,-b,b)*int(1,x,-a,a)*int(N(1)*D(2),y,0,L);
65   K(2,4) =  C11*int(1,z,-b,b)*int(x^2,x,-a,a)*int(D(1)*D(2),y,0,L)...
66            +C44*int(1,z,-b,b)*int(1,x,-a,a)*int(N(1)*N(2),y,0,L);
67   %
68   %
69   K12 = [K(1,3) K(1,4); K(2,3) K(2,4)];
70   %-------------------------------------------------------------
71   %
72   K(3,1) = C44*int(1,z,-b,b)*int(1,x,-a,a)*int(D(1)*D(2),y,0,L);
73   K(3,2) = -C44*int(1,z,-b,b)*int(1,x,-a,a)*int(N(1)*D(2),y,0,L);
74   K(4,1) = -C44*int(1,z,-b,b)*int(1,x,-a,a)*int(D(1)*N(2),y,0,L);
75   K(4,2) = C11*int(1,z,-b,b)*int(x^2,x,-a,a)*int(D(1)*D(2),y,0,L)...
76            +C44*int(1,z,-b,b)*int(1,x,-a,a)*int(N(2)*N(1),y,0,L);
77   %
78   %
79   %
80   K21 = [K(3,1) K(3,2); K(4,1) K(4,2)];
81   %-------------------------------------------------------------
82   %
83   K(3,3) = C44*int(1,z,-b,b)*int(1,x,-a,a)*int(D(2)*D(2),y,0,L);
84   K(3,4) = -C44*int(1,z,-b,b)*int(1,x,-a,a)*int(D(2)*N(2),y,0,L);
85   K(4,3) = -C44*int(1,z,-b,b)*int(1,x,-a,a)*int(N(2)*D(2),y,0,L);
86   K(4,4) = C11*int(1,z,-b,b)*int(x^2,x,-a,a)*int(D(2)*D(2),y,0,L)...
87            +C44*int(1,z,-b,b)*int(1,x,-a,a)*int(N(2)*N(2),y,0,L);
88   %
89   %
90   K22 = [K(3,3) K(3,4); K(4,3) K(4,4)];
91   %
92   %-------------------------------------------------------------
93   %
94   %
95   %
96   K = [K11 K12; K21 K22];
97   %
98   %-------------------------------------------------------------
99   %
100  %
101  A = int(1,z,-b,b)*int(1,x,-a,a); % Cross section area
102  Iz = int(1,z,-b,b)*int(x^2,x,-a,a); % Cross section z Moment of
         Inertia
```

```matlab
103    %
104    %
105    count = 1;
106    %
107    for i=1:2
108        for j=1:2
109    K(1:2,1:2,count) = [C44*A*int(D(i)*D(j),y,0,L), -C44*A*int(N(i)*D(j),
           y,0,L);...
110                -C44*A*int(N(j)*D(i),y,0,L), (C11*Iz*int(D(i)*D(j),y,0,L)...
111                +C44*A*int(N(i)*N(j),y,0,L))];
112        count = count + 1;
113        end
114    end
115    %
116    %
117    K = [K(1:2,1:2,1) K(1:2,1:2,2) ;K(1:2,1:2,3) K(1:2,1:2,4)];
118    %
119    %
120    fprintf(1,'\n\nSymbolic Stiffness Matrix of a Beam under z axis
           bending\n\n\n');
121    %
122    pretty(K)
123    %
124    run INPUT_NUMERIC\INPUT_NUMERIC.m;
125    %
126    K_NUM = subs(K);
127    %
128    %
129    fprintf(1,'\n\n\nNumeric Stiffness Matrix of a Beam under z axis
           bending\n\n\n');
130    %
131    pretty(K_NUM)
132    %
```

In this script the command **diff** is used to compute the symbolic derivatives of the shape functions and the command **int** is used to compute the symbolic integrals, for the calculation of the stiffness matrix **K**.

Once the main code is executed, the following results, symbolic and numeric, are obtained:

```
Symbolic Stiffness Matrix of a Beam under z axis bending


 /      #4,      2 C44 a b,      -#4,      #3 \
 |                                            |
 | 2 C44 a b,        #1,      2 C44 a b, #2 |
 |                                            |
 |     -#4,      2 C44 a b,        #4,      #3 |
 |                                            |
 \      #3,          #2,          #3,      #1 /


where

                    3
          C11 b a  4    C44 L b a 4
    #1 == ---------- + -----------
             3 L            3


                              3
          2 C44 L a b    4 C11 a  b
    #2 == ----------- - ----------
               3            3 L


    #3 == -2 C44 a b


          4 C44 a b
    #4 == ---------
              L




Numeric Stiffness Matrix of a Beam under z axis bending


 /    84000,     21000000,     -84000,    -21000000 \
 |                                                   |
 |  21000000, 7005600000,   21000000, 3494400000 |
 |                                                   |
 |    -84000,    21000000,      84000,    -21000000 |
 |                                                   |
 \ -21000000, 3494400000, -21000000, 7005600000 /

    >>
```

The user can change the input data and verify the variation of the results.

D.3   Evaluation of the Stiffness Matrix via the Recursive Notation Against Theory of Structures

The MATLAB program enables both symbolic and numerical computation of a beam's stiffness matrix using the recursive notation against theory of structures.

To run the program, the user must provide the same INPUT_SYMBOLIC.m as for the matrix notation and the recursive notation against shape functions.

Regarding the INPUT_SYMBOLIC.m input, the user must provide the shape functions ($N(1)$ and $N(2)$) and the expansion function ($F_{\tau y}$ and F_{sy}). The file INPUT_SYMBOLIC.m for is shown below.

(2) INPUT_SYMBOLIC.m

```
1    % ================================================================
2    %                        THIS IS THE INPUT FILE
3    % ================================================================
4    syms y L;
5    syms a b;
6    syms E;
7    syms Fty Fsy;
8    syms Ftz Fsz;
9    % ----------------------------------------------------------------
10   %                        SYMBOLIC VARIABLES
11   % ----------------------------------------------------------------
12   % ================================================================
13   % ================================================================
14   %
15   % 1) Enter shape function in symbolic form and recursive notation:
16   %
17   % Note : shape functions must have the following characterisctics :
18   % - must be a function of y
19   % - the length of the bar must be named L
20   %
21   % The shape functions are entered using Recursive Notation N(i)
22   % and considering Node 1 and Node 2.
23   % N(1) is the shape notation for Node 1 and N(2) is the shape function
          for
24   % Node 2.
25   %
26   % Shape Function for Node 1 :
27   %
28   N(1) = 1-y/L;
29   %
30   % Shape Function for Node 2 :
31   %
32   N(2) = y/L;
33   %
34   % ----------------------------------------------------------------
35   % 2) Enter the Fty(tau) (Ftauy) and Fsy(s) (Fsy) values using
          recursive notation:
36   %     For the problem of a Beam with bending around x axis
37   %     tau = 1 and s = 1 then :
38   %
39   %     Fty(tau) = Fty(1)
40   %     Ftz(tau) = Ftz(1)
41   %
42   %     and
```

```
43   %
44   %      Fsy(s)  =  Fsy(1)
45   %      Fsz(s)  =  Fsz(1)
46   %
47   %
48   Fty(1)  =  -z;
49   Ftz(1)  =  1;
50   %
51   % Similar to the Formula 2.58 of page 15 of REF_1
52   %
53   Fsy(1)  =  -z;
54   Fsz(1)  =  1;
55   %
56   %
57   % NOTE : according with the equation 2.21 of page 10 of REF_1
58   %         the Matrix C = [C11 0; 0 C44]
59   %
60   C = [C11 0; 0 C44];
61   %
62   %============================================================
```

(3) TS_APPROACH.m
The following is the main script to run.

```
1    %
2    syms  N;;
3    syms  D;
4    syms  Bi Bj;
5    syms  x z;
6    syms  y L;
7    syms  E;
8    syms  A;
9    syms  K;
10   syms  a b;
11   syms  Ftx Fty;
12   syms  Fsx Fsy;
13   syms  C11 C44;
14   syms  Iz;
15   syms  Res1 Res2 Res3 Res4 Res5;
16   %
17   %============================================================
18   %============================================================
19   %                     READ INPUT
20   %============================================================
21   %
22   cd ..\
23   run INPUT_SYMBOLIC\INPUT_SYMBOLIC.m;
24   %
25   %============================================================
26   %
27   % Computation of the derivative of N1 and N2 with respect to y
28   %
29   % The derivatives are entered using Recursive Notation D(i)
30   % and considering Node 1 and Node 2.
31   % D(1) is the derivative of the shape notation for Node 1 and D(2) is
         the
32   % derivative of the shape function for Node 2.
33   %
34   D(1)  =  diff(N(1),y,1);
35   D(2)  =  diff(N(2),y,1);
36   %============================================================
37   %
```

```matlab
count1 = 1;
%
for  i=1:2;
    for j=1:2;
BTCB(1:2,1:2,count1) = transpose([ 0 Fsy(1) * diff(N(j),y,1); Fsx(1) *
        diff(N(j),y,1) diff(Fsy(1),x,1) * N(j)])...
            *C*[ 0 Fty(1)* diff(N(i),y,1); Ftx(1) * diff(N(i),y,1) diff(
                Fty(1),x,1) * N(i)];
    count1 = count1 + 1;
    end
end
%
%////////////////////////////////////////////////////////////////////////////
% The previously computed matrices,for the i,j pair,
% are the following 4 matrices:
%
% i =1 , j =1
%
BTCB(1:2,1:2,1);
%
% i =1 , j =2
%
BTCB(1:2,1:2,2);
%
% i =2 , j =1
%
BTCB(1:2,1:2,3);
%
% i =2 , j =2
%
BTCB(1:2,1:2,4);
%
% ===============================================================================
%
%
% The Res(ult)1 is equal to the Cross section AREA
%
Res1 = int(int(Fsx(1)*Ftx(1),z,-b,b),x,-a,a);
%
% The Res(ult)2 is equal to negative value of the Cross section AREA
%
Res2 = int(int(Fsx(1)*diff(Fty(1),x,1),z,-b,b),x,-a,a);
%
% The Res(ult)3 is equal to negative value of the Cross section AREA
%
Res3 = int(int(diff(Fsy(1),x,1)*Ftx(1),z,-b,b),x,-a,a);
%
% The Res(ult)4 is equal to the Cross section Moment of Inerzia Iz
% related to the z axis
%
Res4 = int(int(Fsy(1)*Fty(1),z,-b,b),x,-a,a);
%
% The Res(ult)5 is equal to negative value of the Cross section AREA
%
Res5 = int(int(diff(Fsy(1),x,1)*diff(Fty(1),x,1),z,-b,b),x,-a,a);
%
%
% Equivalence of Res1 to Cross Section Area A
%
A = Res1;
%
%
```

```matlab
98    % Equivalence of Res4 to Cross Section Moment of Inertia Iz
99    %
100   Iz= Res4;
101   %
102   %
103   count2 = 1;
104   %
105   for i=1:2
106       for j=1:2
107   K(1:2,1:2,count2) = [C44*A*int(D(i)*D(j),y,0,L), -C44*A*int(N(i)*D(j),
          y,0,L);...
108             -C44*A*int(N(j)*D(i),y,0,L), (C11*Iz*int(D(i)*D(j),y,0,L)...
109             +C44*A*int(N(i)*N(j),y,0,L))];
110       count2 = count2 + 1;
111       end
112   end
113   %
114   K = [K(1:2,1:2,1) K(1:2,1:2,2) ;K(1:2,1:2,3) K(1:2,1:2,4)];
115   %
116   %
117   fprintf(1,'\n\nSymbolic Stiffness Matrix of a Beam under z axis
          bending\n\n\n');
118   %
119   pretty(K)
120   %
121   run INPUT_NUMERIC\INPUT_NUMERIC.m;
122   %
123   K_NUM = subs(K);
124   %
125   %
126   fprintf(1,'\n\n\nNumeric Stiffness Matrix of a Beam under z axis
          bending\n\n\n');
127   %
128   pretty(K_NUM)
129   %
```

In this script the command diff is used to compute the symbolic derivatives of the shape functions to obtain B_i and B_j and the command **int** is used to compute the symbolic integrals, for the calculation of the stiffness matrix **K**.

Once the MATLAB program is executed the following results, symbolic and numeric, are obtained:

```
Symbolic Stiffness Matrix of a Beam under z axis bending


   /        #4,       2 C44 a b,      -#4,       #3 \
   |                                                 |
   |  2 C44 a b,        #1,       2 C44 a b, #2 |
   |                                                 |
   |     -#4,        2 C44 a b,        #4,      #3 |
   |                                                 |
   \        #3,          #2,            #3,      #1 /


where

                        3
              C11 b a   4    C44 L b a 4
      #1 ==   ---------- + -----------
                 3 L            3

                                        3
              2 C44 L a b    4 C11 a   b
      #2 ==   ----------- - ----------
                  3             3 L

      #3 == -2 C44 a b

             4 C44 a b
      #4 ==  ---------
                 L
```

```
Numeric Stiffness Matrix of a Beam under z axis bending


   /    84000,      21000000,      -84000,     -21000000 \
   |                                                        |
   |  21000000, 7005600000,  21000000, 3494400000 |
   |                                                        |
   |    -84000,    21000000,      84000,      -21000000 |
   |                                                        |
   \ -21000000, 3494400000, -21000000, 7005600000 /

   >>
```

The user can change the input data and verify the variation of the results.

Appendix E
Beam Element Subjected by a Torsion

E.1 Evaluation of the Stiffness Matrix via the Matrix Notation

The MATLAB program enables both symbolic and numerical computation of a beam stiffness matrix using the matrix notation. To run the program, the user must provide two input files for the main script, as listed below.

(1) INPUT_SYMBOLIC.m
In this script the user provides the shape functions.

```
% ==========================================================================
%                         THIS IS THE INPUT FILE
% ==========================================================================
syms y L;
syms a b;
syms C C44;
% --------------------------------------------------------------------------
%                           SYMBOLIC VARIABLES
% --------------------------------------------------------------------------
% 1) Enter shape function in symbolic form:
%
% Note : shape functions must have the following characterisctics :
% - must be a function of y
% - the length of the bar must be named L
% (see Formula 4.23 of page 22 of REF_1)
%
% Shape Function for Node 1 :
N1 = 1-y/L;
%
% Shape Function for Node 2 :
N2 = y/L;
%
%See Formula 4.21 of page 22 of REF_1
%
C = [C44 0;0 C44];
% --------------------------------------------------------------------------
```


E. Carrera et al., *Implementation of Beam-Type Finite Elements Based on Carrera Unified Formulation*, https://doi.org/10.1007/978-3-031-95856-4

(2) INPUT_NUMERIC.m

In this script the user provides the bar cross-section dimensions (a, b), the bar length (L) and the values of the Young Modulus (E) and the Poisson's ratio (ν).

```matlab
%===============================================================
%                    THIS IS THE NUMERIC VALUES INPUT FILE
%===============================================================
%
%===============================================================
%===============================================================
% -------------------------------------------------------------
% 1) Enter Cross Section Dimensions
%
% The dimensions must be expressed in terms a = and b =, without
% dimensions
%
a = 10;
b = 10;
%
% 2) Enter Bar length
%
% The Bar length must be expressed in terms L = a number, without
% dimensions
%
L = 500;
%
% 3) Enter C Matrix values
%
% The dimensions must be expressed in terms of C11 = a number and
% C44 = a number, without dimensions
%
% For isotropic material the elastic constants have the following
%    values :
%
% For Steel material :
%
E = 210000;
%
% The Poisson ratio is set equal to 0 for the BAR ELEMENT
% in order to obtain results corresponding to the theoretical ones.
%
%nu = 0.30;
%
nu = 0.0;
%
%===============================================================
% See pae 30 of the book :
% "FE Anaslysis of Structures Through Unified Formulation"
%===============================================================
%
G = E/(2*(1+nu));
Lambda = (nu*E)/((1+nu)*(1-2*nu));
% ==============================================================
% NOTE :
%           C11 = C22 = C33
%           C44 = C55 = C66
%           C12 = C13 = C21 = C23 = C31 = C32
%
% ==============================================================
C11 = 2*G + Lambda;
C22 = 2*G + Lambda;
C33 = 2*G + Lambda;
```

```
58   %
59   C44 = G;
60   C55 = G;
61   C66 = G;
62   %
63   C12 = Lambda;
64   C13 = Lambda;
65   C21 = Lambda;
66   C23 = Lambda;
67   C31 = Lambda;
68   C32 = Lambda;
69   %
70   % NOTE : according to the equation 4.21 of the REF_1
71   %          the Matrix C = [C44 0; 0 C44]
72   %
```

(3) PVD_APPROACH.m
The following is the main script to run.

```
1    %
2    syms  N1  N2;
3    syms  D1  D2;
4    syms  B  BT;
5    syms  x  z;
6    syms  y  L;
7    syms  E;
8    syms  A  Iz;
9    syms  KK  K;
10   syms  a  b;
11   syms  C  C44;
12   %
13   %=========================================================================
14   %=========================================================================
15   %=========================================================================
16   %
17   %                              READ INPUT
18   %  =======================================================================
19   %
20   cd  ..\
21   run INPUT_SYMBOLIC\INPUT_SYMBOLIC.m;
22   %
23   % Computation of the derivative of N1 and N2 with respect to y
24   %
25   D1 = diff(N1,y,1);
26   D2 = diff(N2,y,1);
27   %
28   %  =========================================================
29   %
30   B = [z*D1  z*D2; -x*D1  -x*D2];
31   %
32   BT = transpose(B);
33   %
34   %
35   BTCB = BT * C * B;
36   %
37   %
38   % Computation of the Stiffness Matrix
39   %
40   %
41   %
42   %
43   KK(1,1) = C44*int(1,z,-b,b)*int(x^2,x,-a,a)*int(D1*D1,y,0,L)...
```

```matlab
44              +C44*int(z^2,z,-b,b)*int(1,x,-a,a)*int(D1*D1,y,0,L);
45 %
46 KK(1,2)  = C44*int(1,z,-b,b)*int(x^2,x,-a,a)*int(D1*D2,y,0,L)...
47              +C44*int(z^2,z,-b,b)*int(1,x,-a,a)*int(D1*D2,y,0,L);
48 %
49 KK(2,1)  = C44*int(1,z,-b,b)*int(x^2,x,-a,a)*int(D1*D2,y,0,L)...
50              +C44*int(z^2,z,-b,b)*int(1,x,-a,a)*int(D1*D2,y,0,L);
51 %
52 KK(2,2)  = C44*int(1,z,-b,b)*int(x^2,x,-a,a)*int(D2*D2,y,0,L)...
53              +C44*int(z^2,z,-b,b)*int(1,x,-a,a)*int(D2*D2,y,0,L);
54 %
55 %
56 %
57 A = int(1,z,-b,b)*int(1,x,-a,a); % Cross section area
58 Ix = int(z^2,z,-b,b)*int(1,x,-a,a); % Cross section z Moment of
        Inertia
59 Iz = int(1,z,-b,b)*int(x^2,x,-a,a); % Cross section x Moment of
        Inertia
60 %
61 %
62 %
63 K(1,1)  = C44*(Ix+Iz)*int(D1*D1,y,0,L);
64 K(1,2)  = C44*(Ix+Iz)*int(D1*D2,y,0,L);
65 K(2,1)  = C44*(Ix+Iz)*int(D1*D2,y,0,L);
66 K(2,2)  = C44*(Ix+Iz)*int(D2*D2,y,0,L);
67 %
68 %
69 fprintf(1,'\n\nSymbolic Stiffness Matrix of a Beam under Torsion\n\n\n
        ');
70 %
71 %
72 K = [K(1,1) K(1,2);K(2,1) K(2,2)];
73 %
74 pretty(K)
75 %
76 run INPUT_NUMERIC\INPUT_NUMERIC.m;
77 %
78 K_NUM = subs(K);
79 %
80 %
81 fprintf(1,'\n\n\nNumeric Stiffness Matrix of a Beam under Torsion\n\n\
        n');
82 %
83 pretty(K_NUM)
84 %
```

In this script the command **diff** is used to compute the symbolic derivatives of the shape functions and the command **int** is used to compute the symbolic integrals, for the calculation of the stiffness matrix **K**.

Once the main code is executed, the following results, symbolic and numeric, are obtained:

```
              Symbolic Stiffness Matrix of a Beam under torsion


              /  #1, -#1 \
              |          |
              \ -#1,  #1 /


              where

                          /     3            3 \
                          | 4 a b      4 a b  |
                     C44  | ------  +  ------ |
                          \    3            3 /
              #1 ==  -----------------------
                                 L
```

```
              Numeric Stiffness Matrix of a Beam under torsion


              /  5600000, -5600000 \
              |                     |
              \ -5600000,  5600000 /

              >>
```

The user can change the input data and verify the variation of the results.

E.2   Evaluation of the Stiffness Matrix via the Recursive Notation Against Shape Functions

The MATLAB program enables both symbolic and numerical computation of a beam's stiffness matrix using the recursive notation against shape functions. To run the program, the user must provide the same input files as for the matrix notation.

(3) FE_APPROACH.m
The following is the main script to run.

```matlab
%
syms N1 N2;
syms D1 D2;
syms B BT;
syms x z;
syms y L;
syms E;
syms A Iz;
syms KK K;
syms K11 K12 K21 K22;
syms a b;
syms C11 C44;
syms M M1 M2 M3 M4;
%
%
%=================================================================
%=================================================================
%
```

```matlab
19   %
20   %                           READ INPUT
21   %=============================================================================
22   %
23   cd  ..\
24   run INPUT_SYMBOLIC\INPUT_SYMBOLIC.m;
25   %
26   % Computation of the derivative of N1 and N2 with respect to y
27   %
28   % The derivatives are entered using Recursive Notation D(i)
29   % and considering Node 1 and Node 2.
30   % D(1) is the derivative of the shape notation for Node 1 and D(2) is the
31   % derivative of the shape function for Node 2.
32   %
33   D(1) = diff(N(1),y,1);
34   D(2) = diff(N(2),y,1);
35   %
36   %=============================================================================
37   %  ========================================================================
38   %
39   B1 = [z*D(1);-x*D(1)];
40   %
41   %
42   B2 = [z*D(2);-x*D(2)];
43   %
44   %
45   B = [B1 B2];
46   %
47   %
48   %
49   % Computation of the Stiffness Matrix
50   %
51   %
52   %
53   %
54   KK(1,1) = C44*int(1,z,-b,b)*int(x^2,x,-a,a)*int(D1*D1,y,0,L)...
55             +C44*int(z^2,z,-b,b)*int(1,x,-a,a)*int(D1*D1,y,0,L);
56   %
57   KK(1,2) = C44*int(1,z,-b,b)*int(x^2,x,-a,a)*int(D1*D2,y,0,L)...
58             +C44*int(z^2,z,-b,b)*int(1,x,-a,a)*int(D1*D2,y,0,L);
59   %
60   KK(2,1) = C44*int(1,z,-b,b)*int(x^2,x,-a,a)*int(D1*D2,y,0,L)...
61             +C44*int(z^2,z,-b,b)*int(1,x,-a,a)*int(D1*D2,y,0,L);
62   %
63   KK(2,2) = C44*int(1,z,-b,b)*int(x^2,x,-a,a)*int(D2*D2,y,0,L)...
64             +C44*int(z^2,z,-b,b)*int(1,x,-a,a)*int(D2*D2,y,0,L);
65   %
66
67   %
68   K = [KK(1,1) KK(1,2); KK(2,1) KK(2,2)];
69   %
70   %-----------------------------------------------------------------------------
71   %
72   %
73   A = int(1,z,-b,b)*int(1,x,-a,a); % Cross section area
74   Iz = int(1,z,-b,b)*int(x^2,x,-a,a); % Cross section x Moment of Inertia
75   Ix = int(z^2,z,-b,b)*int(1,x,-a,a); % Cross section z Moment of Inertia
76   %
77   %
78   % NOTE : in this case the variable i an j on the for statement must be
79   % inverted
80   %
81   for i=1:2
82       for j=1:2
83       K(i,j) = C44*(Ix+Iz)*int(D(i)*D(j),y,0,L);
```

```
84          end
85    end
86    %
87    %
88    K = [K(1,1)  K(1,2)  ;K(2,1)  K(2,2)];
89    %
90    %
91    fprintf(1,'\n\nSymbolic Stiffness Matrix of a Beam under torsion\n\n\n');
92    %
93    pretty(K)
94    %
95    run INPUT_NUMERIC\INPUT_NUMERIC.m;
96    %
97    K_NUM = subs(K);
98    %
99    %
100   fprintf(1,'\n\n\nNumeric Stiffness Matrix of a Beam under torsion\n\n\n');
101   %
102   pretty(K_NUM)
103   %
```

In this script the command **diff** is used to compute the symbolic derivatives of the shape functions and the command **int** is used to compute the symbolic integrals, for the calculation of the stiffness matrix **K**.

Once the main code is executed, the following results, symbolic and numeric, are obtained:

```
      Symbolic Stiffness Matrix of a Beam under torsion


      /  #1,  -#1 \
      |           |
      \ -#1,   #1 /


      where

                        /      3              3 \
                        | 4 a  b       4 a b    |
                  C44 | ------  +  ------ |
                        \      3              3 /
            #1 ==  -----------------------
                              L


      Numeric Stiffness Matrix of a Beam under torsion


      /  5600000, -5600000 \
      |                    |
      \ -5600000,  5600000 /

      >>
```

The user can change the input data and verify the variation of the results.

E.3 Evaluation of the Stiffness Matrix via the Recursive Notation Against Theory of Structures

The MATLAB program enables both symbolic and numerical computation of a beam's stiffness matrix using the recursive notation against theory of structures.

To run the program, the user must provide the same INPUT_SYMBOLIC.m as for the matrix notation and the recursive notation against shape functions.

Regarding the INPUT_SYMBOLIC.m input, the user must provide the shape functions ($N(1)$ and $N(2)$) and the expansion function ($F_{\tau y}$ and F_{sy}). The file INPUT_SYMBOLIC.m for is shown below.

(2) INPUT_SYMBOLIC.m

```
1   %==========================================================================
2   %                         THIS IS THE INPUT FILE
3   %==========================================================================
4   syms y L;
5   syms a b;
6   syms E;
7   syms Fty Fsy;
8   syms Ftz Fsz;
9   % ------------------------------------------------------------------------
10  %                         SYMBOLIC VARIABLES
11  % ------------------------------------------------------------------------
12  %==========================================================================
13  %==========================================================================
14  %
15  % 1) Enter shape function in symbolic form and recursive notation:
16  %
17  % Note : shape functions must have the following characterisctics :
18  % - must be a function of y
19  % - the length of the bar must be named L
20  %
21  % The shape functions are entered using Recursive Notation N(i)
22  % and considering Node 1 and Node 2.
23  % N(1) is the shape notation for Node 1 and N(2) is the shape function for
24  % Node 2.
25  % See Formula 4.23 of page 22 of REF_1
26  %
27  % Shape Function for Node 1 :
28  %
29  N(1) = 1-y/L;
30  %
31  % Shape Function for Node 2 :
32  %
33  N(2) = y/L;
34  %
35  % ------------------------------------------------------------------------
36  % 2) Enter the Ftx(Ftaux), Ftz(Ftauz) and Fsx, Fsz values using recursive
37  %       notation:
38  %       For the problem of a Beam with bending around x axis
39  %       tau = 1 and s = 1 then :
40  %
41  %       Ftx(taux) = Ftx(1)
42  %       Ftz(tauz) = Ftz(1)
43  %
44  %       and
45  %
46  %       Fsx(s) = Fsx(1)
47  %       Fsz(s) = Fsz(1)
48  %
49  % See Formulae 4.48 of page 26 of REF_1
50  %
```

```
51  Ftx(1) = z;
52  Ftz(1) = -x;
53  %
54  % See Formulae 4.53 of page 26 of REF_1
55  %
56  Fsx(1) = z;
57  Fsz(1) = -x;
58  %
```

(3) TS_APPROACH.m
The following is the main script to run.

```
1   %
2   syms N;;
3   syms D;
4   syms Bi Bj;
5   syms x z;
6   syms y L;
7   syms E;
8   syms A;
9   syms K;
10  syms a b;
11  syms Ftx Fty;
12  syms Fsx Fsy;
13  syms C11 C44;
14  syms Iz;
15  syms Res1 Res2 Res3 Res4 Res5;
16  %
17  %=============================================================================
18  %=============================================================================
19  %                              READ INPUT
20  %=============================================================================
21  %
22  cd ..\
23  run INPUT_SYMBOLIC\INPUT_SYMBOLIC.m;
24  %
25  %=============================================================================
26  %
27  % Computation of the derivative of N1 and N2 with respect to y
28  %
29  % The derivatives are entered using Recursive Notation D(i)
30  % and considering Node 1 and Node 2.
31  % D(1) is the derivative of the shape notation for Node 1 and D(2) is
         the
32  % derivative of the shape function for Node 2.
33  %
34  D(1) = diff(N(1),y,1);
35  D(2) = diff(N(2),y,1);
36  %=============================================================================
37  %
38  count1 = 1;
39  %
40  for i=1:2;
41      for j=1:2;
42  BTCB(1:2,1:2,count1) = transpose([ 0 Fsy(1) * diff(N(j),y,1); Fsx(1) *
         diff(N(j),y,1) diff(Fsy(1),x,1) * N(j)])...
43              *C*[ 0 Fty(1)* diff(N(i),y,1); Ftx(1) * diff(N(i),y,1) diff(
                 Fty(1),x,1) * N(i)];
44      count1 = count1 + 1;
45      end
46  end
47  %
48  %///////////////////////////////////////////////////////////////////////////
```

```matlab
49    % The previously computed matrices,for the i,j pair,
50    % are the following 4 matrices:
51    %
52    % i =1 , j =1
53    %
54    BTCB(1:2,1:2,1);
55    %
56    % i =1 , j =2
57    %
58    BTCB(1:2,1:2,2);
59    %
60    % i =2 , j =1
61    %
62    BTCB(1:2,1:2,3);
63    %
64    % i =2 , j =2
65    %
66    BTCB(1:2,1:2,4);
67    %
68    % ==============================================================================
69    %
70    %
71    % The Res(ult)1 is equal to the Cross section AREA
72    %
73    Res1 = int(int(Fsx(1)*Ftx(1),z,-b,b),x,-a,a);
74    %
75    % The Res(ult)2 is equal to negative value of the Cross section AREA
76    %
77    Res2 = int(int(Fsx(1)*diff(Fty(1),x,1),z,-b,b),x,-a,a);
78    %
79    % The Res(ult)3 is equal to negative value of the Cross section AREA
80    %
81    Res3 = int(int(diff(Fsy(1),x,1)*Ftx(1),z,-b,b),x,-a,a);
82    %
83    % The Res(ult)4 is equal to the Cross section Moment of Inerzia Iz
84    % related to the z axis
85    %
86    Res4 = int(int(Fsy(1)*Fty(1),z,-b,b),x,-a,a);
87    %
88    % The Res(ult)5 is equal to negative value of the Cross section AREA
89    %
90    Res5 = int(int(diff(Fsy(1),x,1)*diff(Fty(1),x,1),z,-b,b),x,-a,a);
91    %
92    %
93    % Equivalence of Res1 to Cross Section Area A
94    %
95    A = Res1;
96    %
97    %
98    % Equivalence of Res4 to Cross Section Moment of Inertia Iz
99    %
100   Iz= Res4;
101   %
102   %
103   count2 = 1;
104   %
105   for i=1:2
106       for j=1:2
107   K(1:2,1:2,count2) = [C44*A*int(D(i)*D(j),y,0,L), -C44*A*int(N(i)*D(j),
          y,0,L);...
108           -C44*A*int(N(j)*D(i),y,0,L), (C11*Iz*int(D(i)*D(j),y,0,L)...
109           +C44*A*int(N(i)*N(j),y,0,L))];
```

```
110        count2 = count2 + 1;
111      end
112  end
113  %
114  K = [K(1:2,1:2,1) K(1:2,1:2,2) ;K(1:2,1:2,3) K(1:2,1:2,4)];
115  %
116  %
117  fprintf(1,'\n\nSymbolic Stiffness Matrix of a Beam under z axis
         bending\n\n\n');
118  %
119  pretty(K)
120  %
121  run INPUT_NUMERIC\INPUT_NUMERIC.m;
122  %
123  K_NUM = subs(K);
124  %
125  %
126  fprintf(1,'\n\n\nNumeric Stiffness Matrix of a Beam under z axis
         bending\n\n\n');
127  %
128  pretty(K_NUM)
129  %
```

In this script the command diff is used to compute the symbolic derivatives of the shape functions to obtain B_i and B_j and the command **int** is used to compute the symbolic integrals, for the calculation of the stiffness matrix **K**.

Once the MATLAB program is executed the following results, symbolic and numeric, are obtained:

```
        Symbolic Stiffness Matrix of a Beam under torsion


        /  #1, -#1 \
        |          |
        \ -#1,  #1 /

        where

                        /     3           3 \
                        | 4 a  b    4 a b  |
                   C44  | ------ + ------ |
                        \    3         3   /
             #1 == -------------------------
                              L




        Numeric Stiffness Matrix of a Beam under torsion


        /  5600000, -5600000 \
        |                     |
        \ -5600000,  5600000 /


           >>
```

The user can change the input data and verify the variation of the results.

Appendix F
Beam Element Under Combined Load

F.1 Evaluation of the Stiffness Matrix via the Matrix Notation

The MATLAB program enables both symbolic and numerical computation of a beam stiffness matrix using the matrix notation. To run the program, the user must provide two input files for the main script, as listed below.

(1) INPUT_SYMBOLIC.m
In this script the user provides the shape functions.

```
% ===========================================================================
%                         THIS IS THE INPUT FILE
% ===========================================================================
syms y L;
syms a b;
syms C C11 C44;
% -------------------------------------------------------------------------
%                         SYMBOLIC VARIABLES
% ===========================================================================
% ===========================================================================
% -------------------------------------------------------------------------
% 1) Enter shape function in symbolic form:
%
% Note : shape functions must have the following characterisctics :
% - must be a function of y
% - the length of the bar must be named L
%
% Shape Function for Node 1 :
N1 = 1-y/L;
%
% Shape Function for Node 2 :
N2 = y/L;
%
%
% NOTE : the Matrix C = [C11 0 0; 0 C44 0; 0 0 C44]
%
C = [C11 0 0; 0 C44 0; 0 0 C44];
```


E. Carrera et al., *Implementation of Beam-Type Finite Elements Based on Carrera Unified Formulation*, https://doi.org/10.1007/978-3-031-95856-4

```matlab
28   %
29   %
30   %  ----------------------------------------------------------------------------
```

(2) INPUT_NUMERIC.m

In this script the user provides the bar cross-section dimensions (a, b), the bar length (L) and the values of the Young Modulus (E) and the Poisson's ratio (ν).

```matlab
1    %==========================================================================
2    %                     THIS IS THE NUMERIC VALUES INPUT FILE
3    %==========================================================================
4    %==========================================================================
5    %==========================================================================
6    %  ------------------------------------------------------------------------
7    % 1) Enter Cross Section Dimensions
8    %
9    % The dimensions must be expressed in terms a = and b =, without
10   % dimensions
11   %
12   a = 10;
13   b = 10;
14   %
15   % 2) Enter Bar length
16   %
17   % The Bar length must be expressed in terms L = a number, without
18   % dimensions
19   %
20   L = 500;
21   %
22   % 3) Enter C Matrix values
23   %
24   % The dimensions must be expressed in terms of C11 = a number and
25   % C44 = a number, without dimensions
26   %
27   % For isotropic material the elastic constants have the following
        values :
28   %
29   % For Steel material :
30   %
31   E = 210000;
32   %
33   % The Poisson ratio is set equal to 0 for the BAR ELEMENT
34   % in order to obtain results corresponding to the theoretical ones.
35   %
36   %nu = 0.30;
37   %
38   nu = 0.0;
39   %
40   %==========================================================================
41   % See pae 30 of the book :
42   % "FE Anaslysis of Structures Through Unified Formulation"
43   %==========================================================================
44   %
45   G = E/(2*(1+nu));
46   Lambda = (nu*E)/((1+nu)*(1-2*nu));
47   %  =====================================================================
48   % NOTE :
49   %           C11 = C22 = C33
50   %           C44 = C55 = C66
51   %           C12 = C13 = C21 = C23 = C31 = C32
52   %
53   %  =====================================================================
```

```
54   C11 = 2*G + Lambda;
55   C22 = 2*G + Lambda;
56   C33 = 2*G + Lambda;
57   %
58   C44 = G;
59   C55 = G;
60   C66 = G;
61   %
62   C12 = Lambda;
63   C13 = Lambda;
64   C21 = Lambda;
65   C23 = Lambda;
66   C31 = Lambda;
67   C32 = Lambda;
68   %
69   % NOTE : according to the equation 2.21 the Matrix
70   %          C = [C11 0 0; 0 C44 0; 0 0 C44]
71   %
```

(3) PVD_APPROACH.m
The following is the main script to run.

```
1    %
2    clc
3    %
4    syms N1 N2;
5    syms D1 D2;
6    syms B BT;
7    syms x z;
8    syms y L;
9    syms E;
10   syms A Iz;
11   syms K2 K;
12   syms a b;
13   syms C11 C44;
14   %
15   % ================================================================
16   %                          READ INPUT
17   % ================================================================
18   % ================================================================
19   %
20   cd ..\
21   run INPUT_SYMBOLIC\INPUT_SYMBOLIC.m;
22   %
23   % Computation of the derivative of N1 and N2 with respect to y
24   %
25   D1 = diff(N1,y,1);
26   D2 = diff(N2,y,1);
27   %
28   % ================================================================
29   %
30   B = [D1 D2 0 x*D1 0 x*D2 -z*D1 0 -z*D2 0 0 0;...
31        0 0 -D1 N1 -D2 N2 0 0 0 0 z*D1 z*D2;...
32        0 0 0 0 0 0 -N1 D1 -N2 D2 -x*D1 -x*D2];
33   %
34   BT = transpose(B);
35   %
36   %
37   BTCB = BT*C*B
38   %
39   % ================================================================
40   %
```

```
41   % Computation of the Stiffness Matrix
42   %
43   %
44   K(1,1)  =  C11*int(1,z,-b,b)*int(1,x,-a,a)*int(1/L^2,y,0,L);
45   K(1,2)  =  -C11*int(1,z,-b,b)*int(1,x,-a,a)*int(1/L^2,y,0,L);
46   K(1,3)  =  0;
47   K(1,4)  =  C11*int(1,z,-b,b)*int(x,x,-a,a)*int(1/L^2,y,0,L);
48   K(1,5)  =  0;
49   K(1,6)  =  -C11*int(1,z,-b,b)*int(x,x,-a,a)*int(1/L^2,y,0,L);
50   K(1,7)  =  -C11*int(z,z,-b,b)*int(1,x,-a,a)*int(1/L^2,y,0,L);
51   K(1,8)  =  0;
52   K(1,9)  =  C11*int(z,z,-b,b)*int(1,x,-a,a)*int(1/L^2,y,0,L);
53   K(1,10) =  0;
54   K(1,11) =  0;
55   K(1,12) =  0;
56   %
57   K(2,1)  =  -C11*int(1,z,-b,b)*int(1,x,-a,a)*int(1/L^2,y,0,L);
58   K(2,2)  =  C11*int(1,z,-b,b)*int(1,x,-a,a)*int(1/L^2,y,0,L);
59   K(2,3)  =  0;
60   K(2,4)  =  -C11*int(1,z,-b,b)*int(x,x,-a,a)*int(1/L^2,y,0,L);
61   K(2,5)  =  0;
62   K(2,6)  =  C11*int(1,z,-b,b)*int(x,x,-a,a)*int(1/L^2,y,0,L);
63   K(2,7)  =  C11*int(z,z,-b,b)*int(1,x,-a,a)*int(1/L^2,y,0,L);
64   K(2,8)  =  0;
65   K(2,9)  =  -C11*int(z,z,-b,b)*int(1,x,-a,a)*int(1/L^2,y,0,L);
66   K(2,10) =  0;
67   K(2,11) =  0;
68   K(2,12) =  0;
69   %
70   K(3,1)  =  0;
71   K(3,2)  =  0;
72   K(3,3)  =  C44*int(1,z,-b,b)*int(1,x,-a,a)*int(1/L^2,y,0,L);
73   K(3,4)  =  -C44*int(1,z,-b,b)*int(1,x,-a,a)*int(((y/L-1)/L),y,0,L);
74   K(3,5)  =  -C44*int(1,z,-b,b)*int(1,x,-a,a)*int(1/L^2,y,0,L);
75   K(3,6)  =  C44*int(1,z,-b,b)*int(1,x,-a,a)*int(y/L^2,y,0,L);
76   K(3,7)  =  0;
77   K(3,8)  =  0;
78   K(3,9)  =  0;
79   K(3,10) =  0;
80   K(3,11) =  -C44*int(z,z,-b,b)*int(1,x,-a,a)*int(1/L^2,y,0,L);
81   K(3,12) =  C44*int(z,z,-b,b)*int(1,x,-a,a)*int(1/L^2,y,0,L);
82   %
83   %
84   K(4,1)  =  C11*int(1,z,-b,b)*int(x,x,-a,a)*int(1/L^2,y,0,L);
85   K(4,2)  =  -C11*int(1,z,-b,b)*int(x,x,-a,a)*int(1/L^2,y,0,L);
86   K(4,3)  =  -C44*int(1,z,-b,b)*int(1,x,-a,a)*int(((y/L-1)/L),y,0,L);
87   K(4,4)  =  C44*int(1,z,-b,b)*int(1,x,-a,a)*int((y/L-1)^2,y,0,L)...
88               +C11*int(1,z,-b,b)*int(x^2,x,-a,a)*int(1/L^2,y,0,L);
89   K(4,5)  =  C44*int(1,z,-b,b)*int(1,x,-a,a)*int(((y/L-1)/L),y,0,L);
90   K(4,6)  =  -C11*int(1,z,-b,b)*int(x^2,x,-a,a)*int(1/L^2,y,0,L)...
91               -C44*int(1,z,-b,b)*int(1,x,-a,a)*int((y*(y/L-1)/L),y,0,L);
92   K(4,7)  =  -C11*int(z,z,-b,b)*int(x,x,-a,a)*int(1/L^2,y,0,L);
93   K(4,8)  =  0;
94   K(4,9)  =  C11*int(z,z,-b,b)*int(x,x,-a,a)*int(1/L^2,y,0,L);
95   K(4,10) =  0;
96   K(4,11) =  C44*int(z,z,-b,b)*int(1,x,-a,a)*int(((y/L-1)/L),y,0,L);
97   K(4,12) =  -C44*int(z,z,-b,b)*int(1,x,-a,a)*int(((y/L-1)/L),y,0,L);
98   %
99   %
100  K(5,1)  =  0;
101  K(5,2)  =  0;
102  K(5,3)  =  -C44*int(1,z,-b,b)*int(1,x,-a,a)*int(1/L^2,y,0,L);
```

```
103   K(5,4)  =  C44*int(1,z,-b,b)*int(1,x,-a,a)*int(((y/L-1)/L),y,0,L);
104   K(5,5)  =  C44*int(1,z,-b,b)*int(1,x,-a,a)*int(1/L^2,y,0,L);
105   K(5,6)  =  -C44*int(1,z,-b,b)*int(1,x,-a,a)*int(y/L^2,y,0,L);
106   K(5,7)  =  0;
107   K(5,8)  =  0;
108   K(5,9)  =  0;
109   K(5,10) =  0;
110   K(5,11) =  C44*int(z,z,-b,b)*int(1,x,-a,a)*int(1/L^2,y,0,L);
111   K(5,12) =  -C44*int(z,z,-b,b)*int(1,x,-a,a)*int(1/L^2,y,0,L);
112   %
113   %
114   K(6,1)  =  -C11*int(1,z,-b,b)*int(x,x,-a,a)*int(1/L^2,y,0,L);
115   K(6,2)  =  C11*int(1,z,-b,b)*int(x,x,-a,a)*int(1/L^2,y,0,L);
116   K(6,3)  =  C44*int(1,z,-b,b)*int(1,x,-a,a)*int(y/L^2,y,0,L);
117   K(6,4)  =  -C11*int(1,z,-b,b)*int(x^2,x,-a,a)*int(1/L^2,y,0,L)...
118              -C44*int(1,z,-b,b)*int(1,x,-a,a)*int((y*(y/L-1)/L),y,0,L);
119   K(6,5)  =  -C44*int(1,z,-b,b)*int(1,x,-a,a)*int(y/L^2,y,0,L);
120   K(6,6)  =  C11*int(1,z,-b,b)*int(x^2,x,-a,a)*int(1/L^2,y,0,L)...
121              +C44*int(1,z,-b,b)*int(1,x,-a,a)*int((y^2/L^2),y,0,L);
122   K(6,7)  =  C11*int(z,z,-b,b)*int(x,x,-a,a)*int(1/L^2,y,0,L);
123   K(6,8)  =  0;
124   K(6,9)  =  -C11*int(z,z,-b,b)*int(x,x,-a,a)*int(1/L^2,y,0,L);
125   K(6,10) =  0;
126   K(6,11) =  -C44*int(z,z,-b,b)*int(1,x,-a,a)*int(y/L^2,y,0,L);
127   K(6,12) =  C44*int(z,z,-b,b)*int(1,x,-a,a)*int(y/L^2,y,0,L);
128   %
129   %
130   K(7,1)  =  -C11*int(z,z,-b,b)*int(1,x,-a,a)*int(1/L^2,y,0,L);
131   K(7,2)  =  C11*int(z,z,-b,b)*int(1,x,-a,a)*int(1/L^2,y,0,L);
132   K(7,3)  =  0;
133   K(7,4)  =  -C11*int(z,z,-b,b)*int(x,x,-a,a)*int(1/L^2,y,0,L);
134   K(7,5)  =  0;
135   K(7,6)  =  C11*int(z,z,-b,b)*int(x,x,-a,a)*int(1/L^2,y,0,L);
136   K(7,7)  =  C44*int(1,z,-b,b)*int(1,x,-a,a)*int((y/L-1)^2,y,0,L)...
137              +C11*int(z^2,z,-b,b)*int(1,x,-a,a)*int(1/L^2,y,0,L);
138   K(7,8)  =  -C44*int(1,z,-b,b)*int(1,x,-a,a)*int(((y/L-1)/L),y,0,L);
139   K(7,9)  =  -C11*int(z^2,z,-b,b)*int(1,x,-a,a)*int(1/L^2,y,0,L)...
140              -C44*int(1,z,-b,b)*int(1,x,-a,a)*int((y*(y/L-1)/L),y,0,L);
141   K(7,10) =  C44*int(1,z,-b,b)*int(1,x,-a,a)*int(((y/L-1)/L),y,0,L);
142   K(7,11) =  C44*int(1,z,-b,b)*int(x,x,-a,a)*int(((y/L-1)/L),y,0,L);
143   K(7,12) =  -C44*int(1,z,-b,b)*int(x,x,-a,a)*int(((y/L-1)/L),y,0,L);
144   %
145   %
146   K(8,1)  =  0;
147   K(8,2)  =  0;
148   K(8,3)  =  0;
149   K(8,4)  =  0;
150   K(8,5)  =  0;
151   K(8,6)  =  0;
152   K(8,7)  =  -C44*int(1,z,-b,b)*int(1,x,-a,a)*int(((y/L-1)/L),y,0,L);
153   K(8,8)  =  C44*int(1,z,-b,b)*int(1,x,-a,a)*int(1/L^2,y,0,L);
154   K(8,9)  =  C44*int(1,z,-b,b)*int(1,x,-a,a)*int(y/L^2,y,0,L);
155   K(8,10) =  -C44*int(1,z,-b,b)*int(1,x,-a,a)*int(1/L^2,y,0,L);
156   K(8,11) =  -C44*int(1,z,-b,b)*int(x,x,-a,a)*int(1/L^2,y,0,L);
157   K(8,12) =  C44*int(1,z,-b,b)*int(x,x,-a,a)*int(1/L^2,y,0,L);
158   %
159   %
160   K(9,1)  =  C11*int(z,z,-b,b)*int(1,x,-a,a)*int(1/L^2,y,0,L);
161   K(9,2)  =  -C11*int(z,z,-b,b)*int(1,x,-a,a)*int(1/L^2,y,0,L);
162   K(9,3)  =  0;
163   K(9,4)  =  C11*int(z,z,-b,b)*int(x,x,-a,a)*int(1/L^2,y,0,L);
164   K(9,5)  =  0;
```

```
165  K(9,6)  = -C11*int(z,z,-b,b)*int(x,x,-a,a)*int(1/L^2,y,0,L);
166  K(9,7)  = -C11*int(z^2,z,-b,b)*int(1,x,-a,a)*int(1/L^2,y,0,L)...
167            -C44*int(1,z,-b,b)*int(1,x,-a,a)*int((y*(y/L-1)/L),y,0,L);
168  K(9,8)  = C44*int(1,z,-b,b)*int(1,x,-a,a)*int(y/L^2,y,0,L);
169
170  K(9,9)  = C44*int(1,z,-b,b)*int(1,x,-a,a)*int(y^2/L^2,y,0,L)...
171            +C11*int(z^2,z,-b,b)*int(1,x,-a,a)*int(1/L^2,y,0,L);
172  K(9,10) = -C44*int(1,z,-b,b)*int(1,x,-a,a)*int(y/L^2,y,0,L);
173  K(9,11) = -C44*int(1,z,-b,b)*int(x,x,-a,a)*int(y/L^2,y,0,L);
174  K(9,12) = C44*int(1,z,-b,b)*int(x,x,-a,a)*int(y/L^2,y,0,L);
175  %
176  %
177  K(10,1)  = 0;
178  K(10,2)  = 0;
179  K(10,3)  = 0;
180  K(10,4)  = 0;
181  K(10,5)  = 0;
182  K(10,6)  = 0;
183  K(10,7)  = C44*int(1,z,-b,b)*int(1,x,-a,a)*int(((y/L-1)/L),y,0,L);
184  K(10,8)  = -C44*int(1,z,-b,b)*int(1,x,-a,a)*int(1/L^2,y,0,L);
185  K(10,9)  = -C44*int(1,z,-b,b)*int(1,x,-a,a)*int(y/L^2,y,0,L);
186  K(10,10) = C44*int(1,z,-b,b)*int(1,x,-a,a)*int(1/L^2,y,0,L);
187  K(10,11) = C44*int(1,z,-b,b)*int(x,x,-a,a)*int(1/L^2,y,0,L);
188  K(10,12) = -C44*int(1,z,-b,b)*int(x,x,-a,a)*int(1/L^2,y,0,L);
189  %
190  %
191  K(11,1)  = 0;
192  K(11,2)  = 0;
193  K(11,3)  = -C44*int(z,z,-b,b)*int(1,x,-a,a)*int(1/L^2,y,0,L);
194  K(11,4)  = C44*int(z,z,-b,b)*int(1,x,-a,a)*int(((y/L-1)/L),y,0,L);
195  K(11,5)  = C44*int(z,z,-b,b)*int(1,x,-a,a)*int(1/L^2,y,0,L);
196  K(11,6)  = -C44*int(z,z,-b,b)*int(1,x,-a,a)*int(y/L^2,y,0,L);
197  K(11,7)  = C44*int(1,z,-b,b)*int(x,x,-a,a)*int(((y/L-1)/L),y,0,L);
198  K(11,8)  = -C44*int(1,z,-b,b)*int(x,x,-a,a)*int(1/L^2,y,0,L);
199  K(11,9)  = -C44*int(1,z,-b,b)*int(x,x,-a,a)*int(y/L^2,y,0,L);
200  K(11,10) = C44*int(1,z,-b,b)*int(x,x,-a,a)*int(1/L^2,y,0,L);
201  K(11,11) = C44*int(1,z,-b,b)*int(x^2,x,-a,a)*int(1/L^2,y,0,L)...
202            +C44*int(z^2,z,-b,b)*int(1,x,-a,a)*int(1/L^2,y,0,L);
203  K(11,12) = -C44*int(1,z,-b,b)*int(x^2,x,-a,a)*int(1/L^2,y,0,L)...
204            -C44*int(z^2,z,-b,b)*int(1,x,-a,a)*int(1/L^2,y,0,L);
205  %
206  %
207  K(12,1)  = 0;
208  K(12,2)  = 0;
209  K(12,3)  = C44*int(z,z,-b,b)*int(1,x,-a,a)*int(1/L^2,y,0,L);
210  K(12,4)  = -C44*int(z,z,-b,b)*int(1,x,-a,a)*int(((y/L-1)/L),y,0,L);
211  K(12,5)  = -C44*int(z,z,-b,b)*int(1,x,-a,a)*int(1/L^2,y,0,L);
212  K(12,6)  = C44*int(z,z,-b,b)*int(1,x,-a,a)*int(y/L^2,y,0,L);
213  K(12,7)  = -C44*int(1,z,-b,b)*int(x,x,-a,a)*int(((y/L-1)/L),y,0,L);
214  K(12,8)  = C44*int(1,z,-b,b)*int(x,x,-a,a)*int(1/L^2,y,0,L);
215  K(12,9)  = C44*int(1,z,-b,b)*int(x,x,-a,a)*int(y/L^2,y,0,L);
216  K(12,10) = -C44*int(1,z,-b,b)*int(x,x,-a,a)*int(1/L^2,y,0,L);
217  K(12,11) = -C44*int(1,z,-b,b)*int(x^2,x,-a,a)*int(1/L^2,y,0,L)...
218            -C44*int(z^2,z,-b,b)*int(1,x,-a,a)*int(1/L^2,y,0,L);
219  K(12,12) = C44*int(1,z,-b,b)*int(x^2,x,-a,a)*int(1/L^2,y,0,L)...
220            +C44*int(z^2,z,-b,b)*int(1,x,-a,a)*int(1/L^2,y,0,L);
221  %
222  %
223  fprintf(1,'\n\n\nSymbolic Stiffness Matrix of a Beam ');
224  fprintf(1,'x axis bending, z axis bending and bar behaviour\n\n\n');
225  %
226  K;
```

```
227   pretty(K)
228   %
229   %
230   A  = int(1,z,-b,b)*int(1,x,-a,a); % Cross section area
231   Iz = int(1,z,-b,b)*int(x^2,x,-a,a); % Cross section z Moment of
          Inertia
232   Ix = int(z^2,z,-b,b)*int(1,x,-a,a); % Cross section x Moment of
          Inertia
233   %
234   %
235   pretty(K_NUM)
```

In this script the command **diff** is used to compute the symbolic derivatives of the shape functions and the command **int** is used to compute the symbolic integrals, for the calculation of the stiffness matrix **K**.

F.2   Evaluation of the Stiffness Matrix via the Recursive Notation Against Shape Functions

The MATLAB program enables both symbolic and numerical computation of a beam's stiffness matrix using the recursive notation against shape functions. To run the program, the user must provide the same input files as for the matrix notation.

(3) FE_APPROACH.m
The following is the main script to run.

```
1    %
2    syms N1 N2;
3    syms D1 D2;
4    syms B BT;
5    syms x z;
6    syms y L;
7    syms E;
8    syms A Iz Ix;
9    syms KK K;
10   syms K11 K12 K21 K22;
11   syms a b;
12   syms C11 C44;
13   syms M M1 M2 M3 M4;
14   %
15   %
16   %================================================================
17   %================================================================
18   %
19   %
20   %                          READ INPUT
21   %================================================================
22   %
23   cd ..\
24   run INPUT_SYMBOLIC\INPUT_SYMBOLIC.m;
25   %
26   % Computation of the derivative of N1 and N2 with respect to y
27   %
```

```matlab
28  % The derivatives are entered using Recursive Notation D(i)
29  % and considering Node 1 and Node 2.
30  % D(1) is the derivative of the shape notation for Node 1 and D(2) is
       the
31  % derivative of the shape function for Node 2.
32  %
33  D(1) = diff(N(1),y,1);
34  D(2) = diff(N(2),y,1);
35  %
36  %================================================================
37  %----------------------------------------------------------------
38  %
39  % Computatio of Cross Section Area and
40  % Moment of Inertia around axes x ( Ix ) and z (Iz)
41  %
42  A = int(1,z,-b,b)*int(1,x,-a,a); % Cross section area
43  Ix = int(z^2,z,-b,b)*int(1,x,-a,a); % Cross section x Moment of
        Inertia
44  Iz = int(1,z,-b,b)*int(x^2,x,-a,a); % Cross section z Moment of
        Inertia
45  %
46  %================================================================
47  % Computation of the Stiffness Matrix
48  %================================================================
49  %
50  %
51  count = 1;
52  %
53  %
54  %
55  for j=1:2
56      for i=1:2
57  K(1:6,1:6,count) = [C44*A*int(D(i)*D(j),y,0,L) 0 0 0 0 -C44*A*int(N(i)
        *D(j),y,0,L);...
58                      0 C11*A*int(D(i)*D(j),y,0,L) 0 0 0 0;...
59                      0 0 C44*A*int(D(i)*D(j),y,0,L) -C44*A*int(N(i)*D(j
                        ),y,0,L) 0 0;...                               ...
60  0 0 -C44*A*int(N(j)*D(i),y,0,L) (C11*Ix*int(D(i)*D(j),y,0,L)+C44*A*int
        (N(i)*N(j),y,0,L)) 0 0;...
61                      0 0 0 0 C44*(Ix+Iz)*int(D(i)*D(j),y,0,L) 0;...
62  -C44*A*int(N(j)*D(i),y,0,L) 0 0 0 0 (C11*Iz*int(D(i)*D(j),y,0,L)+C44*A
        *int(N(i)*N(j),y,0,L))];
63      count = count + 1;
64      end
65  end
66  %
67  fprintf(1,'\n\nSymbolic Stiffness Matrix of a Beam x axis bending, z
        axis bending and bar behaviour\n\n\n');
68  %
69  K = [K(1:6,1:6,1) K(1:6,1:6,2) ;K(1:6,1:6,3) K(1:6,1:6,4)];
70  %
71  %
72  pretty(K)
73  %
74  run INPUT_NUMERIC\INPUT_NUMERIC.m;
75  %
76  K_NUM = subs(K);
77  %
78  %
79  fprintf(1,'\n\n\nNumeric Stiffness Matrix of a Beam x axis bending, z
        axis bending and bar behaviour\n\n\n');
80  %
```

```
81   K_NUM
82   %
```

In this script the command **diff** is used to compute the symbolic derivatives of the shape functions and the command **int** is used to compute the symbolic integrals, for the calculation of the stiffness matrix **K**.

Once the main code is executed, the following results, symbolic and numeric, are obtained:

F.3 Evaluation of the Stiffness Matrix via the Recursive Notation Against Theory of Structures

The MATLAB program enables both symbolic and numerical computation of a beam's stiffness matrix using the recursive notation against theory of structures.

To run the program, the user must provide the same INPUT_SYMBOLIC.m as for the matrix notation and the recursive notation against shape functions.

Regarding the INPUT_SYMBOLIC.m input, the user must provide the shape functions ($N(1)$ and $N(2)$) and the expansion function ($F_{\tau y}$ and F_{sy}). The file INPUT_SYMBOLIC.m for is shown below.

(2) INPUT_SYMBOLIC.m

```
1    %=========================================================================
2    %                          THIS  IS  THE  INPUT  FILE
3    %=========================================================================
4    syms  y  L;
5    syms  a  b;
6    syms  C11  C44;
7    syms  Ftx  Fty  Ftz;
8    %  ----------------------------------------------------------------------
9    %                          SYMBOLIC  VARIABLES
10   %  ----------------------------------------------------------------------
11   %=========================================================================
12   %=========================================================================
13   %
14   %  1)  Enter  shape  function  in  symbolic  form  and  recursive  notation:
15   %
16   %  Note :  shape  functions  must  have  the  following  characterisctics :
17   %  -  must  be  a  function  of  y
18   %  -  the  length  of  the  bar  must  be  named  L
19   %
20   %  The  shape  functions  are  entered  using  Recursive  Notation  N(i)
21   %  and  considering  Node  1  and  Node  2.
22   %  N(1)  is  the  shape  notation  for  Node  1  and  N(2)  is  the  shape  function  for
23   %  Node  2.
24   %  Similar  to  the  Formula  2.23  of  page  10  of  REF_1
25   %
26   %  Shape  Function  for  Node  1 :
27   %
28   N(1)  =  1-y/L;
29   %
```

```
30   % Shape Function for Node 2 :
31   %
32   N(2) = y/L;
33   %
34   % -----------------------------------------------------------------------
35   % 2) Enter the Ftx(tau) (Ftaux) , Fty(tau) (Ftauy), Ftz(tau) (Ftauz)
36   %    values using recursive notation:
37   %    For the problem of a Beam with bending around x and z axis in
38   %    combination with a bar, these values are the following
39   %
40   % See Formula 5.5 pf page 33 of REF_1
41   %
42   Ftx(1) = -1;
43   Ftx(2) = z;
44   %
45   Fty(1) = 1;
46   Fty(2) = -z;
47   Fty(3) = x;
48   %
49   Ftz(1) = 1;
50   Ftz(2) = -x;
51   %
52   %
```

(3) TS_APPROACH.m
The following is the main script to run.

```
1    %
2    syms N1 N2;
3    syms D1 D2;
4    syms B BT;
5    syms x z;
6    syms y L;
7    syms E;
8    syms A Ix Iz;
9    syms KK K;
10   syms K11 K12 K21 K22;
11   syms a b;
12   syms C11 C44;
13   syms Ftx Fty Ftz;
14   syms M M1 M2 M3 M4;
15   %
16   %
17   %==============================================================================
18   %==============================================================================
19   %
20   %
21   %                              READ INPUT
22   %==============================================================================
23   %
24   cd ..\
25   run INPUT_SYMBOLIC\INPUT_SYMBOLIC.m;
26   %
27   % Computation of the derivative of N1 and N2 with respect to y
28   %
29   % The derivatives are entered using Recursive Notation D(i)
30   % and considering Node 1 and Node 2.
31   % D(1) is the derivative of the shape notation for Node 1 and D(2) is
           the
32   % derivative of the shape function for Node 2.
33   %
34   D(1) = diff(N(1),y,1);
```

```matlab
35  D(2) = diff(N(2),y,1);
36  %
37  %===============================================================
38  %----------------------------------------------------------------
39  %
40  % Computatio of Cross Section Area and
41  % Moment of Inertia around axes x ( Ix ) and z (Iz)
42  %
43  A = int(1,z,-b,b)*int(1,x,-a,a); % Cross section area
44  Ix = int(z^2,z,-b,b)*int(1,x,-a,a); % Cross section x Moment of
        Inertia
45  Iz = int(1,z,-b,b)*int(x^2,x,-a,a); % Cross section z Moment of
        Inertia
46  %
47  %===============================================================
48  % Computation of the Stiffness Matrix
49  %===============================================================
50  %
51  %
52  count = 1;
53  %
54  %
55  %
56  for j=1:2
57      for i=1:2
58  K(1:6,1:6,count) = [C44*int(int(Ftx(1)*Ftx(1),z,-b,b),x,-a,a)*int(D(i)
        *D(j),y,0,L)...
59  0 0 0 0 C44*int(int(Ftx(1)*diff(Fty(3),x,1),z,-b,b),x,-a,a)*int(N(i)*D
        (j),y,0,L);...
60  0 C11*int(int(Fty(1)*Fty(1),z,-b,b),x,-a,a)*int(D(i)*D(j),y,0,L) 0 0 0
        0;...
61  0 0 C44*int(int(Ftz(1)*Ftz(1),z,-b,b),x,-a,a)*int(D(i)*D(j),y,0,L)...
62  C44*int(int(Ftz(1)*diff(Fty(2),z,1),z,-b,b),x,-a,a)*int(N(i)*D(j),y,0,
        L) 0 0;...
63  0 0 C44*int(int(Ftz(1)*diff(Fty(2),z,1),z,-b,b),x,-a,a)*int(N(j)*D(i),
        y,0,L)...
64  (C11*int(int(Fty(2)*Fty(2),z,-b,b),x,-a,a)*int(D(i)*D(j),y,0,L)...
65  +C44*int(int(diff(Fty(2),z,1)*diff(Fty(2),z,1),z,-b,b),x,-a,a)*int(N(i
        )*N(j),y,0,L)) 0 0;...
66  0 0 0 0 (C44*int(int(Ftx(2)*Ftx(2),z,-b,b),x,-a,a)*int(D(i)*D(j),y,0,L
        )...
67  +C44*int(int(Ftz(2)*Ftz(2),z,-b,b),x,-a,a)*int(D(i)*D(j),y,0,L)) 0;
68  C44*int(int(diff(Fty(3),x,1)*Ftx(1),z,-b,b),x,-a,a)*int(N(j)*D(i),y,0,
        L)...
69  0 0 0 0 (C11*int(int(Fty(3)*Fty(3),z,-b,b),x,-a,a)*int(D(i)*D(j),y,0,L
        )...
70  +C44*int(int(diff(Fty(3),x,1)*diff(Fty(3),x,1),z,-b,b),x,-a,a)*int(N(i
        )*N(j),y,0,L))];
71      count = count + 1;
72      end
73  end
74  %
75  fprintf(1,'\n\nSymbolic Stiffness Matrix of a Beam x axis bending, z
        axis bending and bar behaviour\n\n\n');
76  %
77  K = [K(1:6,1:6,1) K(1:6,1:6,2) ;K(1:6,1:6,3) K(1:6,1:6,4)];
78  %
79  %
80  %
81  pretty(K)
82
83  run INPUT_NUMERIC\INPUT_NUMERIC.m;
```

```matlab
84    %
85    K_NUM = subs(K);
86    %
87    %
88    fprintf(1,'\n\n\nNumeric Stiffness Matrix of a Beam x axis bending, z
          axis bending and bar behaviour\n\n\n');
89    %
90    K_NUM
91    %
```

In this script the command **diff** is used to compute the symbolic derivatives of the shape functions to obtain B_i and B_j and the command **int** is used to compute the symbolic integrals, for the calculation of the stiffness matrix **K**.

Once the MATLAB program is executed the following results, symbolic and numeric, are obtained:

GPSR Compliance
The European Union's (EU) General Product Safety Regulation (GPSR) is a set
of rules that requires consumer products to be safe and our obligations to
ensure this.

If you have any concerns about our products, you can contact us on

ProductSafety@springernature.com

In case Publisher is established outside the EU, the EU authorized
representative is:

Springer Nature Customer Service Center GmbH
Europaplatz 3
69115 Heidelberg, Germany